Packrat Middens

Packrat Middens

The Last 40,000 Years of Biotic Change

Edited by Julio L. Betancourt, Thomas R. Van Devender, and Paul S. Martin

The University of Arizona Press   Tucson

The University of Arizona Press
www.uapress.arizona.edu

Printed in the United States of America
21 20 19 18 17 16   7 6 5 4 3 2

ISBN-13: 978-0-8165-1115-0 (cloth)
ISBN-13: 978-0-8165-3284-1 (paper)

Cover design by Miriam Warren

This book was set in Trump Mediaeval.

Library of Congress Cataloging-in-Publication Data
Packrat middens : the last 40,000 years of biotic change / edited by
  Julio L. Betancourt, Thomas R. Van Devender, and Paul S. Martin.
      p. cm.
    ISBN 0-8165-1115-2 (alk. paper)
    1. Paleoecology—West (U.S.) 2. Paleoecology—Mexico. 3. Wood
rats, Fossil. 4. Paleobotany—Holocene. 5. Paleobotany– West
(U.S.) 6. Paleobotany—Mexico. I. Betancourt, Julio L. II. Van
Devender, Thomas R. III. Martin, Paul S. (Paul Schultz), 1928–
QE720.P3   1990
560'.45'0978—dc20   89-38454
                                                        CIP

British Library Cataloguing-in-Publication data are available.

♾ This paper meets the requirements of ANSI/NISO Z39.48-1992
(Permanence of Paper).

Contents

Packrat Middens

Chapter 1

Introduction

Julio L. Betancourt   Thomas R. Van Devender   Paul S. Martin

The path to the biology building passed an occasional ponderosa pine, the towering Front Range as a backdrop. One of us (JLB) had agreed to preview this book for the departmental seminar, a roomful of environmental, population, and organismic biologists. The afternoon began with a visit to the electrophoresis lab, where a lively discussion ensued about genetic variation in ponderosa pine and its whereabouts during the last ice age. The punchline—that ponderosa's dominance over much of the southern Rockies is only a postglacial phenomenon—was saved for the talk. Afterward, at the reception, the genetics group demanded a full accounting of the evidence. "How reliable are these packrat middens? How much data have you really got? Are you confident relying on negative evidence?" Satisfied with the caveats and explanations, they pressed on. "We thought you guys preached the present as the key to the past; are you turning this around? Are you asking us to set the molecular clock at ten thousand years and manufacture all of this genetic variability overnight? By the way, when did you say the book was coming out?"

This book, which we hope will satisfy our colleagues in population biology and myriad other disciplines, reports on a unique source of fossils and its contribution in reconstructing late Quaternary environments of western North America. The fossils are well-preserved fragments of plants and animals accumulated locally by packrats or woodrats (*Neotoma* spp.) and quite often encased, amberlike, in crystallized urine. Packrat middens, as these deposits have come to be called, are ubiquitous in caves and crevices throughout the arid West. In some cases the deposits represent portions of the animal's house; refuse accumulated along trails or on perch areas also may develop into middens. While long unappreciated for their ultimate worth to ecologists, packrat middens are common enough to have been noticed by early explorers.

Lieutenant J. H. Simpson mentioned packrat middens during a military reconnaissance of the Colorado Plateau in 1849: "Among the escarpment rocks. . . . [we found] in some crevices, also attached to the rock, a dark pitchy substance, agglutinated with the excrement of birds and of animals of the rat species" (Simpson, 1850, p. 72). The Wheeler Expedition of 1872 collected a midden and later subjected it to chemical analysis. H. C. Yarrow and S. F. Baird, prominent zoologists of the time, thought middens were deposited by chuckwallas (*Sauromalus obesus*). E. D. Cope, the renowned paleontologist, had the right idea, "believing them to be the excrement of small mammals, such as *Neotoma*" (Yarrow, 1875, p. 562).

Perhaps the most celebrated early account involves a group of would-be miners in the gold rush of 1849. As they crossed Papoose Lake, about 100 km northwest of Las Vegas, Nevada, they encountered middens in a canyon south of the dry playa: "Part way up we came to a high cliff and in its face were niches or cavities as large as a barrel or larger, and in some of them, we found balls of a glistening substance looking like pieces of variegated candy stuck together. . . . It was evidently food of some sort, and we found it sweet but sickish, and those who were hungry . . . . making a good meal of it, were a little troubled with nausea afterwards" (Manly, 1987, p. 126). Ironically, it was only 50 km south of Papoose Lake that the scientific value of middens first came to be appreciated.

One hundred years after the forty-niners, a huge tract of Mojave Desert was being set aside as the continental proving grounds for atmospheric testing of nuclear devices. In the 1950s, as today, the Nevada Test Site teemed with scientists conducting feasibility and impact studies. Under contract from the U.S. Atomic Energy Commission, the biological effects of fission-type nuclear detonations were being investigated by New Mexico Highlands University, Brigham Young University, and the University of California, Los Angeles.

In 1960 Phil Wells, a plant ecologist fresh from completing a Ph.D. at Duke University, was doing postdoctoral work at New Mexico Highlands. Clive Jorgensen, an entomologist, was the field director of the team from Brigham Young University. Wells and Jorgensen, then in their late twenties, intended to visit Gypsum Cave near Las Vegas to see the deposits of the extinct Shasta ground sloth (*Nothrotheriops shastenses*) made famous by Laudermilk and Munz (1934). Sloth dung had also been reported from Rampart Cave in the western Grand Canyon (Laudermilk and Munz, 1938). In 1956 Dick Shutler of the Nevada State Museum collected radiocarbon samples from an abandoned trench in Rampart Cave. Interspersed in the sloth deposit was a layer of loose packrat debris, including juniper seeds and twigs. From radiocarbon dates Martin et al. (1961) suggested that over twelve thousand years ago junipers grew at the level of Rampart Cave, at least 400 m below their modern lower limit. However, they did not recognize the prevalence and potential usefulness of such debris piles elsewhere in the Southwest.

Wells and Jorgensen never visited Gypsum Cave, but that did not prevent them from seeking fossil deposits in caves and rock shelters at the Nevada Test Site. The real discovery came in spring 1961 on a serendipitous hike up Aysees Peak, the highest mountain around Frenchman Flat. At about 1900 m elevation the mountain seemed high enough to support a woodland of pinyon and juniper. They were disappointed, however, to find only desert shrubs on the summit. They took a different path down the mountain, going by way of a narrow limestone canyon. It was late afternoon when Wells and Jorgensen stopped to rest and muse over their predicament—not much daylight left and quite a way to go still. As they sat and commiserated, Jorgensen looked over his shoulder and something caught his eye about thirty meters away—a dark, shiny mass beneath an overhang. He walked over to it, grabbed a chunk, and broke it open, calling out to Wells, "You've got to see this! This is where all the juniper is!" (Clive Jorgensen, personal communication, 1988). At one stroke a rich new source of fossils had been discovered. Critical determination of the age of the deposits would depend on the effectiveness of radiocarbon dating, another legacy of the atomic age that proved ideally suited to the task. Thus a spinoff of the new technology provided a new source of fossil evidence as well as a way to date it. In another three decades the new source of fossils would prove helpful in selecting the Nevada Test Site for nuclear waste storage.

Given his crucial role in developing the radiocarbon dating method, it is only fitting that W. F. Libby's laboratory at UCLA date that first midden from Aysees Peak. The result, 9320 ± 300 B.P. (UCLA-644, Fergusson and Libby, 1962), supported expectations. Ten other middens dated between 40 and 7.8 ka (thousands of radiocarbon years before the present) indicated that juniper's lower limit in the region was then at least 600 m below where it is now. Wells and Jorgensen's (1964, p. 1172) seminal paper prophesied that middens "may have unique value as a check of the palynological approach to Pleistocene ecology in the arid Southwest." They also

4

might have guessed that middens would soon overtake pollen records as a touchstone for studying Pleistocene climatic and biotic changes in the deserts.

In the Southwest, pollen analysis itself was just getting under way. As recently as 1949, in an extensive review of the continent, E. S. Deevey, Jr., had reported no definitive Pleistocene fossil plant records in the Southwest. Instead, to support geological theory, he made a case for significant ice-age biotic changes based on such tangential evidence as relict fish distributions in Great Basin drainages and chromosome inversions in fruit flies (Deevey, 1949).

In the absence of fossil evidence about desert plants, plant ecologists were encouraged to sing a different tune. Spalding (1909) believed that desert plants such as the stately saguaro had been growing essentially in their modern range since their evolution in the Tertiary. Shreve (1951) concluded that the Sonoran Desert had existed at least since the Miocene. D. T. MacDougal, director of Tuc-

son's Desert Laboratory where both Spalding and Shreve worked, had a different outlook for the Great Basin desert. He wrote that "but few species now inhabit the Great Basin, which grew along the shores or in the tributaries of [Pluvial] Lake Lahontian [sic]" (MacDougal, 1909).

The classic studies of Gilbert (1890) and Russell (1885, 1889) established that pluvial events in the Great Basin could be linked stratigraphically to local glaciation in the Wasatch Mountains and the Sierra Nevada. Eventually, radiocarbon dating supported their case (Flint and Gale, 1958; Broecker and Orr, 1958). Antevs (1954) gave palynologists something to shoot at when he suggested that life zones in New Mexico had been lowered 900–1200 m at the time of maximum glacial extent. He based these estimates largely on glacial cirque depression, the pluvial record at Lake Estancia, and even on fossil records of marmots found at elevations where they no longer occur.

In the absence of peat bogs, pollen analysis

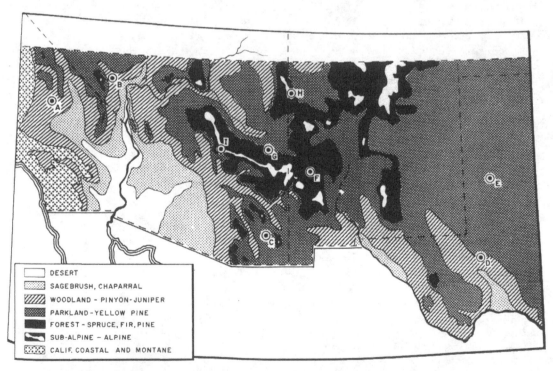

Fig.1.1. Full-glacial vegetation of the Southwest, based on the pollen records that had been studied by the mid-1960s. (A) Searles Lake; (B) Tule Springs; (C) pluvial Lake Cochise; (D) Crane Lake; (E) Rich Lake; (F) San Augustin Plains; (G) Laguna Salada; (H) Dead Man Lake; (I) Potato Lake (from Martin and Mehringer 1965).

in the West got off to a slow start in the 1930s. Laudermilk and Munz (1934, 1938) identified pollen grains as well as macrofossils from coprolites of the Shasta ground sloth, and Sears (1937) studied pollen from one of Antevs's more organic alluvial units in an arroyo cut near Kayenta in northern Arizona. Little else was accomplished and palynology in the arid West did not flourish for another two decades. The discovery of fossil pollen in playa lake sediments (Sears and Clisby, 1952; Clisby and Sears, 1956) opened new doors, with study sites ranging from southern California to west Texas (see reviews by Martin and Mehringer, 1965; Hall, 1985).

One of the first pollen studies of a southwestern playa was made at Searles Lake in California. Roosma's (1958) initial work was supplemented with additional fossil counts by E. B. Leopold and compared with surface samples along an elevational transect in the adjacent Panamint Mountains (Martin, 1964). When pollen in the soil surface was used as a comparison, the best modern analog for the pollen spectra of the Parting Mud unit (dated between 24 and 10 ka) was in pinyon-juniper woodland at least 1200 m higher than the elevation of Searles Lake.

Mehringer's (1965, 1967) landmark studies at Tule Springs also suggested a lowering of vegetation zones by at least 1000 m. Both macrofossil and pollen evidence from spring deposits indicated the presence of junipers in the bottom of the Las Vegas Valley, at about 700 m elevation. Mehringer suggested that ponderosa pine stands may have extended down to 1300 m (900 m below its modern range), but he was understandably cautious. He recognized that over-representation of pine pollen could have come about in various ways and that there were difficulties with inferring paleozonation on the basis of pollen. Mehringer also anticipated problems with the analog approach, particularly because plant associations in the past may have differed radically from modern ones.

A suitable benchmark for demonstrating

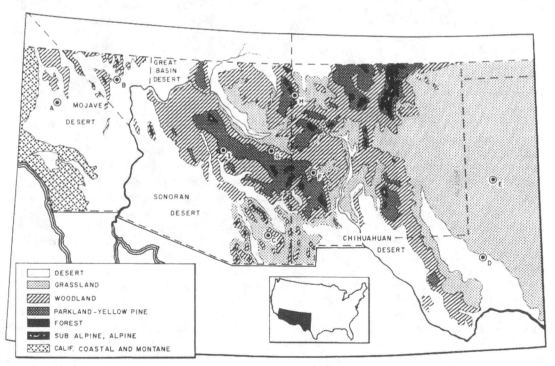

Fig. 1.2. Modern vegetation of the Southwest. Symbols as in Figure 1.1 (from Martin and Mehringer 1965).

the contributions of pollen analysis by the mid-1960s is Martin and Mehringer's (1965) maps of full-glacial (Fig. 1.1) and modern (Fig. 1.2) vegetation in the Southwest. When the full-glacial map was constructed, the Sonoran Desert presented a special problem. In an earlier version Martin (1964) had postulated grassland, based largely on the abundance of horse bones at Ventana Cave (Colbert in Haury, 1950). Fossil pollen in the cave sediments was similar to surface samples in the Sonoran Desert and unlike pollen counts from desert grassland in southeastern Arizona. In addition, the success of feral burros and horses in modern desert environments called into question the validity of using Pleistocene horses as indicators of grassland. The revised map (Fig. 1.1) showed desertscrub only in the lower Colorado Valley, chaparral and sagebrush (*Artemisia tridentata*) over most of the Sonoran Desert, yellow (ponderosa) pine parkland in present desert grasslands, and woodlands in the Chihuahuan, Mojave, and Great Basin deserts. The reader may want to compare Figure 1.1 with a new midden-based map in chapter 21.

This book began as an effort to bring together the most recent findings and views of many of the researchers presently investigating fossil middens. The idea was conceived on Tumamoc Hill, the site of the Carnegie Institution's Desert Laboratory, now affiliated with the University of Arizona and the U.S. Geological Survey. Over the past twenty-five years the editors and many of the authors had their start at packrat midden analysis on "the Hill," inevitably influenced by the ecological legacies of Shreve, Spalding, and MacDougal, among others. The Hill has come to be regarded as an intellectual watering hole, a place of lively debate and shared fondness for the desert.

The chapters that follow open a forum for methodological concerns, update the fossil record of various geographic regions, introduce new applications, and display the vast potential for fossil midden analysis not only in the North American deserts but in arid regions worldwide. Chapters 2–4 represent the first attempts to address how packrat behavior might affect the nature of the fossil midden record. We enlisted the aid of four ecolo-

gists to undertake this novel task for obvious reasons: They have the necessary expertise for studying animal behavior and little stake in actual interpretation of the fossil record.

Packrat dietary specialization introduces a bias that must be acknowledged in fossil interpretations. While packrats appear to sample species richness adequately within their foraging range, the relative abundance of each kind of plant is affected by diet and other factors. Stephens' woodrat (*Neotoma stephensi*), an animal that Vaughan (chapter 2) has studied for some years, is apparently one of only two North American mammals that is a conifer-leaf specialist. Middens produced by this packrat appear to over-represent the abundance of juniper in the fossil community. Finley's (chapter 3) summary of the relevant literature, his own observations over the years, and his prospectus for midden biochemistry will be welcomed by all who have dog-eared copies of his *Woodrats of Colorado: Distribution and ecology* (Finley, 1958). Dial and Czaplewski (chapter 4) take an experimental approach in addressing the problem. At Woodhouse Mesa near Flagstaff, Arizona, they have established a large plot in the range of three sympatric species. They note, much to the chagrin of those seeking simplicity, that a den reoccupied by a different packrat species produces a radically different plant assemblage.

Chapters 5 and 6 extend methodological concerns to the actual nature of fossil middens, analytical strategies, and the biases likely to be reflected in the spatial and temporal distributions of middens reported thus far. Spaulding et al. (chapter 5) show that middens are quite variable in composition. Weights of uriniferous fractions lost in washing, fecal pellets, and the plant matrix do not directly reflect the initial indurated weight. Age and structure of the midden sample may explain some of the variability in the three midden fractions. This suggests that sample size should be expressed only in terms of the actual plant matrix examined. Chapter 5 also explores different methods of quantifying midden macrofossils, age trends in observed diversity, and the relationships between species abundance in the midden and the plant community from which midden contents were derived.

Webb and Betancourt (chapter 6) summarize the spatial and temporal distribution of over 1100 radiocarbon dates from middens, the state of affairs in October 1985. The age frequency of radiocarbon dates reflects a bimodal distribution with peaks at 0–1 ka and 10 ka. A case is made that the older peak reflects the "lure of the Pleistocene," the deliberate search for evidence of glacial-age vegetation. Other observations, such as the fact that 60% of all Pleistocene middens were collected from limestone, a relatively xeric substrate with considerable influence on local plant distributions, also are pertinent to evaluating the midden record.

Six chapters provide regional summaries of late Quaternary vegetation in the Chihuahuan (chapter 7), Sonoran (chapter 8), and Mojave (chapter 9) deserts; the Great Basin (chapter 10); the Grand Canyon (chapter 11); and the Colorado Plateau (chapter 12). Recurrent themes throughout these chapters are the ephemeral nature of existing plant associations and regional vegetational patterns, the distinctive nature and relative stability of glacial-age vegetation, and the individualistic responses of plants to changing climates. Shreve and Spalding would have been intrigued to find that desert plant species did in fact move, much farther and quicker than they would have envisioned. At the end of the Pleistocene the deserts were literally on the march. And in view of their variable rates of arrival, each species seemingly marched to a different drummer.

As will become apparent, each chapter author also marches to a different beat. There is something unique to each of the regional summaries, partly because of differing landscapes and floras, but also due to the varying backgrounds and outlooks of the authors. For those who have witnessed the heated debates at conferences, we are only stating the obvious.

Van Devender (chapter 7) reviews the midden record from the Chihuahuan Desert, a region of limestone sierras and expansive oceans of creosote bush (*Larrea divaricata*). The Pleistocene record is one of pinyon-juniper-oak woodlands, in agreement with Martin and Mehringer's map. Papershell pinyon (*Pinus remota*) and Colorado pinyon (*P. edulis*) apparently switched roles ten thou-

sand years ago. Papershell pinyon was widespread in the Pleistocene, surprisingly so in light of its current relict distribution. Colorado pinyon, now the dominant taxon in the southern Rocky Mountains and Colorado Plateau, was apparently restricted to a small area in the northern Chihuahuan Desert and select localities in the southern part of the Colorado Plateau. Van Devender indicates that glacial-age vegetation was relatively stable until pinyons retreated upslope at 11 ka. Contractions and expansions of plant distributions during glacial-interglacial cycles, combined with decreasing extinction rates at lower latitudes, have yielded high evolutionary rates, a rich endemic flora, and greater species richness toward the Neotropics.

For the Sonoran Desert, an area characterized by low elevations and igneous rocks, Van Devender (chapter 8) summarizes vegetation changes that have been unidirectional since the end of the Pleistocene. He finds the changes analogous to moving down in elevation or latitude. Contrary to the expectations of Shreve and Spalding, the only communities that experienced minimal change were creosote bush–white bursage (*Ambrosia dumosa*) communities of the Gran Desierto and halophytic ones in the Colorado River delta. Chalk one up for Martin and Mehringer's map. However, the distributions of plants like the saguaro and foothills palo verde (*Cercidium microphyllum*) were not stable, and they apparently were reduced in the Pleistocene to small refugia in Sonora. Pinyon-juniper assemblages over most of the Sonoran Desert may not necessarily contrast with the map's claim for chaparral and sagebrush. Pinyon-juniper woodland fuses with interior chaparral south of the Mogollon Rim and in the mountains of the eastern Mojave Desert in southeastern California. Big sagebrush, occasionally found in middens at moderate elevations, may have been more common on fine-grained soils in basins. Numerous pairs of species that no longer grow together or do so only in unusual ecological settings were associated in the late Wisconsin and early Holocene. Van Devender uses these "anomalous" associations to infer equable paleoclimates that led to the relaxation of both the lower elevational limits of woodland and chaparral plants, and the upper limits of desertscrub plants that can utilize winter moisture.

Spaulding's account of the Mojave Desert (chapter 9) draws partly on the rich tradition of midden research at the Nevada Test Site. Mojave Desert middens are the only ones recovered to date that provide evidence for full-glacial desertscrub. A few sites with xeric exposures apparently supported treeless assemblages as high as 900 m elevation, though juniper is the dominant species in most late Wisconsin middens down to 450 m. Juniper woodland was widespread up to 1800 m, where it was replaced by subalpine woodlands with limber pine (*Pinus flexilis*) and bristlecone pine (*P. longaeva*). Pinyon pine (*P. monophylla*) was not as common in these woodlands as it is today. Taking exception to the hypothesis of equability, Spaulding feels that the widespread distribution of subalpine and steppe species indicates that full-glacial winters were colder than now by about 6°C. He also sees vegetation changes coincident with global warming and deglaciation after around 15 ka. During the late Wisconsin, ponderosa pine was absent from areas shaded as yellow pine parkland in Martin and Mehringer's map in the Mojave Desert, the Great Basin, and the Colorado Plateau. This error no doubt resulted from the inability to assign pine pollen to species and the fact that, today, such high pine pollen counts can be found only in ponderosa pine forest.

Thompson finds opportunities to compare fossil pollen and midden records in the Great Basin (chapter 10). The juxtaposition of rocky environments harboring middens and pollen-rich lacustrine deposits allowed him to discern treeless sagebrush communities in alluvial basins, while limber and bristlecone pine dominated rocky substrates. This suggests that, during the late Wisconsin, continuous forest or woodland corridors were not available for migration of boreal mammals into isolated mountains. Pleistocene packrat middens from the northern Winnemucca Lake basin, 100 m below the maximum lake level stage, placed a limiting age on the demise of Lake Lahontan of about 12.1 ka and led to a revision of the lake chronology. Another highlight from Thompson's chapter is the relatively late development of woodlands with singleleaf pinyon and the replacement of sagebrush steppe by shadscale (*Atriplex confertifolia*) communities at 6.6–6.0 ka.

Few would disagree that the Grand Canyon deserves its own venue. Cole (chapter 11) traces paleozonation in the canyon, claims vegetational inertia in explaining lags between departures and arrivals, and introduces new ideas about the stability of ecotones due to physiography. In Hance Canyon today, desertscrub stops and pinyon-juniper woodland begins at the top of the Redwall Limestone, where it grades abruptly to shales of the Supai Group. Similarly, Pleistocene communities dominated by limber pine, Douglas fir (*Pseudotsuga menziesii*), and white fir (*Abies concolor*) descended to the top of the Redwall Limestone but were replaced at lower elevations by Utah juniper (*Juniperus osteosperma*) woodland. Cole uses multivariate ordination techniques to segregate associations in time and space, to show that late Wisconsin and early Holocene communities were in fact unique (rather than a simple vertical displacement of what we see today), and to demonstrate that vegetation zonation was stable from 24 to 12 ka.

Betancourt (chapter 12) covers the remainder of the Colorado Plateau. Much of his discussion takes into account the influence of different rock types in a region where bedrock is exposed almost continuously. The change in physiographic setting with vertical displacement of plants may explain many anomalous associations. In the Pleistocene climbing dunes against a massive sandstone cliff probably supported an aggregate of species that would seem anomalous by present standards. Blue spruce, water birch, and other riparian plants extended well below the base of the mountains into the canyon country. Current disjunctions in well-watered alcoves resulted not from long-distance dispersal but from disintegration of a montane flora that formerly was more widespread. Betancourt suggests that limber pine and Rocky Mountain juniper (*Juniperus scopulorum*) exchanged roles with Colorado pinyon and Utah juniper or one-seed juniper (*Juniperus monosperma*) to create physiognomically equivalent woodlands, which occupied much the same elevational range. He also considers that the main mass of the Rocky Mountains was biologically impoverished and should not be considered the

"Pleistocene mainland" in treatments of insular mountains as ecological islands. The present biota of the assumed mainland is more diverse than that of the isolated mountains because the main mass of the Rockies has suffered fewer extinctions at high elevations while receiving the full complement of Holocene immigrants at low elevations.

Chapters 13–17 develop specific sources of information such as pollen, fossil grasses, vertebrate remains, arthropods, and stable isotopes from plant cellulose in middens. Near the northern limit of fossil middens, Mehringer and Wigand (chapter 13) combine analyses of pollen from a crater lake and middens from nearby lava tubes to address a problem that may be more narrowly viewed by land managers—historic juniper invasion. At Diamond Craters in southeastern Oregon the authors found earlier invasions in the Holocene, long before cattle or sheep arrived, and concluded that today western juniper may also be riding the tides of a changing climate. Mehringer and Wigand's chapter shows how midden and lake pollen can be used to complement plant macrofossil data from packrat middens.

Grass macrofossils have always been more difficult to identify than other taxonomic groups. Van Devender et al. (chapter 14) identify twenty-four species of grasses from sixty-two Sonoran Desert middens. A basic grass flora of ephemerals and short-lived perennials has been in place in the Sonoran Desert during most of the late Quaternary, with minimal invasion by cool-adapted C3 perennials during the late Wisconsin and early Holocene or by large, subtropical C4 perennials in the middle and late Holocene. Drawing on vertebrate remains, Van Devender and Bradley (chapter 15) infer open grassland rather than sagebrush communities in alluvial basins of the Chihuahuan Desert. As with the flora, the numbers of displaced small mammals, from both northern latitudes and mesic mountaintops, decreased to the south in the middle and late Wisconsin. In chapter 16 Hall et al. identify forty-three arthropod taxa from twenty-one middens in Organ Pipe Cactus National Monument. The fauna was apparently more conservative than the flora in that most of these arthropods probably still occur near the midden sites. The number of taxa increased dramatically in the late Holocene

with development of subtropical vegetation. Long et al. (chapter 17) show the great promise implied in the analysis of stable isotopes in plant cellulose, which is abundantly available in middens. The complacency of deuterium ratios between 13 and 8 ka (essentially equal to modern values) is perplexing in light of the dramatic floristic turnovers evident in the same middens. Their suggestion that deuterium chronologies may be useful in discriminating climatic change from vegetational inertia may prove quite valuable in the future.

The final chapters (chapters 18–20) announce that, even in the absence of packrats and porcupines (*Erethizon dorsatum*) (Betancourt et al., 1986), the arid regions of other continents are not lacking in fossil middens. Middens are made by dassie rats (*Petromus* sp.) and hyraxes (*Procavia* sp.) in South Africa (chapter 18), hyraxes (*Procavia* sp.) in the Middle East (chapter 19), and stick-nest rats (*Leporillus* sp.) in Australia (chapter 20). Organic deposits akin to middens also have been found in the Atlas Mountains in Morocco, the Sudan, and the foot of the Andes in Argentina. Some of the research abroad, particularly in the Old World (for example, chapter 19), will undoubtedly track anthropogenic effects on arid landscapes. With mounting interest in global climate change, the pace of research in other arid lands will quicken. We especially hope that this book will offer encouragement to the midden effort abroad.

The chapter authors deserve special thanks for rising to the occasion despite our sometimes erratic instructions and long delays. We also acknowledge the support of individuals who have reviewed, cajoled, collaborated, shared resources, and offered support, moral or otherwise, when needed most: Larry Agenbroad, Platt Bradbury, Jim Brown, Martha Ames Burgess, Bill Critchfield, Owen Davis, Bob Euler, Ben Everitt, Wes Fergusson (deceased), Betty Fink, Hal Fritts, Bill Gillespie, Lisa Graumlich, Don Grayson, Art Harris, Frank Hawksworth, John Hawley, Jim Judge, Jim King, Val La Marche (deceased), Ron Lanner, Estella Leopold, Vera Markgraf, Niall McCarten, Ed McCullough, Chuck Mason, Jim Mead, Kevin Moodie, Bill Peechy, Mary Kay O'Rourke, Art Phillips, Stephen Porter, Jim Scudday, Al Solomon, Watson and Lucy Cranwell Smith, Dave Thomas, Ray Turner, Tom Webb, Phil Wells, Fred Wiseman, Richard Worthington, and Herb Wright. We acknowledge financial support from the National Science Foundation, the National Geographic Society, the National Park Service, the Southwest Parks and Monuments Association, the Big Bend and Grand Canyon Natural History Associations, the U.S. Department of Energy, the U.S. Geological Survey, the University of Arizona, the University of Washington (Quaternary Research Center), Brown University, and the Arizona-Sonora Desert Museum.

REFERENCES

Antevs, E. (1954). "Climate of New Mexico during the last glaciopluvial." *Journal of Geology* 62, 182–91.

Betancourt, J. L.; Van Devender, T. R.; and Rose, M. (1986). "Comparison of plant macrofossils in woodrat (*Neotoma* sp.) and porcupine (*Erethizon dorsatum*) middens from the western United States." *Journal of Mammalogy* 67, 266–73.

Broecker, W. S., and Orr, P. C. (1958). "Radiocarbon chronology of Lake Lahontan and Lake Bonneville." *Bulletin of the Geological Society of America* 69, 1009–32.

Clisby, K. H., and Sears, P. B. (1956). "San Augustin Plains: Pleistocene climatic changes." *Science* 124, 537–39.

Deevey, E. S., Jr. (1949). "Biogeography of the Pleistocene." *Geological Society of America Bulletin* 60, 1315–1416.

Fergusson, G. J., and Libby, W. B. (1962). "UCLA radiocarbon dates. I." *Radiocarbon* 4, 109–14.

Finley, R. B. (1958). "The woodrats of Colorado: Distribution and ecology." *University of Kansas Publications of the Museum of Natural History* 10, 213–552.

Flint, R. F., and Gale, W. A. (1958). "Stratigraphy and radiocarbon dates at Searles Lake, California." *American Journal of Science* 256, 689–714.

Gilbert, G. K. (1890). *Lake Bonneville.* U.S. Geological Survey Monograph 1.

Hall, S. H. (1985). "Quaternary pollen analysis and vegetation history of the Southwest." In *Pollen records of late Quaternary North American sediments.* V. M. Bryant, Jr., and R. G. Holloway, eds., pp. 95–123. Dallas: American Association of Stratigraphic Palynologists Foundation.

Haury, E. W. (1950). *The stratigraphy and archaeology of Ventana Cave, Arizona.* Tucson: University of Arizona Press.

Laudermilk, J. D., and Munz, P. A. (1934). *Plants in the dung of Nothrotherium from Gypsum Cave, Nevada.* Carnegie Institution of Washington Publication 453, pp. 29–37.

Laudermilk, J. D., and Munz, P. A. (1938). *Plants in the dung of Nothrotherium from Rampart and Muav Caves, Arizona.* Carnegie Institution of Washington Publication 487, pp. 271–81.

MacDougal, D. T. (1909). "The origin of desert floras." In *Distribution and movement of desert plants.* V. M. Spalding, ed., pp. 113–19. Carnegie Institution of Washington Publication 113.

Manly, W. L. (1987). *Escape from Death Valley.* Reno: University of Nevada Press.

Martin, P. S. (1964). *Pollen analysis and the full-glacial landscape.* Fort Burgwin (New Mexico) Research Center Publication no. 3, pp. 66–74.

Martin, P. S., and Mehringer, P. J., Jr. (1965). "Pleistocene pollen analysis and biogeography of the Southwest." In *The late Quaternary of the United States.* H. E. Wright, Jr., ed., pp. 433–51. New Haven: Yale University Press.

Martin, P. S.; Sabels, B. E.; and Shutler, D., Jr. (1961). "Rampart Cave coprolite and ecology of the Shasta ground sloth." *American Journal of Science* 259, 102–27.

Mehringer, P. J., Jr. (1965). "Late Pleistocene vegetation in the Mohave Desert of southern Nevada." *Journal of the Arizona Academy of Sciences* 3, 172–88.

———. (1967). "Pollen analysis of the Tule Springs Site area, Nevada." In *Pleistocene studies in southern Nevada.* H. M. Wormington and D. Ellis, eds., pp. 129–200. Nevada State Museum Anthropological Papers 13.

Roosma, Aino. (1958). "A climatic record from Searles Lake, California." *Science* 128, 716.

Russell, I. C. (1885). *Geological history of Lake Lahontan, a Quaternary lake of northwestern Nevada.* U.S. Geological Survey Monograph 11.

———. (1889). *Quaternary history of Mono Valley, California.* U.S. Geological Survey, 8th Annual Report, 1886–1887 (1889), pt. 1, pp. 261–394.

Sears, P. B. (1937). "Pollen analysis as an aid in cultural deposits in the United States." In *Early man.* G. G. MacCurdy, ed., pp. 61–66. London: Lippincott.

Sears, P. B., and Clisby, K. H. (1952). "Two long climatic records." *Science* 116, 176–78.

Shreve, F. (1951). *Vegetation of the Sonoran Desert.* Carnegie Institution of Washington Publication 591.

Simpson, J. H. (1850). *Journal of a military reconnaissance from Santa Fe, New Mexico to the Navajo country . . . in 1849*, pp. 55–168. Thirty-first Congress, 1st session, Senate Executive Document, 64. Washington D.C.: Government Printing Office.

Spalding, V. M. (1909). *Distribution and movements of desert plants.* Carnegie Institution of Washington Publication 113.

Wells, P. V., and Jorgensen, C. D. (1964). "Pleistocene wood rat middens and climatic change in Mohave Desert—a record of juniper woodlands." *Science* 143, 1171–74.

Yarrow, H. C. (1875). "Report upon the collections of batrachians and reptiles made in portions of Nevada, Utah, California, Colorado, New Mexico, and Arizona during the years 1871, 1872, 1873, and 1874." In *Report upon geographical and geological explorations and surveys west of the one hundredth meridian.* G. M. Wheeler, ed., pp. 509–592. Washington, D.C.: Government Printing Office.

PART I
Methods

Chapter 2

Ecology of Living Packrats

Terry A. Vaughan

INTRODUCTION

Packrats (genus *Neotoma*, including twenty-one species) presently occupy many habitats, ranging from near the Arctic Circle in Canada to Nicaragua. Even when desert dwellers, packrats are unable to conserve water and must maintain their water balance by eating moist forage. Leaves and stems are typically low in energy and often contain defensive chemicals that reduce digestibility; selection in packrats has thus favored energy-conserving adaptations such as den building. The den and its enclosed nest of plant fibers markedly reduce the energetic cost of thermoregulation by providing protection from temperature extremes and reducing the rate of heat dissipation from the resting packrat. There are marked differences in feeding and collecting behavior among packrat species. Packrats associated with plant communities with high species diversity tend to be dietary generalists, whereas packrats of communities with low species diversity tend to be specialists. Middens of the former type of packrat are relatively good indicators of the biotic community, while those of the latter type are poor indicators. Despite these complexities, the propensity of today's packrats for discarding an array of rarely eaten plants on the midden and collecting bone-laden carnivore feces and parts of skeletons suggests that fossil middens provide revealing samples of past biotic communities.

While providing a general introduction to the biology of packrats, this chapter deals primarily with those aspects of their ecology that relate to den building and the habit of collecting plants and bones. Because this book is concerned primarily with the southwestern United States, this chapter will stress the ecology of six species that occupy this region now and have probably occupied it, at least intermittently, for the last forty thousand years. Probably no fewer than five species of packrats built the middens useful to workers studying biotic changes during the late Pleistocene and Holocene.

WHAT IS A PACKRAT?

Packrats (*Neotoma*) are compact, long-tailed rodents attaining adult weights of roughly 100–400 g. The eyes are prominent, the ears are large (Fig. 2.1), and the strongly built feet are adapted to grasping branches or rocks when climbing. The molars are flat-crowned, with prismatic patterns of ridges, and the caecum is greatly enlarged. These features in rodents are associated with a diet of fibrous plant material that is low in nutrition. The fur is typically soft and ranges dorsally from pale buff in some desert dwellers to nearly

2.1A

2.1B

Fig. 2.1. (A) Bushy-tailed packrat (*N. cinerea*); long fur on the body and tail and extremely long whiskers are typical of this, the most boreal species of packrat. (B) White-throated packrat (*N. albigula*); note the relatively short pelage and the less furry ears and tail of this mostly desert-dwelling species.

black in some inhabitants of black lava beds. The feet and ventral surfaces are usually white or buffy.

Packrats are members of the cosmopolitan rodent family Muridae, the largest mammalian family (about 1135 species). They belong to the New World subfamily Sigmodontinae, which is largely composed of southern forms: sixty-three of the seventy-three sigmodontine genera live in South America or no farther north than tropical North America. The sigmodontine ancestral stock entered North America from Eurasia early in the Miocene, about 17 mya (million years ago), and under-

went only modest radiation until the late Miocene, some 7 mya (Fahlbusch, 1966; Martin, 1980). The earliest record of *Neotoma* is from the Middle Hemphillian (late Miocene), about 6.6 mya (Dalquest, 1983). The divergence of three living species—*Neotoma floridana*, *N. albigula*, and *N. micropus*—has been dated at from 200 ka (thousands of years before 1950) to 155 ka, during the late Illinoian or early Wisconsin (Birney, 1973; Zimmerman and Nejtek, 1977). The two latter species occur in the Southwest today, and it is likely that all regional species originated well before fifty thousand years ago, the maximum age of the fossil middens reported here.

PACKRAT DISTRIBUTION

The present distribution of the twenty-one living species of packrats is remarkably broad, extending from within two degrees of the Arctic Circle in the Northwest Territories of Canada, at 65°23′ N, to the tropics of Nicaragua, at 13°N (Hall, 1981), with greatest species richness in central Mexico (Table 2.1). In appropriate habitats they occur through much of western Canada, from coast to coast in the United States, and throughout much of Mexico and Central America. Whereas the ranges of some species are restricted, those of others are remarkably extensive. An example of the former is *N. angustapulata*, known from but a few localities in western Tamaulipas, Mexico. An example of the latter is *N. cinerea*, which occupies thirty-two degrees of latitude from the District of Mackenzie in Canada south to Arizona (35°N).

Packrats occupy a variety of climates and habitats. At the northern limit of their range, packrats (*N. cinerea*) are associated with spruce-fir forests and face snowy, boreal winters. At the other extreme, some species (*N. albigula*, *N. devia*, and *N. lepida*) live in low deserts supporting scattered creosote bushes (*Larrea divaricata*) and cacti (*Opuntia* spp.), and survive extreme dryness and searing summer heat. Packrats inhabit alpine tundra, forests, chaparral, grasslands, desert scrub, and tropical thorn scrub, and are known from below sea level near the Salton Sea of California to the top of Pikes Peak in Colorado (4,304 m elevation; Warren, 1942).

PACKRAT ECOLOGY

All packrats share two critically important habitat requirements: succulent plant food (plant material with at least 50% water by weight) and adequate shelter, usually in the form of rocks, shrubs, or trees. Although renowned for their ability to live in desert areas that lack drinking water, packrats void copious urine and depend on succulent or moist forage to maintain their water balance (Schmidt-Nielsen and Schmidt-Nielsen, 1952; Lee, 1963; MacMillen, 1964). The same is true of species that occupy more mesic situations.

Because packrats are not strictly fossorial and are moderately large rodents, adequate shelter consists of rocks, crevices, caves, or dense patches of vegetation. Nearly all spe-

Table 2.1. *Approximate geographic and ecological distribution of living packrats (Neotoma).*
C = central; E = east; N = north; SW = southwest; W = west

Species	Geographic Distribution	Ecological Distribution
N. albigula	SW United States, N Mexico	Deserts, juniper woodland, thornscrub
N. alleni	W and C Mexico	Thornscrub, tropical forest
N. angustapalata	Tamaulipas, Mexico	Thornscrub
N. anthonyi	Todos Santos Island	Chaparral
N. bryanti	Cedros Island, Gulf of California	Brushland
N. bunkeri	Coronados Island, Gulf of California	Desert
N. chrysomelas	Honduras, Nicaragua, Mexico	Highland tropical forest
N. cinerea	W Canada, W United States	Boreal woodland to desert
N. devia	W and NC Arizona	Desert
N. floridana	E and C United States	Woodlands, plains, riparian
N. fuscipes	Pacific Coast of United States	Chaparral, woodlands
N. goldmani	NC Mexico	Desert, thornscrub
N. lepida	SW United States, Baja California	Desert, brushland, juniper woodland
N. martinensis	San Martin Island, Baja California	Brushland
N. mexicana	SW United States, Mexico	Montane woodland
N. micropus	New Mexico, Texas, NE Mexico	Plains grasslands
N. nelsoni	Veracruz, Mexico	Montane woodland
N. palatina	Jalisco, Mexico	Montane woodland
N. phenax	Coastal NW Mexico	Tropical thornscrub and woodland
N. stephensi	N Arizona, W New Mexico	Juniper woodland
N. varia	Turner Island, Gulf of California	Desert

cies of packrats improve such ready-made shelters by building dens of sticks, plant fragments of various sorts, and such detritus as bones and mammal dung. A few species are casual den builders, their presence indicated by inconspicuous deposits of plant debris, whereas some others (most notably *N. fuscipes* of the Pacific Coast chaparral) build monumental dens 2 m or more in height. Except during times of breeding, each den is occupied by a single packrat. Packrats gather plants within a radius of some 30–50 m of their dens, and they discard plant fragments on their middens or cache food plants in chambers in the dens. A nest, about 20–30 cm in diameter, made of soft, shredded plant fibers, and resembling a cup-shaped or spherical bird nest, is built within the heart of the den, in a burrow beneath it, or in an adjacent rock crevice.

Of what value to packrats is the den and under what selective pressures did den-building behavior evolve? Predator avoidance is of prime importance. Packrats are sufficiently large to offer a profitable meal to a coyote or even a black bear, and are small enough to be killed by an owl or a weasel. They are therefore under pressure from a broad array of mammalian and avian predators. The den, by augmenting naturally available cover, generally provides a safe refuge from predators. Selection has clearly favored behavior serving to enhance the impregnability of the den. The signature of several species of packrats is their habit of fortifying dens with cactus. Associated with each den is a deep rock crevice, a burrow among roots, or a cavity in a stump or a living tree. These provide a secure retreat of last resort. To predators, then, assaulting a packrat den is usually unprofitable, energetically wasteful, and perhaps painful. During a six-year study of *N. stephensi* near Flagstaff, Arizona, I observed that only three of some ninety dens were dug into by predators (coyotes). These dens were not protected by cactus but were built against rock crevices or at the bases of large one-seed junipers (*Juniperus monosperma*) whose roots provided secure shelter. Livetrapping after each assault revealed that the resident packrat survived.

Dens also are essential as buffers against temperature extremes. When trees are available in hot, low deserts, *N. albigula* may build dens in the shade of mesquite (*Prosopis* sp.), ironwood (*Olneya tesota*), or palo verde (*Cercidium* sp.). In the heat of the day this packrat retires to a nest in a burrow system beneath the den. During summer days this retreat provides a microclimate that allows the animal to avoid lethal above-ground temperatures (Brown, 1968).

In areas with cold winters the den, or more specifically the insulation of the nest protected by the den, offers an environment with narrow daily temperature fluctuations and a seasonal temperature range as low as 15°C (Brown, 1968). The walls of the nest consist of finely shredded plant material of low thermal conductance and are about 2–8 cm thick, up to five times the thickness of the packrat's pelage, providing an effective insulating layer for the packrat.

Packrats are not known to hibernate, and when foraging is curtailed the den, or a nearby rock crevice, is critically important for food storage. Food storage is especially important for *N. cinerea*, the most boreal species. In northern Arizona both *N. mexicana* and *N. stephensi* remain close to their dens on extremely cold, windy, or snowy nights, and subsist on stored food.

The diet of packrats is often low in energy and high in defensive chemicals. For example, in some desert areas *N. lepida* eats primarily leaves of the creosote bush (Chess and Chew, 1971; Cameron and Rainey, 1972), a plant with potent chemical defenses (Rhoads, 1977). Similarly, *N. stephensi* specializes on the foliage of one-seed juniper (Vaughan, 1980, 1982), which contains defensive terpenoids known to reduce digestibility in deer by inhibiting the microflora in the rumen (Schwartz et al., 1980a, 1980b). The unusually low reproductive rate and slow growth rate of the young in *N. stephensi* are probably associated with its low-energy diet of juniper (Vaughan and Czaplewski, 1985). McNab (1978, 1980) discussed the association between folivory and low reproductive rates in mammals, and careful study of reproduction in *N. floridana* prompted McClure and Randolph (1980) to conclude that packrats are under chronic energy stress. For mammals under such stress, selection clearly favors physiological or behavioral adaptations that lower the energy budget. For packrats, which

Table 2.2. *Other vertebrates inhabiting packrat dens*

Vertebrate Species	Commensal Packrat Species
Terrapene ornata (box turtle)	*N. floridana*
Eumeces fasciatus (five-lined skink)	*N. floridana*
E. obsoletus (plains skink)	*N. floridana*
Sauromalus obesus (chuckwalla)	*N. devia; N. lepida*
Uta stansburiana (side-blotched lizard)	*N. albigula; N. cinerea; N. devia; N. lepida; N. stephensi*
Sceloporus graciosus (eastern fence lizard)	*N. stephensi*
Diadophis punctatus (ring-necked snake)	*N. floridana*
Pituophis melanoleucus (gopher snake)	*N. stephensi*
Sylvilagus floridanus (eastern cottontail)	*N. floridana*
S. audubonii (desert cottontail)	*N. albigula; N. cinerea; N. devia; N. lepida*
Eutamias dorsalis (cliff chipmunk)	*N. stephensi*
Perognathus intermedius (rock pocket mouse)	*N. albigula; N. devia; N. stephensi*
Peromyscus eremicus (cactus mouse)	*N. albigula; N. devia; N. lepida*
P. californicus (California mouse)	*N. fuscipes*
P. maniculatus (deer mouse)	*N. fuscipes*
P. boylei (brush mouse)	*N. stephensi*
P. leucopus (white-footed mouse)	*N. floridana*
P. truei (pinyon mouse)	*N. cinerea; N. stephensi*
Phenacomys longicaudus (red tree vole)	*N. fuscipes*
Mus musculus (house mouse)	*N. fuscipes*
Spilogale gracilis (western spotted skunk)	*N. fuscipes*
Mephitis mephitis (striped skunk)	*N. fuscipes*

Based on observations by English (1923; *N. fuscipes*), Maser et al. (1981; *N. fuscipes*), Pequegnat (1951; *N. fuscipes*), Rainey (1956; *N. floridana*), T. A. Vaughan (unpublished data; *N. albigula, N. cinerea, N. devia, N. lepida, N. stephensi*).

are committed to a diet of moist but low-energy herbage, the den is seemingly of key importance in lowering the energetic cost of thermoregulation.

Packrat dens shelter a variety of other vertebrates (Table 2.2) from predators and thermal extremes. Bones found in fossilized packrat middens may include a variety of species that died there naturally, as well as bones carried in by *Neotoma*.

PACKRATS OF THE SOUTHWEST

The habitat preferences, den-building behavior, food selectivity, and diet of the packrats inevitably determine the contents of a midden. Assuming behavioral consistency, the Pleistocene ecologist can rely on the midden species for guidance in interpreting the fossil record. Excellent den sites such as caves and rock fissures are usually in short supply and tend to be occupied by packrats continuously over long time spans. Contents of fossil middens of these dens may or may not yield proportional representation of all species in the plant communities in which the woodrats foraged, depending on the animals' dietary se-

lectivity and their collecting behavior.

White-throated Packrat, *Neotoma albigula*

This packrat is primarily a desert species, occurring broadly through the Sonoran and Chihuahuan deserts, where it is typically the only packrat species present. Although generally associated with such desert plants as cacti, creosote bush, catclaw acacia (*Acacia greggii*), palo verde, and mesquite, this packrat occurs in juniper woodland and shortgrass prairie at the northern edge of its range, and (locally) in ponderosa pine (*Pinus ponderosa*) and oak (*Quercus* spp.) woods in Mexico (Baker, 1956). It ranges from near sea level in California to roughly 2800 m elevation in Mexico. Over much of its range *N. albigula* seems closely associated with various cacti, especially prickly pear (*Platyopuntia*).

Den-building behavior is relatively strongly developed in this species, and, like most packrats, it is partial to rock outcrops, fissures, and boulder piles. In the absence of rocks, dens are built beneath thickets of catclaw or mesquite, or in patches of cactus or yucca (*Yucca* spp.). Dens are typically untidy

Fig. 2.2. Den of the bushy-tailed packrat built largely of saltbush sticks (21 km west of Winslow, Navaho County, Arizona).

structures (Fig. 2.2) that contain a great variety of debris and are typically festooned with cactus pads, the primary food in many areas (Vorhies and Taylor, 1940; Finley, 1958). Many dens in desert mountain caves are bristling masses of cholla spines with little other building material. The adaptability of *N. albigula* is remarkable. Where cactus is not locally available, as in parts of the Great Basin, dens are built largely of saltbush (*Atriplex* spp.) sticks and cow dung; and Bailey (1931) found a stick den in New Mexico high in a willow tree.

Throughout the period considered in this book, *N. albigula* living in desert areas of Arizona, New Mexico, Texas, and northern Mexico have probably been the most important builders of the dens that have yielded fossilized middens. This species commonly inhabits deep rock fissures, grottos, and caves—the kinds of places where middens are preserved—occupies these sites for many years, and brings a variety of plant material to the den. The strong predilection for cactus as food and den material, however, strongly biases the assemblage of plants that this spe-

cies discards at its middens. That is to say, the preference for cactus is so strong that middens may not contain a proportionately representative sample of the surrounding plant community. This preference does not seriously restrict this packrat's distribution, however, for where cactus is unavailable it will concentrate on other plants. In one area in northern Arizona, yucca (*Yucca baileyi*) and one-seed juniper are major foods (Dial, 1984), whereas in some deserts Mojave yucca (*Yucca schidigera*) is a dietary staple. In addition to plant material, *N. albigula* typically collects bones and mammal dung.

Bushy-tailed packrat, *Neotoma cinerea*

Well able to live under boreal conditions, the bushy-tailed packrat ranges widely from northern Canada to central Arizona, inhabiting alpine talus, spruce-fir forest, pinyon-juniper woodland, Great Basin sagebrush, shortgrass prairie, and Great Basin desert-scrub (Fig. 2.3). *N. cinerea* is primarily a petrean or cliff-dwelling species, but it will also occupy abandoned buildings.

Its large size, equaled among packrats only

Fig. 2.3. One of the habitats occupied by the bushy-tailed packrat: Great Basin desert, 1640 m elevation (21 km west of Winslow, Navaho County, Arizona).

by *N. fuscipes* of the Pacific Coast, and its preference for deep fissures or crevices in cliffs or large rock outcrops contribute to this packrat's ability to occupy seasonally cold areas. Cliffs serve as massive heat sinks with great thermal inertia. Deep-crevice temperatures at dens in Colorado varied but a few degrees daily and not more than 15°C annually, and in winter crevice temperatures generally remained some 10°–15°C above those outside (Brown, 1968). The cliffs are of equal importance as buffers against high temperatures; *N. cinerea* may suffer heat stress at temperatures of only 34°C. The nests of this packrat are large, often globular and with a side entrance, and provide considerable insulation. Brown estimated that *N. cinerea* living near the coast of northern Oregon could save eight kilocalories per winter day by spending three-fourths of the time in its nest. Another adaptation is food caching within or near the den, which is particularly highly developed in this packrat and is seemingly advantageous during winter periods when foraging is restricted. At the Arizona locality shown in Figure 2.3, *N. cinerea* piled cuttings of cliff-rose (*Cowania mexicana*) and shadscale (*Atriplex confertifolia*) in deep pockets in the sandstone, often 3 or 4 m from a den. In Colorado Finley

(1958) observed large stores of food—up to three bushels—stuffed into rock crevices by this packrat.

Den-building behavior is only moderately well developed in *N. cinerea*, and in many areas where this packrat is common one sees relatively little evidence of its presence. Loose piles of sticks, twigs, leaves, and almost invariably some bones accumulate at the base of a crevice or on a ledge protected by overhanging rock. The nest, however, is often deep in a crevice, well away from these debris piles. In Arizona access to a den may be constricted with small rocks instead of sticks.

N. cinerea is a dietary generalist, collecting and eating a variety of plants available near its den. Because it occupies a diversity of habitats, the plants available differ markedly over its range. Major foods in the high country of Colorado include chokecherry (*Prunus virginiana*) and aspen (*Populus tremuloides*); in northern Arizona cliff-rose and shadscale are important food sources. The tendency of *N. cinerea* to collect and eat a variety of plants would suggest that its middens contain a fair representation of the plants growing near the den.

Of particular interest to the paleontologist is its proclivity to collect bones of large and small vertebrates. Finley (1958) noted *N. cinerea* to collect many bones, feces of carnivores, and owl pellets. At one den near Hermosa, Colorado, he recorded the bones of eleven species of vertebrates, including such disparate types as bear (*Ursus americanus*) and dusky grouse (*Dendragopus obscurus*). Dens near Winslow, Arizona, invariably contained vertebrate bones, especially those of cottontails (*Sylvilagus audubonii*) and jackrabbits (*Lepus californicus*), as well as coyote feces, which, in turn, contained rabbit and rodent bones. This packrat is large and strong enough to collect entire, completely articulated axial skeletons or legs of jackrabbits. When raptors use rock outcrops occupied by *N. cinerea* as feeding sites, their castings contribute to the midden ossuary.

The range of *N. cinerea* expanded southward (or to lower elevations) during the Wisconsin full glacial along with boreal and montane plant species. Many, if not all, fossil middens that contain boreal plants were probably made by *N. cinerea* (Harris, 1985).

Desert Packrat, *Neotoma lepida*

Desert packrats range from southeastern Washington and southwestern Idaho through Nevada, the drier parts of Utah, northwestern Colorado, and the deserts and drier chaparral of southern California and Baja California. In Idaho and Utah they are associated with such Great Basin xerophytes as big sagebrush (*Artemisia tridentata*), shadsale, and Utah juniper (*Juniperus osteosperma*). More typically they live in the lower deserts with creosote bush, ironwood, palo verde, brittlebush (*Encelia farinosa*), and many species of cacti. This is the only packrat found in eastern Baja California, the hottest and driest area of North America.

As do most packrats, *N. lepida* prefers rock outcrops for den sites. In open sites dens are often situated in cactus or yucca patches, or at the bases of such desert trees as catclaw or joshuas (*Yucca brevifolia*). On Carmen and Danzante islands in the Gulf of California, *N. lepida* build dens in patches of galloping cactus (*Pitahaya agria, Machaerocereus gummosus*) or in wind-pruned thickets of ironwood; fish bones are common in dens near shorelines (Vaughan and Schwartz, 1980).

Den-building behavior in this species varies from one area to another. Twelve desert packrat dens built at the bases of catclaw bushes in Joshua Tree National Monument averaged 52 cm high and 102 cm wide (Cameron and Rainey, 1972); they were made of interlacing catclaw twigs. Similarly, large and conspicuous houses in this same area were built of cactus, with shrubs such as creosote bush as support. Where huge blocks of rock have lodged against one another and formed cave-like "rooms," dens can be unusually large. In Colorado Finley (1958) recorded such a den measuring 3.7 m long and 1.1 m wide, with a volume of over 2.5 m³. At the other extreme, many dens amid talus or boulder piles are inconspicuous accumulations of sticks and food fragments. Desert packrats commonly gather small bones and carnivore feces.

The desert packrat's variability in den-building behavior is matched by its dietary flexibility. Over its range this species eats a variety of plants; locally it is a dietary specialist. In Joshua Tree National Monument, for example, it eats primarily shrub live oak (*Quercus turbinella*) when available, switches to California juniper (*Juniperus californica*) when living with the larger *N. fuscipes*, eats leaves and seeds of catclaw when available, and depends on creosote bush leaves when in a teddy bear cholla (*Opuntia bigelovii*) and creosote bush community (Cameron, 1971; Cameron and Rainey, 1972). In coastal sage of southern California, white sage (*Salvia apiana*) is the favorite food (Meserve, 1974), whereas on Danzante Island ironwood (*Olneya tesota*) leaves are the mainstay (Vaughan and Schwartz, 1980). Desert populations of this packrat are near the lower size limit for *Neotoma*, with males averaging about 122 g (Lee, 1963). Small size in this case may be an adaptation to a scanty food supply.

Mexican Packrat, *Neotoma mexicana*

Inhabiting a wide variety of plant communities, this species occurs from northern Colorado through parts of Utah, Arizona, and New Mexico, to Central America (El Salvador). In the southwestern United States Mexican packrats occupy montane areas dominated by ponderosa pine and Gambel's oak (*Quercus gambelii*), by pinyon-juniper or juniper woodland, or (rarely) by alpine fir (*Abies lasiocarpa*) and limber pine (*Pinus flexilis*). In Mexico it has an even broader ecological range, occurring in pine-oak forests, tropical thornscrub, tropical deciduous forests, and tropical evergreen forests. Through most of its range in the United States it is a mountain dweller, living at elevations as high as 3300 m on the San Francisco Peaks of Arizona, whereas in Mexico it occurs not only in the mountains (as high as 3075 m in Coahuila), but near sea level along the coasts of Sinaloa and Nayarit.

Throughout its broad range the Mexican packrat is consistently petrean, preferring cliffs, jumbles of huge boulders, or massive outcrops. Because of its rather specialized preferences in den sites and its apparent avoidance of desert conditions, this species has a patchy or even an "insular" distribution in montane refugia. In southern Arizona, for example, Mexican packrats inhabit the higher elevations (mostly above 2000 m) of five mountain ranges; these populations are isolated from one another by broad areas of desert. On a microgeographic scale this species often has a spotty distribution, occupying only the largest outcrops or steepest cliffs, and is completely absent from places offering less imposing cover that are acceptable to other packrat species.

Mexican packrats are indifferent collectors (Baker, 1956; Finley, 1958; Pyc, 1984). In northern Arizona I observed casual scatterings of sticks and plant cuttings associated with packrat feces and blackened deposits of urine to be the usual signs of Mexican packrat presence. Occasionally, conspicuous masses of plant cuttings are cached in rock fissures. This species commonly collects bones, some as large as the pelvis or femur of a mule deer (*Odocoileus hemionus*). Mexican packrats seldom build dens, perhaps because the deep and labyrinthine fissures and crevices that they usually inhabit provide sufficient protection from weather, most predators, and temperature extremes. In such places dens have little purpose.

The Mexican packrat typically eats a variety of plants. Middens in northern Arizona usually contain several species of plants (Table 2.3), with brickellia (*Brickellia californica*), mountain mahogany (*Cercocarpus montanus*), and prickly pear (*Opuntia macrorhiza*) occurring most frequently. Finley (1958) recorded Gambel oak, mountain mahogany, skunkbush (*Rhus trilobata*), and juniper as major food items in Colorado.

Stephens' Packrat, *Neotoma stephensi*

More restricted geographically, ecologically, and in terms of diet than other species in the United States, Stephens' packrat inhabits

Table 2.3. *Plants in twenty-five Mexican packrat middens and within 50 m of each midden; 13 km northeast of Flagstaff, Coconino County, Arizona, 2370 m elevation.*

Plant species	Relative frequency near middens %	Relative frequency in middens %	Rare or absent away from cliffs
Brickellia californica	0.79	0.96	
Cercocarpus montanus	0.73	0.56	
Cowania mexicana	0.48	0.04	x
Opuntia macrorhiza	0.35	0.40	x
Chrysothamnus nauseosus	0.31	0.12	
Ribes cereum	0.22	0.12	
Berberis fremontii	0.18	0.24	x
Yucca baccata	0.18	0.04	x
Pinus edulis	0.09	0.44	x
Datura meteloides	0.04	0.12	
Juniperus deppeana	0.02	0.44	x
Marrubium vulgare	0.02	0.28	
Pinus ponderosa	0.01	0.36	

Data on vegetation partly from Pyc (1984).

parts of northern Arizona, extreme south-central Utah, and part of western New Mexico, in my experience always in association with juniper, usually *Juniperus monosperma* (Vaughan, 1980, 1982). It is commonly found with Colorado pinyon (*Pinus edulis*), Apache plume (*Fallugia paradoxa*), skunkbush, and snakeweed (*Gutierrezia sarothrae*); over much of its range no other packrat species is present.

Located at the bases of junipers or in small rock outcrops, dens are typically built largely of sticks and are rather compact, usually from 1.0 to 1.5 m in diameter. The trademark of Stephens' packrat dens is juniper foliage: freshly cut juniper leaves are present at nearly every occupied den, virtually to the exclusion of other cuttings.

Stephens' packrat is a dietary specialist to a degree unmatched by other species of *Neotoma* that have been studied. Some 90% of its yearly diet is juniper (mostly foliage), and in winter juniper comprises 95% of the diet (Vaughan, 1982). This packrat and the red tree vole (*Phenacomys longicaudus*) are the only two North American mammals that are conifer-leaf specialists.

Because Stephens' woodrats collect and eat mostly juniper, their dens offer strongly biased samples of the plant community. During a five-year study near Flagstaff, Arizona, for example, the only plant other than juniper that I found to be of more or less regular occurrence in the middens was a mistletoe (*Phoradendron juniperinum*) that grew on junipers. The most important perennial in the community in terms of relative frequency (Apache plume) was absent from forty-six middens that I examined carefully. Of all the species of the Southwest, the middens of this packrat give the least-complete picture of the plant community.

My associates and I studied the nocturnal activities of Stephens' packrats by using powdered, fluorescent pigments (See Lemen and Freeman, 1985). The packrats were livetrapped at sundown, dusted with fluorescent powder, and released at their dens. Just before dawn their powder trails were followed with the aid of an ultraviolent light. The mean distance eighteen packrats moved away from their dens during one night was 28.9 m (S.D. = 17.6). Although six individuals moved no far-

ther than 20 m, five moved 40 m or more, and one of these visited a juniper 78 m from its den. Dial (1984) found that some Stephens' packrats fed on junipers over 80 m from their dens. These packrats are strongly arboreal, often moving rapidly and with agility to forage among the topmost interlacing juniper branches.

Western Arizona Packrat, *Neotoma devia*

Previously regarded as representing the species *N. lepida*, populations of packrats from western Arizona were shown by Mascarello (1978) to differ markedly from those of California and Nevada with regard to allozymes, chromosomes, blood proteins, and features of the skull and penis. He named the western Arizona populations *N. devia*, a proposal adopted here. The common name given above is used for convenience; no other is in general use.

This species, probably the smallest of all the packrats, occupies low deserts and desert mountains in western Arizona and parts of the Great Basin desertscrub of northern Arizona. In the low deserts it is associated with such xerophytic plants as creosote bush, brittlebush, palo verde, and various chollas (*Opuntia* spp.), whereas in the Great Basin it occurs with Mormon tea (*Ephedra viridis*), Apache plume, fourwing saltbush, and skunkbush. In its ability to live in hot, barren deserts, *N. devia* resembles *N. lepida*.

Characteristically its dens are skimpy. In the Great Basin desert in northern Arizona, dens were typically built in spots where there was little shelter and much Mormon tea, this packrat's primary food (Dial, 1984). Dens and middens that I observed in the Black Mountains of western Arizona contained cholla (*Opuntia acanthocarpa* and *O. bigelovii*) and fragments of creosote bush and wolfberry (*Lycium* sp.), and were among rock piles or in cliff crevices. Areoles of dead cactus were a favorite item and gave even occupied dens an ancient appearance.

Over much of its restricted range *N. devia* is sympatric with the far more widespread and larger *N. albigula*, and it seems unable to compete effectively with the latter for choice den sites. Probably *N. devia* has consistently occupied "marginal" sites where fossilization of packrat middens seldom occurs.

DISCUSSION

Of central interest in this discussion are factors that affect the contents of packrat middens and their utilization as samples of past
or present biotic communities.

Packrats require large amounts of water and
obtain this from moist leaves. Plants protect
their leaves from folivores by incorporating
defensive chemicals (Rhoads and Cates, 1976;
Rhoads, 1979), and packrats (and other mammalian folivores) have two major strategies
for coping with these chemicals. Some eat a
diverse array of plants and thereby avoid the
ingestion of large amounts of chemicals from
any one plant, while others specialize on one
or a few plants and develop finely tuned
means of dealing with their specific defenses.
The strategy used by a species of packrat is
probably related to habitat quality. In more
mesic habitats packrats can get adequate
water from many species of plants and can
utilize the first (generalist) strategy, whereas
in xeric habitats, with relatively low plant diversity and few plants that remain succulent
and nutritious throughout the year, packrats
must use the second (specialist) strategy.
Among packrats of the western United States,
then, the species that occupy chiefly boreal or
montane areas tend to be dietary generalists,
while those of desert, semiarid chaparral or juniper woodland are dietary specialists. Thus,
although the middens of *N. cinerea* and *N.
mexicana* are relatively good biotic community indicators, those of *N. albigula, N. devia,
N. lepida,* and *N. stephensi* are relatively poor
indicators (Fig. 2.4). Although this broad generalization is probably valid, a number of factors affect the ways in which midden contents reflect plant communities.

In many areas where two or more species of
packrats are sympatric, they are more or less
segregated microgeographically (Fig. 2.5). In
these places the species may use the same
plant community but each has different dietary preferences and, as a result of patterns
of interspecific dominance, each species tends
to occupy den sites of different quality. In
parts of the Black Mountains of northwestern
Arizona, for example, *N. albigula* and *N. stephensi* occur together locally in roughly equal
densities; the former eats largely cactus and
usually occupies the larger rock outcrops,
while the latter eats largely juniper and uses

small outcrops and talus as den sites (personal
observation). The middens of *N. stephensi*
contain mostly juniper; the middens of *N. albigula* largely cactus. Of central importance
here is the fact that each species of packrat
has unique dietary preferences and different
collecting behaviors; a midden contains a distinctive plant assemblage that is characteristic of the packrat species in residence. Within
the same plant community, therefore, the
middens of two packrat species may contain
almost completely complementary groups of
plant species.

The distributions of various species of
packrats have expanded and contracted in association with shifts in the distributions of
plant communities. When a southwestern
cave was surrounded by spruce and fir during
the Wisconsin full glacial, it probably harbored *N. cinerea,* whereas the same cave,
when associated with dry juniper woodland in
the Holocene, may have sheltered *N. stephensi, N. albigula,* or *N. lepida.* The Wisconsinan *N. cinerea* perhaps chose those
plants in the community with the most boreal
or montane affinities; the Holocene *N. albigula,* in contrast, probably selected plants
most nearly typical of deserts. The middens
left by *N. cinerea* might have a "northern"
bias, whereas those of *N. albigula* might have
a "southern" bias.

Cliffs and rock outcrops, where most species of packrats prefer to build their dens,
present special complexities. These places
provide rather unique ecological conditions.
Compared to plant communities away from
cliffs, the near-cliff communities may have
higher species diversity; plants are present
that depend on the more mesic microhabitats
resulting from precipitation runoff from the
cliffs, as are species favored by xeric, warm
microhabitats in south-facing sites. Packrats
that live in cliffs may thus have access to
plants that are absent in plant communities
away from cliffs, and the fossil record will
yield samples only of the former. This problem is illustrated by examples involving cliff-
dwelling *N. mexicana* in northern Arizona.
Frequently encountered in middens of this
packrat are plants that are uncommon or absent away from the cliffs, whereas some of
the most common plants away from cliffs

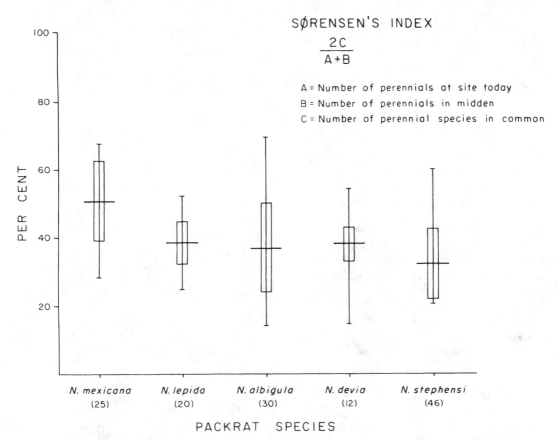

Fig. 2.4. Comparisons of species of plants on packrat middens and species within 50 m of each midden using Sørensen's index of similarity. Horizontal line = x; rectangle = s.d.; vertical line = range. Numbers of middens examined appear in parentheses. Localities for middens: *N. mexicana*, see legend for Table 2.3; *N. lepida*, 8 km northeast of San Bernardino, San Bernardino County, California; *N. albigula*, Plomosa Mountains, La Paz County, Arizona; *N. devia*, see legend for Fig. 2.4; *N. stephensi*, 32 km northeast of Flagstaff, Coconino County, Arizona.

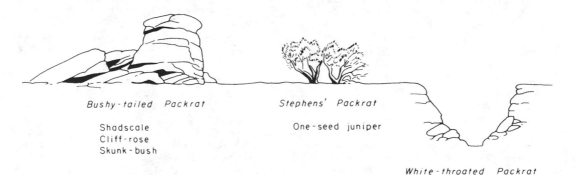

Fig. 2.5. Microhabitat separation in three species of packrats in northern Arizona (see legend of Fig. 2.2B for locality). The plants found most frequently on middens are listed for each species of packrat.

rarely appear in middens. (Table 2.3).

Packrats are seldom such generalized herbivores that their middens contain unbiased plant community samples; their middens, however, clearly offer revealing glimpses of past biotic communities. The importance of middens lies largely in the fact that packrats occupy dens and contribute to middens over spans of time that allow for the fortuitous aggregation of diverse bits of evidence. Packrats occasionally discard fragments of seldom-eaten plants at their dens, they collect bone-laden carnivore feces or parts of prey skeletons discarded by carnivores, and they gather the dung of large herbivores. In time, some of this material is deposited in middens, as are bones of small vertebrates that use packrat dens as shelter. It is this diversity of material, rather than only the food plants brought to the middens by packrats, that makes fossil packrat middens valuable windows to the past.

REFERENCES

Baker, R. H. (1956). "Mammals of Coahuila, Mexico." *Museum of Natural History, University of Kansas* 9, 127–335.

Bailey, V. (1931). "Mammals of New Mexico." *North American Fauna* 53, 1–412.

Birney, E. C. (1973). "Systematics of three species of woodrats (genus *Neotoma*) in central North America." *Museum of Natural History, University of Kansas, Miscellaneous Publications*, 58, 1–173.

Brown, J. H. (1968). "Adaptation to environmental temperature in two species of woodrats, *Neotoma cinerea* and *N. albigula*." *Museum of Zoology, University of Michigan* 135, 1–48.

Cameron, G. N. (1971). "Niche overlap and competition in woodrats." *Journal of Mammalogy* 52, 288–96.

Cameron, G. N., and Rainey, D. G. (1972). "Habitat utilization by *Neotoma lepida* in the Mohave Desert." *Journal of Mammalogy* 53, 251–66.

Carleton, M. D., and Musser, G. G. (1984). "Muroid rodents." In *Orders and families of recent mammals of the world*. S. Anderson and J. K. Jones, Jr., eds., pp. 289–379. New York: John Wiley and Sons.

Chess, T., and Chew, R. M. (1971). "Weight maintenance of the desert woodrat (*Neotoma lepida*) on some natural foods." *Journal of Mammalogy* 52, 193–95.

Dalquest, W. W. (1983). *Mammals of the Coffee Ranch local fauna Hemphillian of Texas*. Pearce-Sellards Series, Texas Memorial Museum, no. 38, 1–41.

Dial, K. P. (1984). "Four sympatric species of *Neotoma*: Ecological aspects of morphology, diets, and behavior." Ph.D. diss., Northern Arizona University, Flagstaff.

English, P. F. (1923). "The dusky-footed woodrat (*Neotoma fuscipes*)." *Journal of Mammalogy* 4, 1–8.

Fahlbusch, V. (1966). "Cricetidae (Rodentia, Mammalia) aus der mittelmiocänen spaltenfullung Erkortshofen bei Eichstatt." Mitt. Bayer. Staatssamml. *Palaont. Hist. Geol.* 6, 109–31.

Finley, R. B. (1958). "The woodrats of Colorado: Distribution and ecology." *Museum of Natural History, University of Kansas* 10, 213–552.

Freeland, W. J., and Janzen, D. H. (1974). "Strategies in herbivory by mammals: The role of plant secondary compounds." *American Naturalist* 108, 269–89.

Hall, E. R. (1981). *The mammals of North America*, vol. 2. New York: John Wiley and Sons.

Harris, A. H. (1985). *Late Pleistocene vertebrate paleoecology of the West*. Austin: University of Texas Press.

Lee, A. K. (1963). "The adaptations to arid environments in woodrats of the genus *Neotoma*." *University of California Publications in Zoology* 64, 57–96.

Lemen, C. A., and Freeman, P. W. (1985). "Tracking mammals with fluorescent pigments: A new technique." *Journal of Mammalogy* 66, 134–36.

McClure, P. A., and Randolph, J. C. (1980). "Relative allocation of energy to growth and development of homeothermy in the eastern woodrat (*Neotoma floridana*) and hispid cotton rat (*Sigmodon hispidus*)." *Ecological Monographs* 50, 199–219.

MacMillen, R. E. (1964). "Population ecology, water relations, and social behavior of a southern California semidesert rodent fauna." *University of California Publications in Zoology* 71, 1–66.

McNab, B. K. (1978). "Energetics of arboreal folivores: Physiological problems and energetic consequences of feeding on an ubiquitous food supply." In *The ecology of arboreal folivores*. G. G. Montgomery, ed., pp. 153–62. Washington, D.C.: Smithsonian Institution.

———. (1980). "Food habits, energetics, and the population biology of mammals." *American Naturalist* 116, 106–24.

Martin, L. D. (1980). "The early evolution of the Cricetidae in North America." *University of Kansas Paleontological Contributions* 102, 1–42.

Mascarello, J. T. (1978). "Chromosomal, biochemical, mensural, penile, and cranial variation in desert woodrats (*Neotoma lepida*)." *Journal of Mammalogy* 59, 477–95.

Maser, C.; Mate, B. R.; Franklin, J. F.; and Dyrness, C. T. (1981). *Natural history of Oregon Coast mammals.* Pacific Northwest Forest and Range Experiment Station, Forest Service, General Technical Report PNW-133.

Meserve, P. L. (1974). "Ecological relationships of two sympatric woodrats in a coastal sage scrub community." *Journal of Mammalogy* 55, 442–47.

Pequegnat, W. E. (1951). "The biota of the Santa Ana Mountains." *Journal of Entomology and Zoology* 42, 1–84.

Pyc, L. R. (1984). "Behavioral ecology of the Mexican woodrat (*Neotoma mexicana*)." Master's thesis, Northern Arizona University, Flagstaff.

Rainey, D. G. (1956). "Eastern woodrat, *Neotoma floridana*: Life history and ecology." *Museum of Natural History, University of Kansas* 8, 535–646.

Rhoads, D. F. (1977). "The anti-herbivore defense of *Larrea.*" In *The biology and chemistry of the creosote bush in the New World deserts.* (T. J. Mabry, J. Hunziker, and D. R. DiFeo, eds., pp. 135–75. Stroudsburg, Penn.: Dowden, Hutchison and Ross.

———. (1979). "Evolution of plant chemical defense against herbivores." In *Herbivores: Their interactions with secondary plant metabolites.* G. A. Rosenthal and D. H. Janzen, eds., pp. 3–54. New York Academic Press.

Rhoads, D. F., and Cates, R. G. (1976). "Toward a general theory of plant anti-herbivore chemistry." In *Biochemical interactions between plants and insects.* J. W. Wallace and R. L. Mansell, eds., pp. 168–213. *Recent Advances in Phytochemistry.* New York: Plenum Press.

Schmidt-Nielsen, K., and Schmidt-Nielsen, B. (1952). "Water metabolism of desert animals." *Physiological Review* 32, 135–66.

Schwartz, C. C.; Nagy, J. G.; and Regelin, W. L. (1980a). "Juniper oil yield, terpenoid concentration, and antimicrobial effects on deer." *Journal of Wildlife Management* 44, 107–13.

Schwartz, C. C.; Regelin, W. L.; and Nagy, J. G. (1980b). "Deer preference for juniper forage and volatile oil treated food." *Journal of Wildlife Management* 44, 114–20.

Vaughan, T. A. (1980). "Woodrats and picturesque junipers." In *Aspects of vertebrate history: Essays in honor of Edwin Harris Colbert.* L. L. Jacobs, ed., pp. 387–401. Flagstaff: Museum of Northern Arizona Press.

———. (1982). "Stephens' woodrat, a dietary specialist." *Journal of Mammalogy* 63, 53–62.

Vaughan, T. A., and Czaplewski, N. J. (1985). "Reproduction in Stephens' woodrat: The wages of folivory." *Journal of Mammalogy* 66, 429–43.

Vaughan, T. A., and Schwartz, S. T. (1980). "Behavioral ecology of an insular woodrat." *Journal of Mammalogy* 61, 205–18.

Voorhies, C. T., and Taylor, W. P. (1940). *Life history and ecology of the white-throated woodrat* Neotoma albigula *Hartley, in relation to grazing in Arizona.* University of Arizona College of Agriculture, Agricultural Experiment Station, Technical Bulletin 86, pp. 453–529.

Warren, E. R. (1942). *The mammals of Colorado: their habits and distribution.* 2d ed. Norman: University of Oklahoma Press.

Zimmerman, E. G., and Nejtek, M. E. (1977). "Genetics and speciation of three semispecies of *Neotoma.*" *Journal of Mammalogy* 58, 391–402.

Chapter 3

Woodrat Ecology and Behavior and the Interpretation of Paleomiddens

Robert B. Finley, Jr.

INTRODUCTION

The conspicuous houses and middens of packrats, or woodrats (genus *Neotoma*), have long attracted attention because of their size and the diversity of their contents. Early mammalogists learned a great deal about the life history and ecology of woodrats by dismantling their houses and analyzing the middens. The structure and materials used to build the house were examined; food litter was analyzed to determine diet; burrow systems beneath houses were excavated to locate nests, food caches, and escape routes. Such studies have been conducted on *N. fuscipes* (Linsdale and Tevis, 1951; Vestal, 1938), *N. albigula* (Vorhies and Taylor, 1940), *N. floridana* (Poole, 1940; Rainey, 1956), *N. lepida* (Stones and Hayward, 1968), and *N. cinerea*, *N. mexicana*, and *N. albigula* (Finley, 1958). These studies were often supplemented by direct observations of woodrat behavior and live-capture–mark-and-release studies of movements and populations.

Interest in packrats and their midden-building habits greatly increased when it was discovered that Pleistocene fossil middens are widespread in parts of the arid West (Wells and Jorgensen, 1964; Van Devender and Spaulding, 1979). Analysis of these ancient midden deposits has raised many questions of interpretation that depend on knowledge of living packrats. I will draw from the woodrat ecology literature and my own field experiences to discuss how the behavior of this animal might affect the nature of the fossil record.

The words used by hunters, naturalists, and scientists to describe "sign" have different meanings to different people. They are used here as follows: *sign*, any indirect physical evidence of animal activity; *den*, any natural or constructed shelter used regularly for protected living space, rest, or rearing of young; *house*, a shelter constructed of various materials to serve as a den and located at a site that may or may not provide some natural shelter; *nest*, a cup or ball of some soft material used as a bed within a house or other den; *midden*, a layer or accumulation of materials brought to or dropped at any location by some animal, usually but not necessarily the den occupant, and including any other materials such as leaves, pollen, and sand that may be deposited there by other means; *paleomidden*, a "fossil" midden that has persisted long enough to survive environmental changes, but has not necessarily been altered over time; *food litter*, material brought to one place for feeding and dropped as unconsumed debris; *fecal deposit*, a layer primarily of fecal droppings that may or may not be indurated by dried urine; *urinary deposit*, a layer of dried urine usually including some fecal droppings and other materials; *post*, a site used repeatedly by an animal for some function, such as a feeding, urinary, or scent post. Posts may give rise to deposits such as food litter either within or away from dens.

Since houses and middens are built by many kinds of animals having very different habits, it is important to identify the agent of deposition. Packrat accumulations can be recognized by size, shape, and non-woody composition of fecal pellets; heavy urine stains or deposits (usually brown or black and hard, smooth, or glossy); presence of many sticks, bones, cactus joints, or other extraneous articles along with intermixed plant cuttings; and tooth marks or materials gnawed by teeth of the right "rat size." Individual features may not be distinguishable from those of some other animal, but the combination of several is unmistakable. Middens of other species occurring in the same area, such as the porcupine (*Erethizon dorsatum*) or marmot (*Marmota flaviventris*), can be recognized by differences in one or more of the above-named features.

Although some other mammals do create middens, were it not for the general habits of the packrat much of the late Quaternary biogeography of the western United States would remain a mystery. Factors that favor paleoecological use of their middens include good preservation in a dry, protected environment; materials incorporated in good, recognizable condition; contents of wide enough variety to yield a good sample of the environment; minimal alteration of the deposit by subsequent intrusion, mixing, or weathering; short transport distance of the material from its source; distinctive features attributable to the animal building the midden; and knowledge of the ecology and behavior of the midden builder.

ADAPTIVE NICHE AND ECOLOGICAL REQUIREMENTS OF *NEOTOMA*

All packrats are terrestrial, scansorial herbivores possessing to some degree the collecting habit and other characteristics that set them apart from nearly all other North American rodents. They are good climbers and can dig burrows but are not specialized for either arboreal or fossorial life. In winter they remain more or less active in or near their dens,

consuming stored food in bad weather. Their house-building and food-storing habits reflect a den-centered mode of life with a rather small home range and strong territoriality. Like most nocturnal mammals they rely on scents for species and sex recognition. They mark their dens and territories and communicate by odors in the urine, feces, and sebum of the ventral sebaceous gland (August, 1978).

Woodrats are unsocial animals and each den is usually occupied by a single individual, except during the mating season and when the female is rearing her litter. In good habitat many clustered houses reflect a large number of individual occupants, each with small, compressed territories, not a true colony. If there is a die-off, the surviving rats expand their territories and occupy adjacent dens as well as their primary ones. At such times the same rat may be live-trapped at more than one den on successive nights.

Houses and other dens provide protection against adverse weather (Brown, 1968) as well as most predators (Fig. 3.1). Owls, hawks, and the medium-sized carnivores are unable to penetrate a substantial stick house or smaller rock crevices. Snakes and weasels can enter packrat dens, but some protection is provided by alternate escape tunnels and exits. Nevertheless, rats do fall prey to all of these enemies when they emerge from the den to forage for food or gather materials to pile on the den (Vestal, 1938). No doubt this danger accounts for the usual caution and speed with which rats usually move from one bit of outdoor cover to another. When a rat has plenty of food plants and collectibles close to the den entrance, it is a considerable advantage.

The great diversity of plants eaten by woodrats has been widely reported, and one can easily gain the impression that their herbivory is unspecialized, but this is not the case. Woodrats eat mostly foliage or soft vegetative parts of woody plants and forbs, but rarely grass. Strong specialization is evident in some cases (Dial and Czaplewski, *this volume*, chap. 4). They also eat mushrooms and some reproductive parts of higher plants (flowers, fruits, and seeds), but these are rarely the primary source of sustenance and are not usually stored in quantity. Woodrats have a large caecum for the microbial breakdown of the large quantities of leaves and

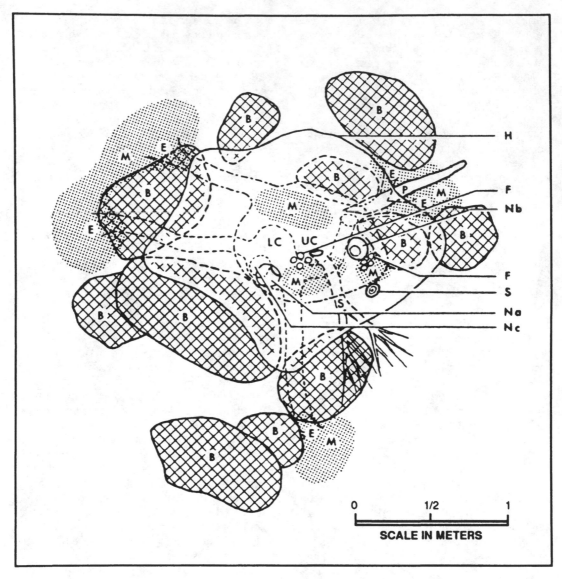

Fig. 3.1. Ground plan of house of *Neotoma albigula* near Gateway, Mesa County, Colorado. *B*, boulder; *E*, entrance; *F*, food; *H*, outline of house; *LC*, lower chamber; *M*, midden; *Na*, nest a; *Nb*, nest b; *Nc*, nest c; *P*, post; *S*, sagebrush stem; *UC*, upper chamber.

stems that form the great bulk of their diet. Their molars are high-crowned with several enamel folds to increase resistance to grinding wear. Their dentition is intermediate in this respect between that of the microtines and the peromiscine rodents, the former being primarily grass eaters and the latter seed and insect eaters. Within the broad adaptive zone of soft foliage eaters, different species of *Neotoma* show some specializations, particularly

for greater use of cactus pulp by the desert dwellers.

In contrast to most rats and mice, woodrats have small litters, usually one to four young, and one to three litters per year. Females have only four mammae. They protect the young, which remain inside the den until they are able to scamper around. If the mother is driven from the den when the young are sucklings, she often flees to other cover dragging one or more nestlings hanging from her nipples (Vestal, 1938). Low productivity and high survival of young are sufficient to maintain relatively stable populations because of strong maternal care and shelter improved by gathering den materials. Even during dispersal the young may benefit from the house building of older rats by moving into an unoccupied or weakly held house or by remaining in the maternal den vacated by the mother (Linsdale and Tevis, 1951).

ECOLOGICAL DIVERGENCE OF *NEOTOMA* SPECIES

The optimum ecological niche of each species of packrat cannot be clearly defined; there seems to be considerable overlap in some, as well as a wide range of geographic variation within a few, such as *N. cinerea* and *N. floridana*. Many of the ecological differences among packrat species deserve mention here because they are reflected in the dens and middens of the species and thus affect interpretations of fossil midden deposits.

Rock Dens and House Building

A significant divergence exists between species denning primarily in rock crevices and those building houses in the open. The rock dwellers have less need to build houses and are less avid collectors. Although sticks and other materials may be carried into a cave, they are not usually piled and organized into a house with passages and chambers. On the other hand, the builders of big houses away from rock crevices are indefatigable packers. Species between these extremes may either occupy rock dens or build houses, depending upon the kinds of habitat available and the presence of competing species of *Neotoma*.

Another important specialization is in climbing ability, either arboreal or on cliffs. The best climbers are *N. cinerea* on cliffs and *N. fuscipes* in trees. *N. cinerea* is only a mediocre house builder (Dixon, 1919) but is fairly industrious in packing and piling sticks, bones, and even flat stones into rock clefts. *N. fuscipes* in California is an outstanding house builder. On the ground it builds stick houses sometimes as much as 2 m in height and 2.3 m in greatest diameter (Vestal, 1938). Its climbing ability is reflected in its frequent construction of large stick houses in live oak trees, sometimes as much as 6 m above the

ground. The tail of each species is somewhat differently specialized for use as a balancing aid in climbing, *N. cinerea* having a bushy tail that serves also as a warm wrap. *N. fuscipes* has a longer tail than any other woodrat, with short hair like other round-tailed woodrats.

N. mexicana is a narrowly specialized rock dweller with the weakest collecting instinct of any packrat. It almost never builds even a poor house. Its shelter needs can be met by almost any horizontal ledge or talus blocks, and it may occupy deep fissures and caves in higher cliffs when these are not already occupied by *N. cinerea*.

N. floridana in the central and eastern United States is a highly successful generalist. It lives in cliffs and caves in the Appalachian region, is a good tree climber, and also can build large, substantial houses. In the sparsely wooded prairie region of the Midwest and Great Plains it also is a fairly good burrower, especially where rocks and woody cover are scarce above ground. *N. albigula*, *N. lepida*, and *N. micropus* are also generalists, using rock shelters in some areas and in others building substantial houses of cactus joints and thorny sticks. Their denning habits seem to depend on the presence or absence of congeneric competitors.

Food Specializations

The kinds of plants most prevalent in the main part of a species' range are usually preferred or utilized in greatest amount, an attribute especially appreciated by paleoecologists. There are both qualitative and quantitative differences in foods consumed and stored. Each kind of woodrat eats a great many kinds of plants in small amounts, but the few kinds of plants and plant parts gathered and consumed in large quantity are the most important for recognition of adaptive differences in food habits.

In general, the species that occur in more boreal or mesic life zones, such as *N. cinerea*, *N. mexicana*, and *N. floridana*, depend primarily on the leaves of woody plants and forbs (Finley, 1958; Rainey, 1956). *N. cinerea* makes much use of conifer needles (*Pseudotsuga*, *Pinus*, *Juniperus*), whereas *N. floridana* makes little use of conifers, preferring instead a wide variety of broad-leaved plants. *N. fus-*

cipes, which lives primarily in the sclerophyl woodlands and chaparral of the Pacific Coast, eats a wide variety of leaves, stems, and flowers, particularly the leaves of the California live oak (*Quercus agrifolia*) and other broad-leaved vegetation (Linsdale and Tevis, 1951; Atsatt and Ingram, 1983).

The species of more austral or xeric life zones, such as *N. micropus*, *N. albigula*, and *N. lepida*, depend much more on the pulp of many kinds of cactus joints, which they consume in large quantity for water as well as food. They also utilize many other desert plants. Some species of prickly pear and cholla (*Opuntia*) are used heavily for food, water, house structure, and defense against intruders. *N. mexicana*, like *N. fuscipes*, consumes very little cactus. *N. cinerea* and *N. floridana* are intermediate in amounts of cactus consumed when available.

SIGNIFICANT FACTORS IN PACKRAT MIDDEN ANALYSIS

Species Differences and Geographic Diversity of Ecological Factors

In the Southwest two or three species of *Neotoma* often occur at the same locality, in either the same or a different habitat. In Grand Canyon National Park there are five species: three confined to the South Rim, one to the North Rim, and one in the canyon on both sides of the Colorado River (Hoffmeister, 1971). Overlapping species usually have differing ecological traits that may or may not be evident by midden examination. Some widespread species live in greatly varying geographical conditions, such as *N. cinerea* and *N. floridana*. Each has differentiated into subspecies that differ ecologically as well as morphologically. Within the species *N. cinerea* the ecological differences between *N. c. orolestes* and *N. c. arizonae* are greater than the differences between the two species *N. albigula* and *N. micropus* (Finley, 1958).

Ecological diversity in packrats influences midden analysis in many ways. If the species responsible for the midden cannot be identified, as is often the case, the interpretation of midden contents must be based on a generalized concept of *Neotoma* ecology. Identification of the packrat species can provide a more reliable interpretation for the presence or absence (or relative abundance) of certain plant species. For instance, while the absence of cactus areoles in a midden of *N. mexicana* says nothing about presence or absence of cactus in the area, it does tell us that the life-zone must be higher than Lower Sonoran, where *N. mexicana* does not occur.

The bushy-tailed woodrat is the most boreal species and one of the most generalized in denning and feeding habits. Throughout most of the northerly reaches of its range it has no congeners, whereas in the southern areas it must divide the woodrat niche with one or more other species. It is primarily a rock, cliff, and cave dweller, preferring vertical rock clefts, where its accumulated sticks and deposits can be seen in fissures and along cliff bases (Fig. 3.2). Cary (1911) reported *N. cinerea* dens in northwestern Colorado in hollow junipers and cottonwoods, in arroyo banks, and in thickets of buffalo-berry (*Shepherdia canadensis*). Grinnell et al. (1930) reported *N. cinerea occidentalis* in the eastern portion of the Lassen Peak region in California living in rocks and in juniper trees (*Juniperus occidentalis*). Nests were located within hollow tree trunks. In addition, houses of juniper were constructed on branches close against the trunk, sometimes as high as 6 m above the ground. Typical habitat in the higher central portion of the region was in rock slides and about boulders.

N. cinerea eats a wide variety of both broad-leaved and conifer foliage. Boreal and mesic species are preferred, but in southern Colorado, northern New Mexico, and northern Arizona the diet includes considerable amounts of cactus joints and other xeric plants. The broad diversities between *N. c. orolestes* and *N. c. arizonae* in den situations and foods eaten were described by Finley (1958). In southwestern Colorado, where dens of the latter occur in the same rock habitat with *N. mexicana* and *N. albigula*, their dens can usually be distinguished comparatively by the lack of cactus spines in middens of *N. mexicana*, the solidly intermeshed bed of spine areoles of *N. albigula*, and the intermediate amounts of spines in *N. cinerea* middens. *N. cinerea* usually drags in more and larger sticks and bones, *N. albigula* more livestock dung, and *N. mexicana* the least material, mostly twigs.

N. mexicana, which adjoins the range of *N.*

Fig. 3.2. *Neotoma cinerea* den 5 km north of Almont, Gunnison County, Colorado, in rock clefts with midden of sticks and litter below.

cinerea at lower elevations in Colorado, has a narrow niche confined to rocky situations, usually ledges and fallen blocks. In southern Arizona, where *N. cinerea* is absent, *N. mexicana* occupies the high mountains and cliffs, while the outcrops and talus of lower elevations are occupied by *N. lepida* or *N. albigula*. At the end of the Wisconsin *N. mexicana* replaced *N. cinerea* in southern New Mexico (Harris, 1985). Because of its poor hoarding and building habits, the contribution of *N. mexicana* to the midden record of Holocene woodlands and forests in Arizona and New Mexico is probably poor.

N. *albigula* is a species of considerable ecological diversity in the Southwest, where it is the most widespread and abundant woodrat in the hot desert. In southern Arizona *N. al-*

bigula constructs imposing houses of cactus joints and sticks of spiny shrubs in large chollas and clumps of prickly pear. In southwestern Arizona the desert flats and bajadas are occupied by *N. albigula* and the rock outcrops are occupied by *N. lepida*. In the mesquite (*Prosopis*) and desert grasslands of southeastern Arizona and southern New Mexico *N. albigula* builds its houses in mesquite, using large amounts of the thorny branches, cactus joints, and cow chips (Vorhies and Taylor, 1940). Mesquite beans are almost as important as cactus for food.

In the desert grassland to the east, *N. albigula* overlaps (and is replaced in New Mexico and Texas) with *N. micropus*, a somewhat larger rat of similar ecological habits but having better digging ability and making more use of underground burrows. At Carlsbad Caverns and on the eastern periphery of its range N. albigula occurs in the higher rocky sites, while N. micropus occupies the lower open country away from rocks (Bailey, 1928; Schmidly, 1977). In southwestern Colorado N. a. laplataensis uses primarily rock dens and feeds on cactus and juniper. In southeastern Colorado N. a. warreni lives in both rock dens and in houses built around the tree cactus (Opuntia arborescens) or in junipers (J. monosperma). Locally these plants are its two most important foods, with soapweed (Yucca glauca) ranking third. At a few localities where cactus was very scarce N. albigula subsisted almost entirely on juniper foliage and yucca blades with a few other shrub and forb cuttings (Finley, 1958).

In southern California Grinnell (1914) found N. albigula venusta only in the mesquite on the outer edges of the riparian zone on both sides of the Colorado River. Rainey (1965) studied this subspecies along Carrizo Creek in San Diego County, California, where it lived entirely in sandy burrows and, in the absence of cactus, subsisted almost entirely on mesquite.

N. stephensi in northern Arizona is a dietary specialist dependent on juniper for 90% of its diet and also for den shelter. "Where juniper densities are high, woodrats usually feed on a tree within 15 or 20 m of the den, but where junipers are widely scattered woodrats occasionally travel up to 50 m from the den to a food tree" (Vaughan, 1982).

In part of its range in Utah N. lepida is also a dietary specialist. In a pinyon-sagebrush community in Juab County, Utah, Stones and Hayward (1968) found 90% of 233 houses associated with live junipers, the needles and berries making up almost all of the food litter and storage. In contrast, in southern California Cameron (1971) found N. lepida to be a generalist where it overlaps with N. fuscipes. Food analyses revealed that "allopatric N. fuscipes and N. lepida both prefer Quercus turbinella, whereas N. lepida switches to Juniperus californica when sympatric with N. fuscipes." In this situation N. fuscipes is the more specialized, being confined to the less xeric coastal areas, whereas N. lepida occupies the extreme deserts as well. These differences may be physiological, with N. fuscipes able to utilize oak leaves and other evergreen vegetation high in fiber, tannins, and related polyphenolics (Atsatt and Ingram, 1983) while N. lepida is able to utilize cactus and other plants containing noxious or toxic compounds. In captive acclimatization experiments Stallone (1979) found that N. lepida and N. fuscipes have about equal capacities to conserve water by increased urine concentration, but N. lepida is superior to N. fuscipes in utilization of cactus for water as well as food.

Although N. fuscipes is usually considered to be a resident of California chaparral, it occurs in some areas without chaparral. The subspecies N. f. riparia along the San Joaquin River lives almost entirely among the branches of trees, including button willow (Cephalanthus), ash (Fraxinus), and especially oak (Quercus). It was suggested that association with one or another kind of oak may be more characteristic than association with chaparral per se (Hooper, 1938).

A matter of taxonomic uncertainty should be mentioned. Joseph Grinnell (1914) collected specimens of N. lepida (which he called Neotoma intermedia desertorum) along the Colorado River only on the rocky desert outside the riparian zone, which was occupied by N. albigula venusta (see above). The populations of N. lepida on opposite sides of the river are ecological equivalents separated by the river and its riparian zones and were described as separate subspecies by E. A. Goldman (1932). Some authors, including Vaughan (this volume, chap. 2) regard N. l. devia in Arizona as a separate species, citing Mascarello's (1978) study on the two sides of the lower Colorado River; but the type locality of devia is Tanner Tank, Painted Desert, in northern Arizona. I follow Hoffmeister (1986), whose morphometric analysis showed that populations from the two sides of the Colorado River in northern Arizona are indistinguishable. The populations in southwestern Arizona are separable from those in California and northern Arizona and "may be part of an incipient species, but the name devia is not available for this group of populations," which Hoffmeister referred to N. l. auripila Blossom.

Woodrat Foraging or Transport Distance

A feature of packrat behavior of special inter-

Table 3.1. *Dens of* Neotoma *in Colorado and occurrences of plant species.*

| | Dens, No. | Occurrences of plant species | | | | | |
| | | In midden | | Within 30 m | | Beyond 30 m | |
		No.	%	No.	%	No.	%
Neotoma cinerea orolestes	42	300	100	211	70.3	89	29.7
cinerea arizonae	11	96	100	72	75.0	24	25.0
cinerea rupicola	4	28	100	24	85.7	4	14.3
Totals	57	424	100	307	72.4	117	27.6
Neotoma mexicana fallax	9	76	100	60	78.9	16	21.1
mexicana scopulorum	5	41	100	30	73.1	11	26.9
mexicana inopinata	37	295	100	242	82.0	53	18.0
Totals	51	412	100	332	80.6	80	19.4
Neotoma albigula brevicauda	7	42	100	35	83.3	7	16.7
albigula laplataensis	13	122	100	88	72.1	34	27.9
albigula warreni	19	131	100	81	61.8	50	38.2
Totals	39	295	100	204	69.2	91	30.8
Neotoma floridana campestris	12	79	100	68	86.1	11	13.9
Neotoma lepida sanrafaeli	9	49	100	39	79.6	10	20.4
Neotoma micropus canescens	8	34	100	27	79.4	7	20.6

est to midden analysts is how far rats forage, or range, from their dens to collect food or sticks. From 176 original den records (partially reported in Finley, 1958) I have retrieved the counts of plant species present at each den, identified either as food litter or as house materials, and not found growing in the vicinity within 30 m of the den. The results for the six species (twelve taxa) of *Neotoma* in Colorado are compiled in Table 3.1. From 62% to 86% of the plant species collected in the midden came from distances of less than 30 m from the den. Although many of the populations are represented by few dens, such as eight for *N. micropus*, the 79% is based on thirty-four species occurrences at those eight dens. All six species rely primarily on plants present within 30 m of the den. The figures also show that the range of species occurrences within 30 m of the den for the three subspecies of *N. albigula* (62%–83%) is almost as great as for all six species, whereas the range for the subspecies of *N. mexicana* is much less (73%–82%).

Thanks to a monumental study of *N. fuscipes* in coastal California, we have more information on the movements of this species than of any other. Linsdale and Tevis (1951) documented the movements of 508 rats in two trapping periods totaling nearly five and a half years. Most rats were trapped at stick houses and the moves recorded were between houses. Of those captured more than once, fewer than 20% were ever caught at points more than 90 m apart, and for nearly half, "all records were in areas less than 100 feet across. Usually this meant less than 50 feet from home." Their data on foods and feeding were obtained by observation of live rats and examination of pellets, food cuttings, and litter at ninety houses in relation to the kinds of plants available within 9 m of each house. An abundance of leaves, berries, and flowers were available nearby at most seasons of the year. Vestal (1938) learned much about the food habits and foraging range of the same species in the Berkeley Hills of California. He gave field evidence for his conclusion that "the forage ranges of a wood rat averages less than 100 feet [30 m] in radius from the nest."

Other studies of living woodrats give some information on the home range and foraging range of other species. In a test of the "tinfoil ball method" of measuring home range of *N. floridana osagensis*, nine hundred such code-labeled balls were placed in a grid along with livetraps. Nine woodrat "nests" (houses) were dismantled after two months and the balls retrieved. Findings were compared with movements obtained from 1260 trap-nights on the live-trap grid. The tinfoil ball method indicated a home range of 52 m from the den, compared with 60 m for the live-trap method (Ireland and Hays, 1969). Wiley (1980) summarized movement data from recaptures of *N. floridana* reported by Lay and Baker, Pearson, Fitch and Rainey, Tate, and Goertz. Normal foraging for food was usually within 21 m of the house. Several investigators recorded longer moves associated with shifts to new territories.

A house of *N. lepida* among lava rocks in Pete's Valley, Lassen County, California, contained juniper branches and fresh foliage from a small tree 30 m away, the only one within hundreds of meters (Grinnell et al., 1930). Stones and Hayward (1968) studied movements of *N. lepida* in Utah by livetrapping. Movements of males averaged 77 m, the longest being 148 m, and those of females averaged 41 m, the longest being 196 m.

Nightly movements of *N. stephensi* monitored in northern Arizona averaged 29 m from the home den (Vaughan, *this volume*, chap. 2). A den of *N. albigula* among junipers in Colorado contained a large amount of tree cactus joints that came from the nearest plant, 82 m from the den (Finley, 1958).

Determination of Midden Age

Modern (active) and ancient middens may not be easily separable in the field. Unindurated deposits can be several thousand years old, and hardened deposits can essentially form from one year to the next, as Thompson (1985) discovered upon revisiting a midden site in Nevada. Fresh (green and undried) plant cuttings are the best indicator of an active, occupied den. Clean entrances and passageways lacking windblown litter also indicate current use. Cobwebs across more than one exit indicate a defunct den. Food litter in dry dens shows little change within the first year or two, but thereafter the greenish color is lost and all leaves and stems turn brown. Fortunately for paleoecologists, woodrats do not dig into their middens to bury food there and dig it up later, as do red squirrels (*Tamiasciurus*), whose middens of cone debris are thoroughly churned each winter in recovering cached cones (Finley, 1969). A defunct den

often has pellets as fresh looking as an active den. Exposed pellets lose color and disintegrate with weathering in a few years; under shelter they become dark brown or grayish without further change.

The presence of extralocal plants in the contents of an ancient midden is a reliable indicator of relative age. Such occurrences may indicate fairly recent local habitat changes or much older climatic changes. Selection of middens that contained extralocal plants yielded a bimodal age distribution with an older peak of about ten thousand years (Webb and Betancourt, *this volume*, chap. 6).

Determination of modernity may be critical in areas of considerable sympatry, as on Woodhouse Mesa (Dial and Czaplewski, *this volume*, chap. 4) and in southwestern Colorado (Finley 1958). Woodrat middens usually remain relatively undisturbed, enough so that layers of different food litter sometimes reveal a change in food habits or occupation by successive species. In southwestern Colorado *N. mexicana* and *N. albigula* sometimes replace each other in taking over the same den, leaving evidence of their different food habits in the stratified midden, the *N. mexicana* stratum being free of cactus areoles. Some appearance of stratification may be seasonal as a result of food storage, but most food stores are stuffed into open spaces and do not appear as a layer until they are consumed.

Rockfalls, erosion, and sedimentation alter middens and eventually destroy them in geological time, but short-term evidence of this is seldom seen. One of the *N. cinerea orolestes* dens I studied and photographed in 1948 in central Colorado was revisited some twenty years later and found to have been greatly altered by rock collapse. Such rockfalls often reveal black urinary stains on vertical rock faces that must have been deposited in a cleft behind the formerly attached rock. Such a stain high on the cliff is visible from the road west of Dolores, Colorado. In the same area an occupied den beneath such a deposit was photographed (Fig. 3.3) and studied in 1948 (Finley, 1958). In 1979 I saw and rephotographed the same stain and picked up a chunk of indurated midden that had fallen with the rock blocks under the overhang. No change is evident in the pattern of the stain over the preceding thirty-one years.

Fig. 3.3. Urinary-fecal stains on a cliff exposed by rockfall 2.5 km west of Dolores, Colorado. Photo taken in 1948 when a den of *N. cinerea* under the fallen blocks was studied. Same exposed stains rephotographed in 1979.

Another midden on the cliff front west of Dolores is clearly ancient, its minimum age datable by rockfall and use by the Anasazi Indians. A large sandstone block that fell about 4 m from the roof of a small rock shelter rests on a deposit of fecal and food litter. On the side of the fallen block at chest height are several deep tool-sharpening grooves such as were made by the Anasazi, whose artifacts and masonry are abundant along the cliff front. The Dolores Archaeological Project found that the area was abandoned by the Anasazi by the year 975 A.D. (Kane, 1984). Thus the rock must have fallen from the roof sometime before that date.

Other archaeological evidence of the age of a deposit was noted by Cary (1911), who saw in the ruins of Spruce Tree House, Mesa Verde, Colorado, large rat fecal deposits that had been blackened by the same smoke that begrimed the interior of many of the rooms. The Anasazi abandoned Mesa Verde by about 1300 A.D. (Euler et al., 1979).

THE NEED FOR STUDIES OF MIDDEN FORMATION, DIAGENESIS, AND BIOCHEMISTRY

The summary of packrat ecology and behavior above, though incomplete, is enough to show how these aspects of biology affect the usefulness and interpretation of ancient middens. Packrats are outstanding collectors and preservers of organic materials that remain in nearly unchanged form for many thousands of years. The study of ancient middens is enormously aided by the fact that, in contrast to fossil bone deposits, the changes with decay and burial are so slight as to permit study by some of the same methods used for the analysis of active middens. Hence the study of midden taphonomy is different from that of bone taphonomy. The processes and biases caused by decay and burial in midden deposits are minimal, but those caused by the animal agent of transport are of major importance. Ecological characteristics of packrats provide many points of generalization and also many sources of bias among different species of *Neotoma* and among different ecological situations. It is the interaction between the rat and its environment that determines what and how organic remains are transported and how the midden is formed by feco-urinary deposition.

Unanswered Questions of Midden Origin

Many questions arise concerning the origin of midden deposits: how they are formed, the sources of the material incorporated, how long a period is required to build up a deposit of any size, how continuously a den or midden is used, the changes in occupancy that may have occurred in the den, etc. Fortunately, these questions of midden origin can be studied directly today, because packrats at presently occupied dens are forming middens and feco-urinary deposits just as they were forty thousand years ago. Opportunities for studies of this kind have great potential for increasing our understanding of both woodrat biology and paleoenvironment.

Paleomiddens, like modern ones, differ greatly in composition, compactness, hardness, and location within the cave. They may consist mainly of unconsolidated sticks and other den materials, of food litter consolidated with feces and urine, or almost entirely of dense, hardened urinary deposits. Indurated ancient deposits of mixed materials, including food litter and feces cemented by urine, are probably the most useful to sample for study because of greater information content and low probability of contamination by later mixing and intrusion. Selection of organic material within the sample for age determination carries with it the question of whether the item dated was laid down simultaneously with other items in the same chunk collected. The same question is even more relevant between different larger strata or sites within the same den shelter. Knowledge of the nature and rates of deposition of different kinds of material at different sites within the shelter would help greatly in field sampling and perhaps reduce the number of samples needed for dating.

Rates of deposition can be approximated by reconstructing or "modeling" the process and using known or estimated values for the elements entering into the formation of a midden. These differ for different constituent materials and kinds of posts, but the theoretical result for a urinary deposit, a fecal deposit, a food-litter deposit, or an accumulation of sticks or other den materials should be usable to reconstruct rates of deposit for mixed materials. The validity of such theoretical rates can be tested by measurements at currently active dens. Such an approach was used to estimate how long it would take red squirrels (*Tamiasciurus hudsonicus*) to accumulate a large, well-delimited cone-storage midden in a blue spruce stand in southern Colorado. By sampling the cone litter of the midden, identifying and counting the kinds of cone cores, and using data on seed biomass and squirrel energy requirements from silvicultural and squirrel literature, it was estimated that the midden could have been produced by twenty-five squirrel-years of cone stripping (Finley, 1969).

Some of the data needed to estimate rates of *Neotoma* midden growth have been reported and others could be measured from captives or field dens. Some data needed are: daily urination under varying heat stress and water availability; grams of urine solids per volume of liquid as above; total kilograms of urine solids produced per year; percentage of total urine on urination posts versus elsewhere; grams of daily food intake, wet and dry weight; total annual food requirement, dry weight; ratio of food consumed to food cuttings in litter for leaves, stems, fruits, cactus joints, etc.; ratios of main foods consumed for the species of rat by season; daily rate of food litter production; total annual food litter production; daily defecation rate; total annual defecation; percentage of annual defecation on midden or defecation posts; percent reduction of feces by drying and compaction; rate of material accumulation for sticks, bones, cactus joints, etc.; and total annual addition of such materials. Information on defecation rates, amounts, and distribution are given for *N. fuscipes* by Vestal (1938) and Linsdale and Tevis (1951). Theoretical reconstructions of this sort obviously contain a wide margin of uncertainty, but they form a starting point for estimating accumulation rates that can be compared with long-term direct measurements of active middens.

In some situations historical evidence can be obtained for rates of accumulation over a period of years or a few decades. Where measurable amounts or depths of deposits have accumulated in old, abandoned buildings, mines, or other structures, the time span of such deposits could be determined from historic or newspaper records. Such urinary deposits made within the past century can be seen on the walls and timbers of the long-abandoned Gold King Mill in La Plata County, Colorado.

An unusual instance of the buildup of urinary posts in approximately two years by the house mouse (*Mus musculus*) was reported by Welch (1953). A laboratory colony of about seventy-five mice was being maintained for testing repellent compounds applied to packaging materials. In the large cage the mice habitually urinated on fecal droppings at many scattered locations, thus building up knob-shaped urinary pedestals, the largest of which measured "1 1/2+ inches high, 1 inch wide at the top with a 3/4 inch base."

Colored aluminum powder like that used in paints can be used as a durable tracer in the diet to mark fecal deposits at outdoor dens. It was used to label bait applied for harvester ant control in New Mexico and was observed in penned birds that had picked up the bait (Finley, 1962). Perhaps the growth of active middens or urinary deposits could be measured by using physical or chemical markers to delimit the deposits at a known time and to measure subsequent changes.

Potential of Midden Analysis for the Identification of *Neotoma* Species

Fortunately, in spite of wide differences in den and food habits, all woodrat species north of Mexico utilize caves and deep rock shelters as dens wherever these are available. The major questions and primary biases concerning middens in such dry, well-protected sites relate to the biology of the animals that made the midden. If the maker can only be identified as *Neotoma*, one must base the interpretation on knowledge of the biology common to all species of *Neotoma* that might have been present at that time; but if one can determine which species made the midden, and the midden is analyzed in appropriate ways, better knowledge of paleoecology and woodrat biogeography can be gained.

The potential for identifying the species of midden builder and gaining more ecological facts by new methods of midden analysis should be investigated. The soluble as well as particulate material in urinary and fecal deposits could be analyzed by chemical, electrophoretic, radioimmunoassay, and other techniques to identify distinctive compounds of plant and animal origin. The proportions of these remaining over long periods may be indicative of various physiological or environmental conditions. Laboratory experiments and field measurements could be made to determine the effects on feces and dried urine of desiccation, bacterial action, oxidation, and other environmental factors. Such studies on the diagenesis of bone have added greatly to our understanding of bone taphonomy. The breakdown of protein such as collagen in bone proceeds slowly by hydrolysis, and the remaining amino acids are slowly leached away, but they can be identified after thousands of years in proportions still indicative of their collagenous origin (Hare, 1980).

Physiological, chemical, and behavioral studies have demonstrated that many mammals produce complex substances that function in scent communication, which is especially important in nocturnal animals (Doty, 1976; Stoddard, 1976; Goodrich and Mykytowycz, 1972). At least forty-five compounds have been identified in castoreum, the anal secretion of the beaver. August (1978) studied scent communication in *Neotoma micropus*. Tests of the responses of males and females to the odors of urine, feces, and sebum from the ventral sebaceous glands of males showed that such odors are important. Although the odors are undoubtedly distinctive for each kind of rat, nothing is known about how long such compounds may remain un-volatilized and stable after being incorporated in a dried urinary deposit.

Proteins, particularly the less soluble ones such as collagen, keratin, and elastin, are the most likely animal molecules to survive and retain their specificity after long periods of burial in dry conditions, but many other proteins are present in small amounts in urine, feces, and glandular secretions. Some methods used to measure the molecular specificity of proteins of living species are sufficiently sensitive to detect the very small amounts remaining in fossil tissues.

Lowenstein (1985) applied a solid-phase double-antibody method of radioimmuno-assay (RIA) to samples of protein from an Egyptian mummy and several kinds of human fossils to show that collagen was detectable in all specimens in amounts decreasing with geological age. The collagen in these fossils bound antiserum to human collagen more strongly than it bound antiserum to rodent or bovine collagen, indicating that some specificity had been preserved for nearly 2 million years. He reviewed the most recent uses of RIA on a wide variety of fossil materials to solve taxonomic problems not resolved on morphological evidence. RIA has been applied to proteins from museum specimens of several extinct animals, including the mammoth (*Mammuthus primigenius*), Steller's sea cow (*Hydrodamalis gigas*), and the quagga (*Equus quagga*). This technique has even been used to identify bloodstains on ancient weapons. Hemoglobin crystals were extracted from arrowheads approximately two thousand years old found in northern Canada, and RIA attributed two of the bloodstains to bison, two to cervids, and one to a human.

Chemical Analysis of Middens for Paleoecological Clues

Other publications also suggest possibilities for midden analysis; I can only scratch the surface here. Atsatt and Ingram (1983) determined the digestibility of oak, sage, and other foliage by *N. fuscipes* and *N. lepida* from the chemical content of feces. The kinds and relative proportions of chemicals in paleomiddens may contain information about past woodrat diets and habitat.

Emerson and Howard (1978) reported on the mineralogy of microcrystalline deposits from urinating posts of *N. cinerea* in northeastern California. The bulk of the deposits were found to be calcite with as much as 45% calcium oxalate. These deposits were visible as white streaks and blotches on some of the rocky bluffs of the Modoc Plateau. Urine samples from a few captive woodrats were analyzed with varying results depending on the diet. This paper also comments on other studies (Shirley and Schmidt-Nielsen, 1967) on oxalate metabolism that relate to water conservation achieved by eating oxalate-rich plants like cactus.

Vernon Bailey (1931) described white streaking or capping of rocky points in New Mexico as calcite deposits up to 2 or 3 mm thick and said that such urine deposits are peculiar to all the rock-dwelling species of *Neotoma*. If this is so, they do not occur everywhere that urinary deposits of rock-dwelling packrats are found. White deposits are relatively scarce in Colorado, even where there are massive dark urinary deposits. I have seen white deposits on some cliffs near woodrat dens but assumed at the time that all were made by birds. Some of these may have been the weathered residues of old packrat urinating posts. If whitish calcite and oxalate deposits are formed only in arid regions where cactus is a prominent part of the diet, I would expect such deposits to be absent from localities occupied only by *N. mexicana* or *N. fuscipes*, which consume very little cactus.

The wealth of data and interpretations about the past obtainable from analyses of plant structure in ancient packrat middens is well presented in other chapters of this volume, but the analysis of chemical content has been neglected. One reason for this may be the expectation that few, if any, large biomolecular compounds remain after initial decay and desiccation. I have tried to show that there is reason to believe that hundreds, if not thousands, of distinctive molecules of biological origin are present in woodrat middens at least in trace amounts sufficient for modern methods of analysis. The intriguing question is whether the conditions of midden deposition are favorable enough for the preservation of large biomolecules for thousands of years and whether many can be identified. The coming decades may show that the analysis of biomolecules in middens can open windows of opportunity in paleoecology and biogeography by way of well-preserved woodrat middens.

REFERENCES

Atsatt, P. R., and Ingram, T. (1983). "Adaptation to oak and other fibrous phenolic-rich foliage by a small mammal, *Neotoma fuscipes*." *Oecologia* 60, 135–42.

August, P. V. (1978). Scent communication in the southern plains wood rat, *Neotoma micropus*." *American Midland Naturalist* 99, 206–18.

Bailey, V. (1928). *Animal life of the Carlsbad Cavern*. American Society of Mammalogists Monograph no. 3. Baltimore: Williams & Wilkins.

———. (1931). "Mammals of New Mexico." *North American Fauna* 53, 1–412.

Brown, J. H. (1968). "Adaptation to environmental temperature in two species of woodrats, *Neotoma cinerea* and *N. albigula*." *Miscellaneous Publications, Museum of Zoology, University of Michigan* 135, 1–48.

Cameron, G. N. (1971). "Niche overlap and competition in woodrats." *Journal of Mammalogy* 52, 288–96.

Cary, M. (1911). "A biological survey of Colorado." *North American Fauna* 33, 1–256.

Dixon, J. S. (1919). "Notes on the natural history of the bushy-tailed wood rats of California." *University of California Publications in Zoology* 21, 49–74.

Dory, R. L., ed. (1976). *Mammalian olfaction, reproductive processes, and behavior*. New York: Academic Press.

Emerson, D. O., and Howard, W. E. (1978). "Mineralogy of woodrat, *Neotoma cinerea*, urine deposits from northeastern California." *Journal of Mammalogy* 59, 424–25.

Euler, R. C.; Gumerman, G. J.; Karlstrom, T. N. V.; Dean, J. S.; and Hevly, R. H. (1979). "The Colorado Plateaus: Cultural dynamics and paleoenvironment." *Science* 205, 1089–1101.

Finley, R. B., Jr. (1958). "The wood rats of Colorado: Distribution and ecology." *University of Kansas Publications, Museum of Natural History* 10, 213–552.

———. (1962). "The effects of harvester ant control on native wildlife." In *The influence of western harvester ant control on re-establishment of range grasses*. S. R. Race, ed., pp. 49–51. Annual Progress Report no. 2, New Mexico State University.

———. (1969). "Cone caches and middens of *Tamiasciurus* in the Rocky Mountain Region." In *Contributions in mammalogy, a volume honoring Professor E. Raymond Hall. University of Kansas Museum of Natural History, Miscellaneous Publications* 51, 1–428.

Goldman, E. A. (1932). "Review of wood rats of *Neotoma lepida* group." *Journal of Mammalogy* 13, 59–67.

Goodrich, B. S., and Mykytowycz, R. (1972). "Individual and sex differences in the chemical composition of pheromone-like substances from the skin glands of the rabbit, *Oryctolagus cuniculus*." *Journal of Mammalogy* 53, 540–48.

Grinnell, J. (1914). "An account of the mammals and birds of the lower Colorado Valley, with special reference to the distributional problems presented." *University of California Publications in Zoology* 12, 51–294.

Grinnell, J.; Dixon, J.; and Linsdale, J. M. (1930). "Vertebrate natural history of a section of northern California through the Lassen Peak region." *University of California Publications in Zoology* 35, 1–594.

Hare, P. E. (1980). "Organic geochemistry of bone and its relation to the survival of bone in the natural environment." In *Fossils in the making*. A. K. Behrensmeyer and A. P. Hill, eds., pp. 208–19. Chicago: University of Chicago Press.

Harris, A. H. (1985). *Late Pleistocene vertebrate paleoecology of the West*. Austin: University of Texas Press.

Hoffmeister, D. F. (1971). *Mammals of Grand Canyon*. Urbana: University of Illinois Press.

———. (1986). *Mammals of Arizona*. University of Arizona Press and Arizona Game and Fish Department.

Hooper, E. T. (1938). "Geographical variation in wood rats of the species *Neotoma fuscipes*." *University of California Publications in Zoology* 42, 213–46.

Ireland, P. H., and Hays, H. A. (1969). "A new method for determining the home range of woodrats." *Journal of Mammalogy* 50, 378–79.

Kane, A. E. (1984). "The prehistory of the Dolores Project Area." In *Dolores Archeological Program: Synthetic Report 1978–1981*, pp. 21–51. Denver: U.S. Department of the Interior, Bureau of Reclamation.

Linsdale, J. M., and Tevis, L. P., Jr. (1951). *The dusky-footed wood rat; a record of observations made on the Hastings Natural History Reservation*. Berkeley: University of California Press.

Lowenstein, J. M. (1985). "Molecular approaches to the identification of species." *American Scientist* 73, 541–47.

Mascarello, J. T. (1978). "Chromosomal, biochemical, mensural, penile, and cranial variation in desert woodrats (*Neotoma lepida*)." *Journal of Mammalogy* 59, 477–95.

Poole, E. L. (1940). "A life history sketch of the Allegheny woodrat." *Journal of Mammalogy* 21, 249–70.

Rainey, D. G. (1956). "Eastern woodrat, *Neotoma floridana*, life history and ecology." *University of Kansas Publications, Museum of Natural History* 8, 535–646.

———. (1965). "Observations on the distribution and ecology of the white-throated wood rat in California." *Bulletin of the Southern California Academy of Sciences* 64, 27–42.

Schmidly, D. J. (1977). *The mammals of trans-Pecos Texas*. College Station: Texas A&M University Press.

Shirley, E. K., and Schmidt-Nielsen, K. (1967). "Oxalate metabolism in the pack rat, sand rat, hamster, and white rat." *Journal of Nutrition* 91, 496–502.

Stallone, J. N. (1979). "Seasonal changes in the water metabolism of wood rats." *Oecologia* 38, 203–16.

Stoddard, D. M. (1976). Mammalian odours and pheromones. Studies in biology no. 73. London: The Institute of Biology.

Stones, R. C., and Hayward, C. L. (1968). "Natural history of the desert woodrat, *Neotoma lepida*." *American Midland Naturalist* 80, 458–76.

Thompson, R. S. (1985). "Palynology and *Neotoma* middens." *American Association of Stratigraphic Palynologists Contribution Series* 16, 89–112.

Van Devender, T. R., and Spaulding, W. G. (1979). "Development of vegetation and climate in the southwestern United States." *Science* 204, 701–10.

Vaughan, T. A. (1982). "Stephens' woodrat, a dietary specialist." *Journal of Mammalogy* 63, 53–62.

Vestal, E. H. (1938). "Biotic relations of the wood rat (*Neotoma fuscipes*) in the Berkeley Hills." *Journal of Mammalogy* 19, 1–36.

Vorhies, C. T., and Taylor, W. P. (1940). *Life history and ecology of the white-throated wood rat,* Neotoma albigula albigula *Hartley, in relation to grazing in Arizona*. University of Arizona, Agricultural Experiment Station, Technical Bulletin 86, 453–529.

Welch, J. F. (1953). "Formation of urinating 'posts' by house mice (*Mus*) held under restricted conditions." *Journal of Mammalogy* 34, 502–3.

Wells, P. V., and Jorgensen, C. D. (1964). "Pleistocene wood rat middens and climatic change in Mojave Desert: a record of juniper woodlands." *Science* 143, 1171–74.

Wiley, R. W. (1980). "*Neotoma floridana*." *Mammalian Species* 139, 1–7.

Chapter 4

Do Woodrat Middens Accurately Represent the Animals' Environments and Diets? The Woodhouse Mesa Study

Kenneth P. Dial   Nicholas J. Czaplewski

INTRODUCTION

Midden studies of woodrats, or packrats (genus *Neotoma*, Fig. 4.1) provide a record of late Pleistocene and early Holocene environments in western North America (Wells and Jorgensen, 1964; Wells, 1976; Van Devender and Spaulding, 1979) that is as yet unparalleled elsewhere in the world. Inevitably, paleoenvironmental reconstructions based on the contents of fossil woodrat middens led to questions about woodrat foraging and other activities. One important question is, what biases can be expected as a result of selectivity by the packrat? For example, the part of a midden suitable for radiocarbon dating has been the subject of disagreement (Van Devender, 1973; Wells, 1976; Mead et al., 1978; Webb and Betancourt, *this volume*, chap. 6). Refined methods of dating are being developed (Van Devender et al., 1985) that permit detection of diachrony or the presence of allochronic contaminants (Wells and Hunziker, 1976) among plant fossils contained in a midden.

Fig. 4.1. *Top*, adult female *Neotoma stephensi*. *Bottom*, one-seeded juniper (*Juniperus monosperma*) on the study area showing extreme damage due to chronic herbivory by *Neotoma stephensi* (see also Vaughan, 1980). Note robustness of normal junipers in background.

Another problem is that of sampling procedures for fossil midden analysis and quantification (Betancourt, 1984). Investigators have used a number of different measures to express the relative abundance of plants in a midden. Usually, heterogeneous samples, possibly including stems, seeds, leaves, and needles must be compared and/or pooled for quantification. Plant parts are sometimes quantified by their mass per kilogram of unwashed indurated midden (including rocks, fecal pellets, amberat; Cole, 1981) and sometimes by frequency of a given taxon relative to the total number of identified specimens (Spaulding et al., *this volume*, chap. 5). Arbitrary measures of relative abundance (e.g., "rare," "uncommon," "common," etc.) are used, and at times no measure of relative abundance is made at all.

Yet another problem has not been fully addressed by those who study fossil middens—inherent biases in the materials collected in a midden that are due to selectivity on the part of the woodrats. Wells (1976), citing a number of natural history studies by woodrat ecologists, noted that "existing wood rats are not only diversely generalist in their herbivorous habits, but also widely indiscriminate in their compulsive collection of transportable objects. . . . The legendary collecting habits, as distinguished from the dietary preferences (which show a somewhat lesser catholicity of taste), have been observed in many species of *Neotoma* (Vorhies and Taylor, 1940; Linsdale and Tevis, 1951; Rainey, 1956; Finley, 1958; Stones and Hayward, 1968)." This view perhaps reflects the failure of woodrat ecologists to carefully quantify dietary preferences and contrast them with resource availability. Van Devender and King (1971) showed that pollen in packrat fecal pellets provides a biased view of the local plant community because it reflects dietary habits and preferences of the rat. In the Mojave Desert Cole and Webb (1985) compared modern plant composition and modern midden composition above and below the *Larrea-Coleogyne* transition, finding Sørensen's index of similarity values of 79% and 80% in presence/absence. They did not know which species of *Neotoma* were involved, and comparisons were not made between midden contents deposited by known species of rats and the ambient plant commu-

nity. On the basis of a comparison between relevés and "modern" middens, Cole (1982) "assume[d] that macrofossil concentration is a rough indicator of past importance of dominant species." As noted above, packrats have been considered generalized herbivores by students of historical and modern woodrat ecology, yet recent and more rigorous studies suggest that at least some are specialists. Cameron and Rainey (1972) suggested that *N. lepida* is a locally facultative dietary specialist concentrating on different plants in different plant communities. Vaughan (1980, 1982) has shown that *N. stephensi* is a strong dietary specialist and occurs only in association with juniper trees. Dial (1984, 1988) found that each of three sympatric woodrat species (*N. albigula*, white-throated woodrat; *N. devia*, desert woodrat; and *N. stephensi*, Stephens' woodrat) specialize on different plant species.

Paleobiologists typically make important assumptions about woodrats and their environments. These assumptions could include: (1) Fossil middens directly reflect the vegetation of past environments, or, stated differently, woodrats accumulate plant species roughly in proportion to plant relative abundance. (2) Whether or not they actually eat them, woodrats sample most, if not all, species of plants within foraging range. (3) The primary mechanism by which modern woodrat middens are accumulated is the diet. (4) One species of woodrat is responsible for midden accumulation at a given locality over a period of time. If several species were involved, no major change in the midden chronosequence need be expected.

This chapter is a case study representing the first attempt to address empirically the validity of the above assumptions. It incorporates data from modern woodrat dens in an area of sympatry of three distinct species of *Neotoma*. We considered ambient vegetation, clippings at the dens, and diets of the respective species in order to understand the interrelationships among these variables. For example, the validity of diet as a mechanism in midden accumulation has not previously been assessed. We have taken an ecological rather than paleontological approach to the analysis of how packrats gather plant remains. At Woodhouse Mesa we were able to

Fig. 4.2. The study area on the east slope of Wood-house Mesa, 45 km northeast of Flagstaff, Arizona. One of the junipers on the slope in the middle dis-tance (right center) shows a lack of foliage due to herbivory by Stephens' woodrat.

assess the effects of differential selectivity by three distinct species of packrats while plant diversity and abundance remained constant and measurable. These factors allow greater environmental control than that commonly available in studies of fossil packrat middens, even when calibrated by modern midden comparisons.

THE WOODHOUSE MESA STUDY AREA

The study area is a 15-ha plot on the Colorado Plateau in an ecotone between cold Great Basin shrub steppe and juniper woodland (Fig. 4.2) near the south border of Wupatki National Monument, 45 km northeast of Flagstaff, Coconino County, Arizona (latitude 35°30′ N, longitude 111°22′30″ W; elevation 1490–1580 m). The substrate consists primarily of a thick layer of volcanic cinders over the slope of an escarpment (the eastern edge of Woodhouse Mesa), with discontinuous, horizontal outcroppings of red sandstone

low on the slope. Higher on the escarpment slope a broken talus of basaltic boulders leads up to a terrace. A 1–5-m-thick basaltic flow caps the mesa and affords den sites for woodrats. The basalt talus, sandstone outcroppings, and juniper trees also provide den sites. The area is quite xeric with no standing water and a mean annual precipitation of 24 cm. Plant species diversity in the area was low, seventeen species of woody plants in 15 ha.

Four species of woodrats lived in the area, in some cases occupying adjacent dens. Three species (*N. albigula*, *N. devia*, and *N. stephensi*) were permanent residents and sympatric (see Mascarello, 1978, for use of the taxon *N. devia* as a full species formerly included in *N. lepida*). Throughout the year these three species had nearly equal population sizes, ranging from approximately twenty-five to fifty individuals per species (Dial, 1988). Four to ten transient individuals of *Neotoma cinerea* (bushy-tailed woodrat) invaded the area each year between August and November. Only the resident species of *Neotoma* were investigated in the dietary and midden studies. The area is not known to harbor fossil middens of Pleistocene age.

MATERIALS AND METHODS

Habitat Analysis

Vegetation was recorded along belt transects to estimate the relative abundance of plant species. Plant species within the belts were listed and crown cover was measured. Relative density, relative cover, and absolute cover were calculated using the methods of Mueller-Dombois and Ellenberg (1974). Because the area was dominated by perennial shrubs, there was little seasonal change in the relative abundance of the plant species. Annuals were of minor seasonal occurrence.

Initially, three sets of five belt transects, each 50 m long by 2 m wide (a total of 1500 m^2) and spaced 10 m apart, were used to characterize the overall vegetation and substrate types of the study area. Subsequently, additional transects were measured within 50 m of each of the fifteen dens (see Fig. 4.3), equalling 200 m^2 per den. This permitted the comparison of both the area-wide and the den-adjacent plant life to the midden contents of each den. Therefore, a total of 4500 m^2 was analyzed on the 15-ha plot. In addition, den-adjacent plant relative abundance was tested for correlation with area-wide plant relative abundance.

Trapping

Livetrapping was usually conducted once a week (minimum twice a month) from September 1980 to September 1984 (Dial, 1984) within a larger 50-ha study area surrounding and including the 15-ha midden study area mentioned above. All active dens were monitored with traps baited with a mixture of

Fig. 4.3. Map of the Woodhouse Mesa study area indicating locations of those dens from which midden materials were sampled. About eighty active dens were located in the area.

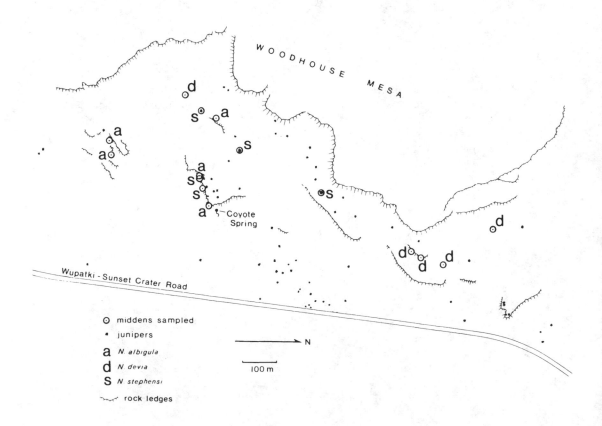

middens sampled
junipers
a *N. albigula*
d *N. devia*
S *N. stephensi*
rock ledges

100 m

N

rolled oats and peanut butter. Dens of pack-rats usually occurred beneath large sandstone or basalt boulders or beneath rock ledges and overhangs; occasionally they were at the bases of junipers.

Diet and midden analysis

Diets of woodrat species were determined (Dial, 1988) by the fecal-pellet-analysis method of Hansen and Flinders (1969). Fecal samples ($N = 372$) from livetrapped, marked (toe-clipped), adult woodrats were studied. Diets were compared with relative plant cover to determine dietary preferences of each resident *Neotoma* species. Diets were compared also with clippings and plant fragments at the dens (midden materials).

Midden is defined in the dictionary as a "dunghill or refuse heap." The midden portion of a modern woodrat den was considered to be those materials accumulated by the woodrat, including unused plant clippings, woodrat feces, and faunal materials (bones and insects). Often the materials were in the process of being coated with or cemented together by rat urine (i.e., they comprised the animal's habitual "toilet" area), but only small portions were consolidated and indurated with urine. Nest-chamber materials (finely shredded grasses or juniper bark fibers) and structural materials (sticks, small rocks, and domestic livestock feces) were not included because they were judged not to comprise part of the midden (see Fig. 4.4). (We realize, nevertheless, that they could become incorporated into a midden through time.) Food caches contributed small amounts to the midden contents in some dens. However, cache provisions usually consisted of clean (completely free of feces), relatively fresh or recently clipped, and partly dried plant materials.

Midden materials were collected as bulk samples from fifteen modern dens (five per woodrat species). A sample amounting to between four and eight liters of midden material was collected from most dens. One species of woodrat, *N. devia*, did not accumulate large amounts of plant materials. Midden samples for this rat usually consisted of about two liters of material. Plant clippings and fragments were identified by macroscopic inspection and direct comparison with plants growing on the study area. Plant parts with a diameter of about 3 mm and larger were sorted by this method. Thus, our method differed from the graduated-screening technique used by most paleoecologists because plant parts, especially seeds, whose greatest diameter was less than 3 mm were not included. Whether this lack of symmetry in technique will strongly affect results (which we doubt) is testable by ascertaining the number of additional plant species added to a floral list represented by seeds or other parts smaller than 3 mm. Plant parts were sorted by species and weighed to the nearest 0.1 g. In an effort to estimate volume, plant parts were crushed using a mortar and pestle and placed in graduated cylinders or beakers. All parts (stems, leaves, flowers, seeds) belonging to one species were measured together. Branches and large sticks were excluded from the analysis because they were considered to be structural elements of the den, not midden elements. Rat fecal pellets and faunal elements also were weighed and their volumes estimated. Feces were not considered a floral component and were left out of analyses of the relative midden-plant weights or volumes.

Percentage relative volumes correlated highly with percentage relative weights (Pearson product correlation coefficients for the three species of packrats: $r = .926-.998$, $p < .001$). Therefore, in the discussion that follows the percentage values given represent percentage relative weights of plant materials.

Removal Experiment

A removal experiment performed as a part of another study (Dial, 1984) bears on the present case. A 50-ha study plot (which surrounded the 15-ha midden study plot) was split in half. The control half contained 55 active dens, (27 of *N. stephensi*, 16 of *N. devia*, and 12 of *N. albigula*), and the treatment half contained 58 active dens (27 of *N. stephensi*, 17 of *N. devia*, and 14 of *N. albigula*). The behaviorally dominant species, *N. albigula* and *N. stephensi*, as determined by Dial (1984), were removed from the treatment plot. Qualitative observations were made on changes in the plant species accumulated at 20 dens where a new species of packrat replaced the former occupant.

Fig. 4.4. Den of *Neotoma devia* before and after dismantling. (**A**) Intact den beneath saltbush with Sherman livetrap for scale. Trap is 25 cm long. Note integration of saltbush stems as structural support and use of saltbush sticks and cattle feces as superstructural materials. (**B**) Dismantled den separated into component parts: *N*, nest chamber of shredded grasses; *C*, urine-consolidated midden material including volcanic cinders (soil substrate), woodrat feces, and unused plant clippings; *M*, un-consolidated midden material consisting mostly of *Ephedra* twigs and woodrat fecal pellets; *S*, sticks removed from the den. No discrete food cache was found in this particular den.

Fig. 4.5. Proportional representation of plant species occurring in the vegetative cover, plant clippings in midden contents, and plants in diets on an area with three sympatric species of *Neotoma*. Plant acronyms are: GRA = grasses; EPH = *Ephedra torreyana* and *E. viridis*; RHU = *Rhus trilobata*; ATR = *Atriplex canescens*; FAL = *Fallugia paradoxa*; JUN = *Juniperus monosperma*; GUT = *Gutierrezia sarothrae*; CHR = *Chrysothamnus nauseosus*; ANN = annuals; YUC = *Yucca baileyi*; ERI = *Eriogonum hookeri*; BRI = *Brickellia californica*; ARF = *Artemisia filifolia*; STA = *Stanleya pinnata*; DAT = *Datura meteloides*; CUC = *Cucurbita foetidissima*.

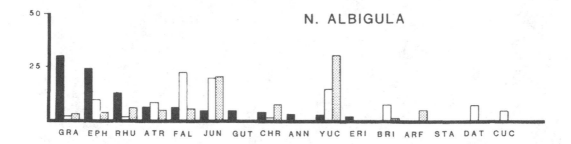

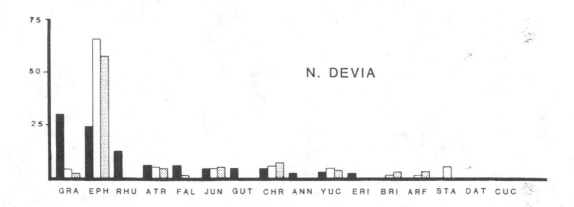

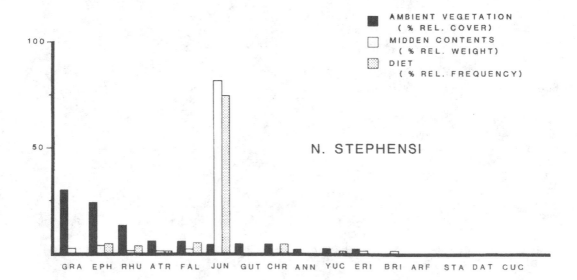

RESULTS

Habitat Characteristics

Suffrutescent perennials and annuals dominate the vegetation of this portion of the Great Basin desertscrub at Woodhouse Mesa. *Juniperus monosperma* occurs in some ecotonal regions. On our study area vegetation was sparse (Fig. 4.2); plants covered only 24% of the ground. Shrubs comprised the majority (48%) of the total vegetative cover. The most important shrubs were Mormon tea (*Ephedra*), skunkbush (*Rhus*), saltbush (*Atriplex*), and Apache plume (*Fallugia*) (Fig. 4.5). Grasses provided 28% of the relative cover but showed a patchy distribution. In most areas where woodrats occur, succulent plants provide them with their major source of water. Other than rare and scattered *Opuntia* cacti, succulent plants were absent on our study area, and evergreens (e.g., *Ephedra, Juniperus,* and *Yucca*) were the source of water (Dial, 1984).

Midden Characteristics

Besides woodrats, other vectors contributed to the accumulation of plant and animal materials in the woodrat middens on Woodhouse Mesa. Some plant materials occurring in the middens almost certainly were blown in by the wind. Such materials included anemochorous seeds (of *Gutierrezia, Chrysothamnus, Atriplex,* and *Andropogon*), propagules of *Salsola*, staminate and ovulate cones of *Ephedra*, and leaflets of *Rhus trilobata*. Porcupine (*Erethizon dorsatum*) activity might have resulted in the accumulation of other plant and faunal materials. At least one woodrat den on Woodhouse Mesa was regularly used by a porcupine, and another midden contained porcupine feces and quills. Still other plant and faunal materials were indirectly contributed by coyotes (*Canis latrans*) whose scats were collected by all three woodrat species. Coyote feces often contained juniper seeds and, in one instance, the undigested red fruits of *Rhus trilobata*. While these seeds and fruits did not occur in the middens of Woodhouse Mesa, Vaughan (1982) found that *N. stephensi* eats fruits of *Juniperus* and *Rhus* elsewhere. Our results do not differentiate among the various agencies of accumulation. Irrespective of their origin, all these materials were incorporated into the midden and all potentially could contribute to environmental reconstruction in the hands of a fossil analyst. Results of the faunal analysis are not presented here but are available from the authors.

Different species of woodrats varied considerably in the plant clippings they accumulated in their middens. Two of the dens examined were occupied by white-throated and Stephens' woodrats, and occurred barely 4 m apart. These two dens emphasized in a dramatic fashion the differences in midden contents collected contemporaneously from the same habitat by different species of rats (Table 4.1). If these same data were to be compared using the five-category relative abundance scale used by paleoecologists (Van Devender, 1973; Betancourt, 1984), these differences in plant abundance might not be as apparent. Our relative abundance data for modern middens, used in calculating the results in Table 4.2, could readily be converted to the 1–5 scale, but the reverse transformation is unlikely.

Table 4.1. *Plants in middens of two woodrat dens occupied by different species located 4 m apart.*
Data are presented as percentage relative mass.

Plant species	Neotoma albigula	Neotoma stephensi
Atriplex canescens	2.5	0.6
Brickellia californica	10.0	4.0
Chrysothamnus nauseosus	0.0	0.1
Datura meteloides	6.3	0.0
Ephedra viridis and *E. torreyana*	3.2	4.7
Eriogonum hookeri	1.2	3.0
Fallugia paradoxa	5.0	4.6
grasses	0.0	0.1
Gutierrezia sarothrae	0.0	0.1
Juniperus monosperma	32.5	80.0
Rhus trilobata	1.2	0.8
Stanleya pinnata	0.0	1.5
Yucca baileyi	32.5	0.0
Unknown	5.0	0.8

Diet Characteristics

Dial (1984, 1988) demonstrated dietary specialization on three different evergreen plants by each of the three species of *Neotoma*. Dietary selectivity is reflected to some extent in

Table 4.2. *Species richness of plants in the midden versus species richness of plants in the vegetation transects adjacent to dens (and percentage representation in the midden) at five individual modern dens for each of three species of packrats.*

Packrat Species	Dens				
	1	2	3	4	5
Neotoma albigula	8/12	7/9	6/8	7/11	6/8
	(67%)	(78%)	(75%)	(64%)	(75%)
Neotoma devia	6/9	3/8	4/8	4/10	4/11
	(67%)	(36%)	(50%)	(40%)	(36%)
Neotoma stephensi	6/9	6/12	5/10	3/8	6/10
	(67%)	(50%)	(50%)	(38%)	(60%)

the plant clippings in middens of different species (Fig. 4.5). Specialization appears to be tied to the rat's ability to track a year-round water source. For example, *Neotoma devia*, apparently the most xeric-adapted species, shaves the green epidermis from *Ephedra* stems (Dial, 1988). Because woodrats require a year-round water source, at least in arid areas, fossil midden hunters can expect to find at least one predominant evergreen or succulent plant in middens. On Woodhouse Mesa *Ephedra*, *Juniperus*, and *Yucca* provided the evergreen sources of water.

Midden and Diet Comparisons

All plant taxa that appeared in the vegetation analyses are listed in Table 4.3 and are contrasted on an area-wide basis with their appearances in woodrat middens and diets. The relative cover percentage of these plants (Fig. 4.5) provides a basis for comparison of their proportional representation in the environment with that of plants occurring in middens and diets. A similar comparison was made with the midden contents and the vegetation around the den from which a particular midden was taken. Data were summed in certain cases where congenerics could not be distinguished in the dietary analysis. Statistical analysis was used to assess selectivity on the part of the woodrats. Plant species richness in the diet exhibited a range from 88% to 100% of plant species richness in the midden. Spearman's rho rank test showed significant correlations between diets and corresponding midden contents in relative abundance of plants common to both for each species of woodrat (Table 4.4).

Midden and Local (Den-adjacent) Vegetation Comparisons

Plants common to both the midden and the local vegetation in the belt transects ranged from 36% to 78% (Table 4.2). Comparisons of individual middens with local vegetation transects gave variable results with respect to relative abundance of plant species. In order to make a conservative comparison, only plants common to both were included in the correlation test. Five of the fifteen dens showed a significant positive correlation between midden contents and adjacent vegetation (Table 4.5). Four of the fifteen dens, however, showed a negative correlation, though not significant at the .05 level. This can occur when a packrat is selective and accumulates plants in the midden that are relatively rare in the area. All *Neotoma albigula* middens showed at least a slight positive correlation with the plant life around their dens.

Midden and Area-wide Vegetation Comparisons

Neotoma albigula was the most eclectic in its foraging. Of twenty-one predominant plant taxa growing in the 15-ha study area, ten (48%) to thirteen (62%) plant taxa were represented per midden. A combined total of eighteen (86%) were present in the five middens of white-throated woodrats (Table 4.3). Proportional representation of midden-plant taxa was most equable across many taxa, but these proportions differed from those of each of the plants growing in the area (Fig. 4.5). Plants predominating in relative abundance in their middens were *Juniperus* (24%), *Fallugia*

Table 4.3. *Plant species growing on the study area, in middens, and in the diets*
of three species of woodrats.
Na = *Neotoma albigula,* Nd = *N. devia,* Ns = *N. stephensi.*

On study area (Sample size)	In Midden			In Diet		
	Na (5)	Nd (5)	Ns (5)	Na (128)	Nd (130)	Ns (114)
Artemisia filifolia		+	+		+	+
A. tridentata		+	+	+	+	+
Atriplex canescens	+	+	+	+	+	+
Brickellia californica	+	+	+	+	+	+
Chrysothamnus nauseosus	+	+	+	+	+	+
Cucurbita foetidissima	+	+		+		
Datura meteloides	+			+		
Ephedra viridis	+	+	+	+	+	+
E. torreyana	+	+	+	+	+	+
Eriogonum hookeri	+	+	+	+	+	+
Fallugia paradoxa	+	+	+	+	+	+
grasses[a]	+	+	+	+	+	+
Gutierrezia sarothrae	+	+	+	+	+	+
Juniperus monosperma	+	+	+	+	+	+
Leucelene sp.			+			
Opuntia whipplei	+			+		
Pinus edulis	+					
Rhus trilobata	+		+	+	+	+
Salsola kali	+	+	+	+	+	+
Stanleya pinnata	+	+	+	+	+	
Yucca baileyi	+	+	+	+	+	+
Total = 21 Maximum number from 5 middens =	18	16	17	18	16	15

[a]Grasses in diets include *Aristida fendleriana, Bouteloua gracilis, Hilaria jamesii,* and *Muhlenbergia minutissima.*

(22%), and *Yucca* (14%). For comparative percentages of these plants in the modern community, see Figure 4.5. In a comparison of den-adjacent to area-wide vegetation relative abundance (at all species' dens), correlation coefficients were .618 or higher for ten of the fifteen dens (Table 4.5).

Neotoma stephensi and *N. devia* were more selective, each accumulating fewer species of plants (Fig. 4.5). Of the twenty-one available pant taxa, from seven (33%) to eleven (52%) were present per midden for *N. stephensi,* with a combined total of seventeen (81%) from five dens. Juniper clippings made up 87% of all plant materials in five *N. stephensi* middens; no other plant species comprised more than 2.7% of its midden contents.

Because three of five *N. stephensi* dens

sampled were under juniper canopies (Fig. 4.3), we considered the possibility that duff from junipers might contribute disproportionately to the midden contents. No significant difference existed in the percentage weight of juniper in *N. stephensi* middens between dens beneath juniper trees (mean = 89.7%) and dens in rocks (mean = 83.2%) (Mann-Whitney U test; $p > .05$; accept equal means). However, the sample size is small, and the 6% higher representation of juniper in middens beneath juniper trees might reflect the contribution of duff. No *N. albigula* dens sampled in this study occurred beneath juniper trees, but one midden that was in rocks next to two juniper trees (Table 4.1) contained 32.5% juniper by weight. The four other *N. albigula* middens studied did not oc-

Table 4.4. *Correlation coefficients[a] of plant relative abundance among area-wide (15 ha) vegetation and midden contents and diets.*
$(N = 13, \text{d.f.} = 11)$

	Neotoma albigula		Neotoma devia		Neotoma stephensi	
	Midden	Diet	Midden	Diet	Midden	Diet
Vegetation	.371	.179	.230	.166	.443	.443
Midden		*.583		*.606		* .478

[a] Computed using Spearman's rho rank correlation test.
* Significant correlation at the 0.05 level.

Table 4.5. *Correlation coefficients[a] between relative abundances of plant midden contents (relative mass) and the corresponding vegetation (relative cover) measured within 50 m of individual modern packrat dens.*
Coefficients were computed conservatively, using *only* relative abundances for those plant species that occurred in both the midden and the vegetation transect.

	Dens				
Packrat Species	1	2	3	4	5
Neotoma albigula	.096	.571	*.941	.306	.543
	(.792)[b]	(.445)	(.850)	(.850)	(.857)
Neotoma devia	*.770	*1.00	.400	−.030	−.400
	(.786)	(.618)	(.643)	(.452)	(.643)
Neotoma stephensi	*.655	*.771	−.785	−.500	.486
	(.452)	(.382)	(.857)	(.900)	(.394)

[a] Computed using Spearman's rho rank correlation test.
[b] Numbers in parentheses are correlation coefficients of den-adjacent vegetation to areawide vegetation.
* Significant correlation at the .05 level.

cur as near to junipers, yet two of these contained larger proportions of juniper clippings (35.8% and 36.0%, respectively) and two contained less (1.6% and 7.1%).

In the case of *N. devia*, from six (29%) to ten (47%) plant taxa occurred per midden, with a combined total of sixteen (76%) from five dens. *Ephedra* clippings made up 66% of all plant materials in five *N. devia* middens; six other plants each comprised between 3.7% and 6.4%. No significant correlations were observed between relative abundance of plant species (on an area-wide basis) and the relative abundance of plant species found in the midden (Table 4.4).

Removal Experiment
After removal of the behaviorally dominant

species (*N. albigula* and *N. stephensi*) from the experimental plot, empty dens were quickly invaded by *N. devia* or other individuals of *N. albigula* or *N. stephensi*. When one species replaced another at a den site, obvious changes began to appear in the accumulated plant clippings. After a period of den occupancy by the new species, the predominant plant species collected by the previous occupant became progressively less obvious. After a year had passed, evidence of the original occupant was completely obscured. For example, a year after *N. devia* replaced *N. stephensi* at den 13 (Fig. 4.6), plant clippings around the den changed from dried juniper twigs and foliage (the hallmark of *N. stephensi*) to mostly *Ephedra* twigs (preferred by *N. devia*).

4.6A

4.6B

Fig. 4.6. Woodrat den no. 13 beneath basalt boulders, before and after removal experiment. Live traps are 25 cm. long. (**A**) Den no. 13 when it was occupied by an adult female *Neotoma stephensi*. Inset shows detail of juniper clippings from midden contents at this den. (**B**) Den no. 13 after removal-trapping of *N. stephensi* and *N. albigula* from the experimental plot. Several weeks later the den was occupied by an adult male *Neotoma devia*. Photo was taken one year after the removal experiment. Inset shows detail of *Ephedra* clippings from this midden.

DISCUSSION

We return to the question raised in the Introduction: Do fossil middens directly reflect the vegetation of past environments? We can never know or directly observe the vegetation of past communities. We can make direct comparisons only in the present, so we must apply uniformitarianism according to the habits of plant selection of living woodrats. Our studies and others (Vaughan, 1982; Atsatt and Ingram, 1983; Dial, 1988) have shown that various woodrat species differ in their plant-collecting behavior. Their dietary preferences emphasize the fact that different species of packrats will accumulate different species of plants from a single vegetative community. Fortunately for the paleoenvironmentalist, these differences are mostly in the relative proportions of plant species collected, not in species richness (see below). Less than half of the plant species present on a 15-ha area may be found in one midden; three-fourths of the species may be found when up to five middens are analyzed. Perhaps our modern middens incorporated material from a shorter interval of time than the modern middens studied by fossil analysts, for example, Cole (1981; 1982) and Cole and Webb (1985).

The time between the original collection of midden materials and the formation of an indurated fossil midden may see numerous events affecting midden composition. Diagenetic enrichment may potentially include at least the following (see also Webb and Betancourt, *this volume*, chap. 6): (1) further addition of midden materials at a given den site by subsequent generations of woodrats; (2) addition of allochthonous material by other animals (especially coyotes) and by wind; (3) bioturbation of the accumulated residues by subsequent generations of woodrats and other animals; (4) differential preservation and erosion.

Strictly applied, uniformitarianism requires that the most closely similar species be used in making inferences about its extinct relatives. During the time period (approximately 40,000 B.P. to the present) in which woodrat middens have been preserved, about twenty-three species of woodrats have existed. Twenty-one of these are extant (Hall, 1981; Mascarello, 1978), and two (*N. pygmaea* and

N. findleyi [Harris, 1984a, 1984b]) are extinct. Probably no more than six of these have produced the majority of fossil middens thus far investigated in North America. All living species are known to build dens and accumulate middens, and therefore all potentially have contributed to the fossil midden record in their respective areas of past geographic distribution. Thus, in work with late Pleistocene and Holocene middens it may be instructive to attempt an identification of the woodrat or woodrats that accumulated the midden in question, since their bones are often preserved in their middens (personal observation). Unfortunately, *Neotoma* species are notoriously difficult to differentiate based on skeletal elements, including teeth.

Table 4.6. *Summary of relationships among diets, midden contents, and vegetation for a three-species community of* **Neotoma** *in northern Arizona.*

	Plant Species Relative Abundance[a]	Plant Species Richness[b]
Diet versus midden contents	100%	88%–100%
Midden versus local (within 50 m of den) vegetation	30%	36%–78%
Midden versus area-wide (15 hectare plot) vegetation	0%	25%–59%

[a]Percentage of fifteen dens found to be significantly correlated.
[b]Range in percentage of plant species common to all.

Do woodrats accumulate plant species in direct proportion to plant relative abundance? This depends on the level of resolution (see Tables 4.5, 4.6). Data from Woodhouse Mesa demonstrate clearly that strong biases exist, depending on the woodrat species. The bias is related to diet (see below). Therefore, the relative abundance of plants in an active midden cannot be taken to represent accurately the relative abundances of plants in the environment. For example, in an area like the Woodhouse Mesa escarpment, where there are few and scattered juniper trees, the contents of a

N. stephensi midden would suggest a dense juniper woodland. In an area of genuine dense juniper woodland, little difference from our data should be expected in Stephens' woodrat midden contents. The analysis of middens of the other two woodrat species involved in the present study would lead to two different habitat reconstructions. We are not convinced that diagenetic alteration or enrichment would be sufficient to eliminate the bias in a fossil midden, and the bias should be sought by paleoecologists.

Do woodrats sample most or all species of plants in their community? Some *Neotoma* species sample the available floral diversity more completely than others. At Woodhouse Mesa the rats incorporated about half of them in a given midden. As stated above, when samples from five or more middens are combined, plant representation is improved. This may be comparable to the expected contribution of additional generations of rats in accumulating what becomes a fossil midden. We recognize that *N. stephensi* may offer a "worst case." Middens of *N. albigula*, a relatively generalized forager, will better record local vegetation. Many fossil midden studies to date, especially in the Chihuahuan, Sonoran, and Mojave deserts (where *N. albigula* is abundant), probably have involved this woodrat. Other taphonomic factors improve the midden representation of floral and faunal diversity. It is conceivable, however, that such agencies as coyotes, foxes, and raptors could act as long-distance dispersers of biota. Since woodrats collect carnivoran feces and raptor pellets, they may be preserving a partly spurious record including extralocal organisms.

Do diets of modern woodrats reflect the contemporaneous vegetation and clippings at a den? This question is of particular importance to those investigators concerned with the mechanism by which plants in the midden initially are accrued. Diets do not reflect the contemporaneous vegetation at Woodhouse Mesa. However, diets do reflect the clippings in the midden. Although the relationship between midden contents and diets is significant, these features are not strongly correlated (Table 4.4).

Is one species of woodrat responsible for midden accumulation at a given locality over a period of time? Probably not. In five years of trapping in dens at Woodhouse Mesa, woodrat species turnover was not uncommon. Additionally, areas of sympatry are common for *Neotoma* species. Thus, in a fossil midden sample thought to represent decades to centuries, the turnover of woodrat species occupying a den may be substantial. For example, at the Grand Canyon of the Colorado River in Arizona, six species of woodrat (*N. cinerea*, *N. mexicana*, *N. stephensi*, *N. albigula*, *N. lepida*, and *N. devia*) occur at various elevations. Changes in the altitudinal distributions of their respective preferred dietary plants could result in complex patterns of synchronous or diachronous overlaps in rat species distributions. In some cases, changing species of woodrats will be difficult or impossible to separate from climatic changes.

Results of the removal experiment demonstrate the implications of a change in the resident rat. Turnovers always involved corresponding changes in the plant clippings accumulated at the den. In studies involving fossil middens, these changes might incorrectly be interpreted as environmental changes. Ideally, we recommend sampling of contemporaneous multiple middens in each region under investigation (as has been the practice of most paleoenvironmentalists when possible). Conversely, it is possible that rat species turnover and continuous occupation over time spans in the range of ten to one hundred years would produce a more representative accumulation of the local flora. Again, empirical data are lacking.

Future investigations could further clarify and enhance the information-retrieval possibilities using fossil middens. In addition to recommendations stated or implied above, we present several more below.

Few dietary studies of living woodrat species are available. Most such studies have been restricted to localized geographic regions, even for species that have broad distributions and probably exhibit some geographic variation in their diets. It is important for ecologists to develop an appreciation for relative specialization or generalization in the trophic roles of *Neotoma* species. More midden studies of living woodrats in a wide variety of regions will help to clarify their selectivity with regard to the available plant species. Long-term studies of modern middens and species-removal experiments would also be instructive.

Techniques for the gathering and analysis of midden constituents should be standardized. As noted above, our methods differed somewhat from those used by analysts of fossil middens. In our case, failure to screen midden materials for particles of very small (< 3 mm) size may bias our data or give incomplete sampling of those species likely to be represented in a midden only by that particle size. This should mainly affect comparisons of species richness, with only a minor impact on comparisons of species relative abundance.

More rigorous means of quantifying fossil midden data are also called for, because the subjective, undefined relative abundance categories often used in studies of fossil middens are not readily repeatable. By virtue of their youth, modern middens are not as susceptible to the biases induced by differential preservation, erosion, et cetera. Techniques should thus be developed which are applicable to fossil middens and reconcilable with the vagaries inherent in the fossil record. Then, similar techniques can be applied to modern middens for more equitable comparisons. Long-term studies (on the order of ten to fifty years) of modern middens are clearly desirable for both ecological and paleontological insights.

Woodrats are not the only agency of accumulation at work in woodrat middens. For example, we have noted usage of woodrat dens by various white-footed mice (*Peromyscus* spp.), ground squirrels (*Spermophilus* and *Ammospermophilus*), lizards, and snakes. Owls at times roost over woodrat dens and regurgitate pellets onto them. Owl pellets and carnivore scats are a potential source of extralocal bones and plants. Porcupines bring in large animal bones (as do Old World porcupines, *Hystrix*; Brain, 1980) and other materials (Betancourt et al., 1986). Wind transports various high-volume–low-density materials such as composite achenes, grass seeds and panicles, and insect parts. Detailed studies are needed of various taphonomic factors other than woodrats that are involved in the accumulation of materials in woodrat middens.

Effective methods for identification of woodrat species from their bones and teeth (Harris, 1984a) are desirable. If the species represented by fossil teeth in a midden can be identified as an extant species, knowledge of the plant-gathering habits of that species can be applied with more confidence to the fossil record.

We conclude that dietary specialization affects plant clippings accumulated at modern woodrat dens, and hence biases the representation of relative abundance of the ambient vegetation over a short time period. Plant clippings accurately reflect the diet of the woodrat species in question. If several middens are sampled, the contemporaneous plant species diversity may be represented fairly. Thus, fossil analysts appear to be justified in their interpretations of species richness, although major uncertainty about interpretations of species abundance persists.

ACKNOWLEDGMENTS

We gratefully acknowledge G. C. Bateman, J. L. Betancourt, S. Cinnamon, G. E. Goslow, Jr., R. H. Hevly, P. S. Martin, J. I. Mead, and T. R. Van Devender for useful discussion and critical comments on the manuscript.

REFERENCES

Atsatt, P. R., and Ingram, T. (1983). "Adaptation to oak and other fibrous, phenolic-rich foliage by a small mammal, *Neotoma fuscipes*." *Oecologia* 60, 135–42.

Betancourt, J. L. (1984). "Late Quaternary plant zonation and climate in southeastern Utah." *Great Basin Naturalist* 44, 1–35.

Betancourt, J. L.; Van Devender, T. R.; and Rose, M. (1986). "Comparison of plant macrofossils in woodrat (*Neotoma* sp.) and porcupine (*Erethizon dorsatum*) middens from the western United States." *Journal of Mammalogy* 67, 266–73.

Brain, C. K. (1980). "Some criteria for the recognition of bone-collecting agencies in African caves." In *Fossils in the making: Vertebrate taphonomy and paleoecology*. A. K. Behrensmeyer and A. P. Hill, eds., pp. 107–30. Chicago: University of Chicago Press.

Cameron, G. N., and Rainey, D. G. (1972). "Habitat utilization by *Neotoma lepida* in the Mojave Desert." *Journal of Mammalogy* 53, 251–66.

Cole, K. L. (1981). "Late Quaternary environment in the eastern Grand Canyon: Vegetational gradients over the last 25,000 years." Ph.D. diss., University of Arizona, Tucson.

———. (1982). "Late Quaternary zonation of vegetation in the eastern Grand Canyon." *Science* 217, 1142–45.

Cole, K. L., and Webb, R. H. (1985). "Late Holocene vegetation in Greenwater Valley, Mojave Desert, California." *Quaternary Research* 23, 227–35.

Dial, K. P. (1984). "Four sympatric species of *Neotoma*: Ecological aspects of morphology, diets, and behavior." Ph.D. diss., Northern Arizona University, Flagstaff.

———. (1988). "Three sympatric species of *Neotoma*: Dietary specialization and coexistence." *Oecologia* 76, 531–37.

Finley, R. B., Jr. (1958). "The wood rats of Colorado: Distribution and ecology." *University of Kansas Publications, Museum of Natural History* 10, 213–552.

Hall, E. R. (1981). *The mammals of North America.* 2d ed. New York: John Wiley and Sons.

Hansen, R. M., and Flinders, J. T. (1969). *Food habits of North American hares.* Range Science Department, Colorado State University, Science Series 1, pp. 1–18.

Harris, A. H. (1984a). "*Neotoma* in the late Pleistocene of New Mexico and Chihuahua." *Special Publications of the Carnegie Museum of Natural History* 8, 164–78.

———. (1984b). "Two new species of late Pleistocene woodrats (Cricetidae: *Neotoma*) from New Mexico." *Journal of Mammalogy* 65, 560–66.

Linsdale, J. M., and Tevis, L. P., Jr. (1951). *The dusky-footed wood rat.* Berkeley: University of California Press.

Mascarello, J. T. (1978). "Chromosomal, biochemical, mensural, penile, and cranial variation in desert woodrats (*Neotoma lepida*)." *Journal of Mammalogy* 59, 477–95.

Mead, J. I.; Thompson, R. S.; and Long, A. (1978). "Arizona radiocarbon dates IX: Carbon isotope dating of packrat middens." *Radiocarbon* 20, 171–91.

Mueller-Dombois, D., and Ellenberg, H. E. (1974). *Aims and methods of vegetation ecology.* New York: John Wiley and Sons.

Nowak, R. M., and Paradiso, J. L. (1983). *Walker's mammals of the world.* 4th ed. Baltimore: Johns Hopkins University Press.

Raincy, D. G. (1956). "Eastern wood rat, *Neotoma floridana*: Life history and ecology." *University of Kansas Publications, Museum of Natural History* 8, 525–646.

Stones, R. C., and Hayward, C. L. (1968). "Natural history of the desert woodrat, *Neotoma lepida*." *American Midland Naturalist* 80, 458–76.

Van Devender, T. R. (1973). "Late Pleistocene plants and animals of the Sonoran Desert: A survey of ancient packrat middens in southwestern Arizona." Ph.D. diss., University of Arizona, Tucson.

Van Devender, T. R., and King, J. E. (1971). "Late Pleistocene vegetational records in western Arizona." *Journal of the Arizona Academy of Sciences* 6, 240–44.

Van Devender, T. R.; Martin, P. S.; Thompson, R. S.; Cole, K. L.; Jull, A. J. T.; Long, A.; Toolin, L. J.; and Donahue, D. J. (1985). "Fossil packrat middens and the tandem accelerator mass spectrometer." *Nature* 317, 610–13.

Van Devender, T. R., and Spaulding, W. G. (1979). "Development of vegetation and climate in the southwestern United States." *Science* 204, 701–10.

Vaughan, T. A. (1980). "Woodrats and picturesque junipers." In *Aspects of vertebrate history.* L. L. Jacobs, ed., pp. 387–401. Flagstaff: Museum of Northern Arizona Press.

———. (1982). "Stephens' woodrat, a dietary specialist." *Journal of Mammalogy* 63, 53–62.

Vorhies, C. T., and Taylor, W. P. (1940). *Life history and ecology of the white-throated woodrat, Neotoma albigula albigula (Hartley), in relation to grazing in Arizona.* University of Arizona College of Agriculture, Agricultural Experiment Station, Technical Bulletin 86, 455–529.

Wells, P. V. (1976). "Macrofossil analysis of wood rat (*Neotoma*) middens as a key to the Quaternary vegetational history of arid America." *Quaternary Research* 6, 223–48.

Wells, P. V., and Hunziker, J. H. (1976). "Origin of the creosote bush (*Larrea*) deserts of southwestern North America." *Annals of the Missouri Botanical Garden* 63, 843–61.

Wells, P. V., and Jorgensen, C. D. (1964). "Pleistocene wood rat middens and climatic change in the Mohave Desert: A record of juniper woodlands." *Science* 143, 1171–73.

Chapter 5

Packrat Middens: Their Composition and Methods of Analysis

W. Geoffrey Spaulding   Julio L. Betancourt   Lisa K. Croft
Kenneth L. Cole

INTRODUCTION

Packrat middens are nondescript masses, gray to dark brown in color and easily overlooked in the darkness of a rock shelter or cave. They are common in many arid to semiarid areas of North America, where a dozen or so researchers routinely exploit their rich store of plant fossils to reconstruct late Quaternary paleoenvironments. In this chapter we discuss the composition of middens, some approaches to their analysis, and examples of vegetational data from middens. Though most midden papers include statements on methods, only four other works (Van Devender, 1973; Wells, 1976; Spaulding, 1985; Thompson, 1985) discuss methods in detail. The reader may wish to consult these references for procedures, assumptions, and insights not discussed in this chapter.

Here we draw on data and research experiences from south-central Nevada, the Colorado Plateau, and the lower Colorado River valley to probe questions encountered when processing midden samples and to assess the qualitative and quantitative attributes of macrofossil assemblages. We examine variability between weights of the plant matrix and other midden constituents (amberat, sediment, and fecal pellets), and discuss how it might influence measures of species richness and abundance adopted by different researchers. Though flexibility in methods may enhance creative and unique applications, lack of standard methods for presenting plant macrofossil data inhibits comparability. Choosing one method over another, however, is no simple matter, as we and others have discovered in exploring the relationship between species representation in modern middens and local vegetation.

The midden-vegetation relationship is of prime concern in paleoecological reconstructions. The correspondence of plant macrofossils to vegetation within a packrat's foraging range (roughly one hectare centered at the midden) is probably complicated by a number of factors that are broadly familiar to taphonomists: (1) packrat dietary preferences (see in this volume Dial and Czaplewski, chap. 4; Vaughan, chap. 2; Finley, chap. 3); (2) the effect of distance to nearest plant and changing packrat selectivity with increasing distance; (3) internal variability in midden composition due to den function; (4) duration of depositional episode; (5) postdepositional history; and (6) the relation of macrofossil amounts (number or weights) to estimates of species importance in the surrounding community (e.g., biomass, percent cover, frequency, or density). The latter is crucial to the goal of reconstructing past communities.

We provide examples by comparing macrofossil abundance in modern middens with
percent cover and density of plants growing
at the midden site.

PHYSICAL PROPERTIES OF MIDDENS

Packrat middens are composed of soluble and
insoluble organic matter, dust, and rocks.
Most ancient middens are indurated, and
many resemble blocks of asphalt with the
consistency and mass of an unfired adobe
brick. Crystallized packrat urine, called amberat, saturates plant fragments, pellets, and
other debris, encasing them in an amberlike
matrix. The mineralogy of the amberat has
been studied in reference to oxalate metabolism in packrats (Emerson and Howard, 1978;
Shirley and Schmidt-Nielsen, 1967). Urine-
soaked, indurated masses appear to be more
resistant to erosion and decay and thus have a
better chance of surviving for tens of millennia than do loose debris piles. However, deposits not cemented by amberat have been
encountered as discrete units in sediments of
some dry caves. An example is unit B of Rampart Cave in the lower Grand Canyon, a stratum composed primarily of loose plant debris
and packrat fecal pellets dating to the last
glacial maximum (Long et al., 1974; Phillips,
1977). Uncemented deposits of Pleistocene
age, retaining some of the original structure
of the packrat's den, also are reported from
caves in southeastern Utah (Betancourt, 1984).

Amberat helps preserve packrat middens in
several ways, including its role as a primary
constituent of the midden rind. Two kinds of
rinds are found on ancient middens: lustrous
coats that resemble solidified asphalt or mo-
lasses, and lusterless rinds that are frequently
convoluted and bear many fine desiccation
cracks. The shiny coats develop where recently liquified amberat has crystallized. A
fresh break on the surface of an indurated
midden will, over a period of months, become
covered with a shiny layer of redeposited amberat. This occurs because amberat is hygroscopic and, during infrequent humid periods,
it will rehydrate. It then flows by capillary action to the surface of any fresh break, where
it recrystallizes to form a glossy layer. Amberat is hard and smooth to the touch when
the air is dry, but sticky and soft when the air
is moist. Thin sections show undulating

layers (about 0.05 mm thick) of microcrystalline minerals (0.005 mm maximum diameter)
and minor amorphous material, with occasionally larger (0.05 mm) detrital inclusions.
The coarser microcrystalline material results
from recrystallization, which fills desiccation
cracks in the weathering rinds (Emerson and
Howard, 1978).

Although the process of rind formation has
not been studied, it is likely that dust and organic debris accumulating on glossy surfaces
over long periods block the free exchange of
moisture between the midden and the atmosphere. Therefore, through the hygroscopic
property of amberat, indurated middens appear to be self-sealing (Spaulding, 1985). Ancient middens are typically encased in a convoluted, weathering rind composed of dust
and degraded organic debris cemented by amberat. These dull gray or brown rinds typify
middens dating to the Wisconsin and early
Holocene (Fig. 5.1).

Saturation by crystallized urine protects
plant macrofossils from decay. Extreme dryness suppresses microbial and chemical activity and favors mummification. Urea is
hypotonic and mummification of plant fragments is likely accomplished in much the
same manner as if they were packed in salt.
Also, most insect reducers (termites, carpet
beetles, dung beetles) probably shun organic
debris saturated in amberat, perhaps because
of its chemical properties (Wells, 1976). Middens depleted of amberat have a punky consistency and are commonly riddled with
termite galleries and show evidence of attack
by dung beetles and moths. Lichens that grow
on the rocks around the deposits do not use
the amberat as substrate (Emerson and Howard, 1978).

The adhesive property of amberat also deserves mention. Finley (*this volume*, chap. 3)
and other researchers have observed massive
indurated middens hanging from cliff walls or
ceilings of rock shelters. Such peculiar situations arise when the rock shelf upon which
the midden accumulated fails, while the midden remains glued to the wall by amberat.
The relative ages of such deposits may yield
approximate rates of exfoliation and cliff retreat (Cole and Meyer, 1982). When middens
are more resistant to degradation than the
host rock, as in some Upper Cretaceous sand-

Fig. 5.1. Photograph of a massive indurated midden from the Eleana Range of south-central Nevada, dating 17.1–10.6 ka (Spaulding, 1985). Note the convoluted surface and many fine cracks that typify the weathering rinds of most ancient packrat middens.

stones of the Colorado Plateau, there is a surplus of "hanging" middens that decompose gradually with exposure to runoff and rain splash.

Variability in the Different Midden Fractions

Samples from a packrat midden must be processed in order to release macrofossils from the impregnating amberat, and this processing allows the different components or fractions of middens to be studied. In the laboratory samples are reduced to the desired size using a hammer and chisel and then are immersed in water to dissolve the amberat. Special attention is given to potential stratigraphic complexity, and questionable portions are deleted by splitting along discernible bedding planes. Presoaking (indurated) sample weights, in our experience, range from 200 to 2000 g. After weighing, the samples are allowed to soak for several days. The disaggre-

gated residue is then wet-screened through a 20-mesh (0.84-mm) soil sieve, oven-dried at about 70°C, and weighed. Some researchers use smaller minimum sieve sizes. Some also remove and weigh fecal pellets prior to initial sorting of the plant matrix.

Observations of hundreds of late Pleistocene middens have led us to suspect that diagenetic changes occur after deposition. If pronounced, these alterations could affect weight and density relations between the different midden fractions. In particular, very young middens are often less saturated with amberat and contain more coarse debris, while older middens are more compact and contain finer debris. The oldest (> 25 ka) middens from the Mojave Desert usually have well-indurated exteriors, but inside they are often depleted of amberat. To better understand these processes, we first examined the relationships between different midden fractions, the amberat, and the sediment. The latter includes dust, gravel, and the ultrafine plant debris lost in washing.

Figures 5.2A and 5.2B compare washed and presoaking weights from middens collected in south-central Nevada and the Colorado Pla-

5.2A

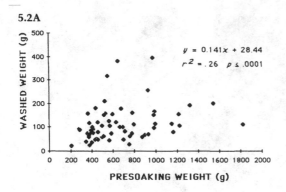

5.2B

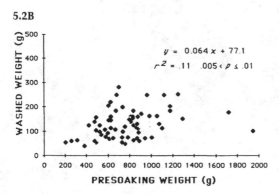

5.2C

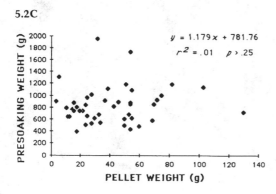

5.2D

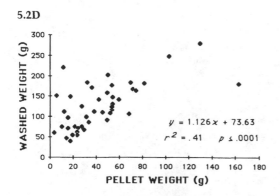

Fig. 5.2. Comparison of the presoaking weight of midden samples with their weight after washing and weight of fecal pellets contained in each sample. (A) Washed weight as a function of presoaking weight for samples from southern Nevada; (B) washed weight as a function of presoaking weight for samples from the Colorado Plateau; (C) comparison of presoaking weight and pellet weight for samples from the Colorado Plateau; (D) comparison of washed sample weight and pellet weight for samples from the Colorado Plateau.

teau. Although statistically significant ($p <$ .01), in both cases the correlations are weak ($r^2 = .26$, $r^2 = .11$). This suggests that the presoaking weight is not a good predictor of the weight of the fraction recovered after soaking. In general, the weight lost in the washing process is 65%–95% of the presoaking weight (Figs. 5.2A, 5.2B). There is little relationship between presoaking weight and pellet weight in the Colorado Plateau data (Fig. 5.2C), a combined effect of variability in amberat saturation, fine sediment content, and fecal pellet abundance. Removal of the amberat and fine sediment fraction results in a more predictable relationship between pellet weight and washed sample weight (Fig. 5.2D).

If amberat saturation increases with age, or if there is gradual breakdown of the organic component of middens, there may be a relationship evident between sample age and percentage of the sample weight lost in the washing process. The "lost fraction" consists of solubles (chiefly amberat) and sediment less than 0.85 mm in size, and is presented as a percentage of the original sample weight compared to sample age in Figures 5.3A and 5.3B. Weight loss is reduced in samples younger than 5 ka, though noticeably less so for the Colorado Plateau samples (60%–95%) than for those from southern Nevada (42%–79%). Although the patterns of dispersion differ slightly for the two data sets, for middens older than 7.5 ka average weight loss for Colorado Plateau samples (83% ± 8%) is quite similar to that for the Nevada samples (85% ± 7%). The trend in percentage loss versus age in the Nevada samples appears statistically valid. We also examined the relationship between age and pellet abundance as the weight percentage of the presoaking and washed weights and found little correlation (Figs. 5.3C, 5.3D).

5.3A

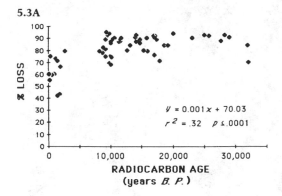

5.3B

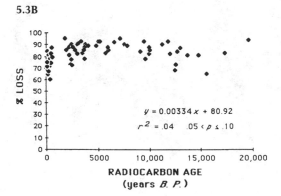

5.3C

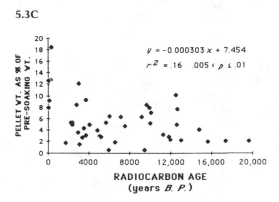

5.3D

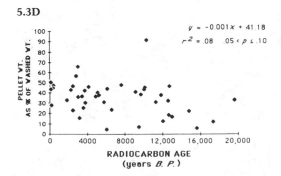

Fig. 5.3. The percentage (by weight) of midden sample lost in the washing process and percentage weight of fecal pellets as functions of radiocarbon age. (A) The percentage weight loss [1 − (weight of washed sample/presoaking weight) × 100] for samples from southern Nevada compared to radiocarbon age; (B) the percentage weight loss for samples from the Colorado Plateau compared to radiocarbon age; (C) pellet weight as a percentage of the presoaking weight compared to radiocarbon age for Colorado Plateau middens; (D) pellet weight as a percentage of the washed weight of the sample against radiocarbon age for Colorado Plateau middens.

Variability in the amount of coarse debris in midden samples could reflect structural differences between middens. There is currently little agreement as to what portions of the rat's den or house become fossilized. Some researchers believe that middens most commonly represent areas near entrances to internal passageways where waste material accumulates from house cleaning (Finley, 1958; Van Devender, 1987). Wells (1976) suggests that middens represent portions of houses that become indurated into a cohesive mass by impregnation with quick-drying urine. We have observed active houses "roofed" with a rind of crystallized urine and pellets (Fig. 5.4). Conceivably, after abandonment the collapsed stick lattice that provides structural support for passageways and chambers combines with urine in the process of rehydration to form middens. We have observed late Pleistocene "houses" in an intermediate stage of collapse in caves on the Colorado Plateau. The loose latticework contained recognizable nest and food chambers subtending an indurated surface. This structure may be inconsistent with the idea that a house "melts" through time to produce a single midden. However, ancient middens frequently have obvious stratigraphy with layers just a few millimeters thick.

Indurated middens can develop from one year to the next, as suggested by the inclusion of a dated newspaper in a well-cemented midden from the Great Basin (Thompson, 1985). Erratic representation of plant organs and disseminules, available to the packrat only seasonally or in mast years, suggests that indurated middens can form during a relatively brief period of a few months to a year. At

Fig. 5.4. Stephens' woodrat (*Neotoma stephensi*) perched atop the indurated "roof" directly above one of the entrances to its den in Chaco Canyon, New Mexico. (Photograph courtesy of Michael Lyon.)

Chaco Canyon, New Mexico, seeds of pinyon and juniper are represented haphazardly in fossil middens containing abundant needles and twigs. However, many middens do contain a diverse array of short-lived perennials and annuals, implying formation over a period of one to several years. In such cases the midden assemblage might average intra- and interannual variability in community composition. Because the present evidence is mainly anecdotal, the duration of midden depositional episodes remains unknown. Radiocarbon dating cannot resolve the issue because, at best, the radiocarbon age of a sample merely brackets the time of deposition.

Wells (1976) and Thompson (1985) believe that in most cases midden samples represent the accumulations of a few years at most. If true, this means that a midden sequence spanning thousands of years, even if from one large midden, is constructed from discontinuous samples, each representing one to several years. These short intervals give no information about the longer, unsampled periods. Where practical, additional middens can be collected to fill time gaps and test for continuity of trends suggested by the initial chronology.

PRESENCE-ABSENCE AND ABUNDANCE OF PLANT TAXA

Quantities of the different macrofossil types (e.g., seeds, twigs, leaves) in a midden assemblage are not always reported in the literature. This is partly a consequence of the sheer number of macrofossils that have to be sorted and quantified. A typical midden sample, commonly weighing 50–200 g after washing, includes many thousands of identifiable plant parts. In tabulation, special note is usually made of single-item occurrences. For example, the presence of a few pine-nut fragments without pine needles suggests caching of nuts by birds or humans, animals that forage over a much larger area than the packrat. Single-item occurrences may also be clues to potential contamination, particularly if the taxon occurs more commonly in older or younger middens from the same locality. If contamination is suspected, accelerator mass spectrometry can discriminate between small plant parts (< 0.1 g) with different ages (Van Devender et al., 1986).

Midden assemblages are autochthonous and, except for very rare occurrences of pine nuts and the like, contain locally occurring plant species only. Therefore, simple presence-absence data can be very useful in midden-based paleoecological reconstructions. Whatever the biases, an uncontaminated assemblage provides a plant species list for a restricted locality. Many analysts have a great interest in the "extralocals," plants in the macrofossils assemblage that do not occur near the site today. Extralocal species can be placed in three categories based on their current distribution: (1) those that presently occur in habitats similar to the fossil site; (2) those that occur in different (usually more mesic) habitats but still exist in the general area; and (3) true extralimitals that have been extirpated from the area.

Similarity Indices

Differences between middens or between a midden assemblage and the modern flora can be quantified using several coefficients of similarity. In recent studies such coefficients have emphasized the number of trees, shrubs, and succulents (N_{ts}) encountered in a sample, or the number of all taxa encountered (N). Use of N_{ts} removes from consideration annual and biennial plants, whose populations can fluctuate widely from year to year (Cole, 1985). It also excludes herbaceous perennials and grasses, which may be difficult to identify or may be avoided by packrats. The mechanics and relative merits of the various indices used in paleontological contexts are discussed by Cheatham and Hazel (1969). The indices behave differently and should not be considered equivalent. For example, in comparing lists of shallow-water marine invertebrates on opposite sides of the Atlantic, trends from the Triassic to the late Tertiary using Simpson's index diverged radically from those computed with Jaccard's index (Flessa and Migasaki, 1978).

Sørensen's index is the most popular method used in midden analyses (Van Devender and Everitt, 1977; Phillips, 1977; Cole,

1985; Spaulding, 1985; Betancourt et al., 1986). It is computed by doubling the number of shared taxa and dividing by the total number of taxa in the two assemblages (Sørensen, 1948). Jaccard's index and Simpson's index also have been used in midden studies. To compute Jaccard's index, the total number of taxa minus the number of shared taxa is divided into the number of shared taxa. Simpson's index is the simple ratio of the number of shared taxa to the number of taxa in the less-rich assemblage. Obviously, each index places differing emphases on the number of shared taxa, the number of unique taxa, and, more critically, the difference in richness between the assemblages. Sørensen's index places emphasis on the number of shared taxa, Jaccard's on the number of taxa that are unique to either sample. Both indices will be affected by a large difference between the total number of taxa in each assemblage, which is avoided by Simpson's index. Because there are differences in the ways in which local vegetation is perceived by the paleoecologist and sampled by the packrat, the use of modern middens to characterize the present flora keeps the biases constant.

Similarity indices have broad applications in midden studies because a fundamental strength of middens lies in presence-absence data. They can be used to compare fossil assemblages to modern floras at each site to reveal whether or not age trends are unidirectional (Fig. 5.5). Another approach would be to plot "running" similarity values, that is, the similarity between a midden assemblage and the next youngest and next oldest assemblage in a time series, to quantify times of relative floral stability or instability.

Species Richness

Interpretations of midden data commonly include inferences based on species-richness values. For example, depauperate assemblages during the Pleistocene-Holocene transition have been cited as evidence for vegetational inertia and migrational lag—a lull between terminal Pleistocene departures and Holocene arrivals (Cole, 1985). One problem with species-richness values is that researchers' abilities to identify plant macrofossils are variable. Species richness also depends on sample size. Figures 5.6A and 5.6B show trends in

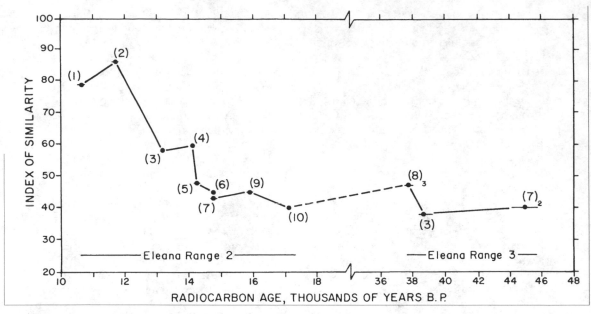

Fig. 5.5. Example of a time-dependent trend in Sørensen's similarity index, comparing macrofossil assemblages from the adjacent Eleana Range-2 and -3 middens with the contents of a modern midden at the Eleana Range-2 site. Note that the oldest samples (ER-3) yield values similar to those for the full-glacial ER-2 assemblages, despite the fact that the middle Wisconsin juniper woodland documented by the ER-3 samples is quite different from the limber pine woodland suggested by the ER-2 assemblages.

numbers of taxa observed with increasing sample size for two middens from southern Nevada.

In Figure 5.7 species richness is expressed as the ratio of number of taxa (N) to the number of identified specimens ($NISP$ or $NISP$-DT) for the south-central Nevada middens. There appears to be little association between N and $NISP$ (Fig. 5.7A), but values of N change significantly with $NISP$-DT. The age relations of $NISP$-DT, N, and N_{ts} (number of taxa for trees, shrubs, and succulents) for these middens are shown in Figures 5.8A, 5.8B, and 5.8C. Note that both $NISP$-DT and N decrease with increasing radiocarbon age. However, when species richness is expressed as the ratio of N to $NISP$-DT, the age trend is reversed. Where the dominant taxon is overrepresented, as in Pleistocene middens containing juniper in south-central Nevada, there are at least two possible explanations for low species richness. It could reflect either low diversity in a community dominated by one taxon with few associates, or simply fewer macrofossils of other species in an assemblage overwhelmingly composed of juniper.

Apparent spatial or temporal trends in species richness might be insignificant if it can be shown that middens from the same plant community exhibit substantial variability. Such variability could be expected from the vagaries inherent in the logarithmic distributions common to plant communities and midden assemblages; a few species occur in abundance, but most species are rare and occur infrequently (Koch, 1987). The probability of sampling a rare species increases with both the number and size (weight, volume, number of plant fragments) of samples. Variability due to small sample size could exceed perceived differences in species richness between middens of different age or from different habitats.

Measures of Macrofossil Abundance
Methods of enumerating macrofossil abundance differ. Thus far, macrofossil abundance has been expressed as: a semiquantitative scale (from 1 = rare to 5 = very abundant) based on actual counts and internal ranking

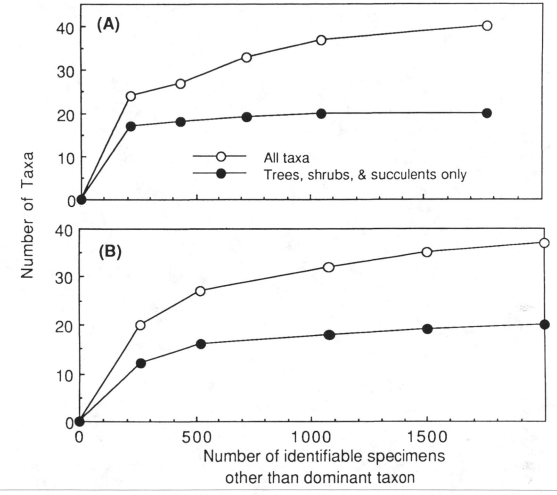

Fig. 5.6. Species richness as a function of sample size for two middens of different ages from southern Nevada. (A) PBP-1(1) was dated at 960 ± 110 B.P., and (B) PBP-3(4) at 6800 ± 110 B.P. Species richness, the ordinate, is measured as the total number of taxa (N), and the total number of trees, shrubs, and succulents (N_{ts}). Sample size is the number of identified specimens minus the number of specimens of the dominant taxon ($NISP$-DT).

(Van Devender, 1987; Thompson, 1985); raw weights of each taxon unrelated to initial sample size (Mehringer and Wigand, 1987); relative weight percentages for selected taxa (as for conifers in Wells, 1983); weight percentages of total plant fragments, including identified and unidentified fractions (Betancourt, 1984); number of plant parts unrelated to sample size (Bright and Davis, 1982); number of plant parts relative to presoaking (Cole, 1981) or washed weights, including pellets (Cole, 1986); and percentage of total number of identified specimens (Cole, 1986), modified in some cases to exclude overrepresented taxa (Spaulding, 1985).

During the initial work of the 1960s and 1970s the abundance of different plant species in a midden sample was often reported in terms of subjectively assigned relative abundance classes (e.g., Wells and Berger, 1967; King, 1976). The use of abundance classes has a long history in vegetation sampling (Braun-

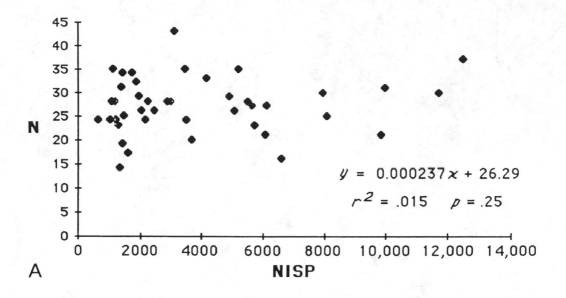

A

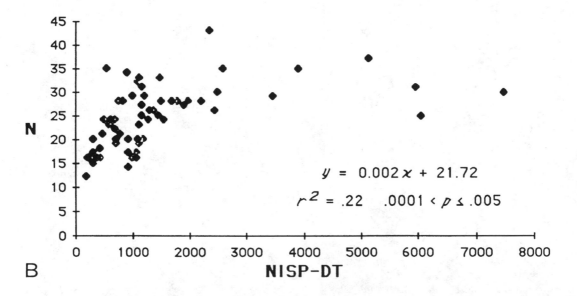

B

Fig. 5.7. Species richness values reflected by *N* (total number of taxa) compared to sample size as reflected by (A) *NISP* (total number of identified specimens), and (B) *NISP-DT* (total number of identified specimens minus the number of specimens of the dominant taxon). Midden data are from south-central Nevada (Spaulding, 1985, 1989).

5.8A

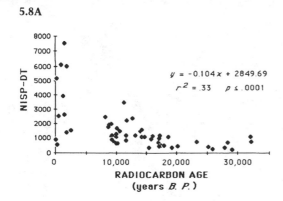

5.8B

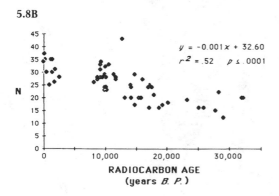

5.8C

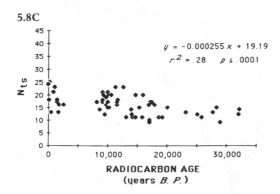

5.8D

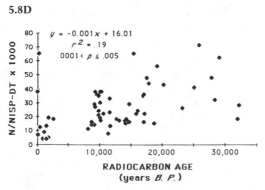

Fig. 5.8. Trends with radiocarbon age for midden samples from south-central Nevada. (A) Sample size as reflected by *NISP-DT* (total number of identified specimens minus the number of specimens of the dominant taxon); (B) species richness values as reflected by *N* (total number of taxa); (C) species richness values as reflected by N_{ts}; (D) the ratio of *N* to *NISP-DT*. Note that, for (D), the age trend appears reversed when sample size is taken into account.

Blanquet, 1965; Daubenmire, 1959). A similar technique is now employed in which the numbers of selected macrofossils are tallied and then the data are converted to a five-part scale based on the tallies (Thompson, 1984). The five-part scale yields a rank order of species similar to that generated by more laborious methods (T. R. Van Devender, personal communication). To be comparable, however, the range of values for each size class must be adjusted to sample weight, volume, or number of identified plant fragments.

Though the five-part scale may abate the common problem of overrepresentation by a single taxon, this also can be accomplished by taking the log of actual concentration or weight values (Cole, 1986), or by deleting the overrepresented taxon to calculate the frequency of other taxa in the midden (Spaulding, 1985). Workers choosing to count or weigh identified plant fragments may convert the tallies or weights to one of several numerical expressions of relative abundance. A simple approach is to use percentage values of either the total number of identified specimens (*NISP*) or the number of identified specimens minus the dominant taxon (*NISP-DT*).

The number or weight of plant fragments for a given taxon can be expressed as a function of the weight or volume of the sample unit. It is important to specify which sizes of fractions are included in the sample unit. Because most middens are composed primarily of material lost in the washing process (Figs. 5.3A, 5.3B), ratios based on presoaking weight would be subject to considerable variability unrelated to the amount of plant fragments. Because the percentage of organic residue recovered after the washing process can vary from 5% to 35% of the presoaking weight, use of the presoaking weight as the denominator in abundance ratios can influence abun-

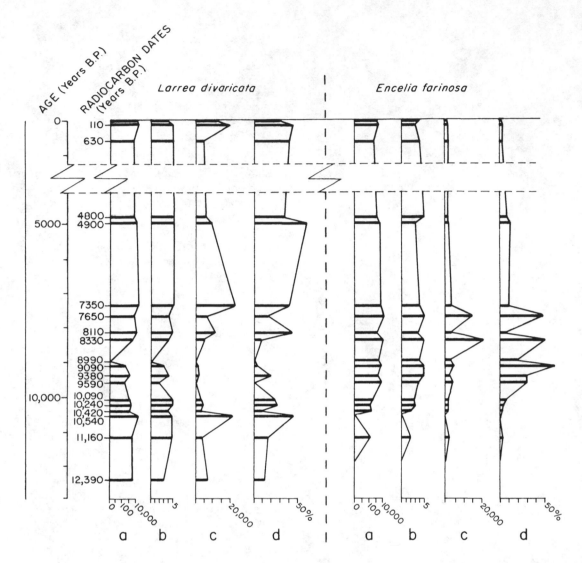

Fig. 5.9. Creosote bush (*Larrea divaricata*) and brittlebush (*Encelia farinosa*) abundance in a series of midden samples from adjacent sites within a 100-m radius at Picacho Peak, California (Cole, 1986). (A) Macrofossil concentration (number of specimens per kilogram washed midden debris, including the weight of fecal pellets) displayed on a logarithmic scale; (B) semiquantitative abundance classes; (C) macrofossil concentration displayed on a normal scale; (D) number of identified specimens as a percentage of *NISP* (total number of identified specimens). The photograph on the opposite page is of Picacho Peak, California, with *Larrea divaricata* (smaller shrubs) in the foreground. Photo courtesy of Steven P. McLaughlin.

dance by a factor of seven. Variability in the amberat, inorganic components, and pellet fraction (Figs. 5.2, 5.3) suggests that the weight or number of the plant fragments recovered after washing is a better measure of sample size.

Comparisons of different measures have not been made since Van Devender (1973) contrasted macrofossil abundances as percentage weight of the identified plant material, *NISP*, and *NISP-DT* for fossil middens from the Sonoran Desert. Data collected by Cole (1986) at Picacho Peak, California, are used to illustrate the effect of different measures of abundance. Twenty fossil middens found within a radius of 100 m record vegetation changes

over the past twelve thousand years in desertscrub of the lower Colorado River valley. The identified macrofossils were counted and weighed and the presoaking and washed sample weights were determined. In Figure 5.9 changes in the abundance of two species, creosote bush (*Larrea divaricata*) and brittlebush (*Encelia farinosa*), as reflected by concentration values (number of specimens per kilogram of washed midden, including pellets) were plotted on logarithmic and normal scales and compared to semiquantitative abundance values and % *NISP*.

For the Picacho Peak data the techniques that appear to be most sensitive to variations in midden content are macrofossil concentration plotted on a normal scale and % *NISP*. These show a pronounced shift from brittle-bush dominance in most early Holocene samples to creosote bush dominance in the middle and late Holocene assemblages. The trend is much less evident in concentration plotted on a log scale and semiquantitative abundance classes (Fig. 5.9). Both log concentration values and semiquantitative abundance classes suppress high values.

THE MIDDEN-VEGETATION RELATIONSHIP

The relationship between macrofossil abundance and local vegetation is both fundamental and less easily studied than it might seem at first. Some workers report reasonable correlations between present vegetation and modern middens (Cole, 1981, 1983; Cole and Webb, 1985; Spaulding, 1985); others do not (Dial and Czaplewski, *this volume*, chap. 4). Similarity values (Sørensen's index) between midden and site floras, which are not weighted toward species abundance, normally range from 75% to 90%. There are usually a few species missing from the midden that are present within 30 m of the site. And, commonly, a few species are encountered in the midden but not in the study plot around the site.

In Death Valley Cole and Webb (1985) compared both species composition (Sørensen's index) and relative abundance between two modern middens and adjacent plant transect data. The abundance data were compared by calculating the Bray-Curtis index (Bray and Curtis, 1957), using % *NISP* and percentage total plant cover. This index is based on the sum of the smaller of the two percentages for all species recorded in both the midden and the plant transect, also expressed as a percentage. Sørensen's index yielded 79% and 89% similarity and the Bray-Curtis index 55% and 64% similarity in these midden-vegetation comparisons. The Bray-Curtis values are actually higher than those obtained when nearby transects were compared to one another, or when cover and density were compared for the same transect.

When comparing percentage plant fragment abundance with cover and density of plants at sites in southern Nevada, Spaulding (1989)

found a less intimate relationship between the modern vegetation and midden contents. The results are recast here to illustrate some of the difficulties encountered in midden-vegetation studies. Both percent total plant cover and density (Curtis and McIntosh, 1950) were measured along three 1 × 30 m transects parallel to slope contours at the sites. The midpoint of the first transect was at the shelter mouth, and the midpoints of the second and third were within 5 m, upslope and downslope of the shelter mouth (topography permitting). In this exercise cover is the extent of a plant's canopy that intercepts the graduated center line of each transect, summed for all individuals of a species and for all three transects. Bare ground is not included in the cover values, nor are cover values divided by the length of the transects. Density is calculated as the number of individuals encountered in all three transects, divided by the area encompassed by those transects (90 m^2). The packrat species was not identified at the sites; Jorgensen and Hayward (1965) report only *Neotoma lepida* in the vicinity of the study sites. We recognize that a total of 90 m^2 may be insufficient to characterize vegetation within the rat's foraging range. By comparison, Cole and Webb (1985) and Dial and Czaplewski (*this volume*, chap. 4) measured 200 m^2 per den area. In this study Sørensen's index is used to compare midden floras and plant lists compiled from plots with a 30-m radius (2827 m^2) centered on the site. The loose debris piles sampled were assumed to be active middens, based on the presence of green leaves and supple twigs.

Cover, Density, and Macrofossil Abundance: Proportionality

Composition and cover of the desertscrub varied at each site considered. Skeleton Hills–1 (Sk-1) is at the foot of a north-facing dolomite cliff, with bedrock, talus slopes, and a small, ephemeral wash providing the available habitat area (Fig. 5.10A). Before relating measures of vegetation to macrofossil abundance (as percentage of number of identified specimens, *NISP*), we first compare percentage cover and density of plants encountered in the local transects (Fig. 5.10B). In the desertscrub at the Sk-1 site large shrubs generally have higher cover values but lower density. This is the case for desert almond (*Prunus fasciculata*), creosote bush (*Larrea divaricata*), and box-thorn (*Lycium* spp.). On the other hand, smaller plants such as bedstraw (*Galium stellatum*) and matchweed (*Gutierrezia microcephala*) are numerous, with consequent high density values, but contribute little cover due to their size. White bursage (*Ambrosia dumosa*) and shadscale (*Atriplex confertifolia*) are intermediate-sized shrubs and also have, proportionately, about the same cover and density (Fig. 5.10B). There are pronounced contrasts in these different measures of species importance, and caution is therefore warranted in judging comparisons with macrofossil frequencies.

Comparison of percent of cover and macrofossil abundance in the Sk-1 midden suggests that only tickweed (*Mortonia utahensis*) appears overrepresented in the midden (Fig. 5.10C). Boxthorn, bursage, desert almond, and shadscale appear underrepresented. Comparison of density values with plant fragment frequencies (Fig. 5.10D) shows overrepresentation of tickweed and creosote bush. Bursage is underrepresented relative to density of plants, as are bedstraw, shadscale, and matchweed.

Point of Rocks–1 (PR-1) is also a rock shelter developed in dolomite, adjacent to a first-order wash (Fig. 5.11A). The site faces southwest, and nearby slopes support a more xeric plant community than the north-facing Sk-1 site. Large shrubs do not dominate. Instead, matchweed provides the greatest cover, while bursage, grasses (primarily fluff grass [*Erioneuron pulchellum*] and three-awn [*Aristida adscencionis*]), and desert rue (*Thamnosma montana*) also contribute considerable relative cover (Fig. 5.11B). Density values reveal that grasses and matchweed are the most numerous plants in the transects. Creosote bush, shadscale, Mormon tea, and rock nettle are more abundant in the PR-1 midden than might be expected from their relative cover or density. The reverse holds true for matchweed, desert rue, grasses, and Indian tobacco (*Nicotiana trigonophylla*) (Figs. 5.11C, 5.11D).

In contrast to these two bedrock sites, Fortymile Canyon–1 (FMC-1) is a shelter in Quaternary alluvium. It is near the bed of a small arroyo (Fig. 5.12A), and the local vegetation is represented by plants typical of the upper ba-

jada as well as the disturbed wash. White bursage, matchweed, Mormon tea, indigo bush (*Psorothamnus fremontii*), and boxthorn dominate percentage cover, but bursage and matchweed dominate density (Fig. 5.12B). Creosote bush was present within 30 m of the midden site but not encountered in the belt transects. It is therefore highly overrepresented in the midden (Figs. 5.12C, 5.12D). Compared to percentage cover, indigo bush and matchweed are underrepresented in the midden, and shadscale is overrepresented (Fig. 5.12C). Relative to their density, Mormon tea and boxthorn are overrepresented in the midden. As with cover values, bursage and matchweed are underrepresented in the midden relative to their density (Fig. 5.12D).

Intersite Comparisons of Selected Taxa

These data also can be studied to show if changes in importance of a given plant from site to site result in corresponding changes in plant fragment abundance in the modern middens. Though based on only five sites, graphs of macrofossil abundance against cover and density are nevertheless instructive. Studies of plant cuticles in fecal pellets (Phillips, 1977) and field observations suggest that four of the taxa selected from this analysis are eaten by packrats. These are creosote bush, saltbush (in these examples both cattle-spinach [*Atriplex polycarpa*] and shadscale are considered), globe mallow (*Sphaeralcea ambigua*), and Mormon tea (*Ephedra nevadensis*, *E. funerea*, and *E. torreyana*). Like the resinous junipers and pines, which usually are the chief constituents of middens in higher-elevation woodlands, packrats preferentially gather and eat creosote bush. It is favored by packrats to the extent that it can comprise more than 90% of the identified fragments in a midden sample, even at sites where the shrub is so scarce as to escape sampling in the transects (Figs. 5.11C, 5.11D; 5.12C, 5.12D). With such a selective bias, in at least some situations, it is not surprising to find little correlation of creosote bush cover and density with macrofossil abundance. Saltbush, represented by much lower frequencies in the middens, also shows no predictable relationship with this range of cover and density values (Figs. 5.13A, B). Globe mallow and Mormon tea frequencies in the middens also

5.10A

5.10B

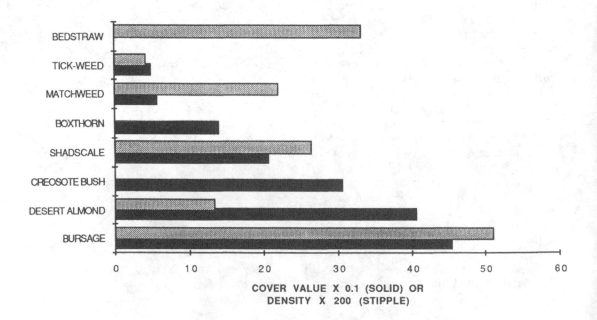

COVER VALUE X 0.1 (SOLID) OR
DENSITY X 200 (STIPPLE)

5.10C

IS, MIDDEN CONTENTS TO RELEVE = 87

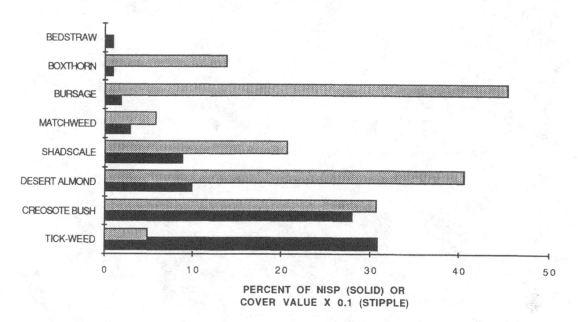

PERCENT OF NISP (SOLID) OR
COVER VALUE X 0.1 (STIPPLE)

5.10D

IS, MIDDEN CONTENTS TO RELEVE = 87

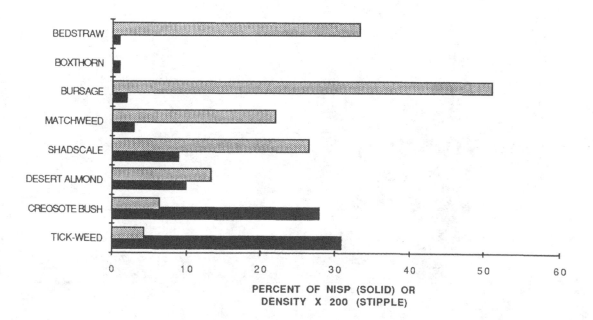

PERCENT OF NISP (SOLID) OR
DENSITY X 200 (STIPPLE)

Fig. 5.10. Midden-vegetation relationships at the Skeleton Hills-1 site, southern Nevada. (A) Photograph of the site; (B) comparison of cover and density values for plant species encountered in transects at the site; (C) macrofossil abundance as percentage of *NISP* (total number of identified specimens) compared to cover values for select species; (D) macrofossil abundance as percentage of *NISP* compared to density for the same species. Note changes in scale.

5.11A

5.11B

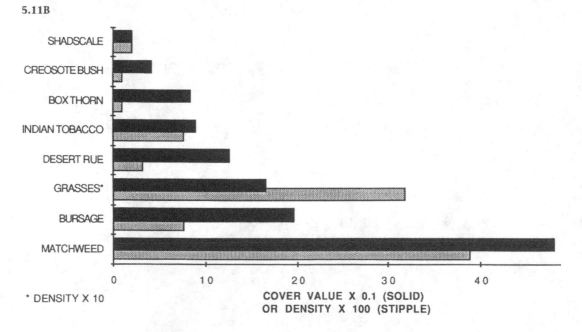

* DENSITY X 10

COVER VALUE X 0.1 (SOLID)
OR DENSITY X 100 (STIPPLE)

Fig. 5.11. Midden-vegetation relationships at the Point of Rocks-1 site, southern Nevada. (A) Photograph of the site; (B) comparison of cover and density values for plant species encountered in transects at the site; (C) macrofossil abundance as percentage of *NISP-DT* (total number of identified

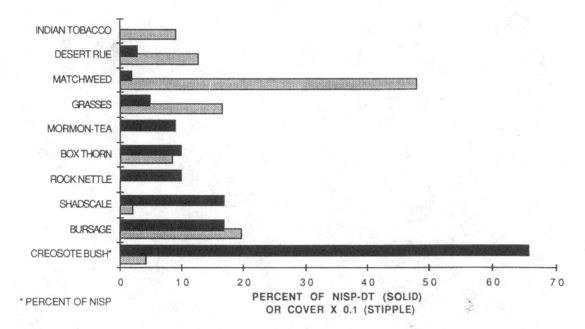

5.11C   IS, MIDDEN CONTENTS TO RELEVE = 88

* PERCENT OF NISP

PERCENT OF NISP-DT (SOLID)
OR COVER X 0.1 (STIPPLE)

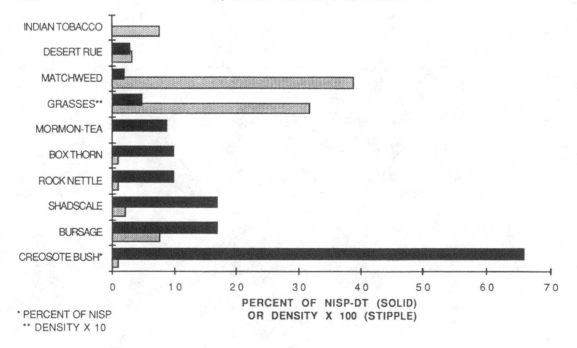

5.11D   IS, MIDDEN CONTENTS TO RELEVE = 88

* PERCENT OF NISP
** DENSITY X 10

PERCENT OF NISP-DT (SOLID)
OR DENSITY X 100 (STIPPLE)

specimens minus the number of specimens of the dominant taxon) compared to cover values for select species; (D) macrofossil abundance as percent-age of *NISP-DT* compared to density for the same species. Note changes in scale.

5.12A

5.12B

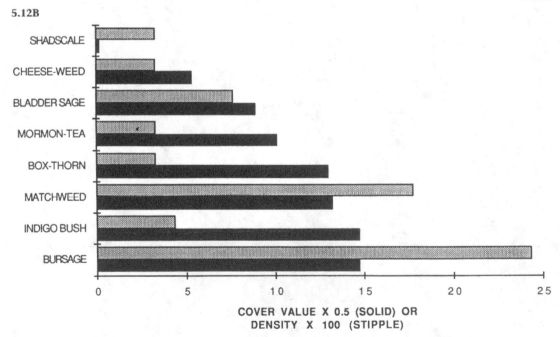

COVER VALUE X 0.5 (SOLID) OR
DENSITY X 100 (STIPPLE)

Fig. 5.12. Midden-vegetation relationships at the Fortymile Canyon-1 site, southern Nevada. (A) Photograph of the site; (B) comparison of cover and density values for plant species encountered in transects at the site; (C) macrofossil abundance as percentage of *NISP* (total number of identified

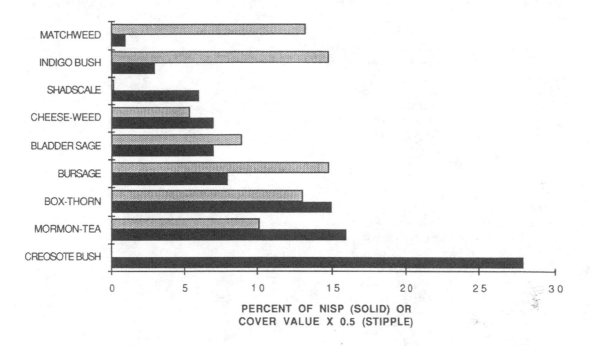

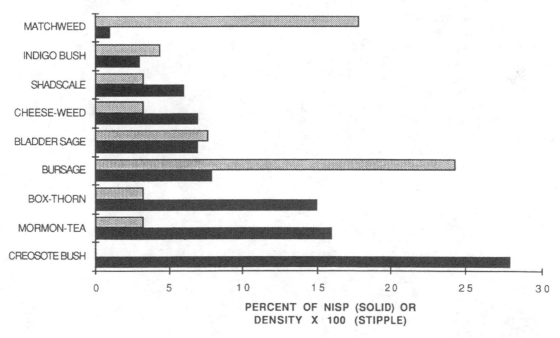

specimens) compared to cover values for select species; (D) macrofossil abundance as percentage of *NISP* compared to density for the same species.

behave erratically relative to their measured importance in the community (Figs. 5.13C, 5.13D).

Four plant taxa of no known dietary value to packrats also were selected for comparison: bursage (in this case white bursage and bur-weed [*Ambrosia acanthicarpa*]), boxthorn (*Lycium pallidum, L. shockleyi,* and *L. andersonii*), matchweed, and perennial grasses (primarily *Erioneuron pulchellum, Aristida* spp., and *Stipa* spp.). Plots of plant fragment frequency versus cover and density values for bursage and boxthorn generate seemingly random scatters (Figs. 5.14A, 5.14B). Density plotted on a logarithmic scale highlights the fact that boxthorn, an uncommon, large shrub, has consistently lower density values than the smaller but more numerous bursages (Fig. 5.14B). Higher percentages of matchweed and grass macrofossils suggest a possible correlation with increasing importance in the local vegetation (Figs. 5.14C, 5.14D), but the relationships are by no means certain. Relatively high density and % *NISP-DT* values for grasses at two sites suggest that threshold values exist for some plant taxa and, when these are exceeded, there is a corresponding increase in percentage of fragments in the midden. This may explain the lack of correlation between percentage values from the middens and modest increases in cover and density, as with globe mallow (Figs. 5.13C, 5.13D).

These results could be affected by the analytical methods employed as much as by packrat selectivity (or lack thereof). There is no assurance that the sampling methods used here are adequate to reveal midden-vegetation relationships. Cover and density values themselves can yield disparate results. It is con-

Fig. 5.13. Comparison of plant fragment abundance in five modern middens with cover and density measured for four plant taxa palatable to packrats in Mojave desertscrub, southern Nevada. The abundance of saltbush (*Atriplex* spp., open square) as percentage of *NISP-DT* and creosote bush (*Larrea divaricata*, closed square) as percentage of *NISP* are compared to (A) cover and (B) density; the abundances of globe mallow (*Sphaeralcea ambigua,* solid square) and Mormon tea (*Ephedra nevadensis, E. torreyana, E. funerea,* open square) are compared to their (C) cover and (D) density. Note changes in scale.

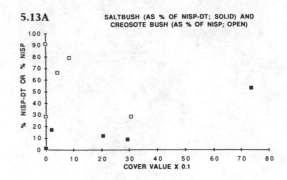

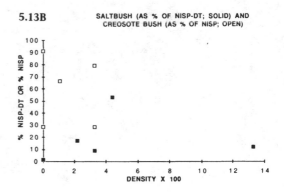

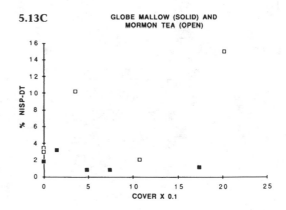

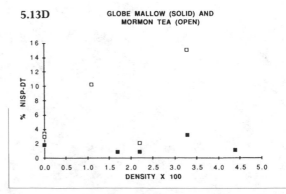

ceivable that other measures, such as the number of plant fragments per sample weight compared with an index expressing the combined importance of each species and the number of potential fragments represented by each plant, may yield different results. At present it seems that proportional representation in middens cannot be relied upon to reflect vegetation conditions. Just as in palynology, some plants are over- and others underrepresented in the fossil assemblage relative to their importance in the local plant community. However, routine measurement of plant cover and density and comparisons with macrofossil abundance can be critical in assessing the midden-vegetation relationship. As long as the discrepancies inherent in the different measures of macrofossil abundance and local vegetation remain unresolved, factors affecting proportional relationships, be they packrat dietary preferences or the effect of distance to nearest plant, cannot be deciphered.

One strategy to consider in future studies is correlating macrofossil abundance with distance to nearest individual of a particular species. McGinley (1984) found that the variance in stick size chosen by *Neotoma floridana* for house construction decreases with increasing distance from the house. Such selectivity with distance, if applicable to dietary as well as construction items, indicates that packrats sample local vegetation in a rather uneven fashion. Studies such as that by Horton and Wright (1944), who examined the contents of *Neotoma fuscipes* houses in southern California, suggest that packrats collect mostly within 10 m of their dens. Furthermore, their analysis shows that the plants at closest

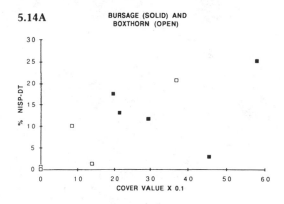

5.14A

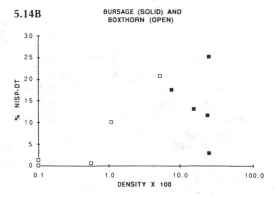

5.14B

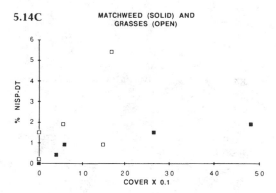

5.14C

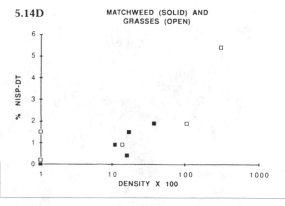

5.14D

Fig. 5.14. Comparison of plant fragment abundance in five modern middens with cover and density measured for four plant taxa of unknown value to packrats in Mojave desertscrub, southern Nevada. The abundances of bursage (*Ambrosia* spp., solid square) and boxthorn (*Lycium* spp., open square) are compared to (A) cover and (B) density. The abundances of matchweed (*Gutierrezia microcephala*, solid square) and grasses (Poaceae, open square) are compared to (C) cover and (D) density. Note changes in scale.

range, those within a few meters, contribute disproportionately to the house contents.

CONCLUSIONS

Plant macrofossil assemblages from packrat middens are unusual. The primary agent of accumulation is an animal, and the fossils themselves are preserved in crystallized urine. The properties of middens, the attributes of macrofossil assemblages, and the collecting behavior of packrats are currently being reevaluated. Roughly one hundred publications deal primarily with the interpretation of fossil midden data, but only nine focus on methods and underlying assumptions (five of these are in this volume).

Our limited analyses raise questions about the nature of middens and methods involved in their analysis. A single midden sample may represent portions of a collapsed den and a depositional episode of a few months to a few years. Some midden samples may have accumulated over longer periods, say several decades to several centuries, but this is not determinable given present dating methods. The standard deviation of a sample's radiocarbon date may bracket but does not define the duration of the depositional episode.

Structural variability within the packrat house and postdepositional alterations probably explain some of the variability seen in the various midden fractions. The presoaking weight is a poor predictor for the fraction lost in the washing process, the fecal pellet weight, or the weight of the plant matrix to be sorted and analyzed.

We offer no standard measure of macrofossil abundance in middens. Even if packrats collect plant debris in proportion to what is present in the local community, relating the number or weight of macrofossils to cover and density values is akin to comparing temperature records from thermistors and thermometers, and even that may be a generous comparison. For now, it appears that neither the relative number (this paper) nor weight (see Dial and Czaplewski, *this volume*, chap. 4) of plant fragments in modern middens accurately reflect percentage cover or density, and, for that matter, the latter two attributes of a plant community are often divergent. Percentage cover and density are only estimates of true species importance and may not

be the most appropriate measures for this kind of comparison. Likewise, there may be little agreement between 100 versus 1000 m^2 of transect data used to characterize vegetation within the same hectare plot, particularly in the heterogeneous environments afforded by cliff sides, rock outcrops, open slopes, and steep canyons. Potential avenues for future research include how the distance between den and individual plants might affect the nature of the midden assemblage and how skewed distributions influence species richness in middens.

Finally, as the level of analysis expands from site to region, using data from more than one researcher, the need for standards will become critical. One possible solution is that each researcher make available the data necessary to convert from one method to another. Ideally, the primary data for each midden includes: an inventory of plants now growing at the site; a macrofossil species list, indicating which species are extralocal or extralimital; the weight of the plant remains that were sorted, including the unidentified portions; the weight of the identified specimens, by taxon; and the number and kinds (e.g., seeds, involucres, leaves, etc.) of identified specimens for each taxon. Such exhaustive lists may be impractical in journal articles and perhaps could be provided only upon request.

Since its discovery in 1960, midden analysis in the desert West, an area with no other major source of plant macrofossils, has relied primarily on empirical methods. Despite this shortcoming, western North America is the first arid region worldwide with a reasonably comprehensive paleoecological record. Midden analysts should continue to probe the strengths and weaknesses of their methods, and to reach for the standards that will empower them to refine our understanding of the area's historical ecology.

REFERENCES

Betancourt, J. L. (1984). "Late Quaternary plant zonation and climate in southeastern Utah." *Great Basin Naturalist* 44, 1–35.
Betancourt, J. L.; Van Devender, T. R.; and Rose, M. (1986). "Comparison of plant macrofossils in woodrat (*Neotoma* sp.) and porcupine (*Erethizon dorsatum*) middens from the western U.S." *Jour-

nal of Mammalogy 67, 266–73.

Braun-Blanquet, J. (1965). *Plant sociology: The study of plant communities*. London: Hafner.

Bray, J. R., and Curtis, J. T. (1957). "An ordination of the upland forest communities of southern Wisconsin." *Ecological Monographs* 27, 225–344.

Bright, R. C., and Davis, O. K. (1982). "Quaternary paleoecology of Idaho National Engineering Laboratory, Snake River Plain, Idaho." *American Midland Naturalist* 108, 21–23.

Cheatham, A. H., and Hazel, J. E. (1969). "Binary (presence-absence) similarity coefficients." *Journal of Paleontology* 43, 1130–36.

Cole, K. L. (1981). "Late Quaternary environments in the eastern Grand Canyon, vegetational gradients over the last 25,000 years." Ph.D. diss., University of Arizona, Tucson.

———. (1983). "Late Pleistocene vegetation of Kings Canyon, Sierra Nevada, California." *Quaternary Research* 19, 117–29.

———. (1985). "Past rates of change, species richness, and a model of vegetational inertia in the Grand Canyon, Arizona." *American Naturalist* 125, 289–303.

———. (1986). "The lower Colorado desert: A Pleistocene refugium." *Quaternary Research* 25, 392–400.

Cole, K. L., and Meyer, L. (1982). "Use of packrat middens to determine rates of cliff retreat in the eastern Grand Canyon, Arizona." *Geology* 10, 597–99.

Cole, K. L., and Webb, R. H. (1985). "Late Holocene vegetation changes in Greenwater Valley, Mojave Desert, California." *Quaternary Research* 23, 227–35.

Curtis, J. T., and McIntosh, R. P. (1950). "The interrelations of certain analytic and synthetic phytosociological characters." *Ecology* 31, 434–55.

Daubenmire, R. F. (1959). "Canopy coverage method of vegetation analysis." *Northwest Science* 33, 43–64.

Emerson, D. O., and Howard, W. E. (1978). "Mineralogy of woodrat, *Neotoma cinerea*, urine deposits from northeastern California." *Journal of Mammalogy* 59, 424–25.

Finley, R. B., Jr. (1958). "The woodrats of Colorado: Distribution and ecology." *University of Kansas Publications in Natural History* 10, 213–552.

Flessa, K. W., and Migasaki, J. M. (1978). "Trends in trans-North Atlantic Phanerozoic invertebrates and plate tectonic events: Discussion and reply." *Geological Society of America Bulletin* 89, 476–77.

Horton, J. S., and Wright, J. T. (1944). "The woodrat as an ecological factor in southern California watersheds." *Ecology* 25, 341–51.

Jorgensen, C. D., and Hayward, C. L. (1965). "Mammals of the Nevada Test Site." *Brigham Young University Science Bulletin, Biological Series* 6, 1–81.

King, T. J., Jr. (1976). "Late Pleistocene–early Holocene history of coniferous woodlands in the Lucerne Valley region, Mohave Desert, California." *Great Basin Naturalist* 36, 227–38.

Koch, C. F. (1987). "Prediction of sample size effects on the measured temporal and geographic distribution patterns of species." *Paleobiology* 13, 100–107.

Long, A.; Hansen, R. M.; and Martin, P. S. (1974). "Extinction of the Shasta ground sloth." *Geological Society of America Bulletin* 85, 1843–48.

McGinley, M. A. (1984). "Central place foraging for nonfood items: Determination of the stick-size value relationship of house building materials collected by eastern woodrats." *American Naturalist* 123, 841–53.

Mehringer, P. J., Jr., and Wigand, P. E. (1987). "Western juniper in the Holocene." In *Proceedings of the Pinyon-Juniper Conference*. R. L. Everett, comp., pp. 109–19. U.S. Department of Agriculture, Forest Service, General Technical Report INT-215.

Phillips, A. M. III. (1977). "Packrats, plants, and the Pleistocene in the lower Grand Canyon." Ph.D. diss., University of Arizona, Tucson.

Shirley, E. K., and Schmidt-Nielsen, K. (1967). "Oxalate metabolism in the pack rat, sand rat, hamster, and white rat." *Journal of Nutrition* 91, 496–502.

Spaulding, W. G. (1981). "The late Quaternary vegetation of a southern Nevada mountain range." Ph.D. diss., University of Arizona, Tucson.

———. (1985). *Vegetation and climates of the last 45,000 years in the vicinity of the Nevada Test Site, south-central Nevada*. U.S. Geological Survey Professional Paper 1329.

———. (1989). *Pluvial climates and paleoenvironments of the Yucca Mountain area, south-central Great Basin, Nevada*. U.S. Geological Survey, Water Resources Investigations Report.

Thompson, R. S. (1984). "Late Pleistocene and Holocene environments in the Great Basin." Ph.D. diss., University of Arizona, Tuscon.

———. (1985). "Palynology and *Neotoma* middens." In *Late Quaternary vegetation and climates of the American Southwest*. B. F. Jacobs, P. L. Fall, and O. K. Davis, eds., pp. 89–112. American Association of Stratigraphic Palynologists Contribution Series 16.

Van Devender, T. R. (1973). "Late Pleistocene plants and animals of the Sonoran Desert: A survey of ancient packrat middens in southwestern Arizona." Ph.D. diss., University of Arizona, Tucson.

———. (1987). "Holocene vegetation and climate in the Puerto Blanco Mountains, southwestern Arizona." *Quaternary Research* 27, 51–72.

Van Devender, T. R., and Everitt, B. L. (1977). "The latest Pleistocene and Recent vegetation of the Bishop's Cap, south-central New Mexico." *Southwestern Naturalist* 22, 337–52.

Van Devender, T. R.; Martin, P. S.; Thompson, R. S.; Cole, K. L.; Jull, A. J. T.; Long, A.; Toolin, L. J.; and Donahue, D. J. (1986). "Fossil packrat middens and the tandem accelerator mass spectrometer." *Nature* 317, 610–13.

Wells, P. V. (1976). "Macrofossil analysis of woodrat (*Neotoma*) middens as a key to the Quaternary vegetational history of arid America." *Quaternary Research* 6, 223–48.

———. (1983). "Paleobiogeography of montane islands in the Great Basin since the last glaciopluvial." *Ecological Monographs* 53, 341–82.

Wells, P. V., and Berger, R. (1967). "Late Pleistocene history of coniferous woodland in the Mohave Desert." *Science* 155, 1640–47.

Chapter 6

The Spatial and Temporal Distribution of Radiocarbon Ages from Packrat Middens

Robert H. Webb   Julio L. Betancourt

INTRODUCTION

Differential preservation and uneven sampling unavoidably hinder interpretation of fossil records, most seriously in analyses of trends over large blocks of time and space. Worldwide Phanerozoic trends in biotic diversity may reflect: (1) sampling bias due to "the pull of the recent," wherein material available for analysis is simply more abundant in younger substrates (Raup, 1972); (2) the relative outcrop area or rock volume of various time units (Raup, 1976); (3) the flux in environmental heterogeneity, or "provinciality" (Signor, 1985); and (4) the intensity of research efforts on special-interest problems (Sheehan, 1977). Not all of these sampling difficulties vanish when study is confined to the last 40,000 years.

In the early enthusiasm for packrat midden research, as with any new technique, special research interests dominated the collection of data. As a result, basic sampling assumptions required for a synthesis of the total midden record have not been explored. In this chapter we describe the spatial and temporal distributions of radiocarbon dates on packrat middens from western North America, identify biases due to the nature of the fossil record, and discuss the influence of special research interests over the past quarter century. We will address how these point samples, each

associated with a particular ^{14}C date, latitude, longitude, elevation, substrate, and aspect, may be collated to reconstruct former plant distributions across the environmental extremes that characterize the region.

METHODS

The individual ^{14}C date, rather than the assigned age of a midden, is the basic unit analyzed here because of problems in mapping and plotting distributions that arise when multiple dates from a single midden conflict with one another. We update earlier summaries by Wells (1974), Mead et al. (1978), and Webb (1985) based on 132, 147, and 910 ^{14}C dates, respectively. Mead's summary (1978) included only dates from the University of Arizona's Laboratory of Isotope Geochemistry. The present study tallies 1113 ^{14}C dates from the western United States and northern Mexico. The dates and associated information were collected from published (see References) and selected unpublished sources (see Acknowledgments). Middens have been collected and analyzed by at least sixteen researchers since the initial announcement of their scientific potential in 1960. The number of published papers and dissertations utilizing ^{14}C dates of packrat middens increased from four in the 1960s to thirty-four in the 1970s and thirty-six between 1980 and 1985 (see References).

We believe that the 1113 ^{14}C dates summarized here represent all published and almost all unpublished dates up to October 1, 1985. Most of the dates were derived using conventional gas-proportional or liquid scintillation techniques; in addition, 45 small samples were recently dated using tandem accelerator mass spectrometry (Van Devender et al., 1985). The University of Arizona's Laboratory of Isotope Geochemistry analyzed 59% of the dates, with the balance analyzed by seventeen other laboratories. Associated sample information that was sought and usually provided in our census includes laboratory number, date of submission, material dated, latitude, longitude, elevation, and substrate. Aspect, important in the local elevational range of plant species, was not available for 35% of the dates.

MATERIAL DATED

The particular fraction of the midden that should be dated has been the subject of considerable debate. King's (1976) systematic excavation and multiple dating of a midden from the Lucerne Valley, California, demonstrate that middens are not always stratigraphically simple (also see Spaulding, 1985). To avoid stratigraphic mixing, Wells (1976) advocates the dating of nonspecific debris (unwashed midden matrix) from a thin, horizontal layer, while Van Devender (1977) suggests dating either a single fragment or a composite sample from a single taxon removed by washing, screening, and sorting the residue from a larger mass. Unfortunately, the large sample size required to secure enough material from a single species for dating increases the risk of stratigraphic mixing. Smaller midden sample size may preclude monospecific dating and may also abbreviate the measurable floral and faunal diversity.

Figure 6.1 shows changes in the types of samples over the past twenty-five years for the categories of unwashed midden matrix; washed midden matrix; miscellaneous, unidentified twigs; fecal pellets (usually of packrats); monospecific plant fragment(s); juniper fragments; and total conifer (including juniper) fragments. The distinction between washed and unwashed midden matrix refers

to the condition of the sample as it was submitted for dating. Both kinds of samples are pretreated with hot HCl and NaOH to remove carbonates, urine, and other soluble contaminants (for clarification, see Wells, 1983). Prior to 1970, over 80% of the middens were dated using unwashed midden matrices. In the 1970s monospecific dates, mainly juniper, provided the bulk of the dates. In the 1980s there has been a significant rise in the dating of fecal pellets. The new emphasis on dating pellets reflects the need for a high mass of contemporaneous material for ^{14}C dates and conservation of bulk, monospecific material for use in phytochemical research (Engel et al., 1978) or paleotemperature and stable-isotope studies (Long et al., *this volume*, chap. 17).

The pattern for material dated varies with age class. For dates older than 10,000 B.P., 65% were monospecific material, 12% were packrat fecal pellets, and 24% were other material (Fig. 6.1). This may reflect interest in extralocal species (e.g., juniper at modern desert elevations) and the fact that conifers, which dominate Pleistocene packrat middens, contribute more bulk material than do other taxa. For middens less than ten thousand years old, only 49% of the dates were monospecific, and fecal pellets contributed 35% with the balance from miscellaneous sources.

AGE DISTRIBUTIONS

A histogram of all ^{14}C dates (Fig. 6.2) indicates a bimodal distribution with sharp peaks at 0–1000 B.P. and 10,000 B.P. This distribution decays asymptotically toward a frequency of near zero at 51,600 B.P., the oldest finite date obtained. "Infinite" dates (beyond the range of ^{14}C analysis) are plotted at the right side of Figure 6.2; these dates have a minimum age that is not necessarily older than 50,000 B.P. The bimodal frequency distribution has not changed shape significantly since 1974 (Fig. 6.2), although numerous samples less than ten thousand years old have been dated since 1980. The frequency distribution also is consistent among the different geographic regions encompassed by midden research (Fig. 6.3).

The temporal distribution of ^{14}C dates can be modeled as a random sample from an underlying probability distribution. The distri-

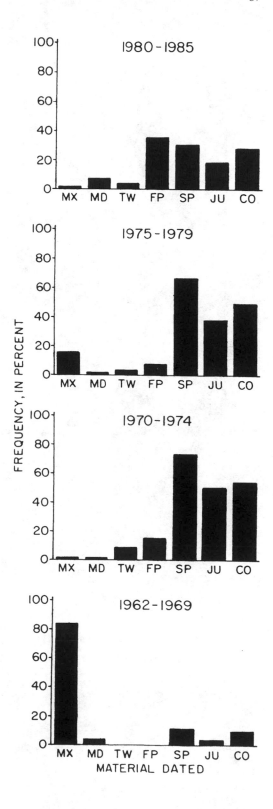

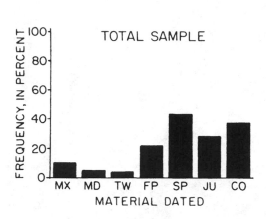

EXPLANATION

MX = UNWASHED MIDDEN

MD = WASHED MIDDEN MATRIX

TW = MISCELLANEOUS TWIGS

FP = FECAL PELLETS

SP = MONOSPECIFIC SAMPLE

JU = JUNIPER

CO = TOTAL CONIFERS

Fig. 6.1. Percentages of material dated relative to submission dates in five-year increments.

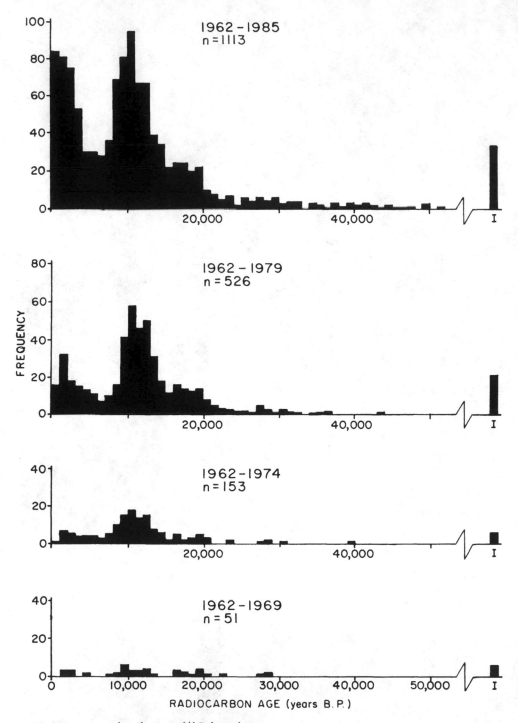

Fig. 6.2. Frequency distribution of ^{14}C dates from the western United States and Mexico. The lower three histograms depict changes in the distribution as the discipline has progressed. See Figures 6.6 and 6.7 for the included area.

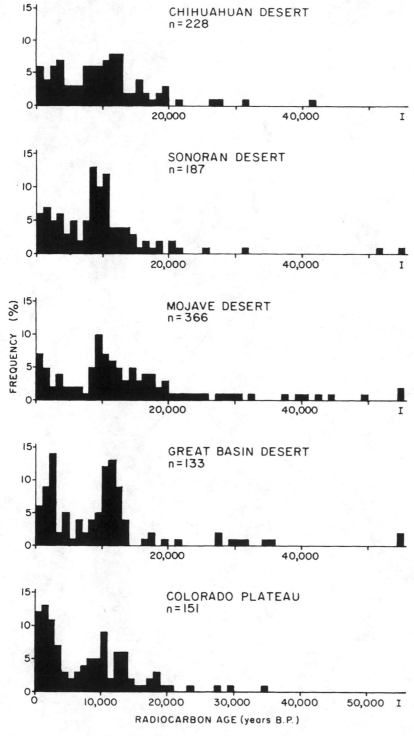

Fig. 6.3. Percentage frequency distributions of ^{14}C dates in five geographic regions of the western United States and Mexico.

90   WEBB AND BETANCOURT

bution for dates greater than 10,000 B.P. is
suggestive of a gamma probability distribution (Haan, 1977). The gamma distribution
is a two-parameter distribution with a shape
factor η and a scale factor λ. The gamma distribution reduces to the exponential probability distribution for $\eta = 1$ and is commonly
used to fit frequency distributions with an apparent exponential decay component. The
probability density function p(x) is given by

$$p(x) = \frac{\lambda^n x^{(n-1)} e^{-\lambda x}}{\Gamma(\eta)} \quad (1)$$

where $\Gamma(\eta)$ is the gamma function (Haan,
1977). The cumulative gamma distribution
P(x) is the integral of p(x) for all x.

A gamma distribution was fit to the dates
greater than 10,000 B.P. (excluding "infinite"
dates) using the method of moments (Haan,

1977). The parameters λ and η are estimated
from

$$\lambda = \frac{\bar{x}}{S^2} \quad (2)$$

and

$$\eta = \frac{\bar{x}^2}{S^2} \quad (3)$$

where $\bar{x}$ and S^2 are the mean and standard deviation of the dates greater than 10,000 B.P.
standardized by

$$x_s = (x - 10,000)/1000 \quad (4)$$

where x and x_s are the real and standardized
[14]C dates, respectively.

A total of 513 finite ^{14}C dates greater than

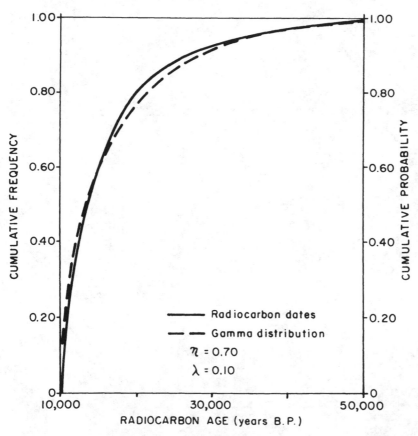

Fig. 6.4. Comparison of cumulative frequency of all
[14]C dates on packrat middens older than 10,000 B.P.
and a cumulative gamma probability distribution
with $\eta = 0.70$ and $\lambda = 0.10$. Solid lines represent
single gamma distributions; dashed lines denote
summary of the two distributions where not concealed by the second distribution curve.

10,000 B.P. were fit to the gamma distribution. Equations (2) and (3) yielded values of
$\lambda = 0.10$ and $\eta = 0.70$. In a similar analysis
on 486 dates, Webb (1985) calculated $\lambda = 0.10$
and $\eta = 0.81$, indicating a slight shift in the
shape of the distribution with the 27 additional dates. The resulting cumulative
gamma distribution closely matches the cumulative frequency of dates greater than
10,000 B.P. (Fig. 6.4).

The shape of the histogram from 0 B.P. to
6000 B.P. is suggestive of the gamma distribution of a lesser number of samples (381)
compared with the number of dates greater
than 10,000 B.P. (513). However, a gamma distribution cannot be fit because a different statistical population of ^{14}C dates begins at about
7000 B.P. If two sampling populations are assumed as a first approximation, then the bimodal distribution can be interpreted as two
gamma probability distributions with different starting age classes at 0 B.P. and approximately 10,000 B.P. (Fig. 6.5).

The bimodal distribution of ^{14}C dates (Fig.
6.2) can be explained best by assuming researcher bias and the fact that midden preservation decreases exponentially with time (the
gamma probability distribution). Researchers
collecting middens can only guess at relative
ages (usually whether a midden is Pleistocene
or Holocene) based on a cursory examination
of the midden exterior. The peak centered on
10,000 B.P. is at least partly the result of selective midden collection, because conifer
needles and cones and other species no longer
occurring at the site can easily be seen in the
midden. The greater interest in middens containing extralocal macrofossils created a collecting bias toward middens older than 8000–
10,000 B.P. (also see Wells, 1974).

Likewise, lack of interest in Holocene middens (because of the greater similarity of midden contents with modern vegetation) caused
an enhancement in the 10,000 B.P. peak (Fig.
6.2). The trough in the distribution between
4000 B.P. and 8000 B.P. may have a physical

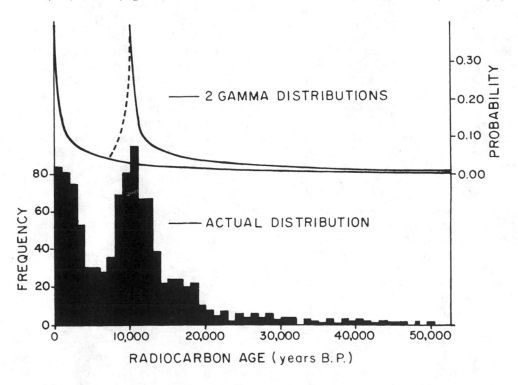

Fig. 6.5. A probabilistic model of the histogram of
^{14}C dates on packrat middens based on two gamma
probability distributions starting at 0 B.P. and
10,000 B.P.

cause in each region (Fig. 6.3), but a single
cause is unlikely for the entire western
United States, given such a wide range of climatic and physiographic conditions. Also,
middens of this age are less restricted spatially than older middens (see below). The
"bias" in the temporal distribution is not necessarily pejorative; a purely random sampling
program, which presumably would have
yielded a single gamma probability distribution, could not have yielded the definitive
record of Pleistocene-Holocene climatic
change as described in this book.

GEOGRAPHICAL DISTRIBUTION

The spatial distribution of ^{14}C dates (Fig. 6.6)
reveals a concentration of research in several
regions. Nevada and Arizona have yielded
more ^{14}C dates on packrat middens (28% and
23%, respectively) than any other states.
Forty-five middens have been dated from
northern Mexico, an increase of thirty-six
dates in two years (Webb, 1985). Middens
have been collected extensively in California's Mojave Desert, but few middens have
been collected elsewhere in the state. Also,
few middens have been collected from central
Arizona, central and eastern New Mexico,
central Utah, and most of the other western
states and northern Mexico.

The ^{14}C dates were divided into age classes
to illustrate temporal patterns in spatial coverage (Fig. 6.7). The 0–4000 B.P. age class (Fig.
6.7A) contains 293 dates and has a wider distribution than any other age class. This illustrates the natural abundance of younger middens. The 4000–8000 B.P. and 8000–11,000
B.P. age classes (Figs. 6.7B and 6.7C), with 124
and 245 dates, respectively, are more spatially
restricted. The numbers of dates in these age
classes have increased 64%, 38%, and 13%,
respectively, between 1984 (Webb, 1985) and
1985, underscoring the current interest in Holocene paleoecology.

The distribution of Pleistocene middens
generally reflects specific study areas in the
Southwest. The 11,000–15,000 B.P. and
15,000–22,000 B.P. age classes (Figs. 6.7D and
6.7E), with 207 and 130 dates, respectively,
were from middens collected predominantly
in southern Nevada (Spaulding, 1981, 1985;
Wells, 1983), the Grand Canyon (Phillips,
1977; Cole, 1981), trans-Pecos Texas (e.g., Van

Devender et al., 1984), and the Colorado River
valley (e.g., Van Devender and Spaulding,
1979). There are 80 dates older than 22,000
B.P., mostly from west Texas (21%) and Nevada (61%) (Fig. 6.7F), with 30% resulting
from replicate dating of two southern Nevada
middens (Spaulding, 1985). Briefly, then, the
number of states in the United States and
Mexico from which middens have been dated
decreases from fourteen in the period 0–4000
B.P. to only eight before 11,000 B.P.

ELEVATION

Vertical plant zonation plays a major role in
paleoecological studies in the western United
States. Although packrats today occur up to
timberline (Vaughan, *this volume*, chap. 2),
the greater moisture of most forested habitats
is unfavorable for fossil midden preservation.
No middens older than 10,000 B.P. have been
collected above 2500 m, although the potential exists in hyperarid ranges such as the
White Mountains of California. Overall, fossil
middens have been collected from sea level to
2680 m, with 90% of all ^{14}C dates collected
between 300 and 2200 m.

Elevation patterns are complicated by geographic variations in base level, specifically
the abrupt transition from the southern, low
deserts (Chihuahuan, Sonoran, and Mojave) to
the higher terrains of the Great Basin and
Colorado Plateau. This transition is defined
by the regional 1200-m contour connecting
the northern extremes of the Chihuahuan,
Sonoran, and Mojave deserts from southeast
(33° N, 106° W) to northwest (38° N, 118° W).
Dates before 10,000 B.P. show no affinity for
either side of the contour, with 49% occurring
above and 51% below 1200 m. The variation in
precipitation with latitude may explain this
parity in midden preservation. In the southern deserts mean annual rainfall at elevations
above 1200 m invariably exceeds 250 mm; in
the Great Basin and Colorado Plateau it is
commonly less than 200 mm. Hence, one
might expect the upper limit for Pleistocene
middens to decrease to the south.

Elevational distributions of midden data indeed follow regional physiographic differences. South of 36° N less than 3% of dates
older than 15,000 B.P. are from elevations
greater than 1500 m, the vertical boundary
between modern desert and woodland. This

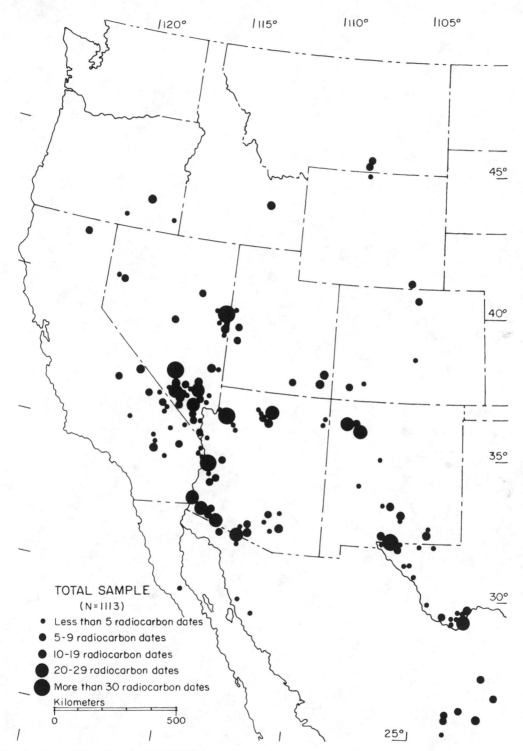

Fig. 6.6. Spatial distribution of 1113 ^{14}C dates on packrat middens in the western United States and northern Mexico.

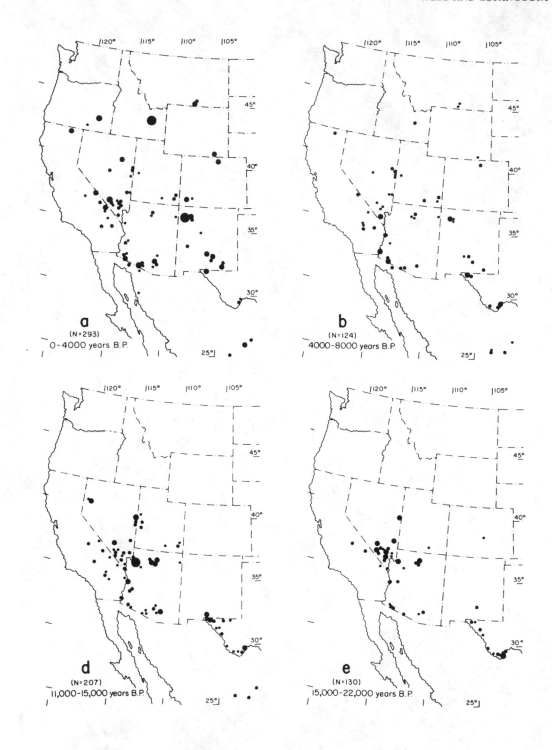

Fig. 6.7. Maps of the western United States and northern Mexico showing the spatial distribution of ¹⁴C dates on packrat middens by six age classes.

(a) 0–4000 B.P.; (b) 4000–8000 B.P.; (c) 8000–11,000 B.P.; (d) 11,000–15,000 B.P.; (e) 15,000–22,000 B.P.; (f) > 22,000 B.P. and "infinite" dates.

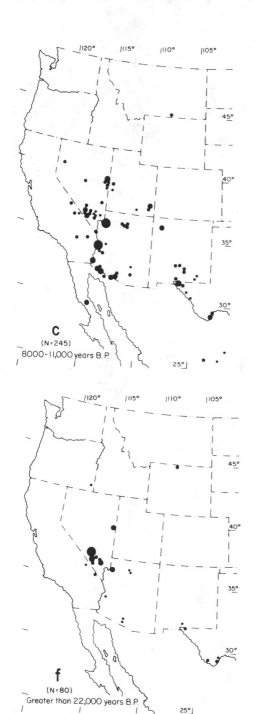

C
(N=245)
8000-11,000 years B.P.

f
(N=80)
Greater than 22,000 years B.P.

• 1 radiocarbon date
• 2-4 radiocarbon dates
● 5-9 radiocarbon dates
● 10-14 radiocarbon dates
● 15 or more radiocarbon dates
Kilometers
0   500

disparity might arise from the interest in demonstrating the relative youth of the modern deserts, the relatively low elevations for most Chihuahuan and Sonoran desert ranges, and the decreasing preservation of older middens with increasing elevation and rainfall.

Figure 6.8 shows the relative frequency of all dates for three elevational ranges (< 1000, 1000–1500, and > 1500 m) in the region as a whole and in the five principal geographic areas. The Mojave Desert has a relatively even distribution in the three elevation classes and is similar to the total sample. The Great Basin and Colorado Plateau contain a disproportionate number of dates from sites above 1500 m, reflecting the higher regional base level. The Chihuahuan Desert is biased toward the elevational range of 1000–1500 m, primarily because of concentrated work near El Paso, Texas (Van Devender, 1977; Van Devender and Everitt, 1977; Van Devender and Riskind, 1979). Most dates from sites above 1500 m in the Chihuahuan Desert come from two studies in southern New Mexico, 100 km north of El Paso (Van Devender et al., 1984; Van Devender and Toolin, 1983). The Sonoran Desert shows the greatest affinity for a given elevational range, with 94% of the dates from sites below 1000 m. Most of the work was done in southwestern Arizona, where peak elevations rarely exceed 1200 m.

In general, the vertical distribution of dates seems to reflect the natural topography. For instance, in the Sonoran Desert the seemingly biased distribution of middens actually does reflect regional physiography. Interregional comparisons of midden floras from north to south must address the problems created by differences in elevational coverage. East-west comparisons at various elevations are best made within transects wherein a common base level and elevation range eliminate the problem (see Betancourt, *this volume*, chap. 12).

PHYSIOGRAPHY AND SUBSTRATE

A significant bias in the midden record stems from the association of midden preservation with rocky environments regardless of substrate. With few exceptions, middens are collected from rocky hillsides, cliff faces, or talus slopes, and alluvial fans, mesas, or floodplains cannot realistically be sampled. Desert moun-

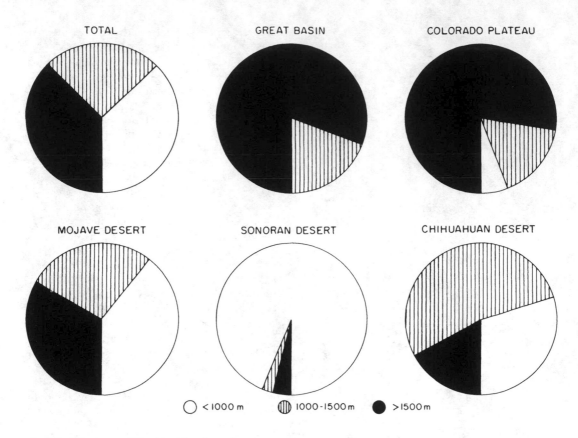

Fig. 6.8. Frequency of ¹⁴C dates for three elevational classes in the five principal geographic areas and the entire western United States and northern Mexico.

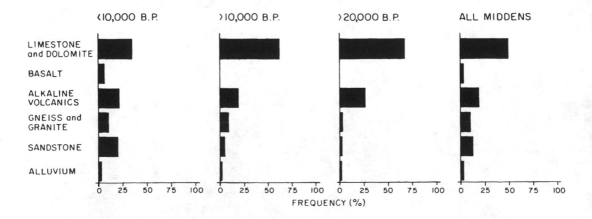

Fig. 6.9. Frequency of ¹⁴C dates in several age classes from different substrates.

tains compose at most 38% of the Mojave and Sonoran deserts (Clements et al., cited in Cooke and Warren, 1973), leaving perhaps 62% of the desert environment unsampled. This problem is perhaps best illustrated by scarp woodlands in the Great Plains, where the midden record would miss the sea of grass for the few clustered trees (Betancourt, 1987).

Preservation of middens varies regionally (Fig. 6.3), partly due to the erodibility of the substrate (Fig. 6.9). The oldest middens from southern Nevada occur in mountains of highly resistant limestone, dolomite, or welded tuffs (Spaulding, 1985). Middens older than 20,000 B.P. are uncommon in the more erodible sandstones of the Colorado Plateau (Betancourt, *this volume*, chap. 12) and the granites of the Sonoran Desert (Van Devender, *this volume*, chap. 8); however, old middens are preserved in the resistant Cambrian Tapeats Sandstone in the Grand Canyon (Cole, 1981). Syenite at Hueco Tanks in west Texas yielded middens no older than 13,500 B.P. (Van Devender and Riskind, 1979), while Paleozoic limestones within 10 km yielded numerous middens older than 20,000 B.P. (Van Devender, *this volume*, chap. 7). Fifty percent of all dates and well over 60% of those older than 10,000 B.P. were collected from limestone and dolomite substrates. Only in the age class of less than 10,000 B.P. is the distribution of dates relatively even among rock types.

Differential preservation by substrate should introduce another bias into the fossil record. Limestone and dolomite weather to calcareous soils with unusual chemical and physical properties, promoting unique local floras with many endemic species (e.g., Beatley, 1976) and vegetation associations that span large elevational ranges irrespective of climate (Peterson, 1984). Vegetation assemblages on limestone tend to be more xeric compared with similar elevations on volcanic or metamorphic substrates (Shreve, 1922; Whittaker and Niering, 1968; Wentworth, 1981, 1985; see Wright and Mooney, 1965, for an exception). The ecotone between the Chihuahuan and Sonoran deserts is a mosaic, with Chihuahuan Desert species common on limestone and Sonoran ones prevalent on less calcic soils. Meyer (1978) considers how the degree of substrate heterogeneity and edaphic

specialization affect migration rates and plant distributions from the Mojave Desert to the Colorado Plateau. Yet edaphic control of the paleoecological distributions of vegetation has been recognized in only a few studies (Van Devender and Riskind, 1979; Cole, 1982; Betancourt, *this volume*, chap. 12).

As with elevation, the severity of a substrate bias depends on the relative outcrop areas of the various rock types. Limestone is the dominant rock type in the Great Basin and Chihuahuan deserts, but is less common in the Mojave and Sonoran deserts. Can substrate be disregarded when comparing Pleistocene floras growing on Paleozoic limestones in the eastern Grand Canyon (Cole, 1981) with similar floras from Mesozoic sandstones in southern Utah (Betancourt, *this volume*, chap. 12)? The problem of potential substrate control on late Quaternary plant distributions must be addressed in future midden studies.

SLOPE ASPECT

The effect of slope aspect on vegetation and microclimate has long been recognized. At middle latitudes in the northern hemisphere a given species is typically found at higher elevations on southern than on northern aspects. Presumably this occurs because the south-facing slopes receive more solar radiation per unit area (e.g., Shreve, 1915). For example, Engelmann spruce and subalpine fir forests are common on the northern slopes but sparse on southern slopes of the laccolithic ranges of southern Utah. The differences in microclimate may be subtle and controlled by meteorology. Haase (1970) suggests that south-southeastern slopes in the Sonoran Desert are the warmest and driest aspects because cloudless mornings in the summer rainy season increase evapotranspiration.

Figure 6.10 shows the frequency of aspects for the 719 dates for which aspect was available, with subsets in the Great Basin, the Mojave Desert, and the Colorado Plateau. Other regions were not analyzed because of substantial missing data. As with elevation, the rose diagram for the Mojave Desert resembles that for the entire region, with a higher frequency of dates from southern, southwestern, northern, and northeastern exposures. The Great Basin shows a deficiency of dates older than

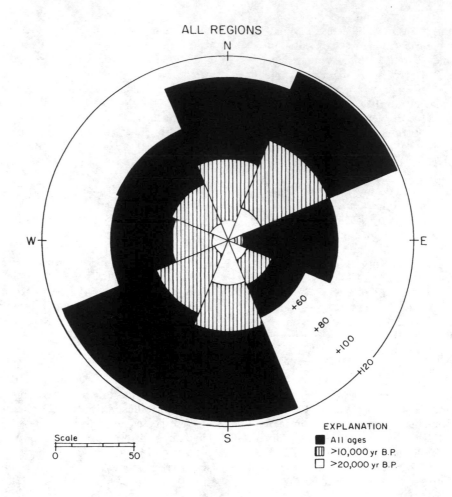

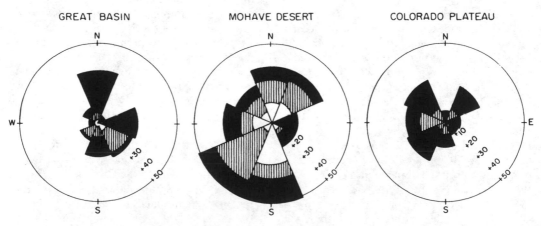

Fig. 6.10. Rose diagrams showing the frequency of ^{14}C dates for different slope aspects in the Mojave Desert, the Great Basin, the Colorado Plateau, and the region as a whole ($N = 719$).

10,000 B.P. on southeastern, southwestern, and southern aspects, while there is a disproportionate number of Holocene dates from northern and eastern aspects. Holocene dates from the Colorado Plateau are predominantly from northwestern, western, southwestern, and northeastern aspects, while Pleistocene ones come primarily from western exposures. If there is an overall bias, it is toward the drier and warmer end of the microclimatic slope gradient, primarily toward southern aspects. The effect of aspect and local moisture gradients has been considered in midden studies in the Sheep Range of southern Nevada (Spaulding, 1981) and the eastern Grand Canyon (Cole, 1981, 1982). Spaulding (this volume, chap. 9) recorded marked variations in full-glacial midden floras from various aspects at the same elevations in the Mojave Desert. A bias for southerly aspects combined with a limestone or dolomite substrate might result in an environmental stress regime that exaggerates the influence of aridity. Future studies should investigate the potential for this type of problem in both local and regional analyses.

DISCUSSION

The temporal and spatial distributions of 1113 ^{14}C dates from packrat middens raise questions about midden preservation, inherent and imposed sample bias, and sampling design for future midden research. Sample bias may be viewed as detrimental and is almost always recognized in hindsight. However, bias may be the only variable that can be held constant in an experimental field approach to paleoecology. Middens are restricted to rocky environments, which, regardless of elevation, substrate, or aspect, share certain characteristics that influence plant-water relations in specific ways. Restriction to rocky environments ensures an unusually rich record of upland vegetation, probably the type of vegetation most sensitive to climate. It also simplifies comparisons in time and space because site variability can be held to a minimum. One drawback is that rocky environments in arid lands generally favor trees in spite of treeless conditions on other nearby landforms (the scarp woodland effect). In general, middens probably underestimate the extent of shrublands and grass-

lands and overlook the mosaic aspects of vegetation distributions induced by landforms.

The curiosity about vegetation patterns of the last glacial, referred to here as the "lure of the Pleistocene," has partly overridden the natural abundance of middens with decreasing age, or "the pull of the modern," to produce a bimodal age distribution with peaks at 10,000 B.P. and 0 B.P. Viewed separately, the Pleistocene and Holocene distributions of ^{14}C dates appear to follow an exponential rate of decay with increasing age, as depicted in a gamma probability distribution (Fig. 6.5). The "lure of the Pleistocene" is common in other Quaternary studies. Earlier North American palynologists devoted more time to the Pleistocene sections of their cores, searching for vegetation changes that occurred as the ice margin retreated.

Overall, the emphasis on the Pleistocene has demonstrated the usefulness of packrat middens in reconstructions of major ecological changes. However, the neglect of the Holocene in midden studies has left significant gaps in our knowledge, most critically between 4000 B.P. and 8000 B.P. This was a time of more subtle changes, when many plants were migrating to fill their present ranges. Some species possibly overextended their northerly ranges compared with their modern distributions (Cottam et al., 1959). Distinct desert floras migrated into sites left vacant by the upward retreat of pygmy conifers (Van Devender and Spaulding, 1979), and upper and lower treelines were diverging in central Colorado between 4000 B.P. and 8000 B.P. (Fall, 1985). The few data available for this period, including alluvial stratigraphy (Haynes, 1968), human settlement patterns (Benedict and Olson, 1978), and preserved pollen (Nichols, 1982), have stirred considerable debate about paleoclimates. Consequently, the middle Holocene should become an important focus of midden research in the future.

We also anticipate a greater interest in the midden record for the last four thousand years through efforts to define the degree of stability of modern ecosystems (e.g., Cole and Webb, 1985) and to furnish a paleoecological backdrop against which the past, present, and future impacts of humans can be evaluated (see Betancourt and Van Devender, 1981). In another decade the histogram of ^{14}C dates may reflect the single gamma probability distribution that one might expect from only the "pull of the modern." We recognize, however, that physical factors may preclude collection of 4000–8000 B.P. middens in certain areas, particularly in communities that now support sparse packrat populations.

The spatial distribution of ^{14}C dates reveals new research opportunities to refine our knowledge of past biogeography and climates. It is now well established that species of pinyon pine vacated desert lowlands by about 11,000 B.P. (Van Devender and Spaulding, 1979) and migrated upward and northward during the Holocene into their present ranges. First arrival at northern sites apparently lagged behind departure from desert lowlands, in some cases by several millennia, but how swift were the pinyons in filling their present elevational ranges above the modern deserts? Similarly, did the desert shrub species that replaced the pinyons migrate at different rates? Is the northward migration of plants ongoing or has migrational equilibrium been achieved? Such questions can be answered as the search for middens continues in areas and at elevations previously neglected.

Historically midden research has been territorial, and new students are usually encouraged to pioneer the voids. The expansion into new field areas will be enhanced by the emphasis on the spatially less-restricted Holocene middens. In another decade spatial distributions of Holocene middens (Figs. 6.7A–C) may be quite different from what they are now. Because they were the focus of early research, the distribution of Pleistocene middens (Figs. 6.7D–E) may not change.

We can only speculate on the amount of sample bias and sampling efficiency, and their effects on a synthesis of the packrat midden record. Raup (1972, 1976) and others questioned elaborate evolutionary scenarios by showing how sample bias could produce the same trends in Phanerozoic diversity. Such high-order syntheses, comparable perhaps to the isochronic, cartographic approach to paleoecology attempted in the eastern and midwestern United States (Davis, 1976; Bernabo and Webb, 1977; Webb et al., 1983), have yet to be attempted with midden data from western North America. Our purpose here is to call attention to the spatial and temporal un-

evenness of the existing record as caveats to a future synthesis of midden data and their biogeographic and climatic significance. A preliminary attempt at synthesis would involve reconstructing modern vegetation from modern middens and then simulating some of the unevenness reported in this chapter to see how it affects the interpretation. How good a climatic proxy would modern vegetation be if 60% of the sample came from patches of limestone or southerly exposures?

At present the midden record reflects the historic collection of data for special-interest projects. The resulting information may be too uneven spatially and temporally to support or refute claims of "vegetational inertia" (Cole, 1985), to reconstruct paleoclimatic gradients across several regions (Wells, 1979), or to properly evaluate Milankovitch forcing of the southwestern monsoon in the early Holocene (Kutzbach, 1983; Spaulding and Graumlich, 1986). Fortunately, a great deal of the unevenness in midden data is artificial and proper planning of future studies will alleviate the problem.

ACKNOWLEDGMENTS

Numerous individuals assisted in all phases of this research. B. Fink and T. Yocum assisted with data collection and processing. T. R. Van Devender and P. S. Martin provided ideas and enthusiasm without which completion of the research would have been difficult. R. C. Bright, K. L. Cole, O. K. Davis, T. R. Van Devender, J. I. Mead, P. J. Mehringer, W. G. Spaulding, and R. S. Thompson generously provided unpublished ^{14}C dates. K. L. Cole, O. K. Davis, K. W. Flessa, P. S. Martin, W. G. Spaulding, and T. R. Van Devender critically reviewed the manuscript. We especially thank C. Sternberg for the drafting.

REFERENCES

Beatley, J. C. (1976). *Vascular plants of the Nevada Test Site and central-southern Nevada.* Springfield, Va.: National Technical Information Center, TID-26881.

Benedict, J. B., and Olson, B. L. (1978). *The Mount Albion complex: A study of prehistoric man and the altithermal.* Center for Mountain Archeology Research Report no. 1.

Bernabo, J. C., and Webb, T. III. (1977). "Changing patterns in the Holocene pollen record of northeastern North America: A mapped summary." *Quaternary Research* 9, 64–96.

Betancourt, J. L. (1987). "Paleoecology of pinyon-juniper woodlands, summary." In *Proceedings of the Pinyon-Juniper Conference*, pp. 129–39. U.S. Department of Agriculture, Forest Service, General Technical Report INT 215.

Betancourt, J. L., and Van Devender, T. R. (1981). "Holocene vegetation in Chaco Canyon, New Mexico." *Science* 214, 656–58.

Cole, K. (1981). "Modern and late Quaternary vegetational gradients in the eastern Grand Canyon." Ph.D. diss., University of Arizona, Tucson.

———. (1982). "Late Quaternary zonation of vegetation in the eastern Grand Canyon." *Science* 217, 1142–45.

———. (1985). "Past rates of change, species richness, and a model of vegetational inertia in the Grand Canyon, Arizona." *American Naturalist* 125, 289–303.

Cole, K. L., and Webb, R. H. (1985). "Late Holocene vegetation changes in Greenwater Valley, Mojave Desert, California." *Quaternary Research* 23, 227–35.

Cooke, R. U., and Warren, A. (1973). *Geomorphology in deserts.* Berkeley: University of California Press.

Cottam, W. P.; Tucker, J. M.; and Drobnick, R. (1959). "Some clues to Great Basin postpluvial climates provided by oak distributions." *Ecology* 40, 361–77.

Davis, M. B. (1976). "Pleistocene biogeography of temperate deciduous forests." *Geoscience and Man* 13, 13–26.

Engel, M. H.; Zumberge, J. E.; Nagy, B.; and Van Devender, T. R. (1978). "Variations in aspartic acid racemization in uniformly preserved plants about 11,000 years old." *Phytochemistry* 17, 1559–62.

Fall, P. L. (1985). "Holocene dynamics of the subalpine forest in central Colorado." *American Association of Stratigraphic Palynologists Contribution Series* 16, 31–46.

Haan, C. T. (1977). *Statistical methods in hydrology.* Ames: Iowa State University Press.

Haase, E. F. (1970). "Environmental fluctuations on south-facing slopes in the Santa Catalina Mountains of Arizona." *Ecology* 51, 959–75.

Haynes, C. V. (1968). "Geochronology of late Quaternary alluvium." In *Means of correlation of Quaternary successions.* R. B. Morrison and H. E. Wright, eds., pp. 591–631. Salt Lake City: University of Utah Press.

King, T. M., Jr. (1976). "Late Pleistocene–early Holocene history of coniferous woodlands in the Lucerne Valley region, Mojave Desert, California." *Great Basin Naturalist* 36, 227–38.

Kutzbach, J. E. (1983). "Monsoon rains of the late Pleistocene and early Holocene: Patterns, intensity, and possible causes of change." In

Variations in the global water budget. A. Street-Perrott, M. Beran, and R. Ratcliffe, eds., pp. 371–90. Dordrecht, Germany: Reidel.

Mead, J. I.; Thompson, R. S.; and Long, A. (1978). "Arizona ¹⁴C dates IX: Carbon isotope dating of packrat middens." *Radiocarbon* 20, 171–91.

Meyer, S. E. (1978). "Some factors governing plant distributions in the Mojave-intermountain transition zone." *Great Basin Naturalist Memoirs* 2, 197–207.

Nichols, H. (1982). "Review of late Quaternary history of vegetation and climate in the mountains of Colorado." *Institute of Arctic and Alpine Research Occasional Paper* 37, 27–33.

Peterson, P. M. (1984). *Flora and physiognomy of the Cottonwood Mountains, Death Valley National Monument, California.* Las Vegas: National Park Service Cooperative Research Unit, University of Nevada, CPSU/UNLV 022/06.

Phillips, A. M. III. (1977). "Packrats, plants and the Pleistocene in the lower Grand Canyon." Ph.D. diss., University of Arizona, Tucson.

Raup, D. M. (1972). "Taxonomic diversity during the Phanerozoic." *Science* 177, 1065–71.

———. (1976). "Species diversity in the Phanerozoic: An interpretation." *Paleobiology* 2, 289–97.

Sheehan, P. M. (1977). "Species diversity in the Phanerozoic: A reflection of labor by systematists?" *Paleobiology* 3, 325–28.

Shreve, F. (1915). *The vegetation of a desert mountain range as conditioned by climatic factors.* Carnegie Institution of Washington Publication 217.

———. (1922). "Conditions affecting vertical distribution on desert mountains." *Ecology* 3, 269–74.

Signor, P. W. III. (1985). "Real and apparent trends in species richness through time." In *Phanerozoic diversity patterns: Profiles in macroevolution.* J. W. Valentine, ed., pp. 129–50. Princeton, N.J.: Princeton University Press.

Spaulding, W. G. (1981). "The Late Quaternary vegetation of a southern Nevada mountain range." Ph.D. diss., University of Arizona, Tucson.

———. (1983). "Late Wisconsin macrofossil records of desert vegetation in the American Southwest." *Quaternary Research* 19, 256–64.

———. (1985). *Vegetation and climate of the last 45,000 years in the vicinity of the Nevada Test site, south-central Nevada.* U.S. Geological Survey Professional Paper 1329.

Spaulding, W. G., and Graumlich, L. J. (1986). "The last pluvial climatic episodes in the deserts of southwestern North America." *Nature* 320, 441–44.

Spaulding, W. G.; Leopold, E. B.; and Van Devender, T. R. (1983). "Late Wisconsin paleoecology of the American Southwest." In *Late-Quaternary en-*

vironments of the United States. Volume 1, The late Pleistocene. S. C. Porter, ed., pp. 259–93. Minneapolis: University of Minnesota Press.

Van Devender, T. R. (1977). "Holocene woodlands in the southwestern deserts." Science 198, 189–92.

Van Devender, T. R.; Betancourt, J. L.; and Wimberly, M. (1984). "Biogeographic implications of a packrat midden sequence from the Sacramento Mountains, south-central New Mexico." Quaternary Research 22, 344–60.

Van Devender, T. R., and Everitt, B. L. (1977). "The latest Pleistocene and Recent vegetation of the Bishop's Cap, south-central New Mexico." Southwestern Naturalist 22, 337–52.

Van Devender, T. R.; Martin, P. S.; Thompson, R. S.; Cole, K. L.; Jull, A. J. T.; Long, A.; Toolin, L. J.; and Donahue, D. J. (1985). "Fossil packrat middens and the tandem accelerator mass spectrometer." Nature 317, 610–13.

Van Devender, T. R., and Riskind, D. H. (1979). "Late Pleistocene and early Holocene plant remains from Hueco Tanks State Historical Park: The development of a refugium." Southwestern Naturalist 24, 127–40.

Van Devender, T. R., and Spaulding, W. G. (1979). "Development of vegetation and climate in the southwestern United States." Science 204, 701–10.

Van Devender, T. R., and Toolin, L. J. (1983). "Late Quaternary vegetation of the San Andres Mountains, Sierra County, New Mexico." In The prehistory of Rhodes Canyon. Survey and mitigation. P. L. Eidenbach, ed., pp. 33–54. Tularosa, New Mexico: Human Systems Research, Inc.

Webb, R. H. (1985). "Spatial and temporal distribution of radiocarbon ages on rodent middens from the southwestern United States." Radiocarbon 28, 1–8.

Webb, T. III; Cushing, E. J.; and Wright, H. E., Jr. (1983). "Holocene changes in the vegetation of the Midwest." In Late-Quaternary environments of the United States. Volume 2, The Holocene. S. C. Porter, ed., pp. 259–93. Minneapolis: University of Minnesota Press.

Wells, P. V. (1966). "Late Pleistocene vegetation and degree of pluvial climatic change in the Chihuahuan Desert." Science 153, 970–75.

———. (1974). "Postglacial origin of the present Chihuahuan Desert less than 11,500 years ago." In Transactions of the Symposium on the Biological Resources of the Chihuahuan Desert Region, United States and Mexico. R. H. Wauer and D. H. Riskind, eds., pp. 67–83. Washington, D.C.: National Park Service Transactions and Proceedings Series no. 3.

———. (1976). "Macrofossil analysis of wood rat (Neotoma) middens as a key to the Quaternary vegetational history of arid America." Quaternary Research 6, 223–48.

———. (1979). "An equable glaciopluvial in the West: Pleniglacial evidence for increased precipitation on a gradient from the Great Basin to the Sonoran and Chihuahuan deserts." Quaternary Research 12, 311–25.

———. (1983). "Paleobiogeography of montane islands in the Great Basin since the last glaciopluvial." Ecological Monographs 53, 341–82.

Wells, P. V., and Berger, R. (1967). "Late Pleistocene history of coniferous woodland in the Mojave Desert." Science 155, 1640–47.

Wells, P. V., and Jorgensen, C. D. (1964). "Pleistocene wood rat middens and climatic change in the Mojave Desert, a record of juniper woodlands." Science 143, 1171–74.

Wentworth, T. R. (1981). "Vegetation on limestone and granite soils in the mountains of Arizona." Ecology 62, 469–82.

———. (1985). "Vegetation on limestone in the Huachuca Mountains, Arizona." Southwestern Naturalist 30, 385–95.

Whittaker, R. H., and Niering, W. A. (1968). "Vegetation of the Santa Catalina Mountains, Arizona IV. Limestone and acid soils." Journal of Ecology 56, 523–44.

Wright, R. D., and Mooney, H. A. (1965). "Substrate-oriented distribution of bristlecone pine in the mountains of California." American Midland Naturalist 73, 257–84.

PART II
Regional Summaries

Late Quaternary Vegetation and Climate of the Chihuahuan Desert, United States and Mexico

Thomas R. Van Devender

Although the North American continental glaciers were distant, the ice sheets affected global thermal regimes and the general circulation of the atmosphere in the Chihuahuan Desert south of the Rocky Mountains and on the Mexican Plateau (Fig. 7.1). Playa lakes such as the San Agustin Plains (Clisby and Sears, 1956; Markgraf et al., 1984) and Lake Estancia (Bachhuber and McClellan, 1977), New Mexico, and the Wilcox Playa in Arizona (Martin, 1963a) were brimming with water and have yielded pollen records suggesting that forest and woodlands grew as much as 1000 m lower than today (Martin and Mehringer, 1965). Montane glaciers were found as far south as the Sacramento Mountains, New Mexico (Richmond, 1962). The Great Plains as far south as the Llano Estacado in Texas supported conifer forests or parklands (Hafsten, 1961). Packrat (*Neotoma* spp.) middens from the limestone sierras of the Chihuahuan Desert are providing detailed vegetation reconstructions for this region of succulent desertscrub and semiarid grassland.

The Chihuahuan Desert is in a sensitive location for paleoclimatic studies, with its moisture-deficient climates and rainfall and freezing winter temperatures imported from other regions. Changes in the position of the polar jet stream have important effects on the distribution of rainfall in the northern Chi-

huahuan Desert (Neilson, 1986). Incursions of arctic air masses from Canada, called "nortes" or "blue northers," can bring intense freezes over the Mexican Plateau as far south as Mexico City (Schmidt, 1986). Increased freeze frequency to the north limits subtropical Chihuahuan plants and favors desert grassland over succulent desertscrub. The present Chihuahuan Desert climate, with over 70% of its rainfall in summer, depends on the development of the Bermuda High to initiate the summer monsoon.

Vegetation changes recorded in Chihuahuan Desert packrat middens can help evaluate various paleoclimatic hypotheses. Views on late Wisconsin climates have ranged from milder and wetter (Van Devender and Spaulding, 1979) to colder and drier (Galloway, 1970; Brakenridge, 1978) than today. Holocene variations in the Earth's orbital parameters have caused changes in the distribution of solar radiation on the planet's surface. In the early and middle Holocene these variations may have raised summer temperatures in much of the Northern Hemisphere, with intensified monsoonal climatic regimes in many parts of the world (Kutzbach, 1983). Based largely on studies in the northern Great Basin, Antevs (1955) invoked a middle Holocene altithermal period characterized by hot and dry conditions across the entire western United States.

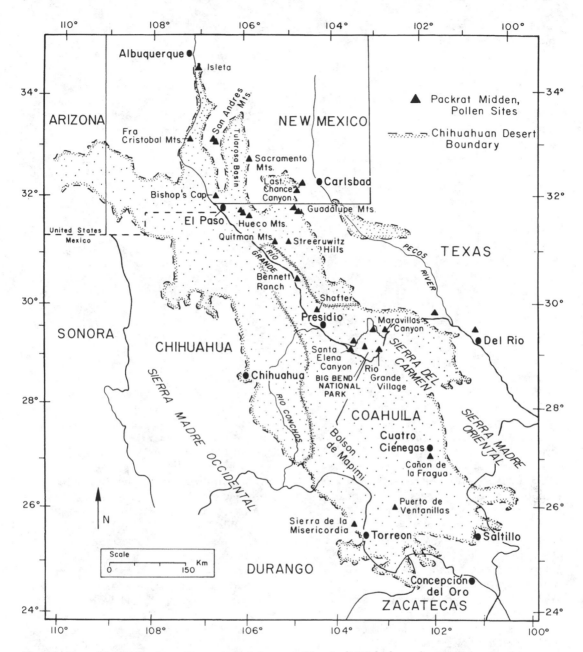

Fig. 7.1. Map of the Chihuahuan Desert (stippled; after Schmidt, 1979). Boundary between northern trans-Pecos and Mapimian subdivisions (double stipple line) is new here. Triangles mark packrat midden or pollen sites discussed in text.

Martin (1963b) hypothesized that in the summer monsoonal portions of the Southwest, the altithermal was a period of enhanced summer precipitation. Desert plants are sensitive to precipitation patterns, thus allowing packrat midden analyses to serve as tests of paleoclimatic models.

A regional bioclimatic chronology for the Southwest based on vegetation changes recorded in packrat midden sequences was proposed by Van Devender and Spaulding (1979) and revised by Van Devender, Thompson, and Betancourt (1987). The end of the late Wisconsin (LW) was placed at about 11 ka (thousands of years ago), the early Holocene (EH) from 11 to about 8.9 ka, the middle Holocene from 8.9 to about 4 ka, and the late Holocene from 4 ka to the present. Vegetation changes through time for different areas are presented in the midden summaries. Age assignment of samples in the broad transition between the early and middle Holocene was based on macrofossil content.

METHODS

Limestone is the prevailing rock type in the mountains of the Chihuahuan Desert. Unless stated otherwise, all fossil packrat middens were collected from dry caves or rock shelters in carbonate rocks. Fossil samples were removed from ledges or cracks and split into their natural layers. Outer weathering rinds were carefully removed to avoid contamination. Sample sizes of about 1000 g or less were soaked in water, screened, dried, and sorted. Plant fragments were identified and ranked in an internal relative abundance ranging from rare to abundant (1–5). A single specimen ranked 1, the most common taxon ranked 5, and the remainder were assigned intermediate values according to quantitative abundance. More elaborate quantitative methods using percentages of identified specimens greatly increase the analytical effort without significantly improving the final result (Betancourt, 1984; Cole, 1981; Spaulding, 1981; Spaulding et al., this volume, chap. 5). Internal relative abundances are readily comparable between samples.

The plant macrofossils in a midden assemblage are well preserved (Fig. 7.2) and provide excellent samples of the local flora within about 30 m of the den with consistent collecting biases. In most cases including samples less than five hundred years old, the midden assemblages contain as many or more plant taxa as occur at the sites today, suggesting that the time represented by a midden layer is long enough to average short-term fluctuations in climate and short-lived plants. I prefer the view that accumulation time is closer to fifty to one hundred years than the idea that rapid accumulation occurs in a few

years (proposed by Spaulding et al., this volume, chap. 5). Packrat midden assemblages cannot directly disclose the height, spacing, coverage, and biomass that enable ecologists to describe, classify, and compare modern vegetation. Density, coverage, biomass, and species composition are variable community parameters. Nevertheless, the species and their relative abundances in the assemblages allow relatively secure inferences about the general heights and structural compositions in a paleocommunity. For example, the relationships between mixed conifers and forest; pinyons and junipers (other than dwarf juniper, Juniperus communis) with woodland; and creosote bush (Larrea divaricata), lechuguilla (Agave lechuguilla), and other succulents with Chihuahuan desertscrub are consistent. In the following discussions inferred vegetation types are used when appropriate. Common and scientific names of plants and animals used in the text are presented in Appendix 7.1.

The fossil middens from the Chihuahuan Desert were predominantly the work of the white-throated packrat (Neotoma albigula; Van Devender, Bradley, and Harris, 1987; Van Devender and Bradley, this volume, chap. 15; Fig. 7.3), reducing potential variability in the fossil record that could be attributed to collecting habits of different species of Neotoma. Although the white-throated packrat has a strong preference for cactus (Vaughan, this volume, chap. 2), it is extremely adaptable and readily shifted its diet as local vegetation shifted from woodland to desert grassland to desertscrub. Several midden samples from Marble Canyon in the Sacramento Mountains, New Mexico, were built by porcupines (Erethizon dorsatum; Van Devender et al., 1984), but they are a relatively minor source of fossils. The differences between the two types of middens are discussed by Betancourt et al. (1986).

Organic materials from packrat middens are ideal for radiocarbon dating, with little hazard of contamination if associational integrity is ensured. Although dates on a single specimen or accumulation of many small specimens of an individual species have the advantage of direct association, dates on packrat fecal pellets and midden debris have proven reliable when taken from well-cleaned samples from

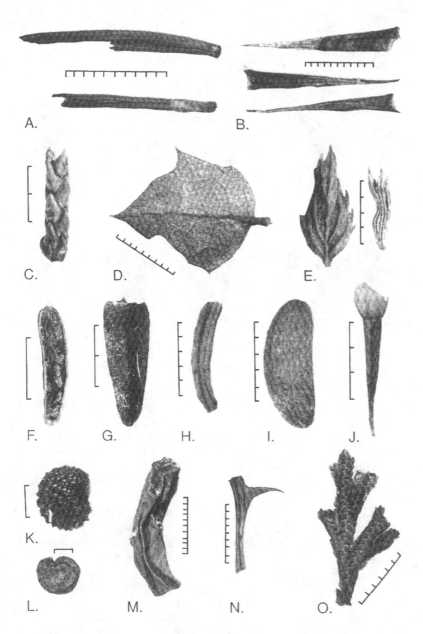

Fig. 7.2. Photographs of plant macrofossils from Maravillas Canyon (MC), Texas, packrat middens. Scale divisions in millimeters; ka = thousand years. (A) papershell pinyon needles; MCTRV#11, 21 ka. (B) Thompson/beaked yucca (*Yucca thompsoniana/rostrata*) leaf tips; MCTRV#11, 21 ka. (C) Juniper twig; MCTRV#11, 21 ka. (D) Shrub oak leaf; MCTRV#1, 20.6 ka. (E) Mock pennyroyal (*Hedeoma plicatum*) leaf and fruit, fruit; MCTRV#1, 20.6 ka. (F) Rock daisy (*Perityle* sp.) achene; MCTRV#1, 20.6 ka. (G) Goldeneye (*Vi-* *guiera* cf. *dentata*) achene; MCTRV#3, 16.2 ka. (H) Winter fat (*Ceratoides lanata*) leaf; MCTRV#3, 16.2 ka. (I) Desert olive (*Forestiera* sp.) seed; MCTRV#16, 6 ka. (J) Gray daisy (*Bahia absinthifolia*) achene; MCTRV#16, 6 ka. (K) False nightshade (*Chamaesaracha* sp.) seed, MCTRV#16, 6 ka. (L) Blind prickly pear seed; MCTRV#13, 5.1 ka. (M) Christmas cactus (*Opuntia leptocaulis*) stem; MCTRV#13, 5.1 ka. (N) Roemer catclaw twig; MCTRV#13, 5.1 ka. (O) Resurrection plant stem; MCTRV#13, 5.1 ka.

Fig. 7.3. The white-throated packrat was responsible for building most of the fossil packrat middens in the Chihuahuan Desert. It apparently has lived in the Hueco Mountains of Texas for the last 42 ka, adapting to *in situ* vegetation changes from Wisconsin woodlands to Chihuahuan desert-scrub (Van Devender, Bradley, and Harris, 1986). Tree cactus (*Opuntia imbricata*) is a commonly eaten succulent. Drawing by Helen A. Wilson.

a single layer or level. Most of the plant taxa in the assemblage are assumed not to be contaminants. Assemblages are arranged in chronological order to produce vegetation sequences through time. With the advent of the tandem accelerator mass spectrometer (TAMS), radiocarbon measurements can be obtained on 10 mg or less of organic carbon. TAMS dates on midden fossils verify the occurrence of ecologically or biographically important species and disclose suspected contaminants (Van Devender et al., 1985). Standard deviations and laboratory numbers for specific dates in years B.P. (before 1950) are cited in the text. Standard deviations are included with dates on the chronological diagrams. Other ages are abbreviated to thousands of years (ka). Additional information on radiocarbon dates is available in references cited in this chapter and in Webb and Betancourt (*this volume*, chap. 6).

Many of the approximately 480 plant taxa identified from Chihuahuan Desert packrat middens were identified to species, allowing their distribution, phenology, and physiology to be used in reconstructing paleoclimates. Plant distributions are available from regional floras (Correll and Johnston, 1970; Martin and Hutchins, 1981), personal notes, and collections. The herbaria at the University of Arizona (ARIZ), the University of New Mexico, New Mexico State University, the University of Texas at El Paso, Sul Ross State University, and the University of Texas at Austin were surveyed for locality records for twenty-nine species of trees, shrubs, and succulents identified in Chihuahuan Desert middens. Phenological responses of plants to climate are available from these sources and other literature. The ephemeral wildflowers that are restricted to summer or winter-spring growing seasons can be indicators of the seasonality of rainfall, although there is more overlap in the Chihuahuan Desert than in the Sonoran and Mojave deserts.

ENVIRONMENTAL SETTING

The Chihuahuan Desert is an interior, continental desert bordered to the west by the Sierra Madre Occidental, to the east by the Sierra Madre Oriental, to the south by the highlands of the Mexican Plateau, and to the north by the Rocky Mountains (Shreve, 1942; Fig. 7.1). The Chihuahuan Desert extends northwest-southeast from Arizona and New Mexico (33° N) south to northern Zacatecas, Mexico (25° N). The vegetation includes xeric succulent desertscrub in the lowlands of the Bolson de Mapimi in Mexico and along the Rio Grande in Texas (600–1675 m; Johnston, 1977; Henrickson and Johnston, 1986). Morafka (1977) proposed three biotic subdivisions of the Chihuahuan Desert based on distributions of animals (mostly reptiles and amphibians) and plants: the Trans-Pecos, the Bolson de Mapimi, and the Saladan. The latter subdivision is south of Schmidt's (1979, 1986) climatic boundary of the Chihuahuan Desert. In all regions creosote bush is important. The Bolson de Mapimi is also a physiographic province that includes the large areas of internal drainage basins east of the Rio Conchos in southeastern Chihuahua and south of the Rio Grande in western Coahuila (Tamayo and West, 1964).

MIDDEN SUMMARIES

A total of 220 packrat middens with 259 associated radiocarbon dates have been studied in the Chihuahuan Desert. The middens were collected in 24 local areas from 25°52' to 34°56' N latitude, and 102°10' W to 107°06' W longitude, ranging from 600 to 2000 m in elevation. Ten study areas have yielded relatively long midden sequences. The following summaries of vegetation history are arranged from south to north. Each section presents the fossil packrat midden record, significant biogeographical results, and important vegetation changes. Summary chronological diagrams illustrate the best plant macrofossil sequences.

Bolson de Mapimi

The Bolson de Mapimi is a large area of internally drained lake basins in southeastern Chihuahua, northeastern Durango, and southwestern Coahuila. Elevations range from 1075 to 2000 m in a landscape of limestone sierras embracing playa lake beds. The vegetation of the intervening rocky slopes is a diverse creosote bush–lechuguilla desertscrub with a rich mixture of shrubs and succulents (Shreve, 1939). Succulent cover and biomass are often very high. Pine-oak woodland and pine forest are found above 1830–2200 m on higher mountains in the area (Henrickson and Johnston, 1986). Tropical influences are evident.

Sierra de la Misericordia. The oldest of fourteen packrat midden samples from Sierra de la Misericordia near Bermejillo, Durango (25°52–56' N, 103°39–41' W; 1250–1310 m elev.), was dated at 13.6 ka. The late Wisconsin assemblages are dominated by juniper and papershell pinyon (*Pinus remota*) associated with Chihuahuan Desert succulents (Van Devender and Burgess, 1985). This community was composed of displaced woodland trees and shrubs, local endemic succulents, and a few subtropical plants from the south. Some time between 11.7 and 9.4 ka the vegetation shifted to a Chihuahuan desertscrub without a transitional woodland. Teeth of meadow vole (*Microtus* cf. *pennsylvanicus*) and a shrew (*Sorex* sp.), small mammals of mesic sedge-grass habitats in montane meadows or lowland cienegas, in samples dated at 7700 ± 180 B.P. and 7770 ± 210 B.P. (A-4126, A-4142) suggest that mesic spring marshes survived into the middle Holocene (Van Devender and Bradley, *this volume*, chap. 15)

Puerto de Ventanillas. The oldest of twelve packrat middens from Puerto de Ventanillas, northeast of Torreon, Coahuila (26°02–03' N, 102°44' W; 1370–1400 m elev.), was dated at 13.2 ka. The site is in one of the driest areas in the Bolson de Mapimi. The late Wisconsin fossils included junipers with abundant succulents (Van Devender and Burgess, 1985). Preliminary analyses of the Holocene samples show that the woodland departed between 12.6 and 8.7 ka. An early Holocene (8700 ± 180 B.P., A-4148) assemblage was rich in Chihuahuan succulents and desertscrub species without woodland plants.

Cañon de la Fragua. The oldest of nine packrat middens from Cañon de la Fragua, Coahuila (26°39′ N, 102°10′ W, 900–930 m elev.), was dated at 13.1 ka. The site is 185 km north-northeast of Puerto de Ventanillas on the southeastern edge of the Cuatro Cienegas Basin (Fig. 7.4). Just to the east, the Chihuahuan Desert gives way to the Sierra Madre Oriental. In the intense rain shadow of surrounding mountains, the Cuatro Cienegas Basin is one of the driest parts of the Chihuahuan Desert. Aquifers support a remarkable system of springs and marshes surrounded by a dense, rich, succulent desertscrub and a highly endemic biota (Minckley, 1969).

Preliminary analyses of midden samples show that the late Wisconsin vegetation on rocky slopes was woodland dominated by papershell pinyon and juniper with some succulents. Evidently here, as elsewhere, glacial climates had an important effect on the vegetation. The woodland withdrew after 12,380 ± 260 B.P. (A-4137). From a study of pollen

from marsh sediments, Meyer (1973) concluded that the vegetation of the Cuatro Cienegas Basin has been stable, with little change in the mountain slope woodlands during the late Wisconsin. Apparently the pollen in the springs predominantly reflects the vegetation of the marsh and does not adequately record the communities of the surrounding uplands.

Big Bend of Texas

Many Chihuahuan succulents reach their northern limits in the Rio Grande lowlands (600–1070 m elev.) from Big Bend National Park to northwest of Presidio, Texas. To the east and northeast on the southern portions of the Stockton and Edwards plateaus, disclimax grasslands abut with many Chihuahuan Desert plants about central Texas woodlands. This area is beyond the limits of the climatic Chihuahuan Desert (Schmidt, 1979, 1986), and is best viewed as part of the broader Chihuahuan biotic province (Blair, 1950). The

Fig. 7.4. Cañon de la Fragua on the south side of the Cuatro Cienegas Basin, Coahuila, Mexico. Today succulents dominate the Chihuahuan desertscrub. A late Wisconsin woodland with papershell pinyon and juniper existed in the area at 12.4 ka. The Cuatro Cienegas marshes studied by Meyer (1973) are about 15 km to the northwest.

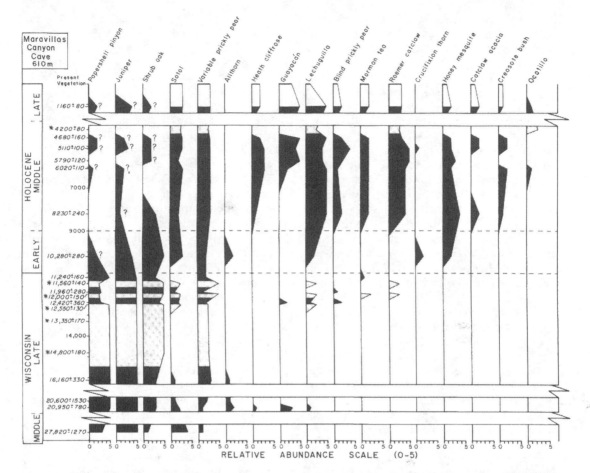

Fig. 7.5. Chronological summary of seventeen important plant macrofossils from Maravillas Canyon Cave packrat middens, Black Gap Wildlife Management Area, Texas. Data for stippled areas and radiocarbon dates marked with asterisks are from Wells (1966). Relative abundances of present vegetation were visually estimated at the site. Relative abundance scale: 0 = absent, 1 = rare, 2 = uncommon, 3 = common, 4 = very common, 5 = abundant; ? = probable contaminant.

Chihuahuan Desert in the Big Bend is bounded on the north by the grasslands and oak woodlands of the Davis Mountains. Well-developed woodlands also are present in the Chisos Mountains in Big Bend National Park.

Maravillas Canyon. The oldest of fourteen packrat midden samples from Maravillas Canyon Cave, Black Gap Wildlife Management Area, Texas (29°33' N, 102°50' W; 610 m elev.), was dated at 27.8 ka (Van Devender, 1986a). Wells (1966) reported five late Wisconsin (14.8–11.6 ka) samples from the cave. A chronological summary of seventeen species of trees, shrubs, and succulents is presented in Figure 7.5. Maravillas Canyon is a major tributary of the Rio Grande just east of Big

Bend National Park. Although the elevations are similar, the vegetation around Maravillas Canyon Cave is not as xeric as the lowlands in the park because of its north-facing aspect and its position near the northeastern edge of the Chihuahuan Desert. Cuatro Cienegas is 275 km to the south-southeast.

Papershell pinyon, shrub oak (*Quercus pungens*), and sotol (*Dasylirion* sp.), a desert-grassland rosette succulent, were common in the oldest sample (27,820 ± 1270 B.P., A-1847). The following sample, dated at 20,950 ± 780 B.P. (A-4230), showed little change. Pinyon-juniper-oak woodland grew at Maravillas Canyon Cave for another 9800 years in the late Wisconsin. Considering the modest differences between middle and late Wisconsin assemblages, the same basic vegetation type seems to have been present for at least 16,600 years.

The fossil content of the middens suggests that the late Wisconsin vegetation was a pinyon-juniper-oak woodland dominated by papershell pinyon, shrub oak, and juniper, including Pinchott juniper (*Juniperus pinchottii*) and rock cedar (*J. ashei*; R. P. Adams, personal communication, 1982). Pinchott juniper is a plant of the lower woodlands and desert grasslands of the southern Great Plains from Oklahoma to Coahuila. Today it occurs at elevations higher than the site in the Big Bend. Rock cedar is the dominant tree of central Texas woodlands to the northeast, although a relict population occurs in the Sierra del Carmen, Coahuila, 50 km to the southeast (Adams, 1977). The presence of shrubs such as honeysuckle (*Lonicera albiflora*) and redbud (*Cercis canadensis*) and the relative scarcity of desert succulents such as lechuguilla suggest that this was a relatively mesic woodland.

The Maravillas Canyon Cave middens contained an unusual number of bones. Van Devender and Bradley (in press) summarize and discuss thirty species of amphibians and reptiles from the middens, including the small, extinct Wilson's tortoise (*Geochelone wilsoni*), the large Bolson tortoise (*Gopherus flavomarginatus*), and many regional endemics. Today the Bolson tortoise is an endangered species restricted to the Bolson de Mapimi in the common corners of Coahuila, Durango, and Chihuahua, 285 km south of Maravillas Canyon. It lives in desert grassland dominated by tobosa grass (*Hilaria mutica*) and honey mesquite (*Prosopis glandulosa*) on the edges of playa lake basins (Morafka, 1982). The fossils were probably transported to the cave by a mammalian predator, a large raptor, or the California condor (*Gymnogyps californiaus*) from the relatively level, grassy margins of Maravillas Creek because tortoises do not climb well and avoid rocky slopes. The slopes above the surviving Bolson tortoise populations support succulent Chihuahuan desertscrub zoned well below any woodland. The late Wisconsin populations in Maravillas Canyon were living in environments quite different from those of today. However, their

range may have been reduced by prehistoric tortoise hunters throughout the Holocene.

Between 11.2 and 10.3 ka, papershell pinyon disappeared while shrub oak increased and a xeric woodland formed. This vegetation change is interesting in comparison with pollen records from caves near the junction of the Pecos and Devil's rivers with the Rio Grande. Pollen of a *Haploxylon* pine that could be papershell pinyon was abundant in cave sediments from 8–9 ka to more than 10.5 ka (Bryant and Holloway, 1985). Today papershell pinyon is found in a number of local relict populations between Maravillas Canyon and Del Rio. In the early Holocene good stands of papershell pinyon apparently survived in wetter areas to the east while it disappeared from the Chihuahuan Desert proper (Lanner and Van Devender, 1981).

A California condor bone from Maravillas Canyon Cave yielded a TAMS radiocarbon date of 10,215 ± 320 B.P. (AA-1276). The only other radiocarbon date for this large scavenger from the Chihuahuan Desert was 12,580 B.P. from Mule Ears Peak in Big Bend National Park (Emslie, 1987). Historically it has been restricted to a small area in California and is presently on the verge of extinction. In the late Wisconsin the California condor was widespread across the Southwest, living on the abundant megafauna (Emslie, 1987). Its survival into the early Holocene in the eastern Chihuahuan Desert may have been due to the regional abundance of bison (*Bison*).

A midden dated at 8230 ± 240 B.P. (A-3098) yielded much lechuguilla, Roemer catclaw (*Acacia roemeriana*), honey mesquite, and variable prickly pear (*Opuntia phaeacantha*-type), and the earliest creosote bush. Later in the middle Holocene (6.0–4.6 ka) species richness increased with 31–54 taxa per assemblage ($\bar{x} = 42$). Subtropical desert shrubs such as guayacán (*Guaiacum angustifolium*), desert olive (*Forestiera angustifolia*), and desert yaupon (*Schaefferia cuneifolia*) were common along with resurrection plant (*Selaginella lepidophylla*) and many grasses.

However, several of the more xeric-adapted plants, including lechuguilla, creosote bush, and ocotillo (*Fouquieria splendens*), in the middle and late Holocene samples do not occur outside the cave but are found in drier, more open areas downslope, 75 m lower. TAMS dates on ocotillo stems from floor fill yielded dates of 2750 ± 120 B.P. (AA-1274–1275). A single late Holocene (1160 ± 80 B.P., A-4229) assemblage was very similar to the middle Holocene samples, with the addition of Big Bend silver leaf (*Leucophyllum candidum*) and an increased abundance of chino grass (*Bouteloua ramosa*), which are dominant on the slope below the cave today. The more mesic character of the vegetation near Maravillas Canyon Cave appears to have developed in the last thousand years.

The middle and late Holocene samples apparently have low levels of systematic contamination by papershell pinyon, juniper, and shrub oak from abundant indurated middens and loose floor fill of Wisconsin age. Juniper twigs from a sample dated 1160 ± 80 B.P. (A-4229) yielded a TAMS date of 15,710 ± 180 B.P. (AA-1595).

Rio Grande Village area. The oldest of thirty-five packrat middens from near Rio Grande Village, Big Bend National Park, Texas (29°11–16′ N, 102°58′–103°01′ W; 600–835 m elev.; Fig. 7.6), was dated at > 36.0 ka (Van Devender, 1986a, 1986b). A chronological summary of fifteen species of trees, shrubs, and succulents is presented in Figure 7.7. The midden sites are 32 km south of Maravillas Canyon Cave. This is the lowest, hottest, and driest area in the Chihuahuan Desert. The slopes are dominated by a mixture of desert shrubs and succulents, especially lechuguilla, blind prickly pear (*Opuntia rufida*), candelilla (*Euphorbia antisyphilitica*), and chino grass (Fig. 7.6).

Six middens contained abundant juniper twigs in what may have been a juniper desert-grassland vegetation in the middle Wisconsin (24.1, 26.4, 30.0, 43.0, > 44.9, 45.6 ka) and again in the early Holocene. Papershell pinyon and other woodland plants increased between 24.1 and 21.8 ka. The late Wisconsin macrofossils of the Rio Grande Village area indicate a pinyon-juniper-oak woodland dominated by papershell pinyon and juniper with Hinckley oak (*Quercus hinckleyi*) rather than shrub oak (Fig. 7.7). Some of the juniper was Pinchott juniper. Today Hinckley oak is a rare endemic found only north of the park in two populations (El Solitario and Shafter; Butterwick and Strong, 1976). Important desertscrub–desert-grassland associates in the woodland include allthorn (*Koeberlinia spinosa*), lechuguilla, and sotol. Allthorn and sotol are rare in the study area today and are most common in desert grassland at 1200–1525 m elevation. Lechuguilla is a dominant species in the present Chihuahuan desertscrub community at the site. Sixteen late Wisconsin samples record woodland assemblages that were remarkably stable for at least 10,300 years (Van Devender, 1986a).

Papershell pinyon disappeared between 11.5 and 10.5 ka. Juniper continued to be important in the middens until after 9870 ± 150 B.P. (A-4237), while Hinckley oak declined. More juniper and oak lingered into the early Holocene in Chihuahuan Desert sites to the north (Van Devender, 1977). Here, a unique Chihuahuan desertscrub dominated by variable prickly pear, lechuguilla, and honey mesquite in association with silver wolfberry (*Lycium puberulum*), Chihuahuan crucifixion thorn (*Castela stewarti*), and creosote bush was present until 8570 ± 170 B.P. (A-4246).

With the disappearance of woodland plants in the middens between 9 and 8.6 ka (or soon after), a basically modern Chihuahuan desert-scrub dominated by blind prickly pear and lechuguilla was established. Younger samples from the middle and late Holocene contain relatively modern assemblages except for very xeric assemblages at 1600 ± 110 B.P. and 1680 ± 80 B.P. (A-4249, A-4252) and 0.7 ka and younger. Judging by the fossils, the modern vegetation is as sparse as at any time in the Holocene.

Big Bend General. Within Big Bend National Park, Wells (1966) analyzed five packrat middens from Dagger Mountain (29°32′ N, 103°06′ W; 850–880 m elev.) and the rhyolitic Burro Mesa (29°16′ N, 103°23′ W; 1200–1210 m elev.), both now supporting Chihuahuan desertscrub. Three late Wisconsin full-glacial samples dated at 16.3–20.0 ka were similar to two middle Wisconsin samples (> 36–> 40 ka) in being dominated by papershell pinyon, juniper, and shrub oak.

Midden assemblages from Santa Elena Canyon (29°09′ N, 103°39′ W; 670 m elev.), Terlingua (15.0 ka; 29°18′ N, 103°41′ W; 910 m elev.), Shafter (29°47′ N, 104°22′ W; 1310 m elev.), and Bennett Ranch (30°37′ N, 105°00′ W; 1035 m elev.) extend the records of late Wisconsin fossils of papershell pinyon, juni-

Fig. 7.6. (A) Rio Grande and limestone cliffs near Rio Grande Village, Big Bend National Park, Texas. This is the lowest (600–610 m), driest area in the Chihuahuan Desert. (B) Succulent desertscrub with ocotillo, blind prickly pear, and chino grass on limestone slopes above the tunnel near Rio Grande Village. Between 22 and 11 ka pinyon-juniper-oak woodland covered the slopes. (C) Close-up of Chihuahuan desertscrub showing lechuguilla, chino grass, and blind prickly pear.

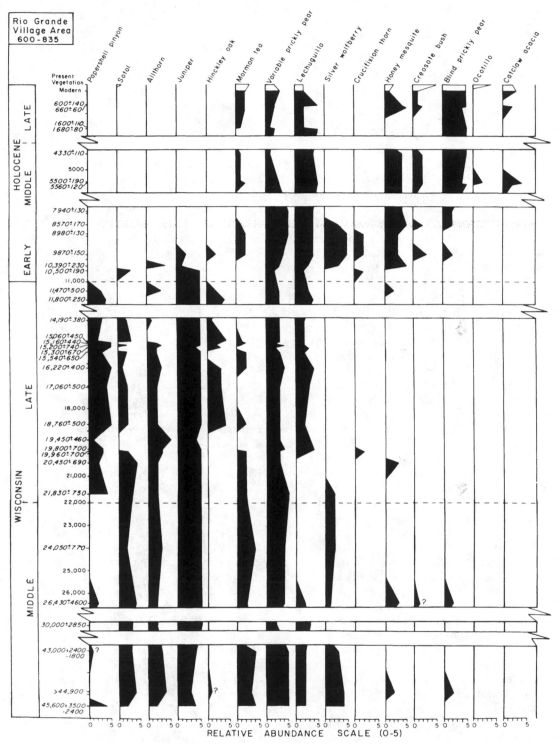

Fig. 7.7. Chronological summary of fifteen important plant macrofossils from the Rio Grande Village area packrat middens, Big Bend National Park, Texas. Relative abundances in present vegetation are composites of visual estimates at each of thirteen subsites or presence in vicinity. Relative abundance scale: 0 = absent, 1 = rare, 2 = uncommon, 3 = common, 4 = very common, 5 = abundant.

per, and shrub oak west and north of Big Bend National Park (Lanner and Van Devender, 1981). The Santa Elena Canyon midden dated at 13,690 ± 460 B.P. (A-1867) yielded juniper and oak with only traces of papershell pinyon at 670 m elevation. With the exception of the juniper samples from Puerto de Ventanillas, Coahuila (Van Devender and Burgess, 1985), this is the most xeric woodland community recorded for the late Wisconsin from the Chihuahuan Desert.

The Shafter samples are interesting because both shrub oak and Hinckley oak were found in full-glacial middens (15.7–16.0 ka; Van Devender et al., 1978). The site is less than 10 km from one of the two relict populations of Hinckley oak, a Chihuahuan desertscrub plant found well below woodlands. Seeds of Chihuahuan crucifixion thorn yielded a TAMS date of 11,460 ± 420 B.P. (AA-430), indicating contamination by younger late Wisconsin material (Van Devender et al., 1985). Today this stout, spinescent shrub is restricted to hot, dry, lowland desert habitats from Big Bend National Park into Coahuila. Its occurrence in a late Wisconsin pinyon-juniper-oak woodland farther north and at higher elevations than it occupies today suggests milder winters.

The Bennett Ranch series of four middens yielded papershell pinyon, juniper, and shrub oak from 11 to 18.2 ka. Again there was little difference between the full- and the late-glacial vegetation. Woodland elements disappeared after 11,000 ± 320 B.P. (A-1801).

Northern Chihuahuan Desert

The vegetation of the northern Chihuahuan Desert is a temperate version of the Bolson de Mapimi and Big Bend areas as cold-sensitive subtropical plants drop out with increasing latitude and elevation. In flatter areas the vegetation is dominated by mixtures of creosote bush, honey mesquite, fourwing saltbush (*Atriplex canescens*), and tarbush (*Flourensia cernua*) in dynamic intermingling with desert grassland. Mixed desertscrub with lechuguilla, creosote bush, mariola (*Parthenium incanum*), ocotillo, etc. are found on rocky slopes. Characteristic Chihuahuan Desert vegetation extends up the Rio Grande from above Presidio to the El Paso area, east to the Guadalupe Mountains, and south to

the Davis Mountains. More temperate desertscrub communities lacking even lechuguilla extend farther north in the Pecos River valley, the Hueco Bolson–Tularosa Basin, and the Rio Grande Valley of New Mexico as well as to the west into southeastern Arizona.

Quitman Mountains–Streeruwitz Hills. Five middens from the Van Horn–Sierra Blanca area, Texas (Lanner and Van Devender, 1981), include three full-glacial (18.1–14.3 ka) samples with papershell pinyon, juniper, and shrub oak from the Streeruwitz Hills (31°07' N, 105°09' W; 1430 elev.). Two samples from the Quitman Mountains (31°08' N, 105°24' W; 1230 m elev.) contain similar woodland assemblages at 12.0–10.9 ka (Van Devender and Wiseman, 1977). Pinyon-juniper-oak woodland was present in the area for at least 7200 years in the late Wisconsin prior to about 10,910 ± 170 B.P. (A-1612) or slightly later. All samples contained papershell pinyon or less diagnostic material that could have been papershell or Colorado pinyon (*Pinus edulis*; Lanner and Van Devender, 1981). A relict population of papershell pinyon exists in the Eagle Mountains (D. K. Bailey 84-01 in ARIZ) about 40 km south of the Streeruwitz Hills and southeast of the Quitman Mountains. The southernmost Colorado pinyons are isolated populations in the Sierra Diablo just north of the Streeruwitz Hills and about 100 km southeast in the Sierra Vieja (L. C. Hinckley collection in ARIZ).

Hueco Mountains. Forty-two packrat middens from the Hueco Mountains of westernmost Texas (31°43'–53' N, 105°59'–106°09' W; 1270–1495 m elev.; Fig. 7.8) provide a rich record of the local flora for the last forty-two thousand years (Van Devender, 1986a, 1986b; Van Devender, Bradley, and Harris, 1987). Two samples were from syenite rock shelters in Hueco Mountains State Park (Van Devender and Riskind, 1979). A chronological summary of sixteen species of trees, shrubs, and succulents is presented in Figure 7.9. Seven assemblages yielded abundant remains of pinyon, juniper, and shrub oak from the late Wisconsin (21.2–12.0 ka). Although many samples could not be identified to species, both Colorado and papershell pinyon were definitely present (Lanner and Van Devender, 1981). For papershell pinyon this is a northwestern range extension of about 160

Fig. 7.8. View of Navar Ranch midden sites, Hueco Mountains, Fort Bliss Army Base, Texas, in the northern Chihuahuan Desert. Vegetation is a northern Chihuahuan desertscrub community domi-nated by creosote bush with reduced abundance of succulents. A middle and late Wisconsin pinyon-juniper-oak woodland grew in the area from 42 to 10.8 ka.

km from the relict stand in the Eagle Mountains, and about 300 km from the extensive populations in the Glass and Del Norte mountains. The juniper was probably oneseed juniper (*Juniperus monosperma*), although some redberry juniper (*Juniperus erythrocarpa*) may have been present.

The dominant, long-lived plants in the midden assemblages changed little in species type or abundance for 9800 years in the late Wisconsin. Moreover, middle Wisconsin samples (32–42 ka) are very similar, suggesting a stable pinyon-juniper-shrub oak woodland for at least 30,000 years. Occasional remains of such plants as big sagebrush (*Artemisia tridentata*-type), winterfat (*Ceratoides lanata*), and Indian rice grass (*Oryzopsis hymenoides*) suggest that the Hueco Mountains were near the southern limit of the Wisconsin expansion of Great Basin plants (Van Devender, Bradley, and Harris, 1987).

Pinyons disappeared sometime between 12

and 10.8 ka, leaving behind a transitional woodland. Caching activities of birds, especially corvids, may account for traces of pinyon nut shells in younger samples. Six early Holocene (10.7–8.3 ka) samples contain abundant shrub oak and juniper. A few relict shrub oak still grow near the sites, and a few junipers linger in protected canyons elsewhere in the Hueco Mountains. Samples dominated by honey mesquite and variable prickly pear appeared between 8.3 and 8.1 ka in a gradual transition to desert grassland.

The shift to a relatively modern desertscrub was completed before 3650 ± 130 B.P. (A-2797). A TAMS date of 4200 ± 600 B.P. (AA-381) for ocotillo marks its first appearance, whereas lechuguilla and creosote bush appear in the 3.7-ka sample. Intense livestock grazing and soil erosion bedrock in the last few centuries have further reduced perennial grasses in favor of shrubs and succulents.

Guadalupe Mountains. The Guadalupe

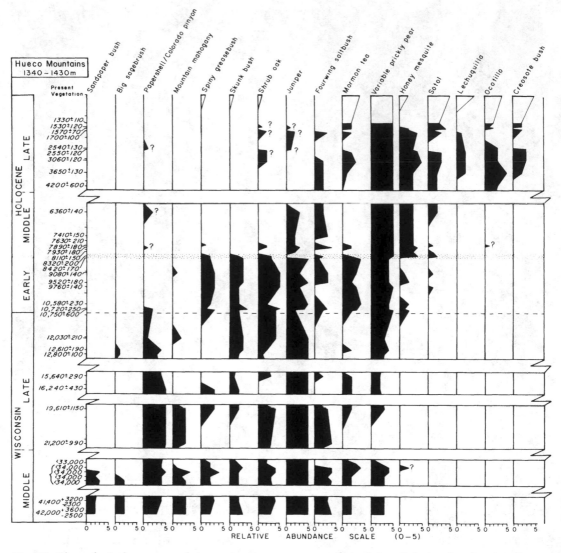

Fig. 7.9. Chronological summary of sixteen important plant macrofossils from Hueco Mountains packrat middens in Texas. Relative abundances in present vegetation are composites of visual estimates at six subsites. Relative abundance scale: 0 = absent, 1 = rare, 2 = uncommon, 3 = common, 4 = very common, 5 = abundant; ? = probable contaminant.

Mountains extend from near the Texas border northward into New Mexico to merge with the lofty Sacramento Mountains. The vegetation at the southern end of the mountains in Guadalupe National Park ranges from a Chihuahuan desertscrub at about 1220 m to a mixed-conifer forest at 2400 m elevation. The dominant trees in the mixed-conifer forest are Douglas fir (*Pseudotsuga menziesii*), southwestern white pine (*Pinus strobiformis*), and ponderosa pine (*P. ponderosa*). To the north,

in the Pecos River valley near Carlsbad, New Mexico, Chihuahuan desertscrub gives way to grasslands as the western edge of the Llano Estacado–southern Great Plains meets the eastern flanks of the Guadalupe Mountains.

A packrat midden from William's Cave at 1500 m elevation at the base of El Capitan (31° N, 104°05′ W), the picturesque peak on the southern end of the Guadalupe Mountains, contained abundant Colorado pinyon, oneseed juniper, and oak at 12,040 ± 210 B.P.

(A-1540; Van Devender, Spaulding, and Phillips, 1979). Traces of Douglas fir, blue spruce (*Picea pungens*), Rocky Mountain juniper (*Juniperus scopulorum*), and New Mexican locust (*Robinia neomexicana*) suggest that the lower edge of the late Wisconsin mixed-conifer forest was not far above. Dung balls from contemporaneous cave fill are attributed to the extinct Shasta ground sloth (*Northrotheriops shastense*), a woodland browser.

The High Sloth Caves on the steep west slope of the Guadalupe Mountains (31°55′ N, 105°50′ W; 2000 m elev.; desert grassland with a few Colorado pinyon and oneseed juniper) also yielded sloth dung balls and late Wisconsin plant macrofossils from both cave fill and packrat middens (Van Devender, Spaulding, and Phillips, 1979). At 13,060 ± 280 B.P. (A-1549) Engelmann spruce (*Picea engelmannii*), blue spruce, southwestern white pine, Douglas fir, Rocky Mountain and dwarf junipers, and Gambel oak (*Quercus gambelii*) grew in the area. These species are typical of the well-developed mixed-conifer forest in the southern Rocky Mountains today. The youngest radiocarbon date on Shasta ground sloth dung is 10,780 ± 140 B.P. (A-1534; Spaulding and Martin, 1979). A TAMS radiocarbon date of 10,850 ± 180 B.P. (AA-1101) on Colorado pinyon probably records its rapid upward expansion at the end of the Wisconsin.

Fifteen packrat middens from Rocky Arroyo (32°29′ N, 104°27′ W; 1140 m elev.) and Last Chance Canyon (32°16′ N, 104°39′ W; 1310 m elev.), New Mexico, in the grasslands west of Carlsbad, provide a history of vegetation for the Holocene (Van Devender, 1980). Juniper dominated samples at 10.0–10.6 ka and continued to do so in thirteen middle and late Holocene middens. The fossils probably represent Pinchott juniper, a common grassland shrub in the area today. Two middle Holocene samples dated at 6.8–7.5 ka reflect a simple juniper grassland community lacking some mesic elements present earlier. By 3.9 ka the species richness in the community increased dramatically with the appearance of many shrubs and succulents, including lechuguilla.

The eastern flanks of the Guadalupe Mountains have probably supported some form of grassland throughout the Holocene. Late Wisconsin plant macrofossils have not been found in the area. Pine dominated pollen profiles on the Llano Estacado to the northeast (Hafsten, 1961). The fossil fauna from nearby Dry Cave included several animals such as sage vole (*Lagurus curtatus*) and sage grouse (*Centrocercus urophasianus*) that suggest an expansion of Great Basin desertscrub with big sagebrush (Harris, 1977). Other animals such as the voles (*Microtus* spp.), bushy-tailed packrat (*Neotoma cinerea*), and yellow-bellied marmot (*Marmota flaviventris*) suggest lowering of cool, moist woodlands or forests.

Bishop's Cap. Twelve packrat middens from Bishop's Cap, south-central New Mexico (32°11′ N, 106°36′ W; 1400–1465 m elev.), provide fossil records from > 28 to 10.3 ka (Van Devender and Everitt, 1977; Thompson et al., 1980). Bishop's Cap is an isolated limestone mountain in a windy pass between the limestone Franklin Mountains to the south and the igneous Organ Mountains to the north. Some of the late Wisconsin midden samples were adhering to dung balls of the extinct Shasta ground sloth and to epidermal scutes of desert tortoise (*Xerobates agassizi*). Late Wisconsin plant fossils on Bishop's Cap reflect a xeric juniper woodland, with Colorado pinyon found only in a sample dated at 11,330 ± 370 B.P. (A-1878). A middle Wisconsin sample dated at 31,250 ± 2200 B.P. (A-2140) also yielded juniper, while three older (34.6, > 39.8, > 43.3 ka) samples yielded substantial amounts of Colorado pinyon with juniper (Lanner and Van Devender, 1981). This suggests the intriguing possibility of more mesic communities in the early or middle Wisconsin than in the late Wisconsin. Juniper woodland continued into the early Holocene (10.3–10.8 ka) with little change (Van Devender and Everitt, 1977).

Sacramento Mountains. The oldest of thirteen packrat and porcupine middens from the Sacramento Mountains, New Mexico (32°45–54′ N, 105°54–55′ W; 1555–1690 m elev.), was dated at 18.3 ka (Van Devender, Betancourt, and Wimberly, 1984). The massive Sacramento Mountains, which form the eastern boundary of the Tularosa Basin, are near the northern edge of the Chihuahuan Desert. The Sacramento Mountains are the southernmost outlier of the Rocky Mountains, with a 2300-m vegetation gradient ranging from Chihuahuan desertscrub to alpine tundra. Moraines on the northeastern slopes of Sierra Blanca mark the southernmost Wisconsin glaciers in the United States (Richmond, 1962). Pluvial Lake Otero was in the Tularosa Basin between the Sacramento and San Andres mountains (Hawley, 1983).

Two Wisconsin full-glacial (16.2–18.3 ka) samples were dominated by Colorado pinyon and Rocky Mountain juniper in association with oneseed juniper, wavyleaf oak (*Quercus* cf. *undulata*), and Douglas fir. The site at 1555 m elevation was apparently near the upper limit of the woodland. In the late Wisconsin Colorado pinyon probably was restricted to a small area in south-central New Mexico and adjacent Texas (31–34° N; Van Devender, 1986c). During the Holocene this tree dispersed several hundred kilometers to the north and west, eventually reaching its northernmost station on the Wyoming-Colorado border at 41° N with a latitudinal range of ten degrees, much expanded from a late Wisconsin range of about three degrees.

Early Holocene (10.6–9.6 ka) samples were dominated by oneseed juniper and oak. The middle Holocene middens were dominated by honey mesquite, variable prickly pear, and sotol. A 7350 ± 240 B.P. (A-2066) sample was transitional with earlier woodlands. A 5430 ± 150 B.P. (A-2096) porcupine midden resembled younger desertscrub assemblages (Betancourt et al., 1986). A relatively modern Chihuahuan desertscrub was established by 3300 ± 110 B.P. (A-2069). Late Holocene immigrants from the south included creosote bush, ocotillo, mariola, and allthorn. Lechuguilla was recovered in a sample dated at 360 ± 90 B.P. (A-2036) from Dog Canyon, a stronghold of the Mescalero Apache. Lechuguilla was an important source of fiber for these desert people. Livestock grazing has shifted the modern vegetation to a sparser desertscrub.

San Andres Mountains. The oldest of thirteen packrat middens from Rhodes Canyon in the San Andres Mountains, New Mexico (33°11′ N, 106°36–39′ W), was dated at 14.9 ka (Van Devender and Toolin, 1983). The study area is on the western side of the Tularosa Basin, 70 km northwest of the Sacramento Mountains midden sites. Although many Chihuahuan Desert plants are present, the vegetation is a shrub-invaded desert grass-

land. A well-developed pinyon-juniper woodland dominated by Colorado pinyon and oneseed juniper is present above 1710 m elevation.

A late Wisconsin sample from Rhodes Canyon at 1705 m elevation dated at 14,920 ± 360 B.P. (A-3134) was dominated by Douglas fir, Rocky Mountain juniper, ponderosa pine, and blue spruce. This is the only Wisconsin record for ponderosa pine east of the Continental Divide. This forest tree is unusually rare in Wisconsin middens, considering that it is one of the most widespread North American conifers today. A TAMS date of 11,670 ± 600 B.P. (AA-616) on fourwing saltbush reveals an unusual association. This desert-grassland–desertscrub shrub is not an expected associate of these conifers. A 11,010 ± 340 B.P. (A-3158) sample from Tip Top Spring at 1465 m elevation contained a juniper assemblage with fourwing saltbush, wild rose (*Rosa stellata*), shrub live oak (*Quercus* cf. *turbinella*), and big sagebrush. The absence of Colorado pinyon from these middens suggests that Rhodes Canyon was beyond its northern limits and that mixed-conifer forest graded directly into juniper woodland.

By 10,290 ± 210 B.P. (A-3139) the mixed-conifer forest in Rhodes Canyon gave way to an unusual woodland community of oneseed juniper, wild rose, fourwing saltbush, and wavyleaf oak without Douglas fir, ponderosa pine, blue spruce, or Colorado pinyon. By 9110 ± 220 B.P. (A-3194) the vegetation had shifted to a more xeric facies with fewer woodland and more desert grassland plants. The 10,900 ± 160 B.P. (A-3210) Tip Top Springs sample records a shift to a more xeric woodland with reduced juniper. The absence of Colorado pinyon suggests that it did not migrate through the 1465–1705-m elevational range in Rhodes Canyon but moved directly into its modern elevational range above 1710 m. Soon after 9.1 ka a desert grassland with oneseed juniper developed. Chihuahuan desertscrub plants, including creosote bush, increased before 4340 ± 150 B.P. (A-3136).

Fra Cristobal Mountains. The Fra Cristobal Mountains, New Mexico (33°18′ N, 107°06′ W, 1675–1740 m elev.), yielded four early Holocene middens (9.3–7.8 ka; Elias, 1987). The Fra Cristobal Mountains are a small range bounded by the Rio Grande to the east and the Jornada del Muerto to the west. The fossils are dominated by oneseed juniper and shrub oak. The vegetation was probably similar to contemporaneous woodlands in the Hueco Mountains. The woodland persisted to the unusually late date of 7830 ± 190 B.P. (A-2636).

Isleta. East of the Continental Divide the northernmost population of creosote bush is near Isleta Pueblo, just south of Albuquerque, New Mexico (34°56′ N, 106°43′ W; 1535 m elev.). In an effort to delimit arrival time, a packrat midden lacking creosote bush from a rock shelter in a basalt lava flow in the midst of the population was dated, yielding an age of 170 ± 80 B.P. (A-4089). The recent invasion of creosote bush in the middle Rio Grande Valley is probably due to the combined influence of overgrazing and dispersal by cattle driven on the Santa Fe Trail.

DISCUSSION

Paleovegetation

In the Chihuahuan Desert the vegetation sequence from the late Wisconsin through the Holocene culminating in relatively modern plant communities was mostly unidirectional. The only reversal occurred from middle to late Wisconsin in the lowest elevations in the Rio Grande Village area. In the Hueco Mountains in the northern Chihuahuan Desert, a four-step vegetation sequence (LW = pinyon-juniper-oak woodland; EH = oak-juniper woodland; MH = desert grassland; and LH = Chihuahuan desertscrub) is similar to both the general regional vegetation from north-central New Mexico south along the Rio Grande into Trans-Pecos, Texas, and the elevational vegetation zonation on nearby mountains. In contrast, desertscrub plants began to increase soon after 9 ka at lower latitudes in the Big Bend and Bolson de Mapimi, where winter freezes are less important and the middle and late Holocene vegetation and climate were not very different. Similarly, the differences between the middle and late Holocene vegetation in areas such as the San Andres Mountains and Last Chance Canyon, New Mexico, which experience more freezes and support desert grassland today, were not marked.

During the late Wisconsin a pinyon-juniper-oak woodland covered the rocky slopes between 600 and 1675 m, the entire elevational gradient of what is now occupied by Chihuahuan Desert in the United States. The importance of succulents and other desert-scrub plants in the woodlands increased to the south. In the Bolson de Mapimi (25°52′ – 26°39′ N) many succulents were found in middens from three sites dated at 11.7–13.6 ka along with juniper or papershell pinyon and juniper (Van Devender and Burgess, 1985). The playas in the Bolson de Mapimi supported enormous pluvial lakes, presumably during the late Wisconsin. The expansion of papershell pinyon and juniper at Cañon de la Fragua, Coahuila, calls for a modification of Meyer's (1973) conclusion (based on fossil pollen from nearby marshes) that there was little change in the Chihuahuan desertscrub on the slopes of the Cuatro Cienegas Basin during the late Wisconsin. The vegetation of special edaphic environments such as the marshes of Cuatro Cienegas and gypsum dunes and flats not inundated by pluvial lakes in the Bolson de Mapimi probably persisted with little modification during the glacial periods. Many endemic plants are found only in these habitats (Powell and Turner, 1977). Chihuahuan Desert plants probably survived the glacial climates as members of equable, xeric woodland communities. Xeric desertscrub communities formed in the early Holocene after pinyon and juniper departed.

Extensive midden sequences from two study areas (Maravillas Canyon and the Rio Grande Village area, 29°11′–33′ N) and isolated records from six other sites, provide an excellent vegetation history for the last forty thousand years in the Big Bend of Texas. A late Wisconsin pinyon-juniper-oak woodland grew on rocky slopes that now support succulent Chihuahuan desertscrub from 22 to 11 ka with little difference between full- and late-glacial vegetation. At the lowest elevation site (Rio Grande Village, 610 m) several important Chihuahuan Desert plants, including lechuguilla and allthorn, were important components in the woodland. A xeric oak-juniper woodland without pinyon persisted in the early Holocene with desertscrub forming after about 8.6 ka. The modern composition of the plant community was reached after 4.3 ka.

A similar vegetation sequence was recorded for the northern Chihuahuan Desert (Hueco

Mountains, Texas; 31°50′ N), with middle and late Wisconsin pinyon-juniper-oak woodland growing on limestone slopes at 1270–1495 m elevation from 42 to 10.8 ka. The woodland contained both papershell and Colorado pinyons without Chihuahuan Desert succulents. An oak-juniper woodland was well developed in the early Holocene and remained until 8.4–8.1 ka. A middle Holocene desert grassland gave way to Chihuahuan desertscrub by about 4.2 ka as more subtropical plants, including lechuguilla and ocotillo, migrated into the area from the Big Bend. The desertscrub corridor across the Continental Divide in southern New Mexico connecting the Chihuahuan and Sonoran deserts was established at that time.

Several sites in Trans-Pecos Texas and south-central New Mexico (Guadalupe, Sacramento, and San Andres mountains; 31°55′–33°11′ N) record the upper boundary of the pinyon-juniper-oak woodland and the lower edge of a mixed-conifer forest with Douglas fir, blue spruce, Rocky Mountain juniper and ponderosa pine at about 1585 m. Ponderosa pine forest was apparently poorly developed or absent from the late Wisconsin vegetation zonation.

In the Hueco Mountains and Maravillas Canyon the vegetation of the middle and late Wisconsin was similar. The middle Wisconsin vegetation of the Rio Grande Village area, the lowest, driest part of the Chihuahuan Desert, was much more xeric than that of the late Wisconsin. The middle Wisconsin climate and vegetation of more southern areas of the Bolson de Mapimi were probably also more xeric than the late Wisconsin, although fossil records are lacking. Throughout the Chihuahuan Desert the vegetation of the late Wisconsin was a relatively stable pinyon-juniper-oak woodland from 22 to 11 ka. There is little difference between full-glacial and late-glacial vegetation in the detailed records from Rio Grande Village, Maravillas Canyon, Bennett Ranch, Streeruwitz Hills–Quitman Mountains, and the Hueco Mountains. The disappearance of Colorado and papershell pinyons from Chihuahuan Desert midden sequences at about 11 ka provides a good estimate of the end of the late Wisconsin (Van Devender, 1986c). Pinyons disappeared rapidly at Rio Grande Village (11.5–10.5 ka),

Maravillas Canyon Cave (11.2–10.3 ka), and in the Hueco Mountains in Texas (12.0–10.8 ka; Fig. 7.10). The velocity of the change on such a wide front suggests climate as a forcing function.

In the Sonoran Desert the vegetation change marking the end of the early Holocene was at about 8.9 ka in three detailed sequences, with survival of juniper to 7.8 and 8.4 ka at two isolated sites (Van Devender, *this volume*, chap. 8). In the Chihuahuan Desert this vegetation change ranged from before 9.4 ka to after 7.8 ka at different localities. A change at 8.9 ka was possible at Maravillas Canyon (10.3/8.2 ka) and Rio Grande Village (9.9/9.0–8.6 ka) of Texas and the Sacramento Mountains (9.6/7.4 ka) of New Mexico. Earlier changes were found in the Bolson de Mapimi (before 8.7 and 9.4 ka) of Mexico and the San Andres Mountains (10.3/9.1 ka) of New Mexico. Younger vegetation transitions were found in the Hueco Mountains (8.3/8.1 ka) of Texas and the Fra Cristobal Mountains (after 7.8 ka) of New Mexico.

Paleoclimates

Packrat midden fossils provide excellent proxy records of the climatic history of the Chihuahuan Desert for the last forty-five thousand years and have proved very useful in the evaluation of various paleoclimatic hypotheses. While the ecophysiology of most plants is not known in detail, their distributions usually follow topographic and climatic gradients, and physiological limits can be inferred (Shreve, 1915). Inferences about paleoclimate are based on each species, and paleoclimatic reconstructions are compilations of these inferences rather than descriptions of the climate of the best analog site (Bryson, 1985).

Middle Wisconsin. Four areas in the Chihuahuan Desert yielded both middle and late Wisconsin middens. Six middle Wisconsin (24.1–45.6 ka) samples from 610–680 m elevation in the Rio Grande Village area (29°13′ N) yielded juniper–desert-grassland assemblages that reflect a relatively drier climate than sixteen late Wisconsin (11.5–21.8 ka) pinyon-juniper-oak samples (Fig. 7.7). Other middle Wisconsin records from higher elevations (880 and 1210 m) in the Big Bend yielded pinyon-juniper-oak assemblages similar to

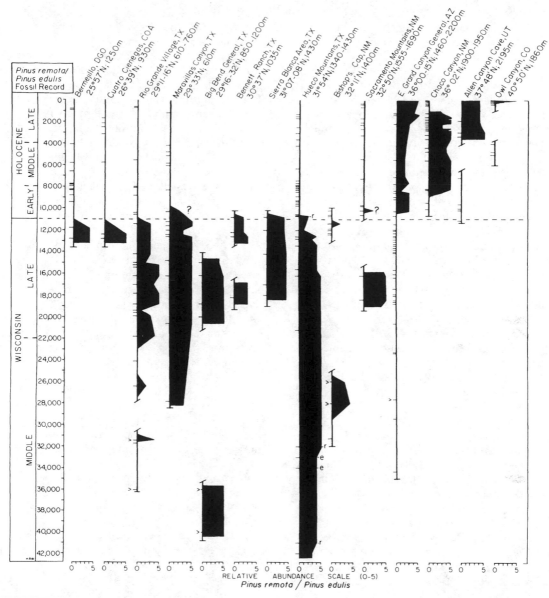

Fig. 7.10. Relative abundances of papershell and Colorado pinyons from packrat midden sequences along a latitudinal gradient from Durango, Mexico, to northern Colorado. Length of baseline indicates midden coverage for each area. Radiocarbon dates marked on baseline. > = infinite date, ? = questionable record. Modified from Van Devender (1986c).

nearby late Wisconsin samples (Wells, 1966). A 27.8-ka sample from Maravillas Canyon Cave (29°33′ N) contained a pinyon-juniper-oak assemblage very similar to those in ten late Wisconsin (11.2–21.0 ka) samples (Fig. 7.5). Apparently the drier conditions in the middle Wisconsin were seen only in the lowest,

hottest elevations in the Chihuahuan Desert. Although fossils are lacking, middle Wisconsin climates were probably relatively dry south of the Big Bend as well.

In the Hueco Mountains of Texas (31°50′ N) pinyon-juniper-oak woodland assemblages were similar in seven middle Wisconsin (> 33–

42 ka) and seven late Wisconsin (12.0–21.2 ka) samples, suggesting relatively stable climates for at least thirty thousand years. Three samples from Shelter Cave, New Mexico (32°11' N), contained abundant Colorado pinyon and oneseed juniper at middle Wisconsin or earlier ages (34.6, > 39.8, > 43.3 ka), while a middle Wisconsin (31.3 ka) sample and three late Wisconsin (11.3–12.3 ka) samples contained only oneseed juniper. A single late Wisconsin sample (11.3 ka) contained modest amounts of Colorado pinyon. These samples present the intriguing possibility of early or middle Wisconsin climates that were wetter than at 31.3 ka in the middle Wisconsin and the late Wisconsin. Indications of more mesic environments in the Wisconsin prior to the beginning of the full glacial at 22 ka have not been found in other areas in the Southwest.

Late Wisconsin. Many of the packrat midden records allow paleoclimatic inferences. The late Wisconsin samples from the Bolson de Mapimi at about 26° N latitude reflect mixed communities composed of displaced woodland trees and shrubs, local endemic succulents, and a few subtropical plants from the south, suggesting a climate characterized by greater precipitation, cooler summers, and few winter freezes.

The pinyon-juniper-oak woodlands at Rio Grande Village and Maravillas Canyon in the Big Bend at about 30° N latitude also indicate greater rainfall and cooler summers. Some of the woodland plants, including papershell pinyon, rock cedar, redbud, and honeysuckle, appear to have expanded into the area from the southern edge of the Edwards Plateau to the west or southwest. Today winter rainfall increases from the Big Bend east to central Texas and the Gulf of Mexico. Both midden series contain abundant C4 perennial grasses, indicating rainfall in late spring or in summer when temperatures were relatively warm. The late Wisconsin woodland in the Rio Grande Village area contained desert grassland and desertscrub elements such as allthorn, lechuguilla, and sotol, which were absent or rare in the Maravillas Canyon samples. This suggests that relatively drier conditions west of the Dead Horse Mountains have been present for at least twenty-eight thousand years. Apparently the Dead Horse Mountains, which reach an elevation of only 1785 m,

mark the northern extension of the massive Sierra Madre Oriental and an important airmass boundary (Neilson, 1987).

The presence of Mexican reptiles near their northern range limits in the late Wisconsin woodlands in the Big Bend (Van Devender and Bradley, in press) is a good indication of a mild, equable climate with relatively few winter freezes, cool summers, and increased rainfall. Important examples include the Texas alligator lizard (*Gerrhonotus liocephalus*), the canyon lizard (*Sceloporus merriami*), the gray-banded kingsnake (*Lampropeltis mexicana*), the Mexican hooknose snake (*Gyalopion canum*), and the blackhood snake (*Tantilla rubra*).

The late Wisconsin plant macrofossils in the Hueco Mountains of Texas (31°54' N) and the Sacramento Mountains of New Mexico (32°50' N) indicate greater rainfall—mostly in the winter—and cooler summers than today. The presence of C4 perennial grasses in the woodland samples suggests that there was some precipitation in the warm season, although typical Chihuahuan Desert plants were absent. Lyre snake (*Trimorphodon biscutatus*) vertebrae were recovered from samples dated at 12,610 ± 190 B.P. (A-1849) and 12,800 ± 100 B.P. (A-2618). This rearfanged tropical snake ranges from Costa Rica northward into the southwestern United States. The Hueco Mountains are near the northern limits of the range of the Texas lyre snake (*T. b. vilkinsoni*). The late Wisconsin records of desert tortoise from Shelter Cave, New Mexico (32°11' N), are of interest because today its eastern range limit is 350 km to the west in Arizona, where it presently is restricted by catastrophic freezes to desertscrub habitats below woodland in the Sonoran and Mojave deserts. The presence of these animals in woodland habitats in the northern Chihuahuan Desert indicates a mild, equable late Wisconsin climate (Van Devender et al., 1976).

The rarity of ponderosa pine in the late Wisconsin and the apparent zonation of pinyon-juniper-oak woodland directly below mixedconifer forest in the Guadalupe Mountains of Texas (31° N) and the San Andres Mountains of New Mexico (33°11' N) have climatic implications. Pearson (1931) found that ponderosa pine needed a longer growing season to set seed than most western forest trees, including Douglas fir, Engelmann and blue spruce, and limber pine (*Pinus flexilis*), a close relative of southwestern white pine. Cooler summers with shorter growing seasons would favor other forest trees over ponderosa pine.

Galloway (1970, 1983) and Brakenridge (1978) suggest that high lake levels and lowering of periglacial conditions on mountaintops during the late Wisconsin could be explained by decreases of 11°–12°C and 7°–8°C for each month of the year, with no increase in precipitation. The undated deposits exhibiting periglacial features in the southwestern mountains, especially the Sacramento Mountains, were assumed to represent full-glacial time and to be solely the result of lowered temperatures. The cold-dry climate models called for a 1300–1400-m timberline depression and a 1000-m snowline depression. Applying an environmental lapse rate of 6°C per 1000 m to the data used by Galloway and Brakenridge, vegetation zones would have been lowered 1750 and 1250 m. These estimates are far in excess of the actual vegetational lowering recorded in the packrat midden records of the Guadalupe and Hueco mountains. There is little evidence of colder winter temperatures (e.g., elevational lowering or southerly displacement of cold-limited plants) that cannot as readily be interpreted as due to cooler summers and/or shifts in rainfall seasonality (e.g., the absence of lechuguilla in the northern Chihuahuan Desert). The small vertebrate faunas from the Guadalupe Mountains with Great Basin elements such as sage vole and sage grouse, and mesic-montane elements such as shrews, voles, bushy-tailed packrat, and yellow-bellied marmot indicate both cooler, drier summers and moister environments at lower elevations than today (Harris, 1977, 1985). Pluvial lake hydrologic budgets can be balanced by adjusting either temperature or precipitation (Snyder and Langbein, 1962). Cooling of lesser magnitude would be insufficient to fill pluvial lakes without increased precipitation. The late Wisconsin biotic record reflects strong summer cooling and substantial rainfall with a strong shift to winter sources over the summer monsoon, and does not support the cold-dry paleoclimatic model.

In contrast, records supporting equable climates with mild winters and cool summers are more convincing; for example, lechuguilla and Chihuahuan crucifixion thorn in pinyon-juniper-oak woodlands at Rio Grande Village and Shafter, Texas (Van Devender et al., 1985), and mixed vertebrate faunas in northern Chihuahuan Desert caves (Harris, 1977, 1985). The most likely explanation of mild winter temperatures in the late Wisconsin is the blocking of arctic air masses by the continental glaciers and virtual elimination of the "blue northers" in the Chihuahuan Desert (Bryson and Wendland, 1967). Equable climates would release the lower and southern range limits of woodland and forest plants and animals but not necessarily extirpate the desertscrub residents, resulting in occasional ecological mixtures that would be nonexistent or very unusual today. The degree of mixing in biotic communities increased to the south as it does today.

The arguments by Kutzbach (1983) for a gradual increase in summer insolation from the full glacial, culminating in warmer-than-modern summer temperatures by 12 ka are not supported. Apparently the effects of the continental glaciers on regional precipitation patterns in the Chihuahuan Desert were much more important than changes in insolation as far south as 26° N. The maximum development of the summer monsoon was in the middle Holocene rather than the latest Wisconsin or the early Holocene (Van Devender et al., 1984).

The late Wisconsin precipitation gradients in the Chihuahuan Desert were similar to those of today, with moisture increasing with elevation and a shift from winter to summer seasonality at lower latitudes. The subtropical high-pressure cells were very weak, with summer rainfall greatly reduced in the northern Chihuahuan Desert. The summer-rainfall-dominated climates of the Chihuahuan Desert and the eastern portion of the Sonoran Desert were greatly reduced in area and displaced into Mexico. The arguments of Wells (1979) for increased summer rainfall in the late Wisconsin based on the expansion of oaks are not supported because shrub live oak was not common in late Wisconsin woodland middens and achieved dominance in early Holocene samples only after pinyons disappeared. The late Wisconsin precipitation was greater than today throughout the Chihuahuan Desert, although the relative increase was greatest in the north and declined at lower latitudes. The Bolson de Mapimi records from 26° N suggest that the northern edge of the Neotropics was not drier than today as suggested by studies farther south in Mexico (Bradbury, 1982; Watts and Bradbury, 1982).

Early Holocene. The climate of the early Holocene was transitional between that of the late Wisconsin and more modern regimes later in the Holocene. Winter rainfall continued to be greater than it is today in the Chihuahuan Desert. The southern edge of the winter storm track moved northward in the southern Chihuahuan Desert, resulting in an early (before 9.4 ka) onset of xeric conditions in the Bolson de Mapimi. An increase in summer temperatures was reflected in a modest increase in the development of the Bermuda High and the summer monsoon. The rapid movement of Colorado pinyon up to 2000 m elevation in the Guadalupe Mountains by 10.9 ka reflects higher summer temperatures and rainfall. The remarkable northward migrations of this tree in the Holocene probably began at this time (Van Devender et al., 1984). In the Sacramento Mountains of New Mexico (32°50′ N) midden fossils suggest that summer rainfall increased to perhaps 30%–40% of the annual total (60%–70% today).

Middle Holocene. With the final demise of the late Wisconsin winter rainfall regime after 8 to 9 ka, a relatively modern climatic regime was established. Dramatically increased summer temperatures produced maximum summer monsoonal rainfall in the Chihuahuan Desert. Fossils of the meadow vole near Sierra de la Misericordia, Durango, suggest that climates of the early portion of the middle Holocene were wet enough to support large, spring-fed sedge marshes in the now-dry Bolson de Mapimi (Van Devender and Bradley, *this volume*, chap. 15). The middle Holocene plant assemblages from Maravillas Canyon, with many subtropical shrubs, lechuguilla, resurrection plants, and many grasses, indicate warm winters and substantial biseasonal rainfall dominated by the summer monsoon. Middle Holocene samples from the nearby Rio Grande Village area re-

flect a similar paleoclimate although relatively drier. However, the abundances of a few riparian shrubs such as honey mesquite and catclaw acacia suggest that the rainfall was somewhat greater than in the driest and lowest part of the Chihuahuan Desert today. Permanent water remained in now-dry playa lakes in New Mexico, although the seasonality and source area of the rainfall shifted dramatically (Van Devender and Worthington, 1977; Markgraf et al., 1984). The enhanced summer monsoonal hypothesis of Martin (1963b) has been substantiated, while Antevs's (1955) model of altithermal drought is retained for winter rainfall areas farther west.

Although summer temperatures were hot, frequencies of severe winter freezes due to southerly incursions of arctic air were apparently much greater than today. Intense freezes may have prevented subtropical desertscrub plants (including lechuguilla) from reaching the Hueco Mountains of Texas and Last Chance Canyon in New Mexico and favored the development of grassland in the present northern Chihuahuan Desert.

Late Holocene. By 4.0 ka the modern climatic regime was established as reduction in the frequency of winter freezes, adequate summer rainfall, and greater variability with more droughts combined to produce the most subtropical climates of the Holocene. In the northern Chihuahuan Desert more subtropical elements of the flora migrated to their present northern ranges as succulent desertscrub displaced desert grassland. The Chihuahuan Desert reached its greatest extent in the late Holocene. As in the Sonoran Desert (Van Devender, *this volume*, chap. 8), the vegetation and climate are as hot and dry today at most sites as at any time in the past.

Biogeography

The fossil packrat midden record yields new information on what happened to plants of the Chihuahuan Desert for species, communities, and vegetation zonation during the last glacial-interglacial cycle. Chihuahuan Desert plants attained their present ranges at different times. There has never been a unified Chihuahuan Desert community with a fixed assemblage of species. From this it appears that plant communities are loosely knit assemblages of individual species with different physiological tolerances to environmental factors, as Shreve (1937) and Gleason (1939) proposed. Community composition has changed continuously over the last eleven thousand years as Wisconsin plants retreated from desert lowlands and woodland, grassland, and desertscrub plants migrated into new areas. Relatively modern compositions of many Chihuahuan desertscrub communities were only attained in the last four to five thousand years in the late Holocene. Community composition continued to vary subsequently during lesser climatic fluctuations, suggesting that differential responses of plant species to climate changes and continuous variation in climate on several scales have resulted in dynamic plant communities that rarely if ever reach equilibrium (Davis, 1986).

Plant fossils in middens revealed a latitudinal gradient in the late Wisconsin of the Chihuahuan Desert with the magnitude of change decreasing to the south (Van Devender, 1986a, 1986b). In the north pinyon-juniper-oak woodlands merged across lowlands that now separate isolated mountain woodlands. Plant and animal species of the Great Basin, including big sagebrush, winterfat, Indian ricegrass, sage grouse, and sage vole, expanded as far south as the northernmost Chihuahuan Desert (Van Devender, Bradley, and Harris, 1987). Ponderosa pine forest appears to have been absent from the late Wisconsin vegetation zonation in the northern Chihuahuan Desert, indicating that mixed-conifer forest directly contacted various types of pinyon-juniper-oak woodland at about 1525 m elevation. Isolated populations of such montane trees as ponderosa pine, southwestern white pine, Douglas fir, and Gambel oak in the Guadalupe, Davis, and Chisos mountains were probably never connected by direct corridors. Long-distance transport of seeds by birds must have been important in establishing these populations. In New Mexico and Texas the modern oak-juniper woodland and juniper grassland appear to have been absent from the late Wisconsin gradient.

Mexican pinyon (*Pinus cembroides*) and alligator bark juniper (*Juniperus deppeana*), dominant trees in the woodlands of the Davis and Chisos mountains today, were absent or relictual in the late Wisconsin. They ex-

panded widely during the Holocene, while papershell pinyon contracted to relict populations east of the Davis Mountains and on mountaintops in northeastern Mexico. At the same time Colorado pinyon began its remarkable Holocene migration toward the Wyoming border.

Woodland plants expanded into the xeric Bolson de Mapimi from the Sierra Madre Oriental to the east in Coahuila and from the Sierra Madre Occidental to the west in Chihuahua and Durango (Van Devender and Burgess, 1985). In the driest areas in Coahuila juniper expanded without papershell pinyon. Shrub oaks were not found in any of the samples. The Pleistocene woodland corridor connecting the two Sierra Madres at about 25° N that was suggested by Martin's (1958) herpetofaunal analyses of the Gomez Farias region, Tamaulipas, was present but much more xeric than predicted.

The associates of the woodland trees in the Bolson de Mapimi midden assemblages are of interest (Van Devender and Burgess, 1985). Chihuahuan Desert succulents, including several local endemics, were abundant. A giant yucca and a legume represent subtropical plants that expanded their ranges to the north, presumably under milder climates. Contractions and expansions of the ranges of plants and animals during many glacial-interglacial cycles combined with low extinction rates in equable climates have resulted in high evolutionary rates, a wealth of endemic species, and high species richness in the southern Chihuahuan Desert.

The Holocene was a dynamic time as Wisconsin plants retreated from desert lowlands and desert grassland and desertscrub plants expanded. In the Bolson de Mapimi the shift to a Chihuahuan desertscrub was rapid and without a transitional early Holocene woodland. Presumably this was because the southern limit of the penetration of winter Pacific frontal storms onto the Mexican Plateau was displaced northward. Early Holocene woodlands were progressively better developed in the Big Bend (29°–30° N) and the Hueco Mountains (32° N). In the latter areas, the final development of relatively modern Chihuahuan desertscrub was very different, with an intermediate desert grassland present in the middle Holocene of the northern Chihuahuan Desert. A greater frequency of winter freezes probably delayed the arrival of desertscrub dominants such as creosote bush, lechuguilla, and ocotillo until about 4.5 ka. In the middle Holocene well-developed grasslands probably reached their maximum expansion and extended from the Great Plains throughout Trans-Pecos Texas into Chihuahua and Durango and across southern New Mexico to Arizona. The northern limits of the Chihuahuan Desert were established at the beginning of the late Holocene. The desertscrub corridor connecting the Chihuahuan and Sonoran deserts across the Continental Divide in southwestern New Mexico and southeastern Arizona is probably of similar antiquity. The warm deserts of North America reached their maximum area in the late Holocene, although historical land uses have further favored them.

Creosote bush. Creosote bush is one of the few dominants in the warm deserts of North America that evolved not here but in South America. Wells and Hunziker (1976) presented a hypothesis for the Holocene history of creosote bush based on the fossil packrat midden record. They proposed that the ancestral diploid stock dispersed from Argentina to the Chihuahuan Desert sometime less than 11.5 ka and dispersed to the western deserts by 10.5 ka, evolving into tetraploid and hexaploid chromosomal races. The absence of creosote bush from middle and late Wisconsin packrat middens from the Big Bend of Texas was interpreted as indicating its absence from the entire Chihuahuan Desert. The youngest Wisconsin midden studied was dated at 11.5 ka (Maravillas Canyon Cave, 610 m elevation).

Recent evidence from 330 m and lower elevations in the Sonoran Desert has disclosed the presence of creosote bush in the late Wisconsin from 11.0 to 18.7 ka (Cole, 1986; Van Devender et al., 1985; Van Devender, *this volume*, chap. 8). The dispersal of creosote bush into North America must have preceded the last glacial maximum.

Creosote bush is usually not a dominant species on rocky slopes in the Chihuahuan Desert, where it is often an uncommon member of a diverse desertscrub community. Creosote bush dominates desertscrub communities on mountain bajadas and in valleys but is

an uncommon plant in the chino grass–desertscrub community on the north-facing slopes below Maravillas Canyon Cave. The few plants below the cave are probably beyond the foraging range of packrats. Therefore the absence of creosote bush from Maravillas Canyon Cave in the Wisconsin need not imply a regional absence.

The oldest Chihuahuan Desert and North American records for creosote bush are low levels from middle and late Wisconsin middens dated at 26,430 ± 4600 B.P. and 20,450 ± 690 B.P. (A-3138 and average of A-3138 and AA 382) from Rio Grande Village, Texas (29°13′ N), but these could be contaminants. However, the plant communities of the middle Wisconsin were more xeric than those of the late Wisconsin, resembling more those of the early Holocene. Several plants, including honey mesquite, Chihuahuan crucifixion thorn, and especially silver wolfberry, have bimodal middle Wisconsin–early Holocene distributions (Fig. 7.7). A TAMS date of 21,300 ± 1250 B.P. (AA-382) verified the presence of honey mesquite in the 20.5-ka creosote bush sample (Van Devender et al. 1985). If these records are valid, creosote bush was probably present in the most xeric habitats in the Big Bend during the last glacial period and was not entirely displaced southward.

In the Holocene creosote bush expanded from its glacial range in the Big Bend or the Bolson de Mapimi and dispersed northward. A TAMS date of 10,000 ± 90 B.P. (AA-1595) verifies the presence of creosote bush in the Rio Grande Village area in the early Holocene. The oldest record for Maravillas Canyon Cave (29°33′ N) on the eastern edge of the Chihuahuan Desert is 8.2 ka (at the beginning of the middle Holocene). Not until the beginning of the late Holocene did creosote bush reach the northern Chihuahuan Desert, arriving in the Hueco Mountains (31°45′ N) by 3.7 ka, the Sacramento Mountains (32°50′ N) by 3.3 ka, and the San Andres Mountains (33°11′ N) by 4.3 ka. These dates mark the establishment of the northern Chihuahuan Desert and probably the desertscrub corridor across the Continental Divide connecting the Chihuahuan and Sonoran deserts. The expansion of creosote bush eastward in the Sonoran Desert may also have occurred in the last five thousand years (Van Devender, *this volume*, chap. 8).

The northern limits of creosote bush apparently have been established in the last few centuries. Creosote bush did not appear in any of a series of eight late Holocene (3.9–0.9 ka) middens from Last Chance Canyon (32°16′ N), although its common associates (including honey mesquite, ocotillo, lechuguilla, fourwing saltbush, and allthorn) were present. The creosote bush population in the canyon just below the midden rock shelters may have arrived in the area historically as desertscrub spread north in the Pecos River valley at the expense of grassland. Evidently the northernmost population of creosote bush in the Chihuahuan Desert near Isleta (34°56′ N) was established in historic times, perhaps aided by livestock.

ACKNOWLEDGMENTS

Midden research in the Chihuahuan Desert has been supported by National Science Foundation grants DEB76-19784 and BSR-8400797 to Thomas R. Van Devender, Fort Bliss Army Base, Human Systems Research, Inc., and the Arizona-Sonoran Desert Museum. Paul S. Martin, Robert S. Thompson, Julio L. Betancourt, Vera Markgraf, and Herbert E. Wright commented on the manuscript at different stages. Helen A. Wilson drafted the figures and drew the packrat. W. Eugene Hall photographed the fossils. Jean T. Morgan typed the manuscript.

REFERENCES

Adams, R. P. (1977). "Chemosystematics—analyses of populational differentiation and variability of ancestral and recent populations of *Juniperus ashei*." *Annals of the Missouri Botanical Garden* 64, 184–209.

Antevs, E. (1955). "Geological-climatic dating in the West." *American Antiquity* 20, 317–55.

Bachhuber, F. W., and McClellan, W. A. (1977). "Paleoecology of marine foraminifera in pluvial Estancia Valley, central New Mexico." *Quaternary Research* 7, 254–67.

Betancourt, J. L. (1984). "Late Quaternary plant zonation and climate in southeastern Utah." *Great Basin Naturalist* 44, 1–35.

Betancourt, J. L.; Van Devender, T. R.; and Rose, M. (1986). "Comparisons of plant macrofossils in woodrat (*Neotoma* sp.) and porcupine (*Erethizon dorsatum*) middens from the western United States." *Journal of Mammalogy* 67, 266–73.

Blair, W. F. (1950). "The biotic provinces of Texas." *Texas Journal of Science* 2, 93–117.

Bradbury, J. P. (1982). "Holocene chronostratigraphy of Mexico and Central America." *Striae* 16, 46–48.

Brakenridge, G. R. (1978). "Evidence for a cold, dry full-glacial climate in the American Southwest." *Quaternary Research* 9, 22–40.

Bryant, V. M., Jr., and Holloway, R. G. (1985). "A late-Quaternary paleoenvironmental record of Texas: An overview of the pollen evidence." In *Pollen records of late Quaternary North American sediments*. V. M. Bryant and R. G. Holloway, eds., pp. 39–70. Dallas: American Association of Stratigraphic Palynologists Foundation.

Bryson, R. A. (1985). "On climatic analogs in paleoclimatic reconstruction." *Quaternary Research* 23, 275–86.

Bryson, R. A., and Wendland, W. (1967). "Tentative climatic patterns for some late glacial and postglacial episodes in central North America. In *Life, land, and water*. S. J. Mayer-Oakes, Jr., ed., pp. 271–98. Winnipeg: University of Manitoba Press.

Butterwick, M., and Strong, S. (1976). "A vegetational survey of the Solitario." In *A Natural Area survey no. 9*, pp. 67–92. Lyndon B. Johnson School of Public Affairs, University of Texas at Austin.

Clisby, K. H., and Sears, P. P. (1956). "San Augustin Plains: Pleistocene climatic changes." *Science* 124, 537–39.

Cole, K. L. (1981). "Late Quaternary environment in the eastern Grand Canyon: Vegetation gradients over the last 25,000 years." Ph.D. diss., University of Arizona, Tucson.

———. (1986). "The lower Colorado River Valley: A Pleistocene desert." *Quaternary Research* 25, 392–400.

Correll, D. S., and Johnston, M. C. (1970). *Manual of the vascular plants of Texas*. Renner, Texas: Texas Research Foundation.

Davis, M. B. (1986). "Climatic instability, time lags, and community disequilibrium." In *Community ecology*. J. Diamond and T. J. Case, eds., pp. 269–84. New York: Harper and Row.

Elias, S. A. (1987). "Paleoenvironmental significance of late Quaternary insect fossils from woodrat middens in south-central New Mexico." *Southwestern Naturalist* 32, 383–90.

Emslie, S. D. (1987). "Extinction of condors in Grand Canyon, Arizona." *Science* 237, 768–70.

Galloway, R. W. (1970). "The full-glacial climate in the southwestern United States." *Annals of the Association of American Geographers* 60, 245–56.

———. (1983). "Full-glacial southwestern United States: Mild and wet or cold and dry?" *Quaternary Research* 19, 236–48.

Gleason, H. A. (1939). "The individualistic concept of the plant association." *American Midland Naturalist* 21, 92–110.

Hafsten, U. (1961). "Pleistocene development of vegetation and climate in the southern High Plains as evidenced by pollen analysis." In *Paleoecology of the Llano Estacado*. F. Wendorf, ed., pp. 59–91. Santa Fe: Museum of New Mexico Press, Fort Burgwin Research Center Publication no. 1.

Harris, A. H. (1977). "Wisconsin age environments in the northern Chihuahuan Desert: Evidence from higher vertebrates." In *Transactions of the Symposium on the Biological Resources of the Chihuahuan Desert Region, United States and Mexico*. R. H. Wauer and D. H. Riskind, eds., pp. 23–52. Washington, D.C.: National Park Service, Transactions and Proceedings Series no. 3.

———. (1985). *Late Pleistocene vertebrate paleoecology of the West*. Austin: University of Texas Press.

Hawley J. W. (1983). "Quaternary geology of the Rhodes Canyon (RATSCAT) site." In *The prehistory of Rhodes Canyon: Survey and mitigation*. P. L. Eidenbach, ed., pp. 17–32. Tularosa, N.M.: Human Systems Research, Inc. Report to Holloman Air Force Base.

Henrickson, J., and Johnston, M. C. (1986). "Vegetation and community types of the Chihuahuan Desert." In *Second symposium on the resources of the Chihuahuan Desert Region*. J. C. Barlow, A. M. Powell, and B. N. Timmermann, eds., pp. 20–39. Alpine, Texas: Chihuahuan Desert Research Institute.

Johnston, M. C. (1977). "Brief résumé of botanical, including vegetational, features of the Chihuahuan Desert region with special emphasis on their uniqueness." In *Transactions of the symposium on the biological resources of the Chihuahuan Desert Region, United States and Mexico*. R. H. Wauer and D. H. Riskind, eds., pp. 335–62. Washington, D.C.: National Park Service, Transactions and Proceedings Series no. 3.

Kutzbach, J. E. (1983). "Modeling of Holocene climates." In *Late-Quaternary environments of the United States*. Volume 2, *The Holocene*. H. E. Wright, ed., pp. 271–77. Minneapolis: University of Minnesota Press.

Lanner, R. M., and Van Devender, T. R. (1981). "Late Pleistocene piñon pines in the Chihuahuan Desert." *Quaternary Research* 15, 278–90.

Markgraf, V.; Bradbury, J. P.; Forester, R. M.; Singh, G.; and Sternberg, R. S. (1984). "San Agustin Plains, New Mexico: Age and paleoenvironmental potential reassessed." *Quaternary Research* 22, 336–43.

Martin, P. S. (1958). "A biogeography of reptiles and amphibians in the Gomez Farias region, Tamaulipas, Mexico." *Museum of Zoology, University of Michigan, Miscellaneous Publication* 101.

———. (1963a). "Geochronology of pluvial Lake Cochise, southern Arizona II. Pollen analysis of a 42-meter core." *Ecology* 44, 436–44.

———. (1963b). *The last 10,000 years*. Tucson: University of Arizona Press.

Martin, P. S., and Mehringer, P. J. (1965). "Pleistocene pollen analysis and biogeography of the Southwest." In *The Quaternary of the United States*. H. E. Wright and D. G. Frey, eds., pp. 433–51. Princeton: Princeton University Press.

Martin, W. C., and Hutchins, C. R. (1981). *A flora of New Mexico*. Vaduz, Licchtenstein: J. Cramer.

Meyer, E. R. (1973). "Late-Quaternary paleoecology of the Cuatro Cienegas basin, Coahuila, Mexico." *Ecology* 54, 982–95.

Minckley, W. L. (1969). Environments of the Bolson of Cuatro Cienegas, Coahuila, Mexico. *University of Texas at El Paso Science Series* (2), 1–65.

Morafka, D. J. (1977). *A biogeographical analysis of the Chihuahuan Desert through its herpetofauna*. The Hague: W. Junk.

Morafka, J. W. (1982). "The status and distribution of the Bolson tortoise (*Gopherus flavomarginatus*)." In *North American tortoises: Conservation and ecology*. R. B. Bury, ed., pp. 71–94. Washington, D.C.: U.S. Department of the Interior, Fish and Wildlife Research Report no. 12.

Neilson, R. P. (1986). "High-resolution climatic analysis and Southwest biogeography." *Science* 232, 27–34.

———. (1987). "Biotic regionalization and climatic controls in western North America." *Vegetatio* 70, 135–47.

Pearson, G.A. (1931). *Forest types in the Southwest as determined by climate and soil*. Washington, D.C.: U.S. Department of Agriculture, Technical Bulletin no. 247.

Powell, A. M., and Turner, B. L. (1977). "Aspects of the plant biology of the gypsum outcrops of the Chihuahuan Desert." In *Transactions of the Symposium on the Biological Resources of the Chihuahuan Desert Region, United States and Mexico*. R. H. Wauer and D. H. Riskind, eds., pp. 315–25. Washington, D.C.: National Park Service, Transactions and Proceedings Series no. 3.

Richmond, G. M. (1962). *Correlation of some glacial deposits in New Mexico*. U.S. Geological Survey Professional Paper 450E, pp. 121–25.

Schmidt, R. H., Jr. (1979). "A climatic delineation of the 'real' Chihuahuan Desert." *Journal of Arid Environments* 2, 243–50.

———. (1986). "Chihuahuan climate." In "*Transactions of the Second Symposium on the Resources of the Chihuahuan Desert Region*." J. C.

Barlow, A. M. Powell, and B. N. Timmermann, eds., pp. 40–63. Alpine, Texas: Chihuahuan Desert Research Institute.

Shreve, F. (1915). *The vegetation of a desert mountain range as conditioned by climatic factors.* Carnegie Institute of Washington Publication no. 217, pp. 1–112.

———. (1937). "Thirty years of change in desert vegetation." *Ecology* 18, 463–78.

———. (1939). "Observations on the vegetation of Chihuahua." *Madroño* 5, 1–48.

———. (1942). "The desert vegetation of North America." *Botanical Review* 9, 195–246.

Snyder, C. T., and Langbein, W. B. (1962). The Pleistocene lake in Spring Valley, Nevada, and its climatic implications. *Journal of Geophysical Research* 67, 2385–94.

Spaulding, W. G. (1981). "The late Quaternary vegetation of a southern Nevada mountain range." Ph.D. diss., University of Arizona, Tucson.

Spaulding, W. G., and Martin, P. S. (1979). "Ground sloth dung of the Guadalupe Mountains." In *Biological investigations in the Guadalupe Mountains National Park.* H. H. Genoways and R. J. Baker, eds., pp. 259–69. Washington, D.C.: National Park Service, Transactions and Proceedings Series no. 4.

Tamayo, J. L., and West, R. C. (1964). "The hydrography of Middle America." In *Natural environment and early cultures.* R. C. West, ed., pp. 84–121. Volume 1 of *"Handbook of Middle American Indians".* Austin: University of Texas Press.

Thompson, R. S.; Van Devender, T. R.; Martin, P. S.; Foppe, T.; and Long, A. (1980). "Shasta ground sloth, *Nothrotheriops shastense* Hoffstetter at Shelter Cave, New Mexico: Environment, diet, and extinction." *Quaternary Research* 14, 360–76.

Van Devender, T. R. (1973). "Late Pleistocene plants and animals of the Sonoran Desert: A survey of ancient packrat middens in southwestern Arizona." Ph.D. diss., University of Arizona, Tucson.

———. (1977). "Holocene woodlands in the southwestern deserts." *Science* 198, 189–92.

———. (1980). "Holocene environments in Rocky Arroyo and Last Chance Canyon, Eddy Country, New Mexico." *Southwestern Naturalist* 25, 361–72.

———. (1986a). "Climatic cadences and the composition of Chihuahuan Desert communities: The Late Pleistocene packrat midden record." In *Community ecology.* J. Diamond and T. J. Case, eds., pp. 285–99. New York: Harper and Row.

———. (1986b). "Pleistocene climates and endemism in the Chihuahuan Desert flora." In *Transactions of the Second Symposium on the Resources of the Chihuahuan Desert.* J. C.

Barlow, A. M. Powell, and B. N. Timmermann, eds., pp. 1–19. Alpine, Texas: Chihuahuan Desert Research Institute.

———. (1986c). "Late Quaternary history of pinyon-juniper-oak woodlands dominated by *Pinus remota* and *Pinus edulis.*" In *Proceedings of the Pinyon-Juniper Conference.* R. L. Everett, comp., pp. 99–103. Reno, Nev.: U.S. Department of Agriculture, Forest Service, Intermountain Research Station.

Van Devender, T. R.; Betancourt, J. L.; and Wimberly, M. (1984). "Biogeographic implications of a packrat midden sequence from the Sacramento Mountains, south-central New Mexico." *Quaternary Research* 22, 344–60.

Van Devender, T. R., and Bradley, G. L. (In press). "The paleoecology and historical biogeography of the herpetofauna of Maravillas Canyon, Texas." *Copeia.*

Van Devender, T. R.; Bradley, G. L.; and Harris, A. H. (1987). "Late Quaternary mammals from the Hueco Mountains, El Paso and Hudspeth Counties, Texas." *Southwestern Naturalist* 32, 179–95.

Van Devender, T. R., and Burgess, T. L. (1985). "Late Pleistocene woodlands in the Bolson de Mapimi: A refugium for the Chihuahuan Desert biota?" *Quaternary Research* 24, 346–53.

Van Devender, T. R., and Everitt, B. L. (1977). "The latest Pleistocene and Recent vegetation of Bishop's Cap, south-central New Mexico." *Southwestern Naturalist* 22, 22–28.

Van Devender, T. R.; Freeman, C. E.; and Worthington, R. D. (1978). "Full-glacial and Recent vegetation of the Livingston Hills, Presidio County, Texas." *Southwestern Naturalist* 23, 289–302.

Van Devender, T. R.; Martin, P. S.; Thompson, R. S.; Cole, K. L.; Jull, A. J. T.; Long, A.; Toolin, L. J.; and Donahue, D. J. (1985). "Fossil packrat middens and the tandem accelerator mass spectrometer." *Nature* 317, 610–13.

Van Devender, T. R.; Moodie, K. B.; and Harris, A. H. (1976). "The desert tortoise (*Gopherus agassizi*) in the Pleistocene of the northern Chihuahuan Desert." *Herpetologica* 32, 298–304.

Van Devender, T. R., and Riskind, D. H. (1979). "Late Pleistocene and early Holocene plant remains from Hueco Tanks State Historical Park: The development of a refugium." *Southwestern Naturalist* 24, 127–40.

Van Devender, T. R., and Spaulding, W. G. (1979). "Development of vegetation and climate in the southwestern United States." *Science* 204, 701–10.

Van Devender, T. R.; Spaulding, W. G.; and Phillips, A. M. (1979). "Late Pleistocene plant communities in the Guadalupe Mountains, Culberson

County, Texas." In *Biological investigations in Guadalupe Mountains National Park*. H. H. Genoways and R. J. Baker, eds., pp. 13–30. Washington, D.C.: National Park Service, Proceedings and Transactions Series no. 4.

Van Devender, T. R.; Thompson, R. S.; and Betancourt, J. L. (1987). "Vegetation history in the Southwest: The nature and timing of the late Wisconsin–Holocene transition." In *North America and adjacent oceans during the last deglaciation*. W. F. Ruddiman and H. E. Wright, Jr., eds., pp. 323–52. Boulder: Geological Society of America.

Van Devender, T. R., and Toolin, L. J. (1983). "Late Quaternary vegetation of the San Andres Mountains, Sierra County, New Mexico." In *The prehistory of Rhodes Canyon: Survey and mitigation*. P. L. Eidenbach, ed., pp. 33–54. Tularosa, N.M.: Human Systems Research, Inc. Report to Holloman Air Force Base.

Van Devender, T. R., and Wiseman, F. M. (1977). "A preliminary chronology of bioenvironmental changes during the Paleoindian period in the monsoonal Southwest." In *Paleoindian lifeways*. E. Johnson, ed., West Texas Museum Association, *Museum Journal* 27, 13–27.

Van Devender, T. R., and Worthington, R. D. (1977). "The herpetofauna of Howell's Ridge Cave and the paleoecology of the northwestern Chihuahuan Desert." In *Transactions of the Symposium on the Biological Resources of the Chihuahuan Desert Region, United States and Mexico*. R. H. Wauer and D. H. Riskind, eds., pp. 85–106. Washington, D.C.: National Park Service, Transactions and Proceedings Series no. 3.

Watts, W. A., and Bradbury, J. P. (1982). "Paleoecological studies at Lake Patzcuaro on the west-central Mexican Plateau and at Chalco in the Basin of Mexico." *Quaternary Research* 17, 56–70.

Wells, P. V. (1966). "Late Pleistocene vegetation and degree of pluvial climatic change in the Chihuahuan Desert." *Science* 153, 970–75.

———. (1979). "An equable glaciopluvial in the West: Pleniglacial evidence of increased precipitation on a gradient from the Great Basin to the Sonoran and Chihuahuan deserts." *Quaternary Research* 12, 311–25.

Wells, P. V., and Hunziker, J. H. (1976). "Origin of the creosote bush (*Larrea*) deserts of southwestern North America." *Annals of the Missouri Botanical Garden* 63, 883–961.

Appendix 7.1. Common and scientific names used in text and figures. Plant nomenclature mostly follows Correll and Johnston (1970).

Alligator bark juniper	*Juniperus deppeana*
Allthorn	*Koeberlinia spinosa*
Big bend silverleaf	*Leucophyllum candidum*
Big sagebrush	*Artemisia tridentata*
Bison	*Bison* sp.
Blackhood snake	*Tantilla rubra*
Blind prickly pear	*Opuntia rufida*
Blue spruce	*Picea pungens*
Bolson tortoise	*Gopherus flavomarginatus*
Bushy-tailed packrat	*Neotoma cinerea*
California condor	*Gymnogyps californianus*
Candelilla	*Euphorbia antisyphilitica*
Canyon lizard	*Sceloporus merriami*
Catclaw acacia	*Acacia greggii*
Chihuahuan crucifixion thorn	*Castela stewarti*
Chino grass	*Bouteloua rufida*
Christmas cactus	*Opuntia leptocaulis*
Colorado pinyon	*Pinus edulis*
Creosote bush	*Larrea divaricata*
Desert olive	*Forestiera angustifolia*
Desert tortoise	*Xerobates agassizi*
Desert yaupon	*Schaefferia cuneifolia*
Douglas fir	*Pseudotsuga menziesii*
Dwarf juniper	*Juniperus communis*
Engelmann spruce	*Picea engelmannii*
False nightshade	*Chamaesaracha* sp.
Fourwing saltbush	*Atriplex canescens*
Gambel oak	*Quercus gambelii*
Goldeneye	*Viguiera dentata*
Gray daisy	*Bahia absinthifolia*

Appendix 7.1. *(continued)*

Gray-banded kingsnake	*Lampropeltis mexicana*
Guayacán	*Guaiacum angustifolium*
Heath cliffrose	*Purshia ericaefolia*
Hinckley oak	*Quercus hinckleyi*
Honey mesquite	*Prosopis glandulosa*
Honeysuckle	*Lonicera albiflora*
Indian ricegrass	*Oryzopsis hymenoides*
Lechuguilla	*Agave lechuguilla*
Limber pine	*Pinus flexilis*
Lyre snake	*Trimorphodon biscutatus*
Mariola	*Parthenium incanum*
Meadow vole	*Microtus pennsylvanicus*
Mexican hooknose snake	*Gyalopion canum*
Mexican pinyon	*Pinus cembroides*
Mock pennyroyal	*Hedeoma plicatum*
Mormon tea	*Ephedra nevadensis*
Mountain mahogany	*Cercocarpus montanus*
New Mexican locust	*Robinia neomexicana*
Ocotillo	*Fouquieria splendens*
Oneseed juniper	*Juniperus monosperma*
Papershell pinyon	*Pinus remota*
Pinchott juniper	*Juniperus pinchottii*
Ponderosa pine	*Pinus ponderosa*
Porcupine	*Erethizon dorsatum*
Redberry juniper	*Juniperus erythrocarpa*
Redbud	*Cercis canadensis*
Resurrection plant	*Selaginella lepidophylla*
Rock cedar	*Juniperus ashei*
Rock daisy	*Perityle* sp.
Rocky Mountain juniper	*Juniperus scopulorum*
Roemer catclaw	*Acacia roemeriana*
Sage grouse	*Centrocercus urophasianus*
Sage vole	*Lagurus curtatus*
Sandpaper bush	*Mortonia sempervirens*
Shasta ground sloth	*Nothrotheriops shastense*
Shrew	*Sorex* spp.
Shrub live oak	*Quercus turbinella*
Shrub oak	*Quercus pungens*
Silver wolfberry	*Lycium puberulum*
Skunk bush	*Rhus aromatica*
Sotol	*Dasylirion* sp.
Southwestern white pine	*Pinus strobiformis*
Spiny greasebush	*Forsellesia spinescens*
Tarbush	*Flourensia cernua*
Texas alligator lizard	*Gerrhonotus liocephalus*
Texas lyre snake	*Trimorphodon biscutatus vilkinsoni*
Thompson/beaked yucca	*Yucca thompsoniana/rostrata*
Tobosa grass	*Hilaria mutica*
Tree cactus	*Opuntia imbricata*
Variable prickly pear	*Opuntia phaeacantha*
Vole	*Microtus* spp.
Yellow-bellied marmot	*Marmota flaviventris*
Wavyleaf oak	*Quercus undulata*
White-throated packrat	*Neotoma albigula*
Wild rose	*Rosa stellata*
Wilson's tortoise	*Geochelone wilsoni*
Winterfat	*Ceratoides lanata*

Chapter 8

Late Quaternary Vegetation and Climate of the Sonoran Desert, United States and Mexico

Thomas R. Van Devender

With its great variety of plants curiously adapted to the arid environment, the Sonoran Desert gives the impression of great antiquity. Forests of giant cacti and trees with minute leaves and chlorophyllous bark reflect the evolutionary radiations of the Miocene that occurred as the North American deserts formed. However, palynological studies in the southwestern dry lakes such as the San Agustin Plains of New Mexico (Clisby and Sears, 1956) and the Willcox Playa of Arizona (Martin, 1963) led to the conclusion that forest and woodland communities grew as much as 1000 m lower in the last glacial period, the Wisconsin, than today leaving little room for desertscrub vegetation (Martin and Mehringer, 1965). Although the areas of continental and montane glaciation were distant during each of many glacial-interglacial cycles, the secondary effects of the ice sheets on global thermal circulation of the atmosphere were readily observed in the subtropical Sonoran Desert. Fortunately, the well-preserved plant macrofossils in packrat (*Neotoma* spp.) middens permit detailed vegetation reconstructions from the deserts themselves for the last glacial-interglacial cycle.

The Sonoran Desert is the subtropical arid area centered around the head of the Gulf of California in western Sonora, southwestern Arizona, southeastern California, and much of the Baja California peninsula (Shreve, 1964; Fig. 8.1). Elevations range from below sea level in the Imperial Valley of California and 278 m at Needles on the Colorado River to about 1000 m at the northeastern margin (Turner and Brown, 1982). All portions of the Sonoran Desert may occasionally experience freezing temperatures, although the duration is rarely more than a single night. In general the frequency and intensity of freezes decrease to the south. Rainfall ranges from a biseasonal regime with strong summer monsoons in Sonora and Arizona to a winter-rainfall regime on the west coast of Baja California. Subdivisions of the Sonoran Desert were proposed by Shreve (1964) and refined by Turner and Brown (1982) based on vegetation and climate. Trees and arborescent cacti are common throughout the Sonoran Desert and reflect its strong affinities with subtropical thornscrub to the south.

The desert is a sensitive area for packrat midden studies because several important climatic features (rainfall, winter freezes) are imported to the Southwest from other regions by large-scale atmospheric circulation features, and are often factors limiting biotic distributions. Seasonal rainfall fluctuations tracking the intensities of the Aleutian Low and the Bermuda High or fluctuations in the frequency of intense freezes during arctic air incursions are reflected in the composition of desertscrub communities. Paleoclimatic reconstructions from Sonoran Desert midden sequences can help evaluate various paleoclimatic hypotheses. Variations in the Earth's orbital parameters over the Holocene have caused changes in the distribution of solar radiation on the planet's surface. In the early and middle Holocene these variations may have raised summer temperatures in much of the Northern Hemisphere, intensifying the summer monsoons in many parts of the world (Kutzbach, 1983). Spaulding and Graumlich (1986) and Spaulding (*this volume*, chap. 9) have suggested that these fluctuations in insolation led to latest Wisconsin and early Holocene maxima in monsoonal precipitation in the Mojave Desert and adjacent Sonoran Desert. Largely on the basis of pluvial lake studies in the northern Great Basin, Antevs (1955) argued that his altithermal period (the middle Holocene) was characterized by hot and dry conditions across the entire western United States. For the summer monsoonal portions of the Southwest, Martin (1963) proposed a period of enhanced summer precipitation.

A regional bioclimatic chronology for the Southwest based on vegetation changes recorded in packrat midden sequences was developed by Van Devender and Spaulding (1979) and revised by Van Devender et al. (1987). The end of the late Wisconsin (LW) was at about 11.0 ka, the early Holocene (EH) from 11 to about 8.9 ka, the middle Holocene from 8.9 to about 4 ka and the late Holocene from about 4 ka to the present. The vegetation of the Sonoran Desert is sensitive to these bioclimatic events (Fig. 8.1). Evidence for vegetation shifts that reflect major climatic changes are presented in the regional sections below.

METHODS

In the Sonoran Desert most midden rock shelters are formed in rhyolite or granite and, less often, in volcanic tuff, gneiss, or limestone. Packrat midden samples were collected in dry rock shelters and carefully separated into natural layers. Outer weathering rinds were removed to avoid contamination. Sample sizes of about 1000 g or less were soaked in water, screened, dried, and sorted. Plant fragments were identified and ranked in an internal relative abundance ranging from rare to abundant (1–5). A single specimen ranked 1, the most common taxon 5, the remainder in between. More elaborate quantitative methods using percentages of identified specimens greatly increase the analytical effort without significantly improving the final result (Van Devender, 1973; Spaulding, 1981; Cole, 1981; Betancourt, 1984; Spaulding et al., *this volume*, chap. 5). Internal relative abundances are readily comparable between samples.

The plant macrofossils in a midden assemblage are well preserved (Figs. 8.2, 8.3) and represent rich samples of the local flora within about 30 m of the fossil site. Any collecting biases should be consistent throughout. In most cases, including samples less than 500 years old, the midden assemblages contain *more* plant taxa than occur at the sites today, suggesting that the time represented by a midden layer encompasses short-term, 50–100-year fluctuations in climate and in short-lived plants. Accordingly, I view the accumulation time of a sample as less rapid than proposed by Spaulding et al. (*this volume*, chap 5).

Packrat midden assemblages cannot directly disclose the height, spacing, coverage, and biomass that enable ecologists to describe, classify, and compare modern vegetation. Density, coverage, biomass, and species composition are variable community parameters. Nevertheless, the species and their relative abundances in the assemblages allow relatively secure inferences about the general heights and structural compositions in a paleocommunity. For example, the relationships between mixed conifers and forest; pinyons and junipers (other than dwarf juniper, *Juniperus communis*) with woodland; and foothills palo verde (*Cercidium microphyllum*) and saguaro (*Carnegiea gigantea*) with Sonoran desertscrub are fairly consistent in modern communities. In the following discussions inferred vegetation types are used when appropriate. Common and scientific names of plants and animals mentioned in the text and figures are presented in Appendix 8.1.

The fossil middens from the Sonoran Desert are predominantly the work of the white-throated packrat (*Neotoma albigula*) and the desert packrat (*N. lepida*, including *N. devia*; Mead et al., 1983). Differences in diets between the two species are probably not a seri-

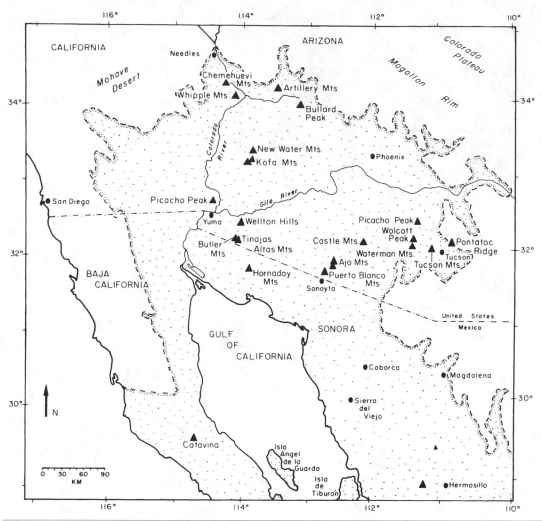

Fig. 8.1. Map of the Sonoran Desert (stippled) as defined by Shreve (1964). Triangles mark packrat midden sites discussed in text.

ous problem for interpretation of the midden record. Although the white-throated packrat has a strong preference for cactus (Vaughan, *this volume*, chap. 2), it has been a dietary generalist for the last forty-five thousand years in the Chihuahuan Desert (Van Devender, *this volume*, chap. 7; Van Devender and Bradley, *this volume*, chap. 15). The desert packrat has probably been the resident generalist in the Mojave Desert. Moreover, a shift in the range of one species at the expense of the other would probably be a reflection of changes in vegetation and climate as well. A middle Wisconsin sample from the

Castle Mountains, Arizona, was deposited by porcupines (*Erethizon dorsatum*; Betancourt, Van Devender, and Rose, 1986). Porcupine middens are not rich in macrofossils from fragile species.

The organic materials from packrat middens are excellent for radiocarbon dating. Without such a dating method, which is reliable to within a few centuries, essential comparisons would be impossible. Although dates on a single specimen or an accumulation of many small specimens of an individual species have the advantage of direct association, dates for packrat fecal pellets and midden de-

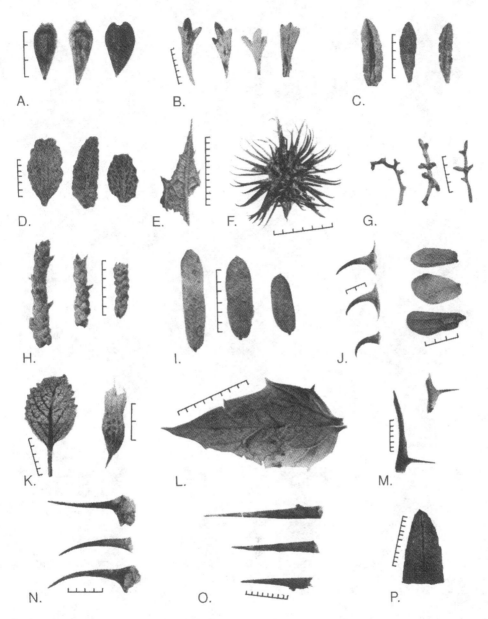

Fig. 8.2. Photos of macrofossils of trees and shrubs from Sonoran Desert packrat middens. AL = Ajo Loop, Puerto Blanco Mountains; and TA = Tinajas Altas Mountains, Arizona. FA = Falling Arches, RP = Redtail Peaks, and TR = Tunnel Ridge in Whipple Mountains, California. Scale divisions in millimeters; ka = thousands of years. (A) Brittle bush achenes; TA#18A, 15 ka. (B) Big sagebrush leaves; TR#5A, 12.7 ka. (C) California buckwheat (*Eriogonum fasciculatum*) leaves; FA#1A, 11.65 ka. (D) Mojave sage leaves; FA#1A, 11.65 ka. (E) Holly-leaf bursage leaf; TA#12A, 10 ka. (F) Hollyleaf bursage bur; TA#16A, 7.9 ka. (G) White bursage leaves; TA#12A, 10 ka. (H) California juniper twigs; RP#6, 9.6 ka. (I) Velvet mesquite leaflets; TA#16A, 7.9 ka. (J) Catclaw acacia thorns and leaflets; TA#16A, 7.9 ka. (K) Desert lavender leaf and fruit; TA#20B, 4 ka. (L) Desert brickell bush (*Brickellia atractyloides*) leaf fragment; TA#20B, 4 ka. (M) Blue palo verde twigs, thorns; TA#20B, 4 ka. (N) Iron-wood thorns; AL#6B, 3.2 ka. (O) Foothills palo verde spinescent branches; AL#6B, 3.2 ka. (P) Mexican jumping bean leaf fragment; AL#2C, 0.1 ka.

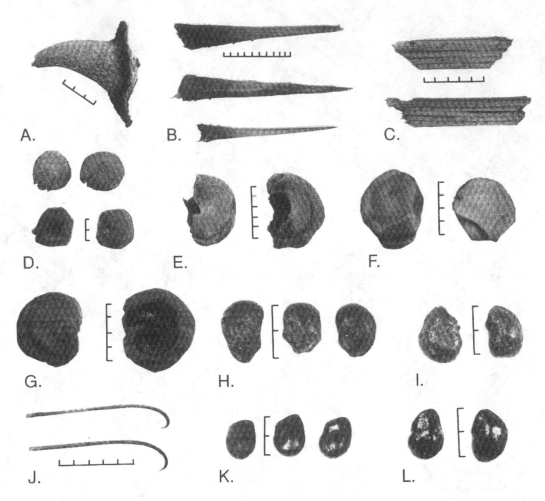

Fig. 8.3. Photos of macrofossils of succulents from Sonoran Desert packrat middens. CP = Cholla Pass, Puerto Blanco Mountains, Arizona. Scale divisions in millimeters; ka = thousands of years. (A) Desert agave (*Agave deserti*) marginal hook; TA#18A, 15.1 ka. (B) Joshua tree leaf tips; TR#5A, 12.7 ka. (C) Bigelow beargrass leaf fragments; RP#6, 9.6 ka. (D) Whipple cholla (*Opuntia whipplei*) seeds; TA#18A. (E) Beavertail cactus (*O. basilaris*) seeds; TA#20B, 4 ka. (F) Staghorn cholla (*O. acanthocarpa*) seeds; CP#2, 3.4 ka. (G) Variable prickly pear (*O. phaeacantha*) seeds; AL#3A, 1 ka. (H) California barrel cactus (*Ferocactus acanthodes*) seeds; FA#1A, 11.7 ka. (I) Coville barrel cactus (*F. covillei*) seeds; AL#6B, 3.2 ka. (J) Fishhook cactus (*Mammillaria microcarpa*) spines; TA#16A, 7.9 ka. (K) Saguaro seeds; TA#16A, 7.9 ka. (L) Organ pipe cactus seeds; AL#2C, 0.1 ka.

bris have proven reliable when taken from well-cleaned samples from discrete layers. Plant taxa in the assemblage are assumed to be associated with the date. Assemblages are arranged in chronological order to produce vegetation sequences through time. With the advent of radiocarbon dating using the tandem accelerator mass spectrometer (TAMS), radiocarbon dates can be run on very small samples. TAMS dates on packrat midden specimens provide direct dates for ecologically important species, verify antiquity of desert plants in woodland assemblages, and identify contaminants (Van Devender et al., 1985). Selected midden radiocarbon dates are presented in Table 8.1. Standard deviations and laboratory numbers for other specific dates are given in the text in years B.P. (radiocarbon years before 1950). Additional information on radiocarbon dates is available in references cited in the text and in Webb and Betancourt (*this volume*, Chap. 6). Other ages are abbreviated to thousands of years (ka).

Many of the plant remains in packrat middens are identifiable to species, allowing their distribution, phenology, and physiology to be invoked in the reconstruction of paleoclimates. Plant distributions and phenological information are available from regional floras (Kearney and Peebles, 1964; Shreve and Wiggins, 1964; Munz, 1974; Wiggins, 1980; Benson and Darrow, 1981; Thorne et al., 1981), atlases of plant distributions (Hastings, Turner, and Warren, 1972; Little 1971, 1976), monographs on cacti (Benson, 1982) and sagebrushes (Beetle, 1960), voucher specimens in the University of Arizona and Arizona State University herbaria, and personal notes and collections. Climatic records for Arizona are available from Sellers and Hill (1974).

MIDDEN SUMMARIES

A total of 199 packrat middens associated with 242 radiocarbon dates have been studied from the Sonoran Desert. The middens were collected in 36 local areas from near sea level to 1555 m elevation, from latitude 29°45' to 34°23' N and longitude 110°53' to 114°45' W. Ten study areas have yielded relatively long midden sequences. The following summaries of the vegetation history are arranged by subdivision of the Sonoran Desert from subtropical Mexico and Arizona west to the lower Colorado River valley and northwest to the edge of the Mojave Desert. Each section summarizes the contents of the fossil packrat middens from which significant biogeographical and vegetational changes are inferred. Summary diagrams illustrate four plant macrofossil sequences that yielded the greatest chronological detail.

Plains of Sonora

In central Sonora, under a biseasonal rainfall regime, a diverse vegetation is dominated by leguminous trees, columnar cacti, and shrubs, especially brittle bush (*Encelia farinosa*) and Sonoran bursage (*Ambrosia cordifolia*). One packrat midden from a rhyolite shelter near Hermosillo in central Sonora (ca. 29° N, 111° W, 457 m elev.) yielded juniper from the middle Wisconsin (> 33 ka; Wells and Hun-

ziker, 1976). Further information and additional records would be of great interest.

Central Gulf Coast

The winter rainfall portion of the Sonoran Desert in Baja California supports a diverse mixture of trees, shrubs, and succulents. Three midden samples from near Catavina, central Baja California (29°45' N, 114°45' W), record early Holocene (10.0–10.2 ka) juniper and chaparral shrubs in an area dominated by boojum tree (*Fouquieria columnaris*; Wells, 1976; Axelrod, 1979). More information on the associated fossils would be desirable.

Arizona Upland

Plant macrofossil records from fossil packrat middens are available from a number of areas in the northeastern Sonoran Desert. The Arizona Upland subdivision forms a narrow, curving border in Arizona from the Buckskin Mountains (35° N, 114° W) southeast to near Phoenix and south to about Altar, Sonora. This area is relatively well watered and supports one of the least "desertlike" of the desertscrub communities in North America. Saguaro, foothills palo verde, and triangleleaf bursage (*Ambrosia deltoidea*) are dominants in association with a diverse mixture of shrubs and succulents. At 1000–1050 m elevation the Arizona Upland merges with desert grassland or interior chaparral. To the west the lower Colorado River valley subdivision begins at about 300 m elevation. The Mojave Desert is to the northwest. The average annual precipitation is 200–425 mm/yr with 30%–60% in the summer (Turner and Brown, 1982).

Pontatoc Ridge. At the upper edge of the Arizona Upland, just above the local limit of saguaro, eight packrat midden assemblages from Pontatoc Ridge in the Santa Catalina Mountains near Tucson, Arizona (32°21' N, 110°53' W, 1555 m elev.), provide a record of an Arizona cypress (*Cupressus arizonica*) forest in the late Wisconsin (18.0, 16.4, 14.5, 14.1, 13.8, 13.2, 12.4, 11.9 ka). Other plants in the community included Douglas fir (*Pseudotsuga menziesii*), ponderosa pine (*Pinus ponderosa*), hybrid singleleaf/Colorado pinyon (*P. monophylla/edulis*), and border pinyon (*P. discolor*). Presumably, the lower edge of mixed-conifer forest dominated by Douglas fir was not far above the site. These are the

Table 8.1. *Radiocarbon and locality information for anomalous species pairs in Sonoran Desert packrat middens listed in Table 8.2.*

Location	Midden	Elev. (m)	Radiocarbon Age (yr B.P.)	Lab. No.	Material Dated
Arizona					
Pima Co.					
Pontatoc Ridge	PR4B	1555	17,950 ± 960	A-1851	*Cupressus arizonica* twigs
	PR81-2		16,400 ± 400	A-2682	*Cupressus arizonica* twigs
	PR4A		14,450 ± 800	WK-163	*Cupressus arizonica* twigs
	PR3A		14,040 ± 600	A-1859	*Cupressus arizonica* twigs
	PR3B		13,670 ± 630	A-1868	*Cupressus arizonica* twigs
	PR7		12,360 ± 280	A-2681	artiodactyl pellets
Wolcott Peak	WP5		12,130 ± 500	A-1287	*Juniperus* sp. twigs
Waterman Mts.	WAM9A1	795	22,380 ± 380	A-4776	*Neotoma* sp. fecal pellets
	WAM2		21,510 ± 240	AA-2497	*Juniperus* sp. twigs
			17,030 ± 380	A-4710	*Neotoma* sp. fecal pellets
			19,270 ± 340	(av. AA-2497 & A-4710)	
	WAM1B		12,690 ± 150	A-4552	*Neotoma* sp. fecal pellets
	WAM1D		12,530 ± 140	A-4551	*Neotoma* sp. fecal pellets
	WAM9A2		11,470 ± 170	A-4777	*Neotoma* sp. fecal pellets
Ajo Mts. (Alamo Canyon)	AC2C	915	32,000 ± 4,400	A-2119	*Juniperus* sp. twigs
	AC1B		14,500 ± 300	A-2120	*Juniperus* sp. twigs
	AC1A1		10,580 ± 220	A-2210	*Neotoma* sp. fecal pellets
			8590 ± 470	AA-539	*Carnegiea gigantea* seeds
			9585 ± 200	(av. A-2210 & AA-539)	
	AC1U		9910 ± 210	A-2211	*Quercus ajoensis* acorns, leaves
			9230 ± 370	AA-533	*Carnegiea gigantea* seeds
			9570 ± 180	(av. A-2211 & AA-533)	
Ajo Mts. (Montezuma's Head)	MH1B	975	21,840 ± 650	A-1696	*Pinus monophylla* needles
	MH1A		20,490 ± 510	A-1695	*Juniperus* sp. twigs
	MH1C		17,830 ± 870	A-1697	*Juniperus* sp. twigs
	MH1D		13,500 ± 390	A-1698	*Juniperus* sp. twigs
Yuma Co.					
Puerto Blanco Mts. (Twin Peaks)	TP1	550	14,120 ± 260	A-3986	*Juniperus californica* twigs
Tinajas Altas Mts.	TA15B1		>43,200	AA-1720	*Juniperus californica* seeds
	TA18B	330	18,700 ± 1,050	A-536	*Larrea divaricata* twig
			16,150 ± 400	A-3407	*Juniperus californica* twigs
			17,425 ± 375	(av. AA-536 & A-3407)	

Location	Sample	Number	Age (yr B.P.)	Lab number	Material
	TA15AL	550	15,680 ± 720	AA-452	*Pinus monophylla* needles
	TA18A	330	15,050 ± 350	A-3459	*Juniperus californica* twigs
	TA2B	365	10,950 ± 90	USGS-958	*Neotoma* sp. fecal pellets
	TA3B	580	10,600 ± 420	AA-781	*Ambrosia dumosa* burs
			10,300 ± 110	USGS-959	*Neotoma* sp. fecal pellets
			10,450 ± 105	(av. AA-781 & USGS-959)	
	TA12A	565	10,070 ± 110	A-3730	*Neotoma* sp. fecal pellets
	TA14	550	9900 ± 200	A-3570	*Juniperus californica* twigs
	TA1B	490	9790 ± 240	A-2494	*Neotoma* sp. fecal pellets
	TA3A	580	9700 ± 100	USGS-957	*Neotoma* sp. fecal pellets
	TA2A	490	9230 ± 140	A-2122	*Juniperus californica* twigs
			8970 ± 75	USGS-956	*Neotoma* sp. fecal pellets
			9100 ± 70	(av. A-2122 & USGS-956)	
Butler Mts.	BM3A	240	11,250 ± 410	AA-469	*Larrea divaricata* twigs
	BM3B		11,060 ± 300	AA-770	*Juniperus californica* twigs
			10,170 ± 610	AA-768	*Larrea divaricata* twigs
			10,615 ± 270	(av. AA-770 & AA-768)	
	BM5	250	10,130 ± 350	AA-537	*Larrea divaricata* twig
	BM1	245	8,160 ± 160	A-2652	*Neotoma* sp. fecal pellets
Kofa Mts. (Brass Cap Point)	BCP1	550	11,450 ± 400	A-1320	*Yucca brevifolia* leaves
	BCP2		9750 ± 470	AA-574	*Castela emoryi* seed
Kofa Mts. (Burro Canyon)	BC1:1	860	14,400 ± 330	A-1315	*Juniperus* sp. twigs
	BC1:6		13,400 ± 250	A-1357	*Juniperus* sp. twigs
La Paz Co. New Water Mts.	NW2	615	12,090 ± 570	A-1445	*Quercus turbinella* acorns, leaves
			11,060 ± 390	A-1353	*Juniperus* sp. twigs
			7870 ± 750	A-1284	*Juniperus* sp. twigs
	NW7	605	11,000 ± 510	A-1295	*Juniperus* sp. twigs
	NW4	615	10,880 ± 900	A-1285	*Juniperus* sp. twigs
California Imperial Co. Picacho Peak area	PP17D2	285	13,380 ± 350	AA-577	*Yucca whipplei* leaf
			12,390 ± 340	AA-578	*Larrea divaricata* twigs
			12,885 ± 245	(av. AA-577 & AA-578)	
	PP17D1		10,850 ± 630	AA-576	*Larrea divaricata* twigs

Table 8.1. *(Continued)*

Location	Midden	Elev. (m)	Radiocarbon Age (yr B.P.)	Lab. No.	Material Dated
San Bernardino Co.					
Whipple Mts. (Redtail Peaks)					
	RP11	520	13,810 ± 270	A-1650	Juniperus californica twigs
	RP8	495	11,520 ± 160	A-1621	Juniperus californica twigs
	RP10A	490	10,840 ± 170	A-1616	Nolina bigelovii leaves
	RP5A	510	10,880 ± 180	A-1661	Juniperus californica twigs
			10,540 ± 140	A-1664	Neotoma sp. fecal pellets
			10,360 ± 350	A-1662	midden debris
			10,600 ± 105	(av. A-1661, A-1662, & A-1664)	
	RP10B	490	10,490 ± 110	A-3944	Neotoma sp. fecal pellets
	RP3	510	10,030 ± 160	A-1620	Juniperus californica twigs
	RP6	495	9600 ± 170	A-1655	Juniperus californica twigs
	RP1A1	520	9160 ± 170	A-1668	Juniperus californica twigs
	RP1A2		8910 ± 380	A-1580	Juniperus californica twigs

only Wisconsin-age records of ponderosa pine from west of the Continental Divide, and the only fossil records for border pinyon. The fossil compositions were quite similar for at least 6100 years, suggesting a relatively stable community in the late Wisconsin. The vegetation change marking the end of the Wisconsin occurred after 11,900 ± 210 B.P. (A-4027). An Arizona cypress forest with composition and relative abundances similar to the fossil assemblages presently occurs in Bear Canyon in the Santa Catalina Mountains.

Waterman Mountains. Seventeen packrat middens from the Waterman Mountains (32°20′30″ N, 111°27′ W; 795 m elev.) provide a rich fossil record from limestone, which is much less frequent in the Sonoran than in the Chihuahuan Desert. The precipitation for Silver Bell, 7 km to the northwest, is 312 mm/yr with 42.3% in the summer (June–August; Sellers and Hill, 1974). The vegetation for the general area is a Sonoran desertscrub dominated by foothills palo verde, saguaro, ironwood (*Olneya tesota*), and triangleleaf bursage. A north-facing slope below one of the midden shelters is locally dominated by mariola (*Parthenium incanum*), a shrub of calcarous soils that is a widespread dominant in the Chihuahuan Desert to the east. The oldest of twenty-five radiocarbon dates (22.5 ka) was associated with a sample that contained abundant juniper and singleleaf pinyon in association with Joshua tree (*Yucca brevifolia*) and Arizona rosewood (*Vauquelinia californica*). Seeds of Nichol's Turk's head cactus (*Echinocactus horizonthalonius* var. *nicholii*), an endemic variety of a common Chihuahuan Desert plant, and an osteoderm of the Gila monster (*Heloderma suspectum*), an ungainly, poisonous lizard that still lives in the area, were also found in the sample. Two late Wisconsin samples dated at 12.5 and 12.7 ka yielded abundant remains of singleleaf pinyon and Utah juniper (*Juniperus osteosperma*) in association with big sagebrush (*Artemisia tridentata*-type) and shrub live oak (*Quercus turbinella*) as well as Joshua tree, striped horsebrush (*Tetradymia argyraea*), rock sage (*Salvia pinguifolia*), Arizona rosewood, and tanglehead grass (*Heteropogon contortus*). A TAMS radiocarbon date of 11,740 ± 110 B.P. (AA-3577) on seeds documents the arrival of velvet mesquite (*Prosopis* *velutina*) in the latest Wisconsin.

An early Holocene sample dated at 9920 ± 160 B.P. (A-4709) yielded abundant juniper and skunkbush (*Rhus* cf. *trilobata*) with Arizona rosewood, catclaw acacia (*Acacia greggii*), and velvet mesquite. A middle Holocene sample dated at 5540 ± 70 B.P. (A-4781) contained abundant saguaro in association with catclaw acacia, velvet mesquite, blue palo verde (*Cercidium floridum*), desert lavender (*Hyptis emoryi*), and elephant tree (*Bursera microphylla*). Catclaw acacia, velvet mesquite, and blue palo verde are mostly restricted to riparian washes below the site today. A relict population of elephant tree occurs in the Waterman Mountains about two kilometers from the site. A TAMS date of 6195 ± 80 B.P. (AA-3578) on twigs is the oldest record for creosote bush (*Larrea divaricata*) in the eastern Sonoran Desert. A 3880 ± 80 B.P. (A-4708) sample was dominated by velvet mesquite, with saguaro and foothills palo verde. Three late Holocene (1.2–2.6 ka) samples contained abundant foothills palo verde and saguaro without riparian plants. On fossil evidence alone one might not realize the substrate was limestone.

Other areas. Late Wisconsin midden assemblages from lower elevations in the Arizona Upland dominated by singleleaf pinyon, juniper, and shrub live oak were widespread in some type of woodland/chaparral. Juniper and shrub live oak without pinyon were found in the early Holocene middens. Important assemblages have been found in the Tucson Mountains, (32°12' N, 111°05' W, 710–890 m elev.; Van Devender, 1973), Wolcott Peak near Silver Bell (32°27' N, 111°28' W; 860 m elev.; Van Devender, 1973), and the Castle Mountains (32°24' N, 112°17' W; 790–855 m elev.; Betancourt et al., 1986). Juniper lingered in the Castle Mountains until 8420 ± 280 B.P. (A-2015). Picacho Peak (32°38' N, 111°24' W; 655 m elev.) middens yielded big sagebrush and Mojave Desert palo verde (*Canotia holacantha*). Singleleaf pinyon disappeared between 11 and 10.3 ka. The youngest date for juniper was 9420 ± 230 B.P. (A-2012).

Ajo Mountains. Additional Arizona Upland midden records are from Organ Pipe Cactus National Monument west of Tucson. Four middens (21.8, 20.5, 17.8, and 13.5 ka) from Montezuma's Head in the Ajo Mountains

(32°07' N, 112°42' W; 975 m elev.) were dominated by singleleaf pinyon, juniper, and Ajo oak (*Quercus ajoensis*). Associates in the woodland included Rocky Mountain juniper (*Juniperus scopulorum*), big sagebrush, shadscale (*Atriplex confertifolia*), Joshua tree, and snowberry (*Symphoricarpos* sp.). The nearest populations of Rocky Mountain juniper today are about 270 km to the northeast below the Mogollon Rim. Big sagebrush, a dominant in Great Basin desertscrub communities, occurs no closer than 390–440 km to the northnortheast above the Mogollon Rim. Joshua tree is a unique arborescent yucca that characterizes modern Mojave desertscrub communities (Fig. 8.4A). The nearest living Joshua trees are near Aguila, Arizona, 225 km to the north-northwest. The composition of the late Wisconsin middens, and presumably the local woodland, changed little for at least 8300 years.

Six middens from Alamo Canyon in the Ajo Mountains (32°05' N, 112°21' W; 915 m elev.) were examined. A sample dated at 14,500 ± 300 B.P. (A-2120) contained a fossil assemblage very similar to those from Montezuma's Head. Moreover, a sample dated at 32 ka extends the record of pinyon-juniper-oak woodland back into the middle Wisconsin. The fossils suggest that a pinyon-juniper-oak woodland was present in the Ajo Mountains for at least 18,500 years. Singleleaf pinyon and Rocky Mountain juniper disappeared in Alamo Canyon before 10.6 ka. Ajo oak and juniper (California juniper, *Juniperus californica*, and the local redberry juniper, *J. erythrocarpa*) associated with the endemic Harrison barberry (*Berberis harrisoniana*) prevailed in Alamo Canyon middens until 8130 ± 370 B.P. (A-2209), although a reduction in juniper occurred earlier. TAMS radiocarbon dates of 8130 ± 430 B.P., 8590 ± 470 B.P., and 9230 ± 370 B.P. (AA-540, AA-539, AA-533) establish that saguaro, the stately arborescent cactus that characterizes the Arizona Upland, arrived in the Ajo Mountains from its Sonoran glacial refugium by the early Holocene (Van Devender et al., 1985).

Puerto Blanco Mountains. Twenty-one packrat midden assemblages yielded radiocarbon ages as old as 14.2 ka, providing a detailed vegetation chronology for the Puerto Blanco Mountains (31°58–59' N, 112°47–48'

Fig. 8.4. (A) Joshua tree, the arborescent yucca characteristic of the Mojave Desert, is widespread in the modern Sonoran Desert. (B) Unique Sonoran desertscrub community dominated by organ pipe cactus and foothills palo verde formed about 4 ka, Puerto Blanco Mountains, Organ Pipe Cactus National Monument, Arizona (Van Devender, 1987).

W; 535–605 m elev.; Van Devender, 1987; Fig. 8.4B). The study area is at the lower edge of the Arizona Upland just west of the Ajo Mountains. Estimated precipitation for the area is 205–215 mm/yr with 40% in the summer. Chronological summaries of the relative abundances of fifteen important trees and shrubs and fifteen succulents are presented in Figures 8.5 and 8.6, respectively.

At 14.1 ka California juniper, Joshua tree, and Whipple yucca (*Yucca whipplei*) were present on a south-facing rhyolitic slope. Many extralocal species departed before 10,540 ± 250 B.P. (A-4276). The early Holocene vegetation was apparently a transitional desertscrub/woodland dominated by brittle bush, catclaw acacia, and saguaro. California juniper was less abundant than before; it disappeared between 9.1 and 8.8 ka. Unlike the Ajo Mountains middens, oak was not found. By 10.5 ka saguaro arrived in the area. Catclaw

and velvet mesquite, now locally restricted to riparian habitats along washes, then lived on rocky slopes as they do today near their northern and upper elevational range limits. After 8.9 ka they were joined by blue palo verde.

Between 5.2 and 3.5 ka foothills palo verde, ironwood, Mexican jumping bean (*Sapium biloculare*), and organ pipe cactus (*Stenocereus thurberi*) became numerous in the middens (Fig. 8.4). With the confinement of catclaw acacia, mesquite, and blue palo verde to washes and the addition of the new arrivals, the desertscrub community approached its modern composition. Creosote bush, appearing in the record at 3.4 ka, was probably the Sonoran Desert tetraploid chromosomal race dispersing northward from Sonora (Van Devender, 1987).

Lower Colorado River Valley

The hottest and driest climate in North Amer-

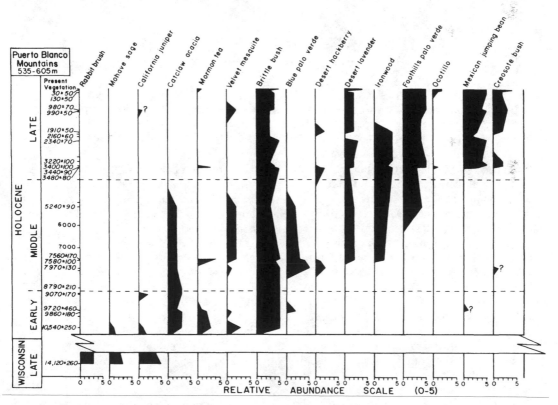

Fig. 8.5. Chronological summary of fifteen important macrofossils of selected trees and shrubs from Puerto Blanco Mountains packrat middens, Arizona (after Van Devender, 1987). Relative abundance in present vegetation is a composite of visual estimates at thirteen subsites. Relative abundance scale: 0 = absent, 1 = rare, 2 = uncommon, 3 = common, 4 = very common, 5 = abundant; ? = probable contaminant.

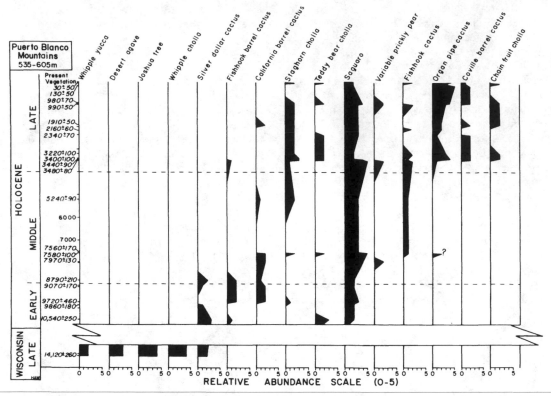

Fig. 8.6. Chronological summary of macrofossils of the succulents from Puerto Blanco Mountains, Arizona, packrat middens (after Van Devender, 1987). Relative abundance in present vegetation is a composite of visual estimates at thirteen subsites. Relative abundance scale: 0 = absent, 1 = rare, 2 = uncommon, 3 = common, 4 = very common, 5 = abundant; ? = probable contaminant.

ica is in the Sonoran Desert lowlands of Sonora, Baja California, Arizona, and California. Precipitation is erratic and in the lower elevations averages less than 100 mm/yr. To the north lies the Mojave Desert, with its colder winters, and to the southeast the Plains of Sonora, with greater summer rainfall (Turner and Brown, 1982). The vegetation is mostly an open desertscrub dominated by creosote bush, white bursage (*Ambrosia dumosa*), brittle bush, and big galleta grass (*Hilaria rigida*).

Hornaday Mountains. Ten midden assemblages from the granitic Hornaday Mountains of northwestern Sonora (31°59′ N, 113°36′ W; 240 m elev.) yielded radiocarbon ages of 10.0–1.7 ka. Only 40 km north of the Gulf of California, the Hornaday Mountains are between the extensive volcanic craters of the Sierra Pinacate and the sands of the Gran Desierto. Preliminary analyses show

that early Holocene fossil assemblages were dominated by brittle bush and Mormon tea (*Ephedra nevadensis*) without the woodland plants or rosette succulents found in other areas of the Sonoran Desert.

Tinajas Altas Mountains. Twenty-one packrat midden assemblages from the granitic Tinajas Altas Mountains, Arizona (32°16–20′ N, 114°02–07′ W; 330–580 m elev.; Fig. 8.7), provide a detailed fossil record from > 43.2 to 1.2 ka. A chronological summary of the relative abundances of sixteen species of trees, shrubs, and succulents is presented in Figure 8.8. The Tinajas Altas Mountains and a small outlier, the Butler Mountains, are the first ranges east of the Colorado River beyond the sandy Yuma Mesa. Estimated precipitation is 125–155 mm/yr with 40% in the summer.

A middle Wisconsin (> 43.2 ka) sample yielded abundant fossils of California juniper and Joshua tree with low levels of singleleaf

8.7A

8.7B

Fig. 8.7. (A) The granitic Tinajas Altas Mountains, Arizona. Singleleaf pinyon grew as low as 430 m in a California juniper–Joshua tree woodland in the late Wisconsin. (B) West base of Tinajas Altas Mountains, at 330 m elevation. Vegetation is a lower Colorado River valley Sonoran desertscrub dominated by creosote bush, white bursage, and ocotillo. Creosote bush was present in an 18.7-ka full-glacial California juniper–Joshua tree woodland.

pinyon. Late Wisconsin samples (15.7–11.0 ka) contained California juniper, singleleaf pinyon, Mojave sage (*Salvia mohavensis*), and Bigelow beargrass (*Nolina bigelovii*) in a more mesic woodland. The 11,040 ± 270 B.P. (A-3393) sample at 460 m is the lowest elevational record for the singleleaf pinyon in the Wisconsin (Van Devender et al., 1985; Fig. 8.7A). A TAMS date of 18,700 ± 1050 B.P. (AA-536) from a California juniper–Joshua tree sample at 330 m elevation is the first Wisconsin full-glacial record for creosote bush, a widespread desertscrub dominant (Fig. 8.7B). The associational context suggests that the Mojave Desert hexaploid chromosomal race of creosote bush was present rather than the tetraploids presently living in the area. Singleleaf pinyon rapidly disappeared from the Tinajas Altas Mountains between 11.04 and 10.95 ka.

California juniper, brittle bush, and catclaw

acacia were abundant in early Holocene middens of the Tinajas Altas Mountains. Hollyleaf bursage (*Ambrosia ilicifolia*), a regional endemic, was an associate. A TAMS date on white bursage establishes its presence by 10,600 ± 420 B.P. (AA-781). California juniper disappeared from the Tinajas Altas Mountains between 9 and 8.9 ka.

In the middle Holocene nearly twice as many species were growing near the rock shelters as occur there today. Now locally found in washes, catclaw and blue palo verde were growing on slopes. A TAMS date on elephant tree (not plotted in Fig. 8.8) revealed its arrival in the area by 5820 ± 310 B.P. (AA-777; Van Devender et al., 1985). Between 6.2 and 4 ka the fossil record was dominated by desert lavender, a shrubby mint that extends into subtropical thornscrub and tropical deciduous forest to the south. After 4010 ± 70 B.P. (A-3535) Bigelow beargrass declined, blue palo

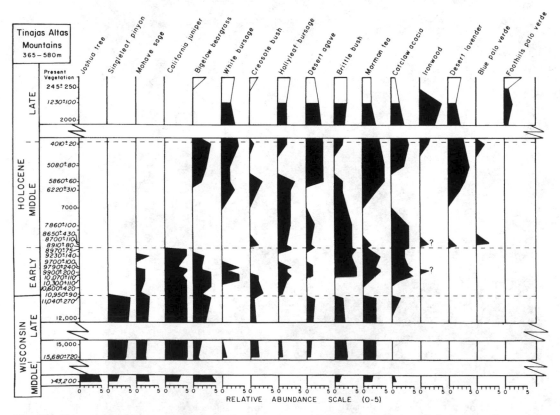

Fig. 8.8. Chronological summary of sixteen important plant macrofossils from Tinajas Altas Mountains packrat middens, Arizona. Relative abundance in present vegetation is a composite of visual estimates at ten subsites. Relative abundance scale: 0 = absent, 1 = rare, 2 = uncommon, 3 = common, 4 = very common, 5 = abundant; ? = probable contaminant.

verde disappeared locally, and foothills palo verde arrived. A 1230 ± 100 B.P. (A-3738) sample reflects relatively modern vegetation.

Butler Mountains. The low, granitic Butler Mountains (32°21' N, 114°12' W; 240–255 m elev.) on the eastern edge of the Yuma Mesa receive an estimated precipitation of 100–105 mm/yr with 35% in the summer. The oldest of nine packrat middens (11.3 and 11.1 ka) were dominated by California juniper, creosote bush, brittle bush, and Mormon tea in association with Mojave sage and white and hollyleaf bursages. These are the oldest records for the bursages and the lowest late Wisconsin or early Holocene elevation records for any juniper (Van Devender et al., 1985). TAMS dates of 11,250 ± 410 B.P., 10,360 ± 430 B.P., and 10,170 ± 610 B.P. (AA-769, AA-537, AA-768) firmly establish the presence of creosote bush in the area in the late Wisconsin and early Holocene. Associated seeds in samples dated at 11.3 and 11.1 ka represent the oldest records for saguaro (if they are not contaminants).

No major changes occurred in the Butler Mountains fossil assemblages between 11.3 and 10.4 ka. California juniper disappeared by 8.6 ka. Middle Holocene middens were dominated by creosote bush, white bursage, and brittle bush, and pygmy cedar (*Peucephyllum schottii*) became important while Mormon tea declined. The presence of catclaw acacia and cheese weed (*Hymenoclea salsola*) growing on rocky slopes indicates wetter conditions; these shrubs are presently confined to washes.

By 3820 ± 70 B.P. (A-3517) the vegetation was relatively modern with ironwood dominant. Catclaw acacia, blue palo verde, and velvet mesquite were again present in a 740 ± 50 B.P. (A-3911) sample. In all cases the modern flora at the midden sites is much impoverished compared to the fossil assemblages.

Picacho Peak area. The oldest of twenty-four packrat midden samples from the rhyolitic Picacho Peak area in southeastern California (32°56–58' N, 114°38–39' W; 240–300 m elev.; Fig. 8.9) was dated at 13.4 ka (Cole, 1986). Picacho Peak is just above the Colorado River, 5 km north of Yuma, Arizona, in one of the hottest and driest areas in North America. A chronological summary of the relative abundances of seventeen species

of trees, shrubs, and succulents is presented in Figure 8.10. Late Wisconsin (11.2, 12.4, and 12.5 ka) samples were dominated by creosote bush in association with Joshua tree, Whipple yucca, Mojave sage, and black bush (*Coleogyne ramosissima*). The assemblage is no different from what one might expect to find in a modern Mojave Desert community. A TAMS date of 12,730 ± 410 B.P. (AA-579B) documents the presence of creosote bush in the late Wisconsin. Joshua tree at 275 m and Whipple yucca and black brush at 285 m elevation are lower than any other records of the species. The nearest living Joshua tree population is 150 km to the northwest in the Cottonwood Mountains of southern Joshua Tree National Monument, California (Rowlands, 1978). The lower elevations in the lower Colorado River valley are one of the few areas in the warm deserts of North America not known to have been invaded by woodland at 11 to 13 ka.

Most extralocal species disappeared after 12.4 ka, possibly close to 11.2 ka. Creosote bush, brittle bush, and pygmy cedar dominated a 10.5-ka midden. These shrubs now form a widespread community on low-elevation rocky slopes from the Gulf of California up the Colorado River into Nevada and in Death Valley, California. Associates in the fossil community such as catclaw acacia, Mojave sage, peachthorn wolfberry (*Lycium cooperi*), Mormon tea, and rabbit brush (*Chrysothamnus teretifolius*) indicate that the early Holocene was transitional between glacial and interglacial communities. Mojave sage and peachthorn wolfberry lingered in the area until 10.1 ka. The youngest records of catclaw acacia and Mormon tea are 8.4 ka and 7.4 ka, respectively.

The middle Holocene vegetation appears to have been a typical creosote bush–brittle bush–pygmy cedar desertscrub until at least 4.8 ka. Sonoran Desert plants such as ironwood, ocotillo, and white bursage arrived in the area subsequently. The late arrival of white bursage is surprising considering its older fossil records in southwestern Arizona and California, notably by 10.2 ka in Death Valley (Woodcock, 1986).

Whipple Mountains. The oldest of thirty packrat midden samples from the rhyolitic Whipple Mountains, southeastern California

Fig. 8.9. The Picacho Peak area, California, just north of Yuma is the hottest, driest desert in North America. Vegetation is a sparse lower Colorado River valley desertscrub dominated by creosote bush. Creosote bush desertscrub remained in the area during the late Wisconsin and possibly much of the Quaternary (Cole, 1986). Photo by S. P. McLaughlin.

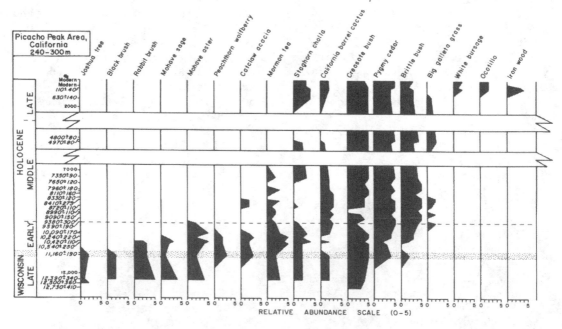

Fig. 8.10. Chronological summary of seventeen important plant macrofossils from Picacho Peak area packrat middens, California (data from Cole, 1986).

Relative abundance scale: 0 = absent, 1 = rare, 2 = uncommon, 3 = common, 4 = very common, 5 = abundant.

(34°13–16′ N, 114°22–25′ W; 320–525 m elev.), was dated at 13.8 ka. The Whipple Mountains are on the northwestern edge of the Sonoran Desert just across the Colorado River from Parker, Arizona. Subtropical Sonoran Desert plants, including ironwood, blue palo verde, and a few saguaro, are found below about 370 m, giving way to a generic lower Colorado River valley desertscrub at higher elevations. Estimated annual precipitation is 140–180 mm with 26% in the summer.

A chronological summary of fifteen trees, shrubs, and succulents is presented in Figure 8.11. Prior to 10.8 ka important midden fossils included California juniper, Joshua tree, singleleaf pinyon, Whipple yucca, Bigelow beargrass, and Mojave sage. Traces of big sagebrush, shadscale, and black brush were found in a few samples. Singleleaf pinyon was uncommon and not found below about 510 m. A TAMS date of 11,360 ± 500 B.P. (AA-380) on

singleleaf pinyon from an 8.9-ka sample establishes it as a contaminant rather than part of the early Holocene woodland (Van Devender et al., 1985). The 11.4-ka date is the youngest record for singleleaf pinyon in the Whipple Mountains. The lowest elevation recorded for California juniper in the Whipple Mountains was 320 m. Wells and Hunziker (1976) reported Utah juniper at 16,900 ± 190 B.P. from 258 m elevation in the Chemehuevi Mountains (ca. 34°42′ N, 114°30′ W), 35 km north of the Whipple Mountains. A TAMS date of 11,015 ± 110 B.P. (AA-1576) on creosote bush twigs records its presence at 320 m in the latest Wisconsin.

In the early Holocene abundant fossils of California juniper, Mojave sage, Mormon tea, and Bigelow beargrass suggest a woodland. However, creosote bush, brittle bush, and pygmy cedar, the modern desertscrub dominants, were present by 10.8 ka. White bursage was present by 10 ka, although it did not be-

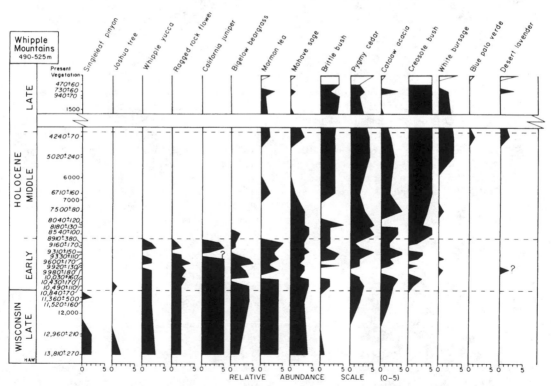

Fig. 8.11. Chronological summary of fifteen important plant macrofossils from Whipple Mountains, California, packrat middens. Relative abundance in present vegetation is a composite of visual estimates at thirteen subsites. Relative abundance scale: 0 = absent, 1 = rare, 2 = uncommon, 3 = common, 4 = very common, 5 = abundant; ? = questionable record.

come common until 5020 ± 240 B.P. (A-2160) in the middle Holocene. California juniper disappeared between 8.9 and 8.5 ka.

The middle Holocene fossils were mainly species one would expect to find in lower Colorado River valley desertscrub. Blue palo verde and desert lavender records at 4240 ± 70 B.P. (A-3649) mark the appearance of more subtropical Sonoran Desert plants. Blue palo verde now occurs in washes within a kilometer of the midden site but at lower elevations. Relatively few differences were seen in the fossils of the middle and late Holocene middens. Fewer plants grow at the midden sites today than in any of the fossil assemblages. With the exception of the blue palo verde record mentioned above, Sonoran Desert plants apparently have not expanded farther into the Whipple Mountains. The present Mojave-Sonoran Desert boundary was probably established soon after 8.9 ka, with final adjustments after 5 ka.

Other Sonoran Desert areas. Middens from the granitic Wellton Hills in the lower Colorado River valley lowlands south of the Gila River, southwestern Arizona (32°36′ N, 114°08′ W; 160–180 m elev.; Van Devender, 1973), included a 10,750 ± 400 B.P. (A-1406) record of creosote bush with Mormon tea and white bursage. The midden yielded no woodland plants or rosette succulents. The site is 80 m lower than the juniper midden sites in the Butler Mountains and the Joshua tree–Whipple yucca sites in the Picacho Peak area; the Wellton Hills probably supported a creosote bush desertscrub in the Wisconsin as well. By 8750 ± 320 B.P. (A-1399) the fossils suggest a shift to dominance by creosote bush, Mormon tea, white bursage, pygmy cedar, ironwood, and catclaw acacia.

Samples dated at 13.4–14.4 ka from Burro Canyon, Kofa Mountains, Arizona (33°24′ N, 114°01′ W; Van Devender, 1973), were dominated by singleleaf pinyon, juniper, and shrub live oak at a higher elevation (860 m) than most of the lower Colorado River valley sites. Associates included big sagebrush, shadscale, and Joshua tree. An 11.5-ka sample from Brass Cap Point at the base of the Kofa Mountains (33°26′ N, 114°06′ W; 550 m elev.; Van Devender, 1973) yielded abundant remains of California juniper and Joshua tree without shrub live oak. A TAMS date of 9750 ± 470

B.P. (AA-574) on a seed of Sonoran crucifixion thorn (*Castela emoryi*) was from a sample containing California juniper, Bigelow beargrass, and Joshua tree (Van Devender et al., 1985). This stout spiny desert tree is not known to grow with these plants presently.

Three middens from the rhyolitic New Water Mountains, Arizona (33°36′ N, 113°55′ W; 605–615 m elev.) also were rich in juniper and shrub live oak from 12.1–10.9 ka (King and Van Devender, 1977). Only the 10.9-ka sample contained singleleaf pinyon, suggesting that its lower elevational limit in the area was close to 615 m. A radiocarbon age of 7.9 ka indicated a late survival of juniper in the New Water Mountains (Van Devender, 1977; Van Devender and Spaulding, 1979).

Three samples from the rhyolitic Artillery Mountains of western Arizona (34°22′ N, 113°38′ W) are of interest. The site is on the northeastern Sonoran–Mojave Desert boundary where Joshua tree populations interfinger with foothills palo verde–saguaro communities. A singleleaf pinyon-juniper-oak assemblage at 725 m elevation dated at older than 30,000 B.P. (A-1100) was very similar to one dated at 18,320 ± 400 B.P. (A-1101) in the Wisconsin full glacial (Van Devender and King, 1971). Dunn oak (*Quercus dunni*) was present in the samples with shrub live oak. Associates included big sagebrush, shadscale, Joshua tree, and catclaw acacia. A 10,250 ± 200 B.P. (A-1099) sample at 615 m elevation yielded a juniper–shrub live oak assemblage with Mormon tea, catclaw acacia, and black brush but no singleleaf pinyon. In the Artillery Mountains the Utah juniper found in Wisconsin woodlands was apparently replaced in the early Holocene by oneseed (*Juniperus monosperma*) or California juniper; the latter is relictual in the area today.

DISCUSSION
Paleovegetation

In the Sonoran Desert the vegetation sequence from the late Wisconsin through the Holocene culminating in relatively modern plant communities was unidirectional. Overall, the sequential changes were similar to moving down in elevation or latitude. In the Arizona Upland (about 550 to 1550 m) there was a four-part sequence: a late Wisconsin woodland dominated by singleleaf pinyon and juniper with shrub live oak and/or Joshua tree; an early Holocene woodland with juniper and, in some cases shrub live oak, with desert riparian trees and Sonoran desertscrub plants; a middle Holocene Sonoran desertscrub with brittle bush, saguaro, and riparian trees; and a late Holocene Sonoran desertscrub with more subtropical desert plants. Below 600 m California juniper woodlands or Mojave desertscrub changed little from the late Wisconsin into the early Holocene. The same can be said for lower Colorado River valley desertscrub communities of the middle and late Holocene. For example, the vegetation sequences at 240–350 m in the Tinajas Altas and Butler Mountains display only two steps, although lesser changes are discernible. Desertscrub communities in the harshest environments may have changed minimally. Potential examples include the creosote bush–white bursage communities of the Gran Desierto and the halophyte communities surrounding the head of the Gulf of California.

During the late Wisconsin a pinyon-juniper-oak woodland with singleleaf pinyon grew in southern Arizona from 1555 down to about 550 m elevation. The lowest fossil record for singleleaf pinyon was 460 m in the Tinajas Altas Mountains, Arizona (Fig. 8.7). Few Sonoran Desert plants could be found in the late Wisconsin middens, and Mojave Desert species, especially Joshua tree, were present instead. Between 600 and 300 m a xeric juniper woodland with California juniper, shrub live oak, Joshua tree, Whipple yucca, and Bigclow beargrass was found. At 330 m elevation in the Tinajas Altas Mountains creosote bush was present in an 18.7-ka full-glacial assemblage of California juniper and Joshua tree (Fig. 8.7). The lowest record for California juniper was 240 m, in a creosote bush–desertscrub assemblage in the Butler Mountains.

From 13.4 to about 11 ka creosote bush, black brush, Joshua tree, and Whipple yucca were present in the Picacho Peak area, California, near Yuma, Arizona (Cole, 1986). Thus desertscrub of Mojavean affinity without woodland plants on rocky slopes in the modern Sonoran Desert was restricted to below 300 m elevation in the xeric trough of the lower Colorado River valley from perhaps 34° N south into the lowlands surrounding the head of the Gulf of California, including

the land exposed by sea level lowering of about 100 m. This area probably has been a core North American desert for much of the Quaternary, although species typical of the Mojave Desert prevailed most of the time with Sonoran Desert plants moving north along the Colorado River and crossing it only during the height of interglacials. The woodland-desertscrub ecotone at 240–300 m at 32°–34° N is much lower than the woodland-semidesert (desertscrub with Joshua tree, Whipple yucca, and shadscale) ecotone at 450 m in Death Valley in the Mojave Desert (Woodcock, 1986).

Between 600 and 1555 m elevation singleleaf pinyon and other mesic woodland plants disappeared at 11 ka. A xeric juniper–shrub live oak woodland or chaparral with California juniper woodland at lower elevations continued into the Holocene. In the early Holocene juniper was present over an elevational range of about 1315 m in southern Arizona. Juniper also grew in a lowland site in central Baja California, Mexico, that now supports the unusual boojum tree (Wells, 1976). The woodlands included Utah, oneseed, redberry, and even Rocky Mountain junipers in the higher areas, and California, and possibly Utah, junipers at lower elevations. The vegetation along the elevational gradient ranged from juniper–shrub live oak woodland/chaparral to creosote bush–desertscrub with California juniper. The disappearance of juniper from the Sonoran Desert at about 8.9 ka (Puerto Blanco, Tinajas Altas, and Whipple mountains) marks the end of the early Holocene as here defined. Survival of juniper until 8.4 ka in the Castle Mountains and 7.8 ka in the New Water Mountains (Van Devender, 1977) may represent persistence of a local population of little regional significance.

In lower areas such as the Wellton Hills, Arizona (160–180 m), the Picacho Peak area, California (240–300 m), and the Hornaday Mountains, Sonora (240 m), desertscrub rather than juniper woodland was already established in the early Holocene. In the Puerto Blanco Mountains, Arizona (535–605 m), California juniper woodland was uncommon in early Holocene midden assemblages while important Sonoran desertscrub dominants, including saguaro and brittle bush, had returned from their glacial refugia to the south

(Van Devender, 1987). In the early Holocene Mojave Desert plants declined or disappeared from what is now the eastern Sonoran Desert while persisting in sites closer to the Colorado River.

After about 8.9 ka desertscrub vegetation was established in southern Arizona. However, the communities in the Waterman, Puerto Blanco, and Tinajas Altas mountains were unlike those found there today. Catclaw acacia, velvet mesquite, and blue palo verde, now restricted to riparian wash habitats, occupied hot, dry, south-facing slopes. To the west the middle Holocene plant communities of the Picacho Peak area and the Whipple Mountains, California, were essentially modern. The transition area between the Sonoran and Mojave deserts has essentially been in the same position for the last 8900 years. After about 4 ka relatively modern vegetation was established throughout the Sonoran Desert as subtropical vegetation reached its northernmost extent since the last interglacial.

Paleoclimates

The fossils in packrat middens provide our best proxy records for deducing the climatic history of the Sonoran Desert and for evaluating various paleoclimatic hypotheses. While their ecophysiology is not known in detail, most plant distributions have their topographic and climatic limits, and physiological limits can be inferred (Shreve, 1915). Thus the modern ranges appear to be controlled by the modern climate, and the fossil ranges reflect past climates. Ephemeral species that are restricted to summer or winter-spring growing seasons are taken as indicators of rainfall seasonality. The past presence of perennial species now confined to riparian washes on exposed, rocky slopes is an indication of greater precipitation. Differences in species richness at a site can reflect climate if sample sizes are not too disparate (Spaulding et al., *this volume*, chap. 5). Inferences about paleoclimates are mostly based on individual species. Paleoclimatic reconstructions are a compilation of these inferences rather than descriptions of the climate of the best analog site, which is unlikely to reflect the true paleoclimate (Bryson, 1985).

Middle Wisconsin. There are relatively few Sonoran Desert middens from the middle

Wisconsin. Middle and late Wisconsin plant assemblages, and presumably climates, were apparently very similar in the Ajo (32, 21.8–13.5 ka) and Artillery (> 30, 18.3 ka) mountains. In the Tinajas Altas Mountains a > 43.2-ka sample contained abundant California juniper and Joshua tree with occasional singleleaf pinyon, while late Wisconsin (15.7, 11 ka) middens lacked Joshua tree and had more singleleaf pinyon. The middle Wisconsin may have been somewhat drier than the late Wisconsin.

Late Wisconsin. Fossils of woodland and desertscrub plants in anomalous or ecotonal associations (Table 8.2) and communities without modern analogs allow various inferences about paleoclimates. In general, there was strong summer cooling and a shift to winter precipitation. Cooling in winter may have been offset by reduced frequencies of catastrophic freezes. Inferences about minimum winter temperatures are difficult to reach because southerly displacement of cold-limited plants can as readily be interpreted as due to cooler summers and concomitant shifts in rainfall seasonality.

Midden fossils of singleleaf pinyon, big sagebrush, and California barrel cactus in the late Wisconsin (13.2 ka) of the Picacho Peak area (655 m) of Arizona reflect a mixed paleoclimate. Means for Eloy (475 m) near Picacho Peak are 10.8°C for January, 32.3°C for July, and 215 mm/yr precipitation, with 34% in summer (Sellers and Hill, 1974). Today, California barrel cactus does not occur higher than 1525 m in central Arizona (Benson, 1982), suggesting that January means would have been warmer than about 4.5°C. The nearest big sagebrush (*Artemisia bigelovii*) is found above 1425 m on the Mogollon Rim and adjacent Colorado Plateau. January means from eleven stations where big sagebrush occurs from the Show Low Snowflake area north of Fredonia are below 1.7°C. The lowest areas with singleleaf pinyons in the Globe-Miami area would have means of about 5°C for January and 27.2°C for July, and about 400 mm/yr precipitation, with 33% in summer. July means for big sagebrush stations are 27°C or less, and 21°–24°C in wooded areas. The late Wisconsin climate of Picacho Peak was probably a combination of these modern climates with means of 5.6°–7.2°C for January (a cooling of 3.6°–5.0°C), about 23.9°C for July (a cooling of 8.4°C), and at least 325 mm/yr precipitation, with 80% in the cool season (October to May; an increase of 50% annually and 110% for the cool season). The late Wisconsin records of singleleaf pinyon at 460–550 m in the Tinajas Altas Mountains indicate an even greater summer cooling of about 19°C.

The expansion of Whipple yucca, a chaparral succulent with limited tolerance for freezing temperatures (Woodcock, 1986), eastward into the present Sonoran Desert (Whipple, Tinajas Altas, and Puerto Blanco mountains) is an indication that late Wisconsin winters were not greatly colder than today. The most likely explanation of mild winter temperatures in the late Wisconsin is the blocking of arctic air masses by the continental glaciers (Bryson and Wendland, 1967). Equable climates would release the lower and southern range limits of woodland and forest plants and animals but not necessarily extirpate the desertscrub residents, resulting in occasional ecological mixtures that would not be found or would be very unusual today.

The arguments by Kutzbach (1983) and Spaulding and Graumlich (1986) for a gradual increase in summer insolation from the full glacial, culminating in warmer-than-modern summer temperatures by 12 ka, are not supported because there was little difference between full-glacial and late-glacial vegetation on Pontatoc Ridge or in the Ajo Mountains. Apparently the effects of the continental glaciers on regional circulation regimes, especially the patterns of precipitation, were much more important than changes in insolation in the Sonoran Desert.

Virtually all the ephemeral herbs and grasses in late Wisconsin middens were winter-spring obligates. Arizona cypress forest similar to the inferred paleocommunity on Pontatoc Ridge (1555 m elev.) occurs in cool, shady canyons in the Santa Catalina Mountains that are much wetter than the exposed Pontatoc Ridge site. Although big sagebrush could live at 655 m elevation on Picacho Peak with today's annual rainfall of 215 mm (127 mm cool season) with cooler summers, singleleaf pinyon could not. In central Arizona singleleaf pinyon occurs in areas with at least 410 mm/yr precipitation (250 mm from October to May).

In the mountains of the eastern Mojave Desert the lower edge of the pinyon-juniper woodland and the upper elevational limit of Joshua tree at about 1600 m receive at least 250 mm/yr with about 190 mm in the cool season (Rowlands, 1978; Thorne et al., 1981). The rainfall in the late Wisconsin at Picacho Peak was perhaps 300 mm/yr (an increase of 40%) with 250 mm in the cool season (an increase of 97%).

The modern climates of the nearest populations of Rocky Mountain juniper below the Mogollon Rim suggest that late Wisconsin rainfall near the midden sites (915–975 m elev.) in the Ajo Mountains was 55%–70% greater than today. Mixed communities of California juniper and Joshua tree similar to the 14.1-ka midden assemblage from the nearby Puerto Blanco Mountains (565 m elev.) presently occur in areas with only 16% greater annual rainfall (Van Devender, 1987). The middle and late Wisconsin records of singleleaf pinyon associated with Joshua tree from 460–550 m elevation in the Tinajas Altas Mountains reflect 40%–60% increases in annual precipitation, with over 100% increase for the cool season. California juniper and Joshua tree presently grow together at 975–1200 m elevation with precipitation of about 140–170 mm/yr (20% in the summer) in the mountains of the eastern Mojave Desert. The association of these plants in 15.1- and 17.4-ka middens from 330 m in the Tinajas Altas Mountains indicates an increase in annual precipitation of 10% to 35%, or 50% to 80% for the cool season. These estimates are minimums based on the modern lower limits of these species or pairs of species in the winter rainfall California portion of their ranges.

Joshua tree and Whipple yucca were found in 12.5-ka samples in a Picacho Peak, California, midden. Joshua tree can maintain itself with as little as 110 mm/yr precipitation (Rowlands, 1978). However, the nearest populations probably receive more than 125 mm/yr with only 19% in summer. To support Joshua trees, precipitation in the Picacho Peak area would probably need to have been 50% greater than today. These results compare favorably with those of Woodcock (1986), who inferred three- to fourfold increases for precipitation for the lower eleva-

Table 8.2. *Plant species found together in Sonoran Desert packrat middens that apparently do not occur together today.*
Information on middens is in Table 8.1. Middens are ordered oldest to youngest. MW = middle Wisconsin, LW = late Wisconsin, EH = early Holocene, MH = middle Holocene.

Plant	Associates	Middens
Trees		
Pinus monophylla	Quercus ajoensis	MW: AC2C
		LW: all AC & MH
	Ambrosia ilicifolia	LW: TA15AL
	Echinocactus horizonthalonius	
	var. nicholii	LW: WAM9A1, WAM9A2
	Tiquila canescens	LW: WAM9A1, WAM1D
Cupressus arizonica	Artemisia tridentata-type	LW: PR6
Pinus discolor	Artemisia tridentata-type	LW: PR6
	Opuntia whipplei	LW: all PR
Juniperus scopulorum	Quercus ajoensis	MW: AC2C
		LW: all AC & MH
	Agave deserti	MW: AC2C
		LW: all AC & MH
Juniperus californica	Quercus ajoensis	EH: AC1A1, AC1U
	Koeberlinia spinosa	LW: BCP1
		EH: BM5
	Ambrosia ilicifolia	LW: TA15AL, TA18A, BM3A
		EH: TA2B, BM3B, TA3B, TA12A, TA14, TA1B, TA3, TA2A
Quercus ajoensis	Artemisia tridentata-type	MW: AC2C
		LW: all AC & MH
	Atriplex confertifolia	LW: MH1B
	Symphoricarpos sp.	LW: MH1D
	Yucca brevifolia	LW: all MH
	Opuntia whipplei	MW: AC2C
		LW: all AC & MH
		EH: AC1A1, AC1U
	Phoradendron coryae	MW: AC2C
		LW: MH1B
Quercus emoryi	Encelia farinosa	LW: WP5
Castela emoryi	Juniperus californica	EH: BCP2
	Yucca brevifolia	EH: BCP2
	Nolina bigelovii	EH: BCP2
Shrubs		
Artemisia tridentata type	Echinocactus horizonthalonius	
	var. nicholii	LW: WAM9A1
	Heteropogon contortus	LW: WAM1D
	Tiquila canescens	LW: WAM9A1, WAM1D
Tetradymia argyraea	Salvia pinguifolia	LW: WAM1B
	Vauquelinia californica	LW: WAM1B
	Heteropogon contortus	LW: WAM1D
Salvia mohavensis	Ambrosia ilicifolia	LW: TA15AL, TA18A
		EH: TA2B, TA3B, BM3B, TA12A, TA14, TA1B
	Yucca whipplei	MW: TA15B1
		LW: TA18B, TA18A, TP1, RP11, PP17D2, RP8
		EH: PP17D1, RP10A, RP5A, RP10B, RP3, RP6, RP1A1, RP1A2

Vauquelinia californica	*Artemisia tridentata*-type	LW:	WAM9A1, WAM2, WAM1B, WAM1D
	Echinocactus horizonthalonius var. *nicholii*	LW:	WAM9A1, WAM9A2
Ambrosia ilicifolia	*Chrysothamnus teretifolius*	LW:	TA15AL, TA18A
	Rhus trilobata	EH:	TA1B
	Monardella sp.	LW:	TA15AL, TA18A
		EH:	TA2B, TA3B, TA12A, TA2A
	Yucca brevifolia	LW:	TA18A
	Bouteloua repens	MH:	BM1
	Vulpia microstachys	EH:	TA2B
	Phoradendron juniperinum	EH:	TA12A, TA1B
	Calycoseris parryi	EH:	TA14
Berberis harrisoniana	*Juniperus californica*	EH:	AC1U
	Opuntia whipplei	EH:	AC1U
	Phoradendron juniperinum	EH:	AC1U
Croton sonorae	*Juniperus californica*	EH:	AC1A1
	Opuntia whipplei	EH:	AC1A1

Succulents

Yucca brevifolia	*Echinocactus horizonthalonius* var. *nicholii*	LW:	WAM9A1

tions in Death Valley, California, based on packrat midden fossils.

Evidence for modest summer rainfall was available from C4 perennial grasses from Pontatoc Ridge and the Ajo Mountains (Soreng and Van Devender, 1989), ponderosa pine and border pinyon from Pontatoc Ridge, and Nichol's Turk's head cactus, oreja de perro (*Tiquilia canescens*), and a few summer ephemerals (*Boerhaavia wrightii, Kallstroemia* sp.) in the Waterman Mountains. Summer rainfall was completely absent from the western Sonoran Desert. The biseasonal rainfall climates of the eastern Sonoran Desert were greatly reduced in area and displaced south into Sonora, Mexico. The arguments of Wells (1979) for increased summer rainfall in the late Wisconsin based on the expansion of oaks, especially shrub live oak, are not supported for several reasons. Shrub live oak is a dominant in winter-rainfall chaparral in Arizona and California rather than an indicator of summer rainfall. Its late Wisconsin range in the present Sonoran Desert was above 600 m elevation, although California juniper grew below 300 m in the lower Colorado River valley.

Today, ecotonal areas at middle elevations in mountains have relatively equable climates because summer temperatures decrease and rainfall increases with elevation, while drainage of cold, dense air on steep slopes ameliorates winter temperatures. At Picacho Peak, Arizona, the difference between the July and January temperature means, an indicator of climatic equability, increased from about 16°–18°C in the late Wisconsin to 21.5°C today. The late Wisconsin climates of the Sonoran Desert were similar to these ecotonal situations in some respects, although the thermal or moisture regimes of many of the anomalous species pairs in Table 8.2 do not overlap today. The paleoclimates differed from any modern analog and combined features of the climates of many surrounding regions.

Early Holocene. The climate of the early Holocene was transitional between that of the late Wisconsin and more modern regimes. The persistence of woodland plants at low elevations suggests that summer temperatures were still cooler than today. The presence of Sonoran crucifixion thorn, a hot desert tree, in a California juniper–Bigelow beargrass woodland at Brass Cap Point (550 m elev.) at 9.8 ka indicates an equable environment. By 9.9 ka catclaw acacia, velvet mesquite, and summer ephemerals were common in juniper woodland in the Waterman Mountains (795 m), suggesting greater precipitation than today with reliable summer rainfall.

In the Puerto Blanco Mountains (560–605 m elev.) the presence of catclaw acacia and velvet mesquite on south-facing slopes suggests that precipitation was 30% greater in the early Holocene (Van Devender, 1987). A total of 71.4% of the ephemeral herbs (N = 11) and grasses (N = 3) identified in the samples were winter-spring obligates, indicating continued dominance by winter rainfall.

California juniper in the early Holocene of the Tinajas Altas Mountains at 550 m elevation does not necessarily indicate an increase in total annual precipitation but does suggest that winter rainfall was 20% higher than today. California juniper in the late Wisconsin and early Holocene of the Butler Mountains (240–255 m) suggests precipitation increases of 40% annually or 70% for the cool season.

Most of the ephemerals (N = 21) from the last thirteen thousand years in the Picacho Peak area (240–300 m elev.) middens are winter-spring plants (Cole, 1986). Only a few plants that typically grow in the summer rainy season elsewhere (*Boerhaavia* sp., *Aristida* cf. *adscensionis*, *Bouteloua barbata*) were found in eight samples from 10,240 ± 220 B.P. (A-3406) to the present. However, these plants also respond to winter-spring rains if temperatures are warm (Van Devender et al., *this volume*, chap. 14). Indeed, with pan evaporation greater than 3650 mm/yr, relative humidity of 10% or less, average monthly temperatures of 24°–36°C, and maximum temperatures of 46°C or higher from May to September, a 50-mm rainfall, equal to the entire annual average for the Picacho Peak area, would evaporate in a few days. Summer monsoonal rainfall has probably never been effective for inducing plant growth in this core desert area.

In the Whipple Mountains (490–525 m elev.), which receive about 27% summer rainfall today, a rich ephemeral flora in thirteen early Holocene (8.9–10.9 ka) middens did not contain any summer-obligate species. While a modest increase in summer temperatures can be inferred for the early Holocene from the midden fossils, reliable summer rainfall was present only in the eastern Sonoran Desert. The Bermuda High was apparently not well developed enough for summer monsoons to reach the Whipple Mountains. The maximum development of the monsoon and its penetration into the Mojave Desert in the early Holocene proposed by Spaulding and Graumlich (1986) and Spaulding (*this volume*, chap. 9)

are not supported. Their suggestion that California juniper was maintained in the desert lowlands in the early Holocene is unlikely in view of the above records, considering its modern distribution in winter rainfall areas and its early Holocene distribution in the lower Colorado River valley. Its relative abundance in middens increased from, at most, low levels in Puerto Blanco Mountains (565–605 m) samples (Fig. 8.5; Van Devender, 1987) to abundant in the Tinajas Altas (365–580 m; Fig. 8.8), Butler (240–250 m), and Whipple (365–510 m; Fig. 8.11) mountains in an east-west gradient toward the source of winter precipitation.

Middle Holocene. With the final demise of the late Wisconsin winter rainfall regime after 9 ka, a relatively modern climatic regime was established with hot summers. In most areas middle Holocene midden assemblages reflect desertscrub communities with different compositions and higher species richness than today. The lack of organ pipe cactus, Mexican jumping bean, limber bush, pink felt plant (*Horsfordia newberryi*), Coville barrel cactus (*Ferocactus covillei*), and chain fruit cholla (*Opuntia fulgida*) in the Puerto Blanco Mountains fossils suggests a greater frequency of catastrophic severe winter freezes due to southerly incursions of arctic air, preventing more subtropical Sonoran desertscrub plants from reaching the area (Van Devender, 1987).

Rainfall continued to be greater than today throughout the Sonoran Desert. In the Waterman and Puerto Blanco mountains catclaw acacia, velvet mesquite, and blue palo verde grew on hot, dry, south-facing slopes but now grow locally in washes (Fig. 8.5). Similar desertscrub communities currently occur at the upper elevational and northern latitudinal limits of the Sonoran Desert, where freezes are more frequent and rainfall is about 30% greater today (Van Devender, 1987). Only 37.5% of the ephemeral herbs (N = 4) and grasses (N = 4) in the Puerto Blanco Mountains samples were winter-spring obligates, suggesting that perhaps 60% of the annual rainfall was in the summer (June–August), with additional warm-season precipitation in September and early October.

In the Tinajas Altas Mountains, ephemerals (N = 11) reached a maximum of 46% summer-rainfall obligates in the middle Holocene. Riparian trees grew near the sites and species richness was nearly twice that of today, indicating greater rainfall. A sample dated at 8.2 ka from the Butler Mountains yielded florets of slender grama (*Bouteloua repens*), a subtropical grass, 160 km west-northwest of its nearest current population (Van Devender et al., *this volume*, chap. 14). Comparison of area climatic records suggests that summer rainfall was 20% greater than today.

In the Whipple Mountains increases in species richness and in summer ephemerals and grasses suggest a modest increase in summer rainfall between 8 and 4.2 ka. A sample dated at 5020 ± 240 B.P. (A-2160) yielded florets of hook threeawn (*Aristida hamulosa*), a C4 perennial grass no longer found in the lower Colorado River valley. Apparently it was able to expand from western Arizona to the Whipple Mountains due to greater summer rainfall at that time (Van Devender et al., *this volume*, chap. 14). The middle Holocene was probably the only time when summer rainfall was of some importance in the area.

Thus summer rainfall was considerably greater than today in the eastern Sonoran Desert but only slightly so near the Mojave Desert boundary. Monsoonal storms probably did not penetrate into the Mojave Desert and Great Basin much farther than they do today. Summer rainfall was probably significantly greater than today only in the mountains of the eastern Mojave Desert and west of the lower Colorado River valley in California, and not in the desert lowlands. The only important consequence of this increase in summer rainfall to the plants in this area may have been that conditions for seedling establishment were met more often. Warmer sea-surface temperatures may have increased the frequencies of tropical storms and hurricanes reaching the Sonoran and Mojave deserts from the Pacific Ocean and the Gulf of California in late summer–early fall (Huning, 1978).

The monsoonal maximum for the Holocene occurred after 9 ka, later than proposed by Spaulding and Graumlich (1986) and Spaulding (*this volume*, chap. 9). Arizona's middle Holocene summers seem to have been wetter, not drier (Martin, 1963). A dry altithermal may apply only to winter rainfall areas to the

west and northwest (Antevs, 1955).

Late Holocene. By 4 ka the modern climatic regime was established. Reduction in the frequency of winter freezes and adequate summer rainfall combined to produce more subtropical climates in the Sonoran Desert as more frost-sensitive elements of the flora migrated to their present northern ranges. Resurgence of velvet mesquite, increased species richness in long-lived perennials, and the abundance of ephemerals in samples dated at 0.98 and 0.99 ka suggest that there was a short climatic fluctuation with greater rainfall in both seasons in the Puerto Blanco Mountains (Van Devender, 1987). A 0.7-ka sample from the Butler Mountains that contained catclaw acacia, blue palo verde, and velvet mesquite also indicates wetter conditions than today. In the Puerto Blanco, Tinajas Altas, Butler, and Whipple mountains, impoverished modern floras at the midden sites suggest that the present climate is as hot and dry today as at any time in the Holocene.

Biogeography

The fossil packrat midden record provides information on what happened to plant species, communities, and vegetation zonation in the Sonoran Desert during the last glacial-interglacial cycle. The packrat middens have yielded a wealth of records of plants that no longer grow together. Table 8.2 presents 46 "anomalous" species pairs found in Sonoran Desert packrat middens. Most of the pairs were found in late Wisconsin (32 pairs, 76 records) or early Holocene (19 pairs, 44 records) samples. Nine pairs were in samples of both ages. Seven anomalous associations were found in the two middle Wisconsin middens. The only anomalous association after 8.9 ka was the record of slender grama grass in association with hollyleaf bursage in the Butler Mountains at 8.2 ka. Associations at the species level were essentially modern in the middle Holocene, although relatively modern composition of plant communities was not seen until the late Holocene.

These anomalous associations included Great Basin desertscrub plants (big sagebrush, shadscale) with Sonoran desertscrub (Nichol's Turk's head cactus), desert grassland (Arizona rosewood, tanglehead grass), and woodland (Arizona cypress, border pinyon, Ajo oak) plants. Other pairs were woodland plants (singleleaf pinyon, Rocky Mountain juniper, and snowberry) with Sonoran desertscrub (hollyleaf bursage, Nichol's Turk's head cactus, Sonoran crucifixion thorn) and relict woodland (Ajo oak, Harrison barberry) species. Mojave desertscrub plants (Joshua tree, Whipple cholla, rabbit brush, Mojave sage, striped horsebrush) formerly grew with Sonoran desertscrub (hollyleaf bursage, rock sage), desert grassland (Arizona rosewood, tanglehead grass), and relict woodland (Ajo oak) in various combinations. And finally, Mojave sage, a plant of the dry woodlands and adjacent desertscrub, commonly occurred with Whipple yucca, a characteristic plant of chaparral and coastal sage scrub in southern California and northern Baja California.

Besides the anomalous associations presented in Table 8.2, the middens have yielded many additional species pairs that occur only in very local areas, special habitats, or in ecotones between communities. Whipple yucca has disjunct, isolated populations in the western Grand Canyon in Arizona and the Sierra Pinacate area in northeastern Sonora where it associates with Whipple cholla, cottontop cactus (*Echinocactus polycephalus*), hollyleaf bursage, and teddy bear cholla. A relictual population of Mojave sage in the Sierra Estrella near Phoenix, Arizona, may bring it in contact with the fishhook barrel cactus. Joshua tree, California barrel cactus, and Whipple cholla, which are mostly in desertscrub today, range into the lower edge of pinyon-juniper woodlands in associations that were more common in the late Wisconsin.

A number of common desert reptiles have been found in late Wisconsin or early Holocene woodland middens, including the desert tortoise (*Xerobates agassizi*), chuckwalla (*Sauromalus obesus*), desert spiny lizard (*Sceloporus* cf. *magister*), leopard lizard (*Crotaphytus wislizeni*), banded sand snake (*Chionactis occipitalis*), desert rosy boa (*Lichanura trivirgata*), and spotted leaf-nosed snake (*Phyllorhynchus decurtatus*; Van Devender and Mead, 1978). Although these are mostly inhabitants of desertscrub, they can occasionally be found in chaparral or xeric juniper woodland, especially where California juniper occurs below 900 m. The banded sand snake, an inhabitant of hot, sandy desert areas be-

low 460 m, would be an unusual associate of even California juniper. These animals were apparently more typical woodland residents in the past.

Thus the late Wisconsin and early Holocene communities of the Sonoran Desert were mixtures of these unusual pairs and more common associates. Today these plants can be found in Great Basin, Mojave, and Sonoran desertscrubs, desert grassland, chaparral, and woodlands in Arizona, California, Sonora, and Baja California. Beyond the general community level, many of the late Wisconsin and early Holocene communities of the Sonoran Desert do not have modern analogs. Partial analogs occur in the relict areas mentioned above, in the interior chaparral and singleleaf pinyon woodlands of central Arizona, in the California juniper woodlands and Joshua tree communities that surround the Hualapai Mountains of Arizona and the mountains of the eastern Mojave Desert in southeastern California (Thorne et al., 1981), and where Sonoran desertscrub mingles with woodland on the interior mountain slopes in southeastern California and northern Baja California. The paleocommunities of southwestern Arizona were mixtures of plants now found in all of these areas.

In the Holocene Sonoran Desert plants attained their present ranges at different times. There has never been a unified Sonoran Desert community with a fixed assemblage of species. The packrat midden records strongly support the view that plant communities are loosely knit assemblages of individual species with different physiological tolerances to environmental factors, as Shreve (1937) and Gleason (1939) proposed. Community composition has changed continuously for the last eleven thousand years as Wisconsin woodland and Mojave desertscrub plants retreated from lowlands, and desertscrub plants migrated into new areas. Relatively modern compositions of many Sonoran desertscrub communities were only attained in the last four to five thousand years. Community composition continued to vary subsequently during lesser climatic fluctuations, suggesting that differential responses of plant species to climate changes and continuous variation in climate on several scales have resulted in dynamic plant communities that rarely, if ever, reach equilibrium (Davis, 1986). Only simple communities in extremely harsh environments or substrates may have been more stable. Relict populations in the modern Sonoran Desert are not only stranded elements of ice-age woodlands (e.g., Mojave sage in the Sierra Pinacate) but also of middle Holocene warm, wet climates (e.g., elephant tree in the Waterman Mountains) and local endemics that are tolerant to glacial and interglacial climates (e.g., Nichol's Turk's head cactus in the Waterman Mountains).

Late Wisconsin cooling and winter rains had a profound effect on the region now occupied by the subtropical Sonoran Desert as woodland or Mojave desertscrub expanded. Middens record singleleaf pinyon, Utah juniper, shrub live oak, big sagebrush, and Joshua tree reaching the Waterman Mountains in the eastern Sonoran Desert. Lowland woodlands with California juniper and Joshua tree have been found close to the Sonoran border in the Puerto Blanco and Tinajas Altas mountains (Van Devender, 1987). Rocky Mountain juniper grew at 905–975 m elevation in the Ajo Mountains, suggesting the potential for other mesic montane trees such as Douglas fir and ponderosa pine to have occurred at higher elevations on isolated mountains, as they did on Pontatoc Ridge in the Santa Catalina Mountains. The dispersal of these plants would have been via long-distance transport of seeds by birds, along lowland riparian corridors if the climate were cool and wet enough, or possibly through pinyon-juniper-oak woodland not typically seen today. Montane forests did not occur low enough to connect isolated forests in the Graham and Chiricahua mountains with the highlands of the Mogollon Rim and the White Mountains. In Mexico woodland expanded to lower elevations from the peninsular ranges in Baja California and the Sierra Madre Occidental in Sonora. While the upper and northern limits of the Sonoran Desert shrank, new land was exposed around the Gulf of California as sea level was lowered one hundred meters or so.

What was the Sonoran Desert like in the late Wisconsin? Biotic distributions were very different from today, although the virtual absence of fossil records from Mexico makes inferences speculative. The winter-rainfall plants and animals of Baja California probably expanded into a sizable area on the Sonoran mainland, leaving many relict populations, including the Punto Cirio boojum trees near Puerto Libertad. The Sonoran Desert plants such as the saguaro and foothills palo verde that live in biseasonal regimes with important summer rainfall were probably reduced to a small area in central Sonora. If rainfall was less in the more tropical areas to the south, thornscrub may have retreated southward, yielding area to the Sonoran Desert. If rainfall increased to the south, the biseasonal Sonoran Desert would have been restricted to a small area.

In the Holocene the expansion of the Arizona Upland subdivision into Arizona probably represents the development of a new vegetation type composed of the more cold-tolerant Sonoran Desert plants. Sonoran plants requiring winter rainfall mostly withdrew from the Sonoran mainland to the Baja California peninsula, where winter rainfall desertscrub expanded as chaparral and juniper woodland retreated to higher elevations. Relict summer-rainfall thornscrub and desertscrub on the southern end of Baja California expanded northward. The boundary between the Mojave and Sonoran deserts can be detected by about 8900 years ago with the maximum expansion of the Sonoran Desert species in the late Holocene. Presumably the modern Sonoran desertscrub communities resemble those of the warmest portions of previous interglacials and of the original late Miocene–Pliocene Sonoran Desert (Axelrod, 1979). Desertscrub communities of the lower Colorado River may have presented an even more barren aspect before the immigration of creosote bush from the deserts of South America (Wells and Hunziker, 1976; Van Devender, *this volume*, chap. 7).

ACKNOWLEDGMENTS

Packrat midden research was supported by National Science Foundation grants DEB 79-23840 to P. S. Martin and BSR 82-14939 to P. S. Martin and T. R. Van Devender, the Southwest Parks and Monuments Association, Raymond M. Turner of the U.S. Geological Survey, and the Arizona-Sonoran Desert Museum. Paul S. Martin, Julio L. Betancourt, Vera Markgraf, and Herbert E. Wright provided helpful editorial comments. Tony L.

Burgess helped evaluate anomalous associations using plant distributions. Richard S. Felger provided information about the Sonoran population of Whipple yucca and its associates. George L. Bradley provided information on the distributions of reptiles in chaparral and woodland in Arizona. Diagrams were drafted by Helen A. Wilson. W. Eugene Hall photographed the fossils. Jean T. Morgan typed the manuscript.

REFERENCES

Antevs, E. (1955). "Geological-climatic dating in the West." *American Antiquity* 20, 317–55.

Axelrod, D. I. (1979). "Age and origin of Sonoran Desert vegetation." *California Academy of Sciences Occasional Papers* 132, 1–74.

Beetle, A. A. (1960). *A study of sagebrush: The section Tridentatae of Artemisia.* Laramie: University of Wyoming Agricultural Experiment Station Bulletin no. 368.

Benson, L. (1982). *The cacti of the United States and Canada.* Stanford: Stanford University Press.

Benson, L., and Darrow, R. A. (1981). *Trees and shrubs of the southwestern deserts.* Tucson: University of Arizona Press.

Betancourt, J. L. (1984). "Late Quaternary plant zonation and climate in southeastern Utah." *Great Basin Naturalist* 44, 1–35.

Betancourt, J. L.; Van Devender, T. R.; and Rose, M. (1986). "Comparison of plant macrofossils in woodrat (*Neotoma* spp.) and porcupine (*Erethizon dorsatum*) middens from the western United States." *Journal of Mammalogy* 67, 266–73.

Bryson, R. A. (1985). "On climatic analogs in paleoclimatic reconstruction." *Quaternary Research* 23, 275–86.

Bryson, R. A., and Wendland, W. (1967). "Tentative climatic patterns for some late glacial and postglacial episodes in central North America." In *Life, land, and water.* S. J. Mayer-Oakes, Jr., ed., pp. 271–98. Winnipeg: University of Manitoba Press.

Clisby, K. H., and Sears, P. P. (1956). "San Augustin Plains: Pleistocene climatic changes." *Science* 124, 537–39.

Cole, K. L. (1981). "Late Quaternary environment in the eastern Grand Canyon: Vegetation gradients over the last 25,000 years." Ph.D. diss., University of Arizona, Tucson.

———. (1986). "The lower Colorado Valley: A Pleistocene desert." *Quaternary Research* 25, 392–400.

Davis, M. D. (1986). "Climatic instability, time lags, and community disequilibrium." In *Community ecology.* J. Diamond and T. J. Case, eds., pp. 269–84. New York: Harper and Row.

Gleason, H. A. (1939). "The individualistic concept of the plant association." *American Midland Naturalist* 21, 92–110.

Hastings, J. R.; Turner, R. M.; and Warren, D. K. (1972). An atlas of some plant distributions in the Sonoran Desert. *University of Arizona Institute of Atmospheric Physics Technical Reports on the Meteorology and Climatology of Arid Regions* (21), 1–255.

Huning, J. R. (1978). *A characterization of the climate of the California desert.* Riverside, Calif.: U.S. Bureau of Land Management Report.

Kearney, T. H., and Peebles, R. H. (1964). *Arizona Flora.* Berkeley: University of California Press.

King, J. E., and Van Devender, T. R. (1977). "Pollen analysis of fossil packrat middens from the Sonoran Desert." *Quaternary Research* 8, 191–204.

Kutzbach, J. E. (1983). "Modeling of Holocene climates." In *Late-Quaternary environments of the United States.* Volume 2, *The Holocene.* H. E. Wright, ed., pp. 271–77. Minneapolis: University of Minnesota Press.

Little, E. L., Jr. (1971). *Atlas of United States trees.* Volume 1. *Conifers and important hardwoods.* Washington, D.C.: U.S. Forestry Service Miscellaneous Publication no. 1146.

———. (1976). *Atlas of United States trees.* Volume 1. *Minor western hardwoods.* Washington, D.C.: U.S. Forestry Service Miscellaneous Publication no. 1314.

Martin, P. S. (1963). *The last 10,000 years.* Tucson: University of Arizona Press.

Martin, P. S., and Mehringer, P. J. (1965). "Pleistocene pollen analysis and biogeography of the Southwest." In *The Quaternary of the United States.* H. E. Wright and D. G. Frey, eds., pp. 433–51. Princeton: Princeton University Press.

Mead, J. I.; Van Devender, T. R.; and Cole, K. L. (1983). "Late Quaternary small mammals from Sonoran Desert packrat middens, Arizona and California." *Journal of Mammalogy* 64, 173–80.

Munz, P. A. (1974). *A flora of southern California.* Stanford: University of California Press.

Rowlands, P. G. (1978). "The vegetation dynamics of the Joshua tree (*Yucca brevifolia* Engelm.) in the southwestern United States of America." Ph.D. diss., University of California, Riverside.

Sellers, W. D., and Hill, R. H. (1974). *Arizona climate, 1931–1972.* Tucson: University of Arizona Press.

Shreve, F. (1915). *The vegetation of a desert mountain range as conditioned by climatic factors.* Carnegie Institute of Washington Publication 217, pp. 1–112.

———. (1937). "Thirty years of change in desert vegetation." *Ecology* 18, 463–78.

———. (1964). "Vegetation of the Sonoran Desert." In *Vegetation and flora of the Sonoran Desert.* F.

Shreve and I. L. Wiggins, eds., pp. 259–93. Stanford: Stanford University Press.

Shreve, F., and Wiggins, I. (1964). *Vegetation and flora of the Sonoran Desert*. Stanford: Stanford University Press.

Soreng, R. J., and Van Devender, T. R. (1989). "Late Quaternary fossils of *Poa fendleriana* (muttongrass): Holocene expansions of apomicts." *Southwestern Naturalist* 34, 35–45.

Spaulding, W. G. (1981). "The late Quaternary vegetation of a southern Nevada mountain range." Ph.D. diss., University of Arizona, Tucson.

Spaulding, W. G., and Graumlich, L. J. (1986). "The last pluvial climate episodes in the deserts of southwestern North America." *Nature* 320, 441–44.

Thorne, R. F.; Prigge, B. A.; and Henrickson, J. (1981). "A flora of the higher ranges and the Kelso Dunes of the eastern Mojave Desert in California." *Aliso* 10, 71–186.

Turner, R. M., and Brown, D. E. (1982). "Sonoran desertscrub." *Desert Plants* 4, 181–221.

Van Devender, T. R. (1973). "Late Pleistocene plants and animals of the Sonoran Desert: A survey of ancient packrat middens in southwestern Arizona." Ph.D. diss., University of Arizona, Tucson.

———. (1977). "Holocene woodlands in the southwestern deserts." *Science* 198, 189–92.

———. (1987). "Holocene vegetation and climate in the Puerto Blanco Mountains, southwestern Arizona." *Quaternary Research* 27, 51–72.

Van Devender, T. R., and King, J. E. (1971). Late Pleistocene vegetational records in western Arizona. *Journal of the Arizona Nevada Academy of Sciences* 6, 240–44.

Van Devender, T. R.; Martin, P. S.; Thompson, R. S.; Cole, K. L.; Jull, A. J. T.; Long, A.; Toolin, L. J.;

and Donahue, D. J. (1985). "Fossil packrat middens and the tandem accelerator mass spectrometer." *Nature* 317, 610–13.

Van Devender, T. R., and Mead, J. I. (1978). "Early Holocene and late Pleistocene amphibians and reptiles in Sonoran Desert packrat middens." *Copeia* 1978, 464–75.

Van Devender, T. R., and Spaulding, W. G. (1979). "The development of vegetation and climate in the southwestern United States." *Science* 204, 701–10.

Van Devender, T. R.; Thompson, R. S.; and Betancourt, J. L. (1987). "Vegetation history in the Southwest: The nature and timing of the late Wisconsin-Holocene transition." In *North America and adjacent oceans during the last deglaciation*. W. F. Ruddiman and H. E. Wright, Jr., eds., pp. 323–52. Boulder: Geological Society of America.

Wells, P. V. (1976). "Macrofossil analysis of woodrat, *Neotoma*, middens as a key to the Quaternary vegetational history of arid America." *Quaternary Research* 6, 223–48.

———. (1979). "An equable glaciopluvial in the West: Pleniglacial evidence of increased precipitation on a gradient from the Great Basin to the Sonoran and Chihuahuan deserts." *Quaternary Research* 12, 311–25.

Wells, P. V., and Hunziker, J. H. (1976). "Origin of the creosote bush (*Larrea*) deserts of southwestern North America." *Annals of the Missouri Botanical Garden* 63, 833–61.

Wiggins, I. L. (1980). *Flora of Baja California*. Stanford: Stanford University Press.

Woodcock, D. (1986). "The late Pleistocene of Death Valley: A climatic reconstruction based on macrofossil data." *Palaeogeography, Palaeoclimatology, Palaeoecology* 57, 273–83.

Appendix 8.1. Common and scientific names used in the text and figures.

Ajo oak	*Quercus ajoensis*
Allthorn	*Koeberlinia spinosa*
Arizona cypress	*Cupressus arizonica*
Arizona rosewood	*Vauquelinia californica*
Banded sand snake	*Chionactis occipitalis*
Beavertail cactus	*Opuntia basilaris*
Bigelow beargrass	*Nolina bigelovii*
Big galleta grass	*Hilaria rigida*
Big sagebrush	*Artemisia tridentata*
Black brush	*Coleogyne ramosissima*
Blue palo verde	*Cercidium floridum*
Boojum tree	*Fouquieria columnaris*
Border pinyon	*Pinus discolor*
Brittle bush	*Encelia farinosa*
California barrel cactus	*Ferocactus acanthodes*
California buckwheat	*Eriogonum fasciculatum*
California juniper	*Juniperus californica*
Catclaw acacia	*Acacia greggii*
Chain fruit cholla	*Opuntia fulgida*
Cheese weed	*Hymenoclea salsola*
Chuckwalla	*Sauromalus obesus*
Colorado pinyon	*Pinus edulis*
Cottontop cactus	*Echinocactus polycephalus*
Coville barrel cactus	*Ferocactus covillei*
Creosote bush	*Larrea divaricata*
Desert agave	*Agave deserti*
Desert brickell bush	*Brickellia atractyloides*
Desert hackberry	*Celtis pallida*
Desert lavender	*Hyptis emoryi*
Desert packrat	*Neotoma lepida*
Desert rosy boa	*Lichanura trivirgata*
Desert spiny lizard	*Sceloporus magister*
Desert tortoise	*Xerobates agassizi*
Douglas fir	*Pseudotsuga menziesii*
Dwarf juniper	*Juniperus communis*
Elephant tree	*Bursera microphylla*
Fishhook barrel cactus	*Ferocactus wislizenii*
Fishhook cactus	*Mammillaria microcarpa*
Foothills palo verde	*Cercidium microphyllum*
Gila monster	*Heloderma suspectum*
Harrison barberry	*Berberis harrisoniana*
Hollyleaf bursage	*Ambrosia ilicifolia*
Hook threeawn	*Aristida hamulosa*
Ironwood	*Olneya tesota*
Joshua tree	*Yucca brevifolia*
Leopard lizard	*Crotaphytus wislizeni*
Limber bush	*Jatropha cuneata*
Mariola	*Parthenium incanum*
Mexican jumping bean	*Sapium biloculare*
Mojave aster	*Xylorhiza tortifolia*
Mojave Desert palo verde	*Canotia holacantha*
Mojave sage	*Salvia mohavensis*
Mormon tea	*Ephedra nevadensis*
Nichol's Turk's head cactus	*Echinocactus horizonthalonius* var. *nicholii*
Ocotillo	*Fouquieria splendens*
Oneseed juniper	*Juniperus monosperma*
Oreja de perro	*Tiquilia canescens*

Appendix 8.1. (continued)

Organ pipe cactus	*Stenocereus thurberi*
Peachthorn wolfberry	*Lycium cooperi*
Pink felt plant	*Horsfordia newberryi*
Ponderosa pine	*Pinus ponderosa*
Porcupine	*Erethizon dorsatum*
Pygmy cedar	*Peucephyllum schottii*
Rabbit brush	*Chrysothamnus teretifolius*
Redberry juniper	*Juniperus erythrocarpa*
Rock sage	*Salvia pinguifolia*
Rocky Mountain juniper	*Juniperus scopulorum*
Saguaro	*Carnegiea gigantea*
Shadscale	*Atriplex confertifolia*
Shrub live oak	*Quercus turbinella*
Silver dollar cactus	*Opuntia chlorotica*
Singleleaf pinyon	*Pinus monophylla*
Slender grama	*Bouteloua repens*
Skunkbrush	*Rhus trilobata*
Snowberry	*Symphoricarpos* sp.
Sonoran bursage	*Ambrosia cordifolia*
Sonoran crucifixion thorn	*Castela emoryi*
Spotted leaf-nosed snake	*Phyllorhynchus decurtatus*
Staghorn cholla	*Opuntia acanthocarpa*
Striped horsebrush	*Tetradymia argyraea*
Tanglehead grass	*Heteropogon contortus*
Teddy bear cholla	*Opuntia bigelovii*
Triangleleaf bursage	*Ambrosia deltoidea*
Utah juniper	*Juniperus osteosperma*
Variable prickly pear	*Opuntia phaeacantha*
Velvet mesquite	*Prosopis velutina*
Whipple cholla	*Opuntia whipplei*
Whipple yucca	*Yucca whipplei*
White bursage	*Ambrosia dumosa*
White-throated packrat	*Neotoma albigula*

Chapter 9

Vegetational and Climatic Development of the Mojave Desert: The Last Glacial Maximum to the Present

W. Geoffrey Spaulding

INTRODUCTION

Past climatic changes profoundly affected the biogeography and ecology of the Mojave Desert and, during the Quaternary (the last 1.7 million years), the pace of these changes has been rapid and their impact pronounced. These observations are applicable to most terrestrial ecosystems, but only recently has a detailed description of the response of Mojavean ecosystems to climatic change become possible, largely due to the region's rich pack-rat midden record. Ancient middens from the Mojave were the first to be brought to the attention of the scientific community (Wells and Jorgensen, 1964).

In the southern Mojave Desert south of latitude 36° N, vast bajadas and low valleys separate many small mountain ranges (Fig. 9.1). Basal elevations of the southern valleys are generally below 500 m, and the surrounding mountains seldom exceed 1500 m in height. Warm desertscrub, typified by creosote bush (*Larrea divaricata*; botanical nomenclature in this chapter follows Munz [1974], with exceptions after Cronquist et al. [1972] and Beatley [1976]), white bursage (*Ambrosia dumosa*), and desert spruce (*Peucephyllum schottii*), occupies much of the southern Mojave. To the north basal valley elevations often exceed 1000 m, and many mountain ranges have maximum elevations above 2000 m. As Beatley (1975) points out, increasing elevations to the north exaggerate the latitudinal decline in thermal regimes. Although still present, warm desertscrub is less widespread in the northern Mojave. Habitats too cold and/or moist to be dominated by warm-desert species are occupied by temperate desertscrub. Blackbrush (*Coleogyne ramosissima*), Joshua tree (*Yucca brevifolia*), ground-thorn (*Menodora spinescens*), boxthorn or wolf-berry (*Lycium* spp.), and goldenbushes (*Haplopappus* spp.) are plants that typify Mojavean cold-temperate desertscrub (Brown et al., 1979). On the valley flanks of the northern Mojave Desert, Great Basin desertscrub is an important vegetation type. Zoned immediately below pinyon-juniper (*Pinus monophylla–Juniperus osteosperma*) woodland, these steppe communities are dominated by sagebrushes (*Artemisia tridentata, A. nova*), horsebrushes (*Tetradymia glabrata, T. canescens*), and rabbitbrushes (chiefly *Chrysothamnus nauseosus* and *C. viscidiflorus*). Rowlands et al. (1982) present a summary of the phytogeography of the region, while Beatley (1976) provides an excellent summary of the floristics and ecology of the north-central Mojave Desert.

Topographic variation in this area is ex-

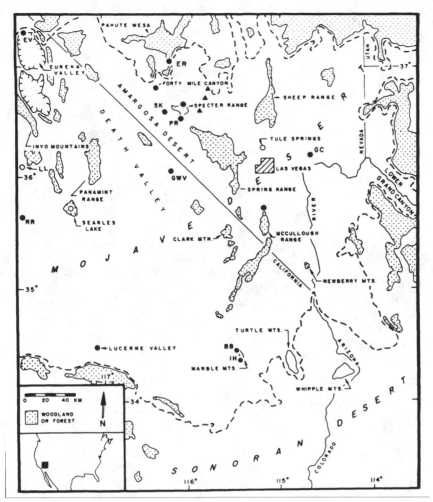

Fig. 9.1. The Mojave Desert, showing the current distribution of woodland and localities mentioned in the text. The bold dashed line represents the boundary of the Mojave Desert (after Cronquist et al., 1972; Brown and Lowe, 1980). Open dots are pollen sites, closed dots are packrat midden and cave sites. Solid triangles are midden sites from the Spotted Range and vicinity. Abbreviations: BB, Brown Buttes; ER, Eleana Range; EV, Eureka View; GC, Gypsum Cave; GWV, Greenwater Valley; IH, Iron Hat; LL, Little Lake; PR, Point of Rocks and Specter Range; RR, Robbers Roost; Sk, Skeleton Hills; TM, Turtle Mountains.

treme. From Death Valley west to the Sierra Nevada, regional relief of the Mojave Desert exceeds 4000 m. There are many mountain ranges in the region, but only two, the Spring and Sheep ranges of Nevada, are large enough and far enough east to support ample forest vegetation. These ranges support a full complement of vegetation zones: desertscrub, usually restricted to elevations below 1800 m, gives way to singleleaf pinyon–Utah juniper woodland, which in turn gives way to fir-pine forest (*Abies concolor*–*Pinus ponderosa*)

above about 2200 m. Above about 2700 m subalpine woodland dominated by Great Basin bristlecone pine (*Pinus longaeva*) occurs, frequently in a topographically and edaphically induced vegetation mosaic of woodland and black sagebrush (*Artemisia nova*) scrub. Farther west in the Inyo and Panamint ranges (Fig. 9.1) fir-pine forest is lacking. Instead, a treeless vegetation zone occurs at middle elevations between pinyon-juniper and subalpine woodland. This vegetation is characterized by black sagebrush scrub or interior chaparral commonly dominated by little-leaf mountain mahogany (*Cercocarpus intricatus*). To the east, where average annual and summer precipitation are greater, these communities are restricted to xeric aspects, as well as to windswept plateaus and ridges above about 2100 m (Bradley and Deacon, 1967; Spaulding, 1981).

Time-Stratigraphic Terms

For brevity, radiocarbon ages are rounded to the nearest century and expressed as thousands of years before the present (ka). Terms applied to intervals of the late Quaternary follow the global chronology of climatic changes and regional responses reflected in the well-dated sedimentary record of Searles Lake (Smith, 1979). Briefly, the "middle Wisconsin" lasted from about 65 to 22.6 ka. The end of the "late Wisconsin" is coeval with the end of the Pleistocene, placed by international convention (Olausson, 1982) at 10 ka. The conventional date of the full glacial (or "Wisconsin maximum") is 18 ka. The "latest Wisconsin" is an informal term used to refer to the warming phase of the last glacial age, from about 12 to 10 ka. In this treatment the "early Holocene" spans the period from 10 to 8 ka, the "middle Holocene" from 8 to 4 ka, and the "late Holocene" from 4 ka to the present (Spaulding, 1985).

OBJECTIVES

In this chapter I use fossil packrat middens from the Mojave Desert to examine responses of plant species and associations to climatic change since the last glacial maximum. The fossil samples include isolated middens providing only a point or two in radiocarbon time as well as suites of dated samples from single localities (chronosequences). Consistent variations (trends) in the relative abundance of plant species in a chronosequence of macrofossil assemblages presumably reflect real changes in local vegetation. I assume that these plant macrofossil assemblages allow reasonable characterization of local vegetation at the community-type or "series" level. Despite some infidelity in the midden-vegetation relationship (see *this volume*, chapters 2, 4, and 5), different community types "distinguished primarily on taxa that are distinctive climax plant dominants" (Brown et al., 1979) leave different plant macrofossil records. Vari-

ations in these records, then, are the basis for inferring variations in the type of vegetation occurring at a given site. My goal is to present a detailed view of vegetation change in the Mojave Desert and its mountains, a view previously unavailable to the ecologist and biogeographer. Regional boundaries and the study areas to be discussed are shown in Figure 9.1.

RECORDS FROM THE SOUTHERN MOJAVE

Packrat midden studies have been limited in the southern Mojave Desert. King (1976) conducted research in the Lucerne Valley of California, while Leskinen (1975) surveyed midden sites in the Newberry Mountains of southern Nevada. New midden records from the Marble Mountains east of Lucerne Valley and the McCullough Range northwest of the Newberry Mountains are discussed below.

Marble Mountains

The bulk of the Marble Mountains (MM) lies below 900 m elevation, with Castle Peak, 1170 m elevation, the highest point. While volcanic rocks form most of the range, isolated limestone outcrops yield distinct habitats of calcareous soils. At the Iron Hat locality a low-elevation midden site provides a macrofossil record of sparse desertscrub dating to the latest Wisconsin. Rock shelters that developed in rhyolitic breccia at the Brown Buttes yield a more comprehensive record of desertscrub and woodland. All the sites are from xeric southeast- to west-facing slopes.

Iron Hat (450–475 m elevation). The sere south- to southwest-facing limestone slopes near the Iron Hat Mine support widely scattered shrubs; only brittle-bush (*Encelia farinosa*), Death Valley sage (*Salvia funerea*), and creosote bush are common. Death Valley sage occurs here about 275 km southwest of its main range in the tributaries of Death Valley (Munz, 1974; Benson and Darrow, 1981). On the alluvial fan and in canyon bottoms vegetation is less sparse and more diverse and includes white bursage, desert lavender (*Hyptis emoryi*), sweet bush (*Bebbia juncea*), and desert spruce.

The Marble Mountains-2 midden, dated at 10.5 ka, reveals desertscrub vegetation different from that of today (Spaulding, 1983).

Macrofossils of extralocal Mormon tea (*Ephedra* cf. *californica*) and wolf-berry (*Lycium* sp., perhaps *L. californicum*) are abundant. Death Valley sage appears to be the only representative of the modern flora. Also missing are woodland and high-desert plants that typify contemporaneous middens from lower elevations in the Whipple Mountains, about 120 km to the east-southeast (Fig. 9.1; Van Devender, *this volume*, chap. 8). These species are present, however, in assemblages from the Brown Buttes locality.

Brown Buttes (840–900 m elevation). The perennial vegetation of the Brown Buttes is denser and more diverse than that at Iron Hat, a consequence of the higher elevation and more favorable edaphic setting. The talus slopes and cliff bases support well-developed desertscrub dominated by creosote bush and brittle-bush with white bursage, sweet bush, desert lavender, desert spruce, and wolf-berry common to occasional. Mesophytes largely restricted to the washes, away from the midden sites, include catclaw acacia (*Acacia greggii*) and Mojave yucca (*Yucca schidigera*).

As at the Iron Hat locality, desertscrub without juniper occupied the Brown Buttes at the close of the Wisconsin glacial age. Extralocal shrubs typical of higher elevations above 1400 m dominate the record from 10.2 ka to 7.9 ka, including turpentine bush (cf. *Haplopappus laricifolius*), wedgeleaf goldenweed (*H. cuneatus*), Mojave sage (*Salvia mohavensis*), and Mojave yucca (Fig. 9.2) (Spaulding, 1983). The oldest macrofossil assemblages, dated at 10.2 and 9.5 ka, contain no woodland species. The next two middens, dated to 8.9 ka, suggest a brief woodland incursion. They contain not only juniper (*Juniperus* sp.) but also sagebrush (*Artemisia* sec. Tridentatae), Whipple yucca (*Yucca whipplei*), and Joshua tree (Fig. 9.2). Samples that contain no juniper or yucca provide bracketing dates on this woodland period of 9.5 and 7.9 ka. Thus it appears that the development and demise of woodland at the Brown Buttes took less than 1600 radiocarbon years. Alternatively, woodland existed at the adjacent MM-7 and MM-9 sites, but not at the MM-4 and MM-5 sites. The latter two sites are from a locality about 1 km distant with somewhat different habitat characteristics.

Desertscrub modernization was under way before the brief reversal to woodland vegetation. Brittle-bush was the first of the modern dominants to arrive. A single seed in the oldest sample is a possible contaminant, but in the next-youngest sample, dated at 9.5 ka, brittle-bush macrofossils are common. Apparently the shrub was an important component of vegetation during the woodland period. Creosote bush, its common associate today, is present in trace amounts only. Neither creosote bush nor white bursage are well represented in the fossil record until about 7.9 ka (sample MM-6), when the modern dominants co-occurred with the last of the extralocals (Mojave yucca, goldenbush, Mojave sage; Fig. 9.2).

More than three thousand years elapsed between deposition of the MM-6 macrofossil assemblage and accumulation of MM-7(2) and MM-5(2), dated at 4.5 and 4.4 ka respectively. These two middle Holocene samples suggest a vegetation similar to that of today. In contrast, the late Holocene MM-4(3) assemblage lacks the white bursage and desert spruce found in older samples and in the modern vegetation.

McCullough Range

Located in the southern tip of Nevada close to the northwestern boundary of the Sonoran Desert, the McCullough Range is a moderate-sized mountain mass supporting pinyon-juniper woodland at elevations above 1800 m (Fig. 9.1; Bostick, 1973). The flora is less xerophytic than that of the Marble Mountains, with catclaw acacia (*Acacia greggii*) and succulents common on south-facing slopes and in washes. The midden sites are two adjacent rock shelters at 1045 m elevation developed in rhyolitic substrate similar to that at Brown Buttes. The modern vegetation is mixed desertscrub dominated by Mormon tea (*Ephedra torreyana*), catclaw acacia, creosote bush, white bursage, and roundleaf rabbitbrush (*Chrysothamnus teretifolius*).

The McCullough Range time series discussed here consists of five macrofossil assemblages dated from 6.5 to 1 ka, with a modern midden as a control (Fig. 9.3). All assemblages contain abundant creosote bush although, at 53% to 64% of *NISP* (number of identified specimens), it is more frequent in

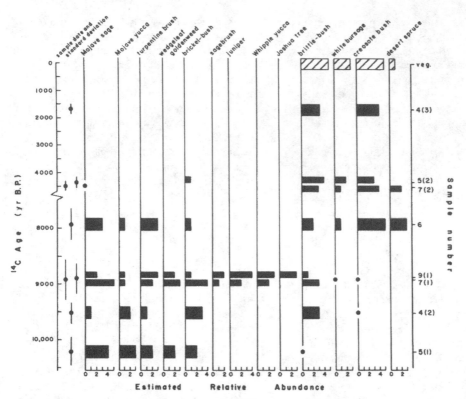

Fig. 9.2. The estimated relative abundance of selected plant taxa in midden samples from the Brown Buttes locality, northern Marble Mountains, California. Half-width bars indicate assemblages with ¹⁴C ages that overlap at 1 standard deviation. Hachured bars indicate estimate of species abundance in the current vegetation (data from Spaulding, 1983, and unpublished).

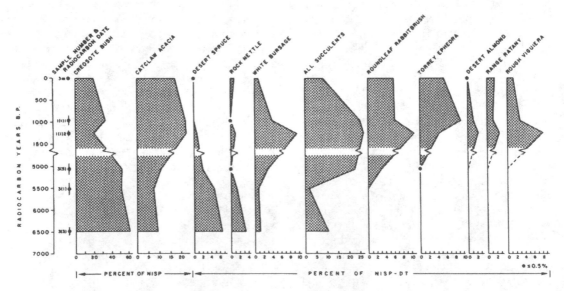

Fig. 9.3. Changes in percentage values of selected plant taxa in packrat midden samples from the McCullough Range, Nevada. Note break and changes in scales. Solid dots indicate ≤ 1%. Abbreviations are *NISP*, number of identified specimens; *DT*, dominant taxa.

the three middle Holocene samples than in the late Holocene and modern assemblages (21% to 37% of *NISP*). Catclaw acacia, white bursage, and succulents are present in the older samples, but they are more abundant in those dating to the late Holocene. Desert spruce is present in quantity only in the two oldest samples, dated at 6.5 and 5.5 ka (Fig. 9.3).

A number of shrubs appeared at the site between the deposition of the youngest middle Holocene assemblage and the oldest late Holocene assemblage. Species richness, as measured by the number of taxa (N), is low ($\bar{x} = 26 \pm 4$) in the three oldest samples compared with the two youngest and modern sample ($\bar{x} = 38 \pm 2$). The immigrants were relatively mesophytic, and included Mormon tea, roundleaf rabbitbrush, rayless brittle-bush (*Encelia virginensis*), range ratany (*Krameria parvifolia*), desert almond (*Prunus fasciculata*), and rough viguiera (*Viguiera reticulata*). The apparent trend toward more diverse vegetation may have begun early; Mormon tea and rabbitbrush first appear in the youngest middle Holocene assemblage (Fig. 9.3). Interestingly, with the possible exception of desert spruce there were no local extirpations; the change from middle to late Holocene environmental conditions led to apparent enrichment of the local flora. The depauperate middle Holocene flora and its xerophytic character indicate increased aridity relative to the last twelve hundred years.

Other Sites in the Southern Mojave Desert

The oldest packrat midden records from the southern Mojave Desert are from near its perimeter. To the south and east, sites in the Turtle Mountains (Wells and Berger, 1967) and Newberry Mountains (Leskinen, 1975; Fig. 9.1) reveal pinyon-juniper woodland at low elevations (700 to 850 m) at about 19.5 ka and from 13.9 to 12.6 ka. To the west, pinyon-juniper woodland dating from 13.8 to 12.8 ka is recorded at Robbers Roost (Van Devender and Spaulding, 1979). The Robbers Roost middens also provide the oldest dates on oak (*Quercus turbinella*) from the Mojave Desert (Van Devender and Spaulding, 1979).

Latest Wisconsin and early Holocene midden records from the Newberry Mountains and the Lucerne Valley are woodland charac-

terized by juniper (*Juniperus* sp.) and, in extreme southern Nevada, live oak (*Quercus* cf. *chrysolepis* and *Q. dunnii*). Unlike the Marble Mountains, these middens provide no evidence for desertscrub until about 7.8 ka, and it may be that woodland prevailed at these localities until the end of the early Holocene. Causes for the persistence of woodland may lie in the higher elevation of the Lucerne Valley sites and in their more mesic, granitic substrate. Also, the Newberry Mountains lie in the Colorado River trough (Fig. 9.1). This region and areas to the east supported widespread woodland during the latest Wisconsin and early Holocene, while most areas to the west and north were apparently drier (Spaulding, 1983).

RECORDS FROM THE NORTHERN MOJAVE

Quaternary paleoecological studies in the northern Mojave Desert began in the first half of the twentieth century with excavations at Gypsum Cave and Tule Springs in southern Nevada (Fig. 9.1; Harrington, 1933; Laudermilk and Munz, 1934; Simpson, 1933). Along with subsequent pollen-stratigraphic studies (Mehringer, 1965, 1967; Mehringer and Sheppard, 1978), a rich packrat midden record has been recovered from the region's calcareous mountains (Mehringer and Ferguson, 1969; Phillips, 1977, 1984; Spaulding, 1981, 1985; Wells and Berger, 1967).

Eureka Valley

Eureka Valley is the northernmost intermountain basin in the Mojave Desert. The lofty Inyo Mountains form its west wall, and immediately west of the Inyos, across the narrow Owens Valley, rises the Sierra Nevada. The rain shadow in the lee of these two massive mountain ranges is severe. For example, at the north end of the Eureka Valley creosote bush–white bursage desertscrub reaches 1500 m, an elevation that typically supports oak woodland in the Sonoran Desert. The roughly 1000-m altitudinal gain from the Marble Mountains to the north end of Eureka Valley reflects the general rise in basin elevation with increasing latitude. The northern limit of creosote bush, and by definition the northern boundary of the Mojave Desert, is in the hills that separate Eureka Valley from the next higher tier of valleys to the north.

The Eureka View locality (EV; 1430–1510 m elevation) is an area of rugged south-facing cliffs and talus slopes (Fig. 9.4) named for the spectacular view of the Eureka Valley afforded from the midden sites. Despite the sere aspect of the blackened dolomite slopes, diverse desertscrub vegetation occurs here. Chaff bush (*Amphipappus fremontii*) and California buckwheat (*Eriogonum fasciculatum*) are subordinate to the dominating shadscale (*Atriplex confertifolia*) and creosote bush. Other common perennials include brickel-bush (*Brickellia arguta*), fluff grass (*Erioneuron pulchellum*), needle grass (*Stipa arida*), false brickel-bush (*Haplopappus brickellioides*), and Indian tobacco (*Nicotiana trigonophylla*).

The oldest macrofossil assemblage, dated at 14.7 ka (EV-5A), was recovered from a small cave at 1430 m elevation (Fig. 9.4). Macrofossils of Utah juniper and shadscale prevail, with joint-fir (*Ephedra* cf. *viridis*) and

Fig. 9.4. View north to the Eureka View locality, California. Arrow points to the EV-5 site. In the foreground is a golden cholla (*Opuntia echinocarpa*), one of the species that once occurred on the hills in the background but today are restricted to the deeper soils of the valley.

the endemic Mojave prickle-leaf (*Hecastocleis shockleyi*) as common associates. Utah juniper and joint-fir occur today about 500 m upslope in the Last Chance Range, 10 km to the east. Mojave prickle-leaf still grows here but is restricted to north-facing habitats. A few limber pine (*Pinus* cf. *flexilis*) needles suggest that this subalpine conifer occurred nearby. An early Holocene assemblage dated at 8.3 ka (EV-5B) postdates the extirpation of Utah juniper and other woodland plants, and marks the arrival of a few thermophilous species such as false brickel-bush and Indigo bush (*Psorothamnus fremontii*; Fig. 9.5).

The oldest middle Holocene assemblage, dated at 6.8 ka (EV-4[2]), shows the advent of boxthorn (*Lycium* sp., perhaps *L. andersonii*), rayless brittle-bush, beavertail prickly pear (*Opuntia basilaris*), and white bursage. More apparent additions to the local flora were made in the fifteen hundred years between the accumulations of EV-5B and EV-4[2] than in the preceding six thousand years separating EV-5A from EV-5B (Fig. 9.5). Some continuing immigration is recorded in the two younger middle Holocene samples dated at 5.6 and 5.4 ka. However, until sometime after 3.9 ka (EV-2C) the vegetation was relatively stable. EV-2C is the last macrofossil assemblage to document the co-occurrence of rayless brittle-bush, boxthorn, cacti, and white bursage (Fig. 9.5). Boxthorn and cacti are presently restricted to more mesic soils of the upper bajada (Fig. 9.4), while brittle-bush occurs in the vicinity on exposures of highly calcified soil.

Creosote bush arrived at Eureka View between 3.9 and 5.6 ka, 1400 to 2900 years after the immigration of white bursage, its common associate in modern deserts. Its tardiness may trouble biogeographers using creosote bush to define the northern limits of the Mojave Desert. The abundance of creosote bush in the Eureka View time series after 3.9 ka, and the disappearance of boxthorn and rayless brittle-bush, signal the end of middle Holocene vegetation conditions (Fig. 9.5).

There were some distinct alterations of the local flora during the late Holocene. Chaff bush and California buckwheat do not appear until 1.6 ka. Desert spruce appears earlier (3.9 ka), although it is absent from EV-1 (2.6 ka) *and* the local vegetation (Fig. 9.5). Apparently, beavertail prickly pear, cholla (*Opuntia echi-*

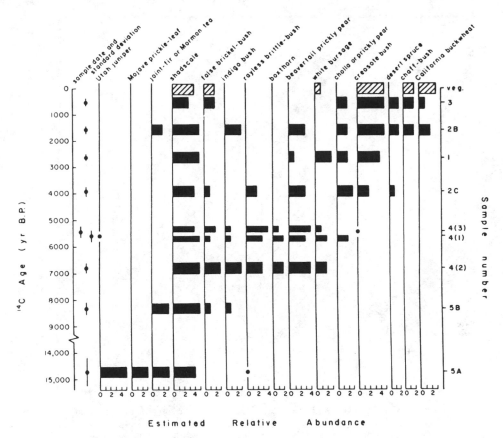

Fig. 9.5. The estimated relative abundance of selected plant taxa in midden samples from the Eureka View locality. Note break in time scale (see Fig. 9.2 for explanation of format; data from Spaulding, 1985).

nocarpa), and desert spruce withdrew from the sites during the last five hundred radiocarbon years. The cacti persist within 500 m (Fig. 9.4), and desert spruce occurs in habitats similar to the fossil sites about 1 km to the west.

Amargosa Desert and Vicinity

The next intermountain basin east of Death Valley is known as the Amargosa Desert (Fig. 9.1). Lying about 150 km southeast of the Eureka Valley, midden sites from the Amargosa Desert are also rock shelters developed in Paleozoic dolomite at habitats 340–600 m lower in elevation than the Eureka View sites.

Specter Range (1100–1190 m elevation). The Specter Range-2 site (Spc-2) is on the north side of a high ridge (Fig. 9.6). The current vegetation is mixed Mojave Desert scrub in which chaff bush, shadscale, and matchweed (*Gutierrezia microcephala*) are impor-

tant. Shadscale, associated with juniper, also appears in the middle Wisconsin assemblages dated as young as 23.3 ka (Fig. 9.6). The Spc-2(2) assemblages indicate local extirpation of shadscale by about 19 ka, immigration of pinyon pine, and the increase of shrubs such as little-leaf mountain mahogany, snowberry (*Symphoricarpos* sp.), and cliff-bush (*Fendlerella utahensis*). This full-glacial expansion of woodland trees and shrubs marks the only recorded period in which steppe shrubs or desertscrub species were not common in the vegetation of the Specter Range.

In contrast to the ridgetop location of the Spc-2 site, the north-facing Specter Range-3 shelter is at the head of a stabilized talus slope that stretches out to an alluvial fan. The relative percentage of sagebrush in the Spc-3B sample, dated at 28.5 ka, is the highest value for this species in a Mojave Desert mid-

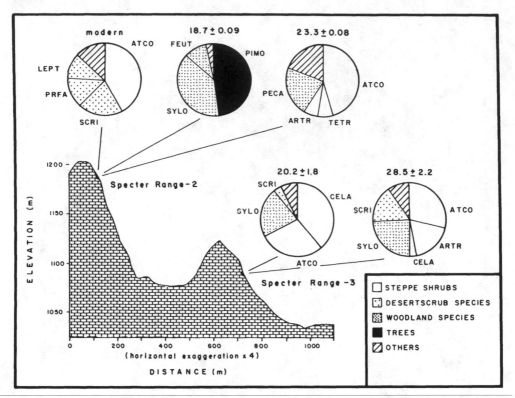

Fig. 9.6. Topography of the Specter Range midden locality and secondary components of selected macrofossil assemblages from the Specter Range-2 and -3 packrat middens. Utah juniper is the dominant macrofossil type in all fossil samples. Percentages for the modern sample are of the total number of identified species (*NISP*), while those of the fossil assemblages are of *NISP* minus the number of juniper fossils. Abbreviations for plant species are: ARTR, *Artemisia* subgen. Tridentatae (sagebrush); ATCO, *Atriplex confertifolia* (shadscale); CELA, *Ceratoides lanata* (winter fat); FEUT, *Fendlerella utahensis* (cliff-bush); LEPT, *Leptodactylon* sp. (spiny-phlox); PECA, *Petrophytum caespitosum* (rock spirea); PIMO, *Pinus monophylla* (pinyon pine); PRFA, *Prunus fasciculata* (desert almond); SCRI, *Scopulophila rixfordii* (rock matchweed); SYLO, *Symphoricarpos longiflorus* (snowberry); TETR, *Tetradymia* sp. (horsebrush). Data from Spaulding (1985).

den. Winter fat (*Ceratoides lanata*) occurs infrequently in the regional fossil record but is relatively abundant in the Spc-3A assemblage, dated at 20.2 ka (Fig. 9.6). Thus, although Utah juniper in the Spc-3 assemblages indicates a woodland, the fine-grained soil of the valley flank may have favored shrubs such as sagebrush and winter fat.

Glacial-age Records from the Skeleton Hills (910–940 m). Isolated on the north flank of the Amargosa Desert, the Skeleton Hills are stark, blackened, dolomite inselbergs partially draped with sand sheets. Of the midden sites here, two are from mesic habitats that record both glacial-age woodland and postglacial desertscrub, while a third is in

a xeric setting and provides an early Holocene chronosequence to be discussed later. Despite their fanciful name and barren appearance from a distance, the Skeleton Hills support a diverse desert flora. And, like the Specter Range, they supported woodland during the last glacial age.

The Skeleton Hills-1 (Sk-1) rock shelter faces north-northwest, and mixed desertscrub in the vicinity is dominated by white bursage, creosote bush, shadscale, and desert almond. The oldest assemblage from the site is dated at 36.6 ka and reflects vegetation in which juniper, sagebrush, shadscale, and snowberry appear to have been important (Fig. 9.7). The relative abundance of Utah juniper is com-

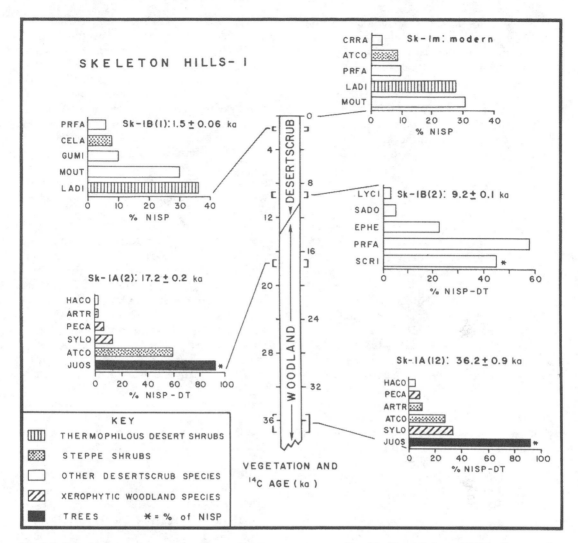

Fig. 9.7. Principal constituents of packrat midden samples and inferred vegetation chronology at the Skeleton Hills-1 site, Amargosa Desert, southern Nevada. Abundance of taxa marked with an asterisk is expressed as a percentage value of *NISP*, rather than *NISP-DT* (number of identified specimens minus number of specimens of the dominant taxon). Radiocarbon dates are expressed in millennia before the present (ka). Abbreviations for plant taxa not listed in Figure 9.6 are: CRRA, *Cryptantha racemosa* (nievitas); EPHE, *Ephedra* sp. (Mormon tea); GUMI, *Gutierrezia microcephala* (matchweed); HACO, *Haplopappus cooperi* (Cooper's goldenbush); JUOS, *Juniperus osteosperma* (Utah juniper); LADI, *Larrea divaricata* (creosote bush); LYCI, *Lycium* sp. (boxthorn); MOUT, *Mortonia utahensis* (tick-weed); SADO, *Salvia dorrii* (sage). Data from Spaulding (1989).

parable to that of the subsequent full-glacial assemblage, dated at 17.2 ka. However, the mesophytes snowberry and sagebrush are reduced in this sample and supplanted by abundant shadscale (60% of *NISP-DT*; Fig. 9.7).

Two samples from the Sk-1 site record vegetation after the demise of the local juniper woodland. The Sk-1B(2) assemblage, dated at 9.2 ka, contains abundant rock matchweed (*Scopulophila rixfordii*; Carophyllaceae), a petrophyte with a stout, woody stalk. Like Mojave prickle-leaf, it is a monotypic genus endemic to the calcareous mountains of the northern Mojave and southern Great Basin deserts. Its abundance in this sample suggests that it was common to abundant on the cliffs around the site. Desert almond and Mormon tea may have been locally abundant on stony soils at this time. Shadscale apparently disappeared by 9.2 ka, only to reappear in trace quantities in the sample dated 1.5 ka. It is again common (13% of *NISP-DT*) in the modern control sample (Fig. 9.7). Rock matchweed, Mormon tea, and desert almond declined in importance in the late Holocene, while matchweed, creosote bush, and tickweed (*Mortonia utahensis*) gained their present dominance.

While the Sk-4 site is on a southeast-facing slope that affords much less shade than the Sk-1 site, it also supported juniper woodland during the late Wisconsin (17.9–15.5 ka; Fig. 9.8). A larger suite of plants is present in the Sk-4 samples compared with the contemporaneous Sk-1A(2) assemblage. Chief among these are ball-flower (*Buddleja utahensis*), cliff-rose (cf. *Cowania mexicana*), and Joshua tree. Cliff-rose is most abundant in the two youngest Sk-4 samples, dated at 16.1 and 15.5 ka. The relatively xerophytic rubber rabbitbrush (*Chrysothamnus nauseosus*), tickweed, and rock matchweed attain their highest percentages in Sk-4A(1), dated at 17.9 ka (Fig. 9.8). Two samples contain a few pinyon pine needles. Assuming the species grew in the vicinity, the record constitutes the lowest (925 m) fossil occurrence of *Pinus monophylla* in the northern Mojave Desert, and the northernmost (36°38' N) full-glacial record of the species.

Point of Rocks (900–910 m elevation). On the xeric southwest-facing side of a ridge at 910 m elevation, the Point of Rocks-3 site (PR-3) yielded six midden assemblages dating from 17.5 to 10 ka (Fig. 9.9). The local vegetation is now dominated by white bursage; creosote bush is common, along with many species tolerant of calcareous soils that also occur at the Eureka View and Specter Range localities. The PR-3 fossil assemblages also reflect a treeless vegetation but are dominated by shrubs no longer found here. Macrofossils of steppe and xerophytic woodland shrubs (shadscale, rubber rabbitbrush, Mormon tea, and snowberry) comprise between 70% and 81% of all identified specimens (*NISP*) in assemblages dating from 17.5 to 12.7 ka. Only four minute twigs of Utah juniper were encountered in these samples, unlike the abun-

dance of juniper found in glacial-age samples from the Skeleton Hills and Specter Range. Evidently there existed a mosaic of vegetation types, with scrub on xeric aspects and woodland on more mesic slopes (Fig. 9.10).

The PR-3 chronosequence appears relatively complacent until between 12.8 and 10 ka, when a major vegetation change occurred. Steppe and woodland shrubs typical of the older PR-3 samples decline at this time (Fig. 9.9). Grasses (Poaceae) are surprisingly abundant in PR-3(8), also the only assemblage to contain appreciable quantities of succulent macrofossils (*Opuntia* spp. and *Yucca brevifolia*).

The PR-1 site (900 m elevation) is adjacent to a large wash, a habitat more mesic than the PR-3 site. The site now supports thermophilous shrubs, including creosote bush, white bursage, matchweed, and boxthorn (*Lycium pallidum*). The nearby wash supports a disturbance-adapted community dominated by cheeseweed (*Hymenoclea salsola*). Of six macrofossil assemblages from this site, three have statistically indistinguishable dates ranging from 11.7 to 11.4 ka and record latest Wisconsin vegetation conditions; two record early Holocene vegetation from 9.8 to 9.4 ka; and one late Holocene sample provides a measure of vegetation change after 3.7 ka (Fig. 9.11).

Around 11.5 ka vegetation at PR-1 apparently was characterized by scale-broom (*Lepidospartum latisquamum*), greasebrush (*Forsellesia nevadensis*), sagebrush, rubber rabbitbrush, and cactus (*Opuntia* spp.). This vegetation changed by 9.8 ka to one in which antelope bitterbrush (*Purshia glandulosa*) was common (Fig. 9.12). The late Holocene sample reveals the modern dominants, indicating that they immigrated between 9.3 and 3.7 ka. The relative abundance of desert rabbitbrush (*Chrysothamnus paniculatus*) in this assemblage is curious; it was found neither in the modern sample nor in the surrounding vegetation. The nearest population is about 50 m away on the shoulders of a highway. Phreatophytic plants are missing from the PR-1 assemblages, suggesting that the wash in front of the site was ephemeral during the period of record.

Holocene-Age Records from the Skeleton Hills (940 m elevation). The Skeleton Hills-2 site (Sk-2) is a xeric, southwest-facing rock

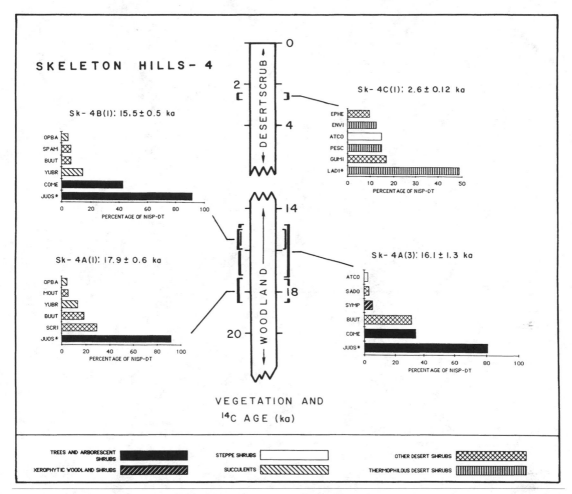

Fig. 9.8. Principal constituents of midden samples and inferred vegetation at the Skeleton Hills-4 site in the Amargosa Desert. Abbreviations follow Figures 9.6 and 9.7 except for BUUT, *Buddleja utahensis* (ball-flower); COME, *Cowania mexicana* (cliff-rose); ENVI, *Encelia virginensis* (rayless brittle-bush); OPBA, *Opuntia basilaris* (beavertail prickly pear); PESC, *Peucephyllum schottii* (desert spruce); SPAM, *Sphaeralcea* cf. *ambigua* (globemallow); SYMP, *Symphoricarpos* sp. (snowberry); YUBR, *Yucca brevifolia* (Joshua tree). Data from Spaulding (1989).

slope near the head of an alluvial fan (Fig. 9.12). Local vegetation is dominated by white bursage in association with creosote bush, shadscale, and desert spruce. The macrofossil assemblages from the Sk-2 midden date from 10 to 8.2 ka and reflect desertscrub apparently dominated by tick-weed (Fig. 9.13). This shrub is now restricted to less xeric habitats in the vicinity.

The two oldest macrofossil assemblages, dated at 10 and 9.9 ka, contain a distinctive species array that includes rubber rabbitbrush, Mormon tea, Joshua tree, ground-thorn, and desert-rue (*Thamnosma montana*). By 9.2 ka these taxa were reduced, with the exception of ground-thorn. Several currently important species first appeared at this time; trace amounts of white bursage, desert spruce, and fluff grass are recorded, as well as common Indian tobacco. The marked increase of fluff grass, desert spruce, and white bursage in Sk-2(1), dated at 8.8 ka, is accompanied by the disappearance or drastic decline of rabbitbrush, Joshua tree, ground-thorn, and Mormon tea (Fig. 9.13). Between 8.8 and 8.2 ka shadscale and rayless brittle-bush immigrated and as-

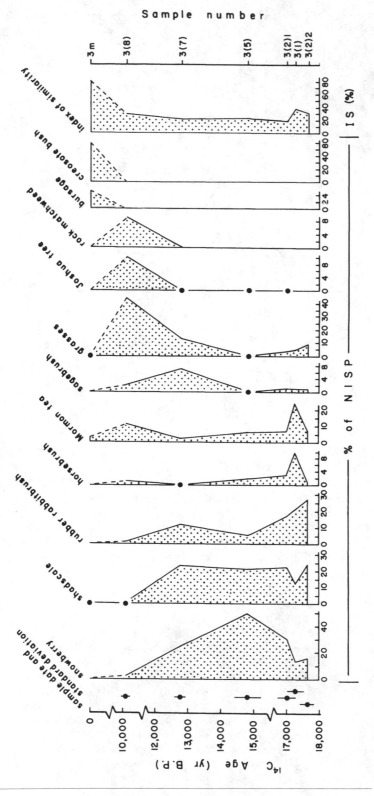

Fig. 9.9. Changes in percentage values of selected plant taxa in midden samples from the Point of Rocks-3 site, Amargosa Desert. Note breaks and changes in scales. Abbreviations follow Fig. 9.3. IS, Sørensen's index of similarity (see chapter 5). Data from Spaulding (1989).

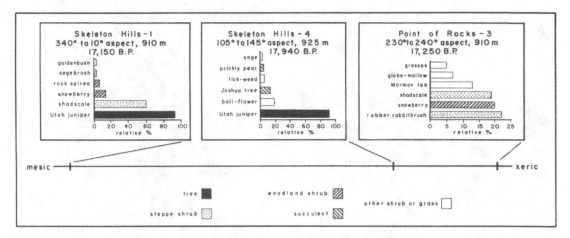

Fig. 9.10. Summary percentage diagrams for late full-glacial assemblages from sites of similar elevation and substrate but with different slope aspects in the Amargosa Desert. The Point of Rocks-3 diagram presents the averaged percentages and dates of PR-3(1), PR-3(2)1, and PR-3(2)2. Percentage values are those of *NISP*, except for the succulents and shrubs in the Skeleton Hills-1 and -4 samples (*NISP-DT*). Data from Spaulding (1989).

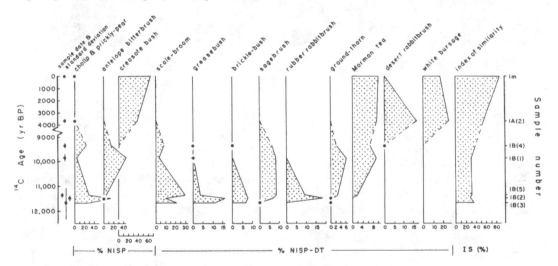

Fig. 9.11. Changes in percentage values of selected plant taxa in midden samples from the Point of Rocks-1 site, Amargosa Desert. Note breaks and changes in scale. Abbreviations follow Fig. 9.3. Data from Spaulding (1989).

sumed importance in the local vegetation. Occasional achenes of the latter in some of the older assemblages are probably contaminants.

This chronosequence shows an episode of increased flux in the local vegetation dating to between 9.9 and 8.8 ka. Plant species now found at elevations exceeding 1000 m, such as Joshua tree, desert-rue, and ground-thorn, were replaced by more xerophytic taxa. Some

of these early Holocene immigrants were in turn reduced or extirpated. Among the many changes that occurred after 8.2 ka, tick-weed became very rare in the present plant association and creosote bush arrived and assumed its modern dominance (Fig. 9.13).

Fortymile Canyon (925–1050 m elevation). Unlike most fossil packrat middens, those from Fortymile Canyon (FMC) occur in Quaternary alluvium and reflect vegetation

Fig. 9.12. View north to the Skeleton Hills-2 site, Amargosa Desert, Nevada. Arrow points to the small cave that contains the midden. The dark shrub in the foreground is creosote bush, a species that was absent from the site until after 8.2 ka.

conditions in a large wash. Despite their youth (< 1.9 ka), they disclose marked late Holocene vegetation change. The desert riparian vegetation at the sites is typified by saltbushes (*Atriplex polycarpa, A. canescens*), desert rabbitbrush, cheeseweed, and creosote bush. Modern middens from the locality indicate that, at present, local packrats prefer creosote bush to other shrubs (Spaulding et al., *this volume*, chap. 5). In contrast, two samples dating to 1.8 and 1.5 ka contain only trace amounts of creosote bush, 2% and 7% of *NISP-DT* relative to 59% to 91% of *NISP* in subrecent to modern samples (Fig. 9.14). They also contain the remains of an extralocal shrub, antelope bitterbush.

The present upper limit of creosote bush–dominated desertscrub in Fortymile Canyon is above 1050 m elevation (Fig. 9.14–9.15), and antelope bitterbrush has been observed no lower than 1170 m in the canyon. The scarcity of creosote bush and presence of bitterbrush in the oldest samples suggest a downslope movement of mesic vegetation between 1.8 and 1.5 ka. A contemporaneous midden 6.4 km farther down Fortymile Canyon yielded abundant creosote bush and but a single bitterbrush leaf, suggesting that, at about 1.8 ka, the upper limit of creosote bush desertscrub was closer to 1000 m elevation (Fig. 9.14). Two subrecent samples show that it returned to near its modern upper limit by 0.5 ka.

Other Records in the Northern Mojave Desert

Spotted Range and Vicinity (1100–1830 m elevation). Packrat middens from the desert ranges of south-central Nevada (Fig. 9.1) show woodland vegetation persisting to the end of the early Holocene at elevations as low as 1250 m (Van Devender, 1977; Wells and Berger, 1967), long after desertscrub developed in the adjacent Amargosa Desert. The youngest record of juniper woodland in the region is dated at 7.8 ka and comes from Mercury Ridge, a spur of the Spotted Range. Conflicting radiocarbon dates of > 40 and 9.4 ka have been reported for the Spotted Range-2 midden, which provides a rare record of mesophytic pinyon-juniper woodland (Wells and Jorgensen,

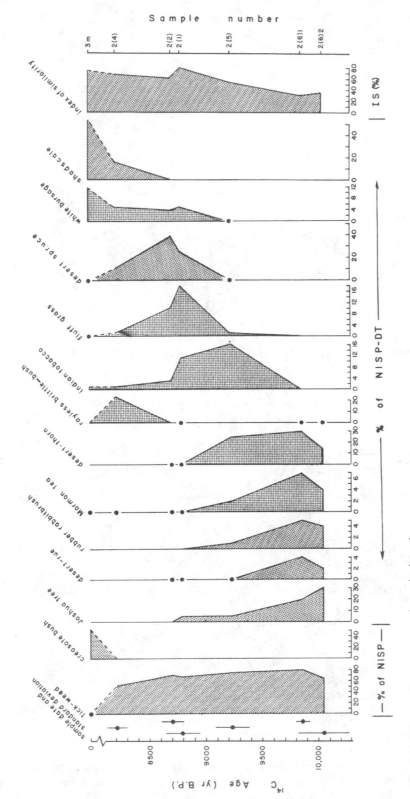

Fig. 9.13. Changes in percentage values of selected plant taxa in midden samples from the Skeleton Hills-2 site, Amargosa Desert. Note breaks and changes in scale. Abbreviations follow Fig. 9.3. Data from Spaulding (1989).

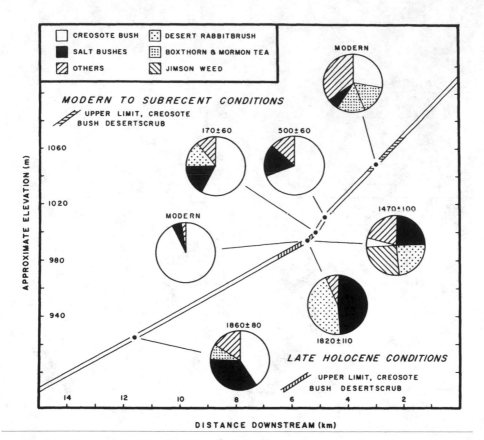

Fig. 9.14. Principal constituents of selected plant macrofossil assemblages from midden sites in the Fortymile Canyon drainage (represented by the diagonal double line). Modern to subrecent samples (upper left) indicate creosote bush dominance to at least 1040 m elevation. Late Holocene samples (lower right) suggest an upper limit below 1000 m for creosote bush. Irregularity of canyon "profile" is due to varying height of sites on canyon wall and varying elevation of tributaries relative to the thalweg of Fortymile Wash. Radiocarbon dates are in years B.P. Data from Spaulding (1989).

1964; Wells and Berger, 1967; Spaulding, 1985). The postglacial replacement of woodland by desertscrub on these small mountain ranges led to the local extinction of Utah juniper and pinyon pine. They persist in high-elevation refugia on larger massifs, such as the Sheep Range.

Sheep Range (1500–2400 m elevation). Late Quaternary vegetation zone changes in the Sheep Range of southern Nevada (Spaulding, 1981) were studied using macrofossil assemblages spanning a 900-m elevational gradient. The vertical displacements of species displayed individualistic responses to climatic change (Fig. 9.15). For example, Utah juniper and limber pine occurred more than 1100 m below their present lower limits, while the

observed depression of white fir and bristlecone pine did not exceed 650 m. Reversing the trend, shadscale ascended to higher altitudes than it occupies today to coexist with subalpine shrubs and conifers (Fig. 9.16; Spaulding et al., 1983). This plant association is not known at present, but it appears repeatedly in midden assemblages from the Mojave and Great Basin deserts (Spaulding, 1981; Thompson, 1984). Release from competition in the context of a cold, relatively dry climate can explain this anomaly (see Spaulding, 1985).

As a consequence of differential displacement, Wisconsin-age vegetation zonation was not analogous to that of the present. While today's pinyon-juniper woodland grades up-

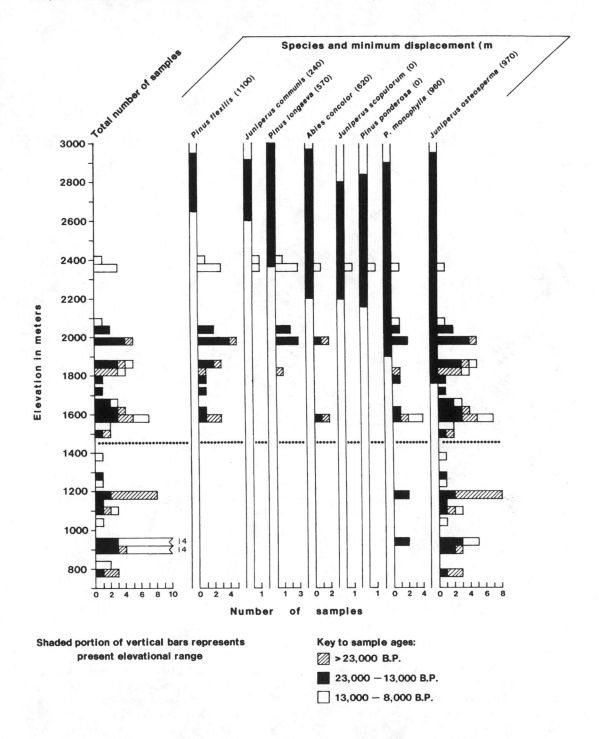

Fig. 9.15. The present elevational range (solid vertical bars) and fossil records (histograms to the right of each bar) of conifers in south-central Nevada (bounded by latitude 36°20′N and 36°40′ N, and by longitude 115° W and 116°20′ W). The modern ranges are for calcareous soils of the Sheep Range (Spaulding, 1981). Records plotted above the dotted line are primarily from the Sheep Range; below they are from smaller mountains and inselbergs that no longer support trees.

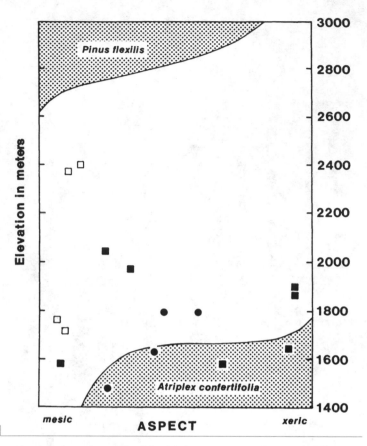

Fig. 9.16. Fossil records of shadscale (*Atriplex con-fertifolia*) and one of its anomalous glacial-age associates, limber pine (*Pinus flexilis*), from packrat middens in the Sheep Range. Solid circles represent midden records of shadscale alone; solid squares represent records of shadscale and limber pine together; and open squares, those of limber pine alone. All fossil records are older than 15 ka.

ward to fir-pine (*Abies concolor–Pinus pon-derosa*) forest, late Wisconsin juniper woodland gave way to subalpine woodland. There is no evidence for the fir-pine zone that currently separates xerophytic pygmy conifer and subalpine conifer woodland (Fig. 9.17). Below 1800 m Utah juniper was widespread while pinyon pine seems to have been an infrequent associate near its glacial-age north-ern limit (Fig. 9.16). There also were dif-ferences in the composition of ice-age sub-alpine woodland. Today limber pine is very rare near the top of the Sheep Range, which it shares with the more abundant bristlecone

pine. The middens indicate it was widespread and more common than bristlecone pine dur-ing the late Wisconsin (Fig. 9.15; Spaulding, 1981).

Eleana Range (1810 m elevation). The Ele-ana Range-2 (ER-2) midden allows reconstruc-tion of late Wisconsin vegetation on the south flank of a mesa just north of the Mojave Desert boundary (Figs. 9.1, 9.18). The modern vegetation is a mosaic of pinyon-juniper woodland and Great Basin desertscrub on tuffaceous rocks and alluvial sediments. In a chronosequence spanning 17.1 to 10.6 ka, samples as young as 13.2 ka disclose abundant

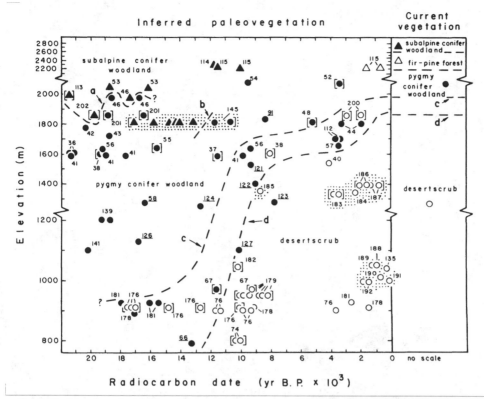

Fig. 9.17. Zonation of major late Wisconsin and Holocene vegetation types in central-southern Nevada. All sites are from calcareous rocks except for those encompassed by the stippled zones. Brackets indicate samples from xeric (south-east to west aspect) sites; underlined sample numbers (from Wells and Jorgensen, 1964; Wells and Berger, 1967) are not controlled for aspect. Dashed lines indicate approximate ecotone (vegetation transition) positions and represent (a) the ecotone between late Wisconsin subalpine and pygmy conifer woodland in the Sheep Range; (b) a similar ecotone in the Eleana Range; (c) the woodland-desertscrub ecotone at mesic to intermediate sites; (d) the woodland-desertscrub ecotone at xeric sites. Site numbers are those listed in Table 2 of Spaulding (1987). Sample area follows Figure 9.15. Solid triangles indicate samples containing abundant limber pine; solid circles indicate abundant juniper; open circles lack conifer remains or contain rare juniper.

limber pine with steppe shrubs (*Artemisia* sec. Tridentatae, *Chrysothamnus* spp., *Tetradymia* sp.). Between 13.2 and 11.7 ka this community gave way to pinyon-juniper woodland (Figs. 9.18, 9.19). Although pinyon pine and Utah juniper persist today, changes occurred in the understory after 10.6 ka. Although rare today, prickly pear cactus (*Opuntia* cf. *erinacea*) was apparently common during the latest Wisconsin (Fig. 9.19). Four-wing saltbush, presently dominant on this south-facing slope, occurs in the youngest assemblage only and in only trace amounts. Gambel oak (*Quercus gambelii*), currently important in nearby mesic woodland, is absent from the ER-2 assemblages. Presumably it withdrew from the northern Mojave during glacial times and had yet to arrive at this site by 10.6 ka.

The Eleana Range record provides the opportunity to compare a detailed chronology of vegetation changes with two other well-dated records from the Mojave Desert, those from Searles Lake in the west-central Mojave (Smith, 1979; Smith and Street-Perrott, 1983) and Rampart Cave in the lower Grand Canyon (Long et al., 1974; Fig. 9.1). Between about 23

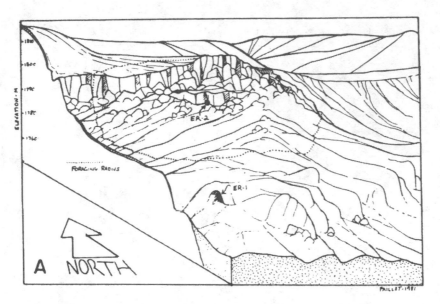

Fig. 9.18. Reconstruction of conditions at the Eleana Range-2 site, Nevada. (A) The physiography of the site, dotted line indicates the 30-m distance arbitrarily chosen to represent the extent of packrat foraging activity; (B) the present vegetation at the site; (C) reconstructed limber pine–steppe shrub woodland recorded in samples dating from 17.1 to 13.2 ka; (D) reconstructed pinyon-juniper woodland recorded in samples dating from 11.7 to 10.6 ka. Note the inferred continuity of physiognomy of vegetation from (C) to (D), despite the fact that the tree species are different. Drawings by F. L. Paillet, U.S. Geological Survey, Denver. From Spaulding et al. (1984).

and 16 ka the Searles basin held a large freshwater lake. Desiccation occurred by 14 ka and left a shallow, saline lake in the Searles basin (Smith, 1979). At about the same time (14.8–13.1 ka), ground sloths (*Nothrotheriops shastense*) reoccupied Rampart Cave. The cave stratigraphy reveals that they were present during the middle Wisconsin interstadial, but not during full-glacial times (Martin et al., 1961; Long et al., 1974). The only contemporaneous vegetation change apparent at the Eleana Range was a gradual increase in two thermophiles, bitterbrush (*Purshia triden-*

tata) and dwarf goldenweed (*Haplopappus nánus*). In other words, the climate change that forced the partial desiccation of Searles Lake and allowed the return of ground sloths to Rampart Cave did not appear to cause a pronounced vegetation change at the Eleana Range. That change is evident only in the increasing abundance of thermophiles; there was no attendant turnover in dominants at the site. The apparent persistence of subalpine conifers and steppe shrubs after the decline in Searles Lake appears to indicate lack of major vegetation response to increasing aridity (i.e.,

vegetational inertia; see Cole, 1985).

The resurgence of Searles Lake to full-stage conditions at about 12 ka is correlated with the failure of limber pine and steppe shrubs and the development of pinyon pine–juniper-cactus vegetation at the Eleana Range (Fig. 9.19). While a conventional explanation would attribute this vegetation change to increasing aridity, the Searles Lake record seems to invalidate that hypothesis. Instead, Spaulding and Graumlich (1986) have proposed that it was due to the inception of a climatic regime typified by increased monsoons

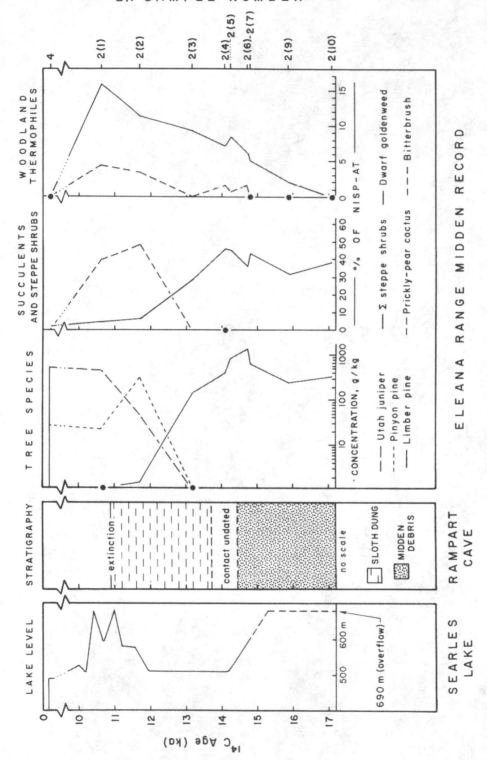

Fig. 9.19. The Eleana Range-2 chronosequences compared to the late Wisconsin lake-level history of pluvial Searles Lake (left; Smith, 1979), and the stratigraphy of Rampart Cave in the Lower Grand Canyon (second from left; Long et al., 1974). Figure modified from Spaulding et al. (1984).

and annual temperatures near the modern mean. This second, monsoonal pluvial was forced by enhanced solar radiation during the summer half-year—the Milankovitch effect discussed by Kutzbach and others (Kutzbach, 1983; Kutzbach and Street-Perrott, 1985; Ritchie et al., 1985). It was this radical change in precipitation seasonality that likely favored pinyon pine, juniper, and cactus, and forced the vegetation turnover at the Eleana Range site (Fig. 9.19). The disappearance of ground sloths from Rampart Cave at around 11 ka would be difficult to attribute to this type of climatic change.

Lower Grand Canyon (425–675 m elevation). In northwestern Arizona the Mojave Desert penetrates the lower Grand Canyon (Fig. 9.1), where it is enriched by a number of species absent from the rest of the region. These include ocotillo (*Fouquieria splendens*), typical of the Sonoran and Chihuahuan deserts, and Whipple yucca (*Yucca whipplei* ssp.

caespitosa). Mojave Desert occurrences of Whipple yucca are restricted to this area and to the mountain flanks of the Mojave's western boundary, hundreds of kilometers to the west in California (Benson and Darrow, 1981). Ocotillo and Whipple yucca are absent from the fossil record of the lower Grand Canyon, as are many other elements of the modern desert flora.

The lower Grand Canyon is an area of high relief (Fig. 9.20), and its midden record (Phillips 1977; Mead and Phillips, 1981) includes thirty dated samples from sixteen sites (Fig. 9.21) located primarily in side canyons that empty into the Colorado River. Sampling strategy differed from that used at most other localities discussed here in that only middens containing the remains of extralocal juniper were collected. Perhaps not surprisingly, then, juniper is abundant in most of the samples from the lower Grand Canyon, but other species have more variable records. Currant

Fig. 9.20. View northeast from Rampart Cave in the lower Grand Canyon of Arizona. Photograph by T. R. Van Devender.

(*Ribes* sp.) and hop-hornbeam (*Ostrya knowl-toni*) commonly appear only in full-glacial samples (Fig. 9.21). The broad pattern of late-glacial change appears consistent with the Eleana Range record. Phillips's (1977) full-glacial vegetation stage II ends at about 14 ka, and stage III vegetation is characterized by more records of single-leaf ash (*Fraxinus anomala*) and desert species such as desert al-mond and Mormon tea. Shadscale, the only shrub of Great Basin affinity common in this record, is first absent from sample MG-1, dated at 12.4 ka. Its last dated record is from VC-14 (11.9 ka; Fig. 9.21), making its decline coincident with that of the steppe shrubs at the Eleana Range (13.2–11.7 ka; Fig. 9.19). Other species lingered a bit longer; black-brush and prickly pear (*Opuntia basilaris, O. erinacea*) last appeared at about 11 ka (the end of stage III). Phillips's (1977) stage IV is the last period in which extralocal species such as juniper and single-leaf ash occurred, but at this time they shared dominance in the mid-dens with many modern desert species.

The lower Grand Canyon midden record (Fig. 9.21) reflects paleovegetation in the im-mediate vicinity of Rampart Cave. A marked change in that record could be expected at about 11 ka if ground sloth extinction was precipitated by major vegetation changes. However, a pronounced change in community type, such as that which occurred at the Ele-ana Range between 11.7 and 13.2 ka (Fig. 9.19), is not evident (Fig. 9.21). Vegetation change was under way well before ground sloth extinction and continued for several millennia thereafter (Phillips, 1984). Because there is an absence of evidence for substantial vegetation change correlated with extinction, and because ground sloths relied on plant spe-cics still widespread in the lower Grand Can-yon (Hansen, 1978), it is difficult to relate extinction of this herbivore to climate-induced vegetation changes.

REGIONAL VEGETATION SINCE THE LAST GLACIAL MAXIMUM

Mojave Desert packrat middens have been used to trace the expansion of desert vegeta-tion since the last glacial maximum and the restriction of woodland to isolated highlands. Within the context of broad altitudinal and latitudinal shifts in species's ranges, changes

in plant associations at individual sites pro-vide examples of local alterations in commu-nity composition. It is not my intention to define an exact time when glacial age gave way to interglacial, or to propose time bound-aries for the late Quaternary of the Mojave Desert. Biotic changes were largely time-transgressive, and apparently they varied from locality to locality in their magnitude and pace. Glacial-age vegetation conditions did not end precisely at 12 ka (or, for that matter, at 11 ka), nor did the transitional phase of Mojave Desert vegetation end at exactly 8 ka. These divisions merely prove useful in group-ing the fossil data for assessment and discussion.

Glacial-Age Vegetation Conditions (18–12 ka)

Below 1000 m full-glacial middens in the Mo-jave Desert record xerophytic juniper wood-land and, at sites in Death Valley and the Amargosa Desert, desertscrub. At intermedi-ate elevations (1000–1800 m) records of juni-per and pinyon-juniper woodland prevail (Figs. 9.15, 9.17). The highest sites, on the flanks of the Sheep, Eleana, and Spring ranges, and at Clark Mountain (Mehringer and Ferguson, 1969; Fig. 9.1), reveal subalpine woodland characterized by the xerophytic, cold-tolerant conifers limber pine and bristlecone pine (Thompson and Mead, 1982; Wells, 1983).

Desertscrub of the Last Glacial Age. Until recently, evidence for glacial-age desertscrub was lacking. Data now indicate that, at least in the northern Mojave region, some low-elevation sites may have always been treeless. On xeric exposures near the close of the full glacial (about 17.5 ka; Fig. 9.17) the lower limit of woodland was *above* 900 m clcva-tion. Contemporaneous middens from differ-ent habitats in the Amargosa Desert reveal a dichotomy in vegetation type depending on slope aspect. Juniper woodland occupied north- to southeast-facing slopes, while des-ertscrub occurred on southwest-facing slopes (Fig. 9.10).

The oldest records of desertscrub presently known date to 17.5 ka. A midden from Death Valley dated at 19.6 ka documents Utah ju-niper on south-facing slope at 425 m eleva-tion (Wells and Woodcock, 1985), 500 m lower than the PR-3 site. The next youngest sample

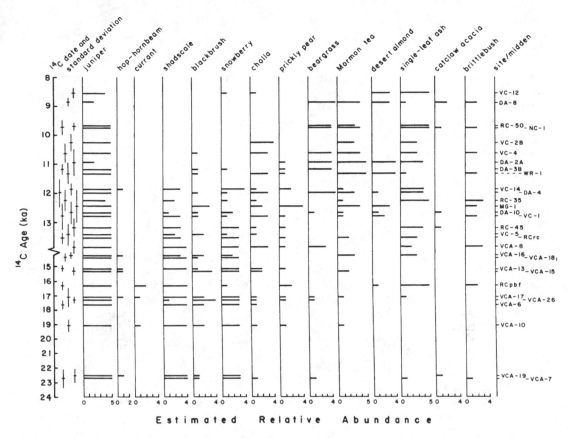

Fig. 9.21. The estimated relative abundance of selected plant taxa in late Wisconsin and early Holocene middens from the lower Grand Canyon, Arizona. Site abbreviations are: DA, Desert Almond Canyon; MG, Muav Gate; NC, Needle-eye Canyon; RC, Rampart Cave; VC, Vulture Canyon; VCA, Vulture Cave; WR, Window Rock. Data from Phillips (1977), Mead and Phillips (1981).

from the site, dated at 17.1 ka (contemporaneous with the oldest PR-3 records; Fig. 9.9), contains no juniper. It is therefore possible that woodland was more widespread prior to 17.5 ka, and the oldest desertscrub middens from Death Valley and the Amargosa Desert postdate woodland retreat near the close of the glacial maximum.

Glacial-age desertscrub lacked the thermophiles that presently abound in the Mojave Desert. Instead, cold-desert and woodland shrubs like Mormon tea, shadscale, rubber rabbitbrush, and snowberry were important. In terms of the classification of Brown et al. (1979), fossil assemblages from the PR-3 site (910 m elevation) reflect cold-temperate desertscrub, in contrast to current warm-temperate desertscrub. Indices of similarity between desertscrub assemblages older than 10

ka and modern midden samples range from 19% to 38%.

At lower elevations (425 m) in Death Valley, Wells and Woodcock (1985) report glacial-age "semi-desert" in which Whipple yucca apparently played an important role. Whipple yucca also was present at the Marble Mountains (Fig. 9.2) and at the Whipple Mountains in the Colorado River trough (King and Van Devender, 1977). These records are of interest because, as mentioned previously, Whipple yucca is presently restricted to western California and the Grand Canyon. Isolation of the Grand Canyon population of this species could have occurred at the close of the last glacial age, as might be implied by Wells and Woodcock's (1985) treatment. However, the macrofossil record from the lower Grand Canyon reveals an enigma. Whipple yucca is not present in fossil middens from that area, and it is not known when it first became established there.

Glacial-Age Woodlands of the Mojave Desert. In the last glacial age woodland extended across the lowlands of the Mojave Desert (Mehringer, 1965, 1967; Van Devender and Spaulding, 1979; Wells, 1983). Just as thermophilous desertscrub typifies the current warm deserts of the Southwest, pygmy conifer woodland apparently typified the region during the Wisconsin glacial age. Regional differences in this paleowoodland are poorly understood, and ice-age phytogeographic boundaries have not been determined. Five low-elevation (< 1200 m) full-glacial middens from the Sonoran region (Van Devender, 1973; King and Van Devender, 1977) yielded the remains of seven tree and arborescent shrub species, fourteen succulents (Cactaceae and Agavaceae), and two steppe shrub species. Eleven full-glacial Mojave samples yielded eight steppe shrub taxa, with only three tree and arborescent shrub species (Utah juniper, pinyon pine, and cliff-rose) and seven succulents (Spaulding, 1985; Wells and Woodcock, 1985). Then (as now) the Sonoran region harbored more species of trees and succulents and fewer steppe shrubs. Oaks (*Quercus turbinella, Q. dunnii, Q. gambelii*) apparently were absent from the Mojavean region until the last millennia of the late Wisconsin. Full-glacial middens from the lower Grand Canyon (Mead and Phillips, 1981; Phillips, 1984) have revealed an increase in succulent diversity toward the southeastern periphery of the present Mojave Desert (Fig. 9.21). Beargrass (*Nolina microcarpa*), yuccas (*Yucca schidigera, Y. baccata*), and barrel cacti (*Echinocactus polycephalus, Ferocactus acanthodes*) are recorded here but nowhere else in Mojavean woodlands of the last glacial maximum (Spaulding, 1985; Wells and Woodcock, 1985). The differences apparently reflect the higher latitudinal position and consequent drier and colder climate of most of the Mojave region (Spaulding, 1985; Spaulding and Graumlich, 1986). North of the Mojave, pygmy conifer woodland apparently was absent from the central Great Basin (Thompson, 1984; *this volume,* chap. 10; Wells, 1983). The northernmost full-glacial record of pinyon pine is from the Sk-4 site, while Utah juniper oc-

curred at least as far north as the Eureka Valley (Fig. 9.1). The discovery of pinyon pine in glacial-age packrat middens from the *west* side of the Sierra Nevada (Cole, 1983; *this volume,* chap. 11) indicates that elements of Mojavean woodland extended well into cismontane California during the late Wisconsin.

Vegetation in Transition (12–8 ka)

Worldwide deglaciation began between 12 and 16 ka (Ruddiman and Duplessy, 1985). In the southern part of the continent vegetation changes brought on by global climatic warming may have started by 15 ka (Delcourt and Delcourt, 1984; Spaulding et al., 1984). Pollen-stratigraphic data from Tule Springs in the Las Vegas Valley (Mehringer, 1967) and the chronosequence from the ER-2 site (Fig. 9.19) indicate major vegetation changes by about 12 ka. By 10 ka desertscrub was widespread below 1000 m elevation (Fig. 9.17). This vegetation has been termed "anomalous desertscrub" because extralocal shrubs are abundant and most modern elements, such as creosote bush, are lacking (Spaulding, 1985). Latest Wisconsin vegetation changes at most sites involved reduction of steppe shrubs such as shadscale, rabbitbrush, and sagebrush, and increases in succulents and grasses (Figs. 9.9, 9.13, 9.19). Figure 9.22 summarizes changing relative abundances of steppe shrubs, grasses, succulents and thermophiles in midden samples from the Amargosa Desert. Steppe shrubs, dominant in the 18- and 15-ka age classes, give way to succulents and grasses in the 12-, 10-, and 9-ka age classes. The early Holocene advent of thermophiles is evident in a comparison of the 9- and 8-ka classes.

Dynamics of Desertscrub Modernization. Plant community turnover in response to late-glacial climatic change was not a simple combination of local extinction of ice-age elements and immigration of modern species. At most sites there were latest Wisconsin and early Holocene immigrants that subsequently underwent local extinction. At the Marble Mountains juniper, sagebrush, Joshua tree, and Whipple yucca arrived in the early Holocene but did not survive the middle Holocene (Fig. 9.2). At the Eureka View locality, a depauperate early Holocene flora was supplemented by the arrival of rayless brittle-bush

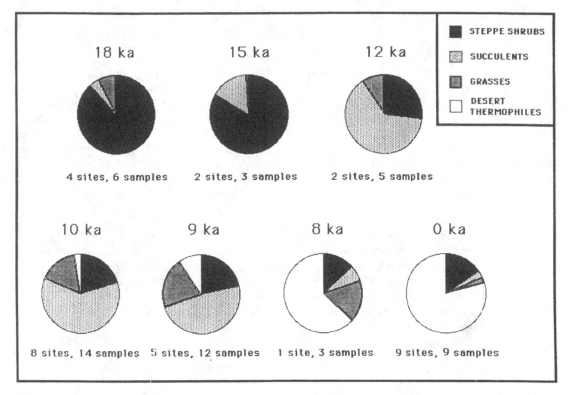

Fig. 9.22. Averaged relative abundance of selected plant types in macrofossil assemblages from below 1200 m elevation on the periphery of the Amargosa Desert. All assemblages dating within 1000 years of the specified age are considered in averages for that age class. Steppe shrubs are *Artemisia* subgen. Tridentatae, *Atriplex confertifolia, Ceratoides lan-* *ata, Chrysothamnus nauseosus, C. viscidiflorus,* and *Tetradymia* spp.; succulents are all Cactaceae and Agavaceae; grasses, all Poaceae; and desert thermophiles are *Larrea divaricata, Ambrosia dumosa, Eucnide urens, Encelia virginensis, Peucephyllum schottii,* and *Tidestromia oblongifolia.* From Spaulding and Graumlich (1986).

and boxthorn, species that disappeared at the end of the middle Holocene (Fig. 9.5). The appearance and subsequent extirpation of rock matchweed and Joshua tree at PR-3, and rayless brittle-bush at Sk-2 (Figs. 9.9, 9.13) are further examples of the singular nature of latest Wisconsin and early Holocene vegetation.

Patterns of immigration cannot be predicted from the present distributions of the plant species involved. For example, brittle-bush (*Encelia farinosa*) is restricted to the southern Mojave Desert, while creosote bush and white bursage occur much farther north (Hastings et al., 1972). Since creosote bush and bursage tolerate a wider range of environmental conditions, one might expect them to appear first in the southern Mojave Marble Mountains time series. Instead, they are not reliably recorded until 7.9 ka, while brittle-bush appeared by 9.5 ka (Fig. 9.2).

Immigration of desert thermophiles from their refugia was staggered; some plants arrived much later than their common modern associates. A case in point is provided by the contrasting histories of white bursage and creosote bush. At Marble Mountains white bursage and creosote bush are reliably recorded at 7.9 ka, concordant with the first record of white bursage farther west in the Lucerne Valley. However, creosote bush does not appear in the Lucerne Valley sites until the next-youngest sample, dated at 5.9 ka (King, 1976). Farther north, white bursage was present at Death Valley by 10 ka (Wells and Woodcock, 1985), and it was well established at the Skeleton Hills by 8.8 ka. Creosote bush, on the other hand, did not arrive there until after

8.2 ka (Fig. 9.13). At the northernmost Eureka View locality, bursage first appeared at 6.8 ka, while creosote bush was first found (in trace amounts) in a sample dated 5.4 ka (Fig. 9.5). White bursage appears to have been more rapidly dispersed throughout the Mojave than creosote bush, indicating that common associates may have had different migrational histories.

Changes toward Modern Woodland. Extinction of low-elevation conifer populations and progressive restriction of woodland to isolated mountain masses accompanied the expansion of desertscrub. Unfortunately, it has been impossible to firmly establish the time of woodland extirpation at any one site. The transitions are lost in the substantial time gaps that appear in midden sequences at about the time of local woodland extirpation. The composite record for southern Nevada shows that woodland persisted longer at higher elevations and in less xeric habitats (Fig. 9.17). In the north-central Mojave Desert packrat middens as young as 7.8 ka record lingering woodland on treeless desert mountain ranges. However, on inselbergs in the adjacent valleys woodland gave way to desertscrub by at least 10 ka.

If woodland persisted longer at higher elevations and less xeric sites, one might well expect that it persisted longer at higher latitudes. A comparison of Mojave Desert middens with those from the south and east shows that this is not the case. Widespread desert vegetation appears to have developed earlier in the Mojave than in the Sonoran and Chihuahuan deserts (Spaulding, 1983). Middens from as close as the lower Grand Canyon (Fig. 9.21) show that juniper woodland persisted until about 8.5 ka at elevations more than 400 m lower than those providing contemporaneous records of desertscrub in the core of the Mojave Desert.

As mentioned, oaks are not recorded in the full-glacial Mojavean woodlands, although today Gambel oak occurs as far north as Pahute Mesa, and shrub live oak reaches the Spring Range (Fig. 9.1). There is some evidence for northward spread of oaks during the latest Wisconsin and early Holocene, prior to the final demise of low-elevation woodlands. Older middens from the Newberry Mountains, dating from 19.6 to 12.6 ka, contain pinyon pine and juniper without oak. The younger two, dated at about 9.5 ka, contain the remains of oak (Leskinen, 1975). Records of oak from the Robbers Roost locality, dating from 13.8 to 12.8 ka, are contemporaneous with a marked increase in oak pollen percentages in sediments from Searles Lake dating from about 14 to 10 ka (Spaulding et al., 1983; Figs. 9.13–9.16). This may reflect the increase of oak in woodlands of upwind, pollen-source areas like Robbers Roost. Northward migration of oak after the close of the full glacial was recorded by Davis et al. (1984) at Bechan Cave in southeastern Utah, where a rise in its pollen percentages is dated at about 12.9 ka.

Oak was not alone in migrating from southerly latitudes (or expanding from relictual populations). Other mesophytic trees and shrubs that expanded during the postglacial include ponderosa pine, Rocky Mountain juniper (*Juniperus scopulorum*), skunk bush (*Rhus trilobata*), and (farther north) pinyon pine. These invasions are clearly separate from the progressive restriction of woodland to isolated, high-elevation habitats. There may have been but a brief interval of time—after the initiation of latest Wisconsin warming and before the final demise of the pluvial precipitation regime—when moisture-loving, montane-temperate species could have spread northward through woodlands that were nearly continuous. After that, long-distance dispersal would necessarily be the only mechanism by which species such as oak or ponderosa pine could reach isolated montane habitats that were inhospitable during the last glaciation, and hospitable but isolated during the current interglaciation.

Vegetation Changes in the Last Eight Thousand Years

The McCullough Range and Eureka View time series conflict (Figs. 9.3, 9.5). At the former, middle Holocene vegetation was noticeably impoverished, while at Eureka Valley middle Holocene vegetation appears as rich in desertscrub species as late Holocene vegetation. The transition to modern vegetation at both localities involved the immigration of species important today. At Peanut Butter Pass the new floristic elements were not as strongly drought-adapted, and only one local extirpation accompanied their arrival (Fig.

9.3). At Eureka View the immigrants were xerophytes (creosote bush, desert spruce) and some relative mesophytes (California buckwheat); and there were some losses (indigo bush, prickly pear) (Fig. 9.5).

The McCullough Range record of a richer and less drought-tolerant late Holocene community agrees with studies from the Sheep Range and Fortymile Canyon, and with pollen-stratigraphic evidence from Little Lake (Mehringer and Sheppard, 1978; Fig. 9.1). In the Sheep Range, middens dating from 3.5 ka to 2 ka revealed pinyon-juniper woodland 100–200 m below its current lower limits (Spaulding, 1977, 1981); its consequent upward retreat is undated. At Fortymile Canyon, middens suggest a depression of the upper limit of creosote scrub between 1.8 and 1.5 ka, and its upward expansion between 1.5 and 0.5 ka. At Little Lake, northwest of Searles Lake, a transition to less arid conditions occurred at about 3 ka. A return to somewhat drier conditions took place within the last millennium (Mehringer and Sheppard, 1978). These data do not entirely agree with reconstructed late Holocene vegetation changes in the Greenwater Valley (Fig. 9.1; Cole and Webb, 1985), where middens demonstrate a downward shift in the lower elevational limit of blackbrush desertscrub between A.D. 1435 and A.D. 1795, a relaxation of aridity more recent than those recorded at Fortymile Canyon and at the McCullough and Sheep ranges.

SUMMARY AND CONCLUSIONS

Plant macrofossils from ancient packrat middens offer a detailed view of local vegetation changes that took place as present-day Mojave Desert ecosystems developed. These changes were pronounced over the past eighteen thousand years, and their chronologies provide a complex record of ecological dynamics in response to shifting climatic controls.

Vegetation Dynamics

Just as modern plant associations differ from site to site, sequential changes in local vegetation differed at all sites studied. Important contrasts occurred in both the timing and components of shifting plant associations. Given the present record, the varied histories of immigrations and extirpations appear

highly individualistic. While two species of desert shrubs may have arrived together at one site, at another site one may have preceded the other by thousands of years. New studies on a neighboring inselberg are necessary to replicate, say, the Sk-2 record. Did the arrival and departure times (Cole, 1985) of species vary substantially among sites in a restricted area such as the Amargosa Desert? This may be expected if heterogeneity of habitat types, which is pronounced in rugged terrain, dictated substantial site-to-site differences in the chronology and components of vegetation changes. Alternatively, species such as creosote bush and white bursage might have occupied all available habitats in the valley within several centuries. If this were the case, we may expect good correlation between arrival and departure times at different sites in the Amargosa Desert. Differences in the timing of immigrations (but not necessarily local extinctions) would be noted only when comparing contemporaneous time series from regions with significantly different climates (i.e., the Marble Mountains and Skeleton Hills).

During the last glacial age, juniper and pinyon-juniper woodlands were widespread in the current Mojave Desert. In the northern interior desertscrub was probably widely distributed on xeric sites. That desertscrub was present at all during the last glacial age is a revelation; its subsequent expansion during the latest Wisconsin and early Holocene contrasts with the idea of widespread woodland persistence until around 7.8 ka (Van Devender and Spaulding, 1979). Desertscrub development during the latest Wisconsin and early Holocene may have been punctuated by brief reversals. Mehringer (1967) notes two moist episodes at Tule Springs, dated at 10.5–10.0 ka and 8.5–8.0 ka. Another possible reversal is the Marble Mountains woodland period of 8.9 ka.

The regional expression of immigrations and extinctions after the end of the full glacial is seen in the expansion of the coverage and altitudinal range of desertscrub. Woodland retreated to higher altitudes and became restricted to isolated, montane islands. The northward migration of mesophytic, relatively thermophilous trees and shrubs was probably hindered by increasing aridity in the

lowlands that separate montane islands. Desertscrub development progressed from low to
high elevations, and apparently began much
earlier in the interior Mojave Desert than in
regions to the south and east.

Paleoclimates and Bioclimatic Phenomena

During the full glacial the northern Mojave
Desert was colder and drier than that called
for in most models of a "pluvial" climatic regime. Current reconstructions call for a decline in average annual temperature of at
least 6°C, and not more than a 40% increase
in average annual precipitation (Dohrenwend,
1984; Spaulding, 1985). The changes herald-
ing an end to this climatic regime began after
16 ka, were well under way by about 11 ka,
and involved alterations of seasonal temperature and precipitation regimes. At the ER-2
site increases in bitterbrush and dwarf goldenweed began well before the final demise of
subalpine woodland (Fig. 9.19) and may reflect
increasing temperatures (Spaulding et al.,
1984). That and the failure of the intensified
westerly airflow typical of the last glacial
maximum (Spaulding and Graumlich, 1986)
account for the decline in pluvial Lake
Searles by around 14.5 ka (Fig. 9.20; Smith,
1979).

   Warming temperatures during the latest
Wisconsin and early Holocene forced the
northward migration of many biotic components. There also were responses to changing
seasonal precipitation regimes, which in turn
were forced by changes in boundary conditions and global temperature patterns (Kutzbach and Guetter, 1986). These probably were
not restricted to plant species. The re-
appearance of the Shasta ground sloth at
Rampart Cave (Fig. 9.19) may have been related to warmer winters (Spaulding et al.,
1984). While the low-elevation Gypsum and
Rampart Cave sites provide middle Wisconsin
dates on *Nothrotheriops* (Long et al., 1974;
Thompson et al., 1980), there are no sloth
records from the Southwest reliably dated
to the full glacial (broadly, 21–15 ka). Most
megafauna remains from well-dated cave
strata, other than those of cervids (deer and
elk) and caprines (sheep and goats), actually
begin at about 13 ka (e.g., Davis et al., 1984;
Hansen, 1980; Long et al., 1974; Spaulding

and Petersen, 1980; Ven Devender et al.,
1979).

   The advent of pinyon-juniper woodland in
the Eleana Range (Fig. 9.19) and replacement
of steppe shrubs by grasses and succulents in
the Amargosa Desert (Fig. 9.22) occurred between 13 and 12 ka, when Searles Lake returned to full-stage conditions. Increased
precipitation accompanying warmer temperatures is required to account for all these phenomena (Spaulding et al., 1984). The increase
in effective moisture brought about by a postulated latest Wisconsin pluvial (Spaulding
and Graumlich, 1986) was insufficient to re-
tard desertification in the interior Mojave
Desert. But on the periphery of the Mojave,
closer to the Pacific Ocean and the Gulf of
Mexico, woodland persisted millennia longer.
Intensified summer monsoons in response to
increased summer insolation from 12 to 9 ka
(Kutzbach and Street-Perrott, 1985) account
for low-elevation woodlands in the Sonoran
Desert and trough of the Colorado River
(Spaulding, 1983), the return of Searles Lake
to high-stand conditions, and the abundance
of succulents in Mojavean desertscrub at this
time (Spaulding and Graumlich, 1986). These
enhanced monsoons, and not increasing
aridity, may have forced the major vegetation
change at the Eleana Range site (Fig. 9.19).

   The advent of the thermophilous shrubs in
the Mojave Desert, disappearance of succulents, and elimination of woodland from all
low-desert habitats by 7.8 ka mark the collapse of the latest Wisconsin to early Holocene pluvial (Spaulding and Graumlich, 1986).
The slow spread of creosote bush to reach its
northern limit by about 5.5 ka may indicate
either lower winter temperatures or a slow
dispersal rate. Middle Holocene aridity is
clearly reflected by the McCullough Range
middens. Other records from farther north in
the Great Basin also indicate minimum effective moisture between 7.5 and 5.5 ka (e.g.,
Currey and James, 1982; Thompson and
Kautz, 1983). On the other hand, middle Holocene aridity is not obvious in the packrat
midden record from Eureka Valley.

   Dendroclimatic evidence from the White
Mountains of California shows that reduction
of growing-season temperatures occurred in
the transition from the middle to late Holo-

cene (La Marche, 1973). Increased effective moisture, relative to the present, is evident at most Mojave Desert localities between 3.8 and 1.5 ka. Modest increases in winter precipitation may have accompanied decreasing temperatures during this period. Movement of vegetation zones upslope by as much as 100–200 m between 1.5 and 0.5 ka marks the end of the late Holocene "neopluvial."

Future Directions

Research in the Mojave Desert over the past two decades has provided a wealth of paleoecological information. The richness of the data itself becomes burdensome because its analysis involves multidimensional comparisons and correlations. Readers lacking field experience in the Mojave Desert will be at a disadvantage. Computer-based data-retrieval systems will be essential to utilize the growing packrat midden record effectively. Site-specific chronosequences and quantitative-analysis techniques also are necessary to resolve the details of late Quaternary vegetation and climatic change.

Much remains to be done in order to perfect our picture of paleoenvironmental dynamics of the Mojave Desert. The list of "unknowns" is large. Refined analyses will improve our understanding of the dynamics of plant community alteration, shifts in vegetation zone composition and position, plant species migration rates and driving mechanisms, paleophytogeographic patterns, the nature of paleoclimates and climatic change and, not least, the limitations and biases of the midden method.

ACKNOWLEDGMENTS

Funding for this research was provided by the U.S. Geological Survey, the National Science Foundation, the California Bureau of Land Management, and Applied Conservation Technology, Inc. My thanks to J. L. Betancourt, P. S. Martin, D. K. Grayson, and T. R. Van Devender for reviewing earlier versions of this manuscript, and to L. J. Graumlich and J. R. Sharp for their assistance in computer analyses. Data-base analyses were made possible with additional funds provided by the Department of Botany and the Graduate School of the University of Washington. Research was done at the Laboratory of Arid Lands Paleoecology, Quaternary Research Center, University of Washington.

REFERENCES

Beatley, J. C. (1975). "Climates and vegetation pattern across the Mojave/Great Basin Desert transition of southern Nevada." *American Midland Naturalist* 93, 53–70.

———. (1976). *Vascular plants of the Nevada Test Site and central-southern Nevada.* National Technical Information Service, TID-26881.

Benson, L., and Darrow, R. A. (1981). *Trees and shrubs of the southwestern deserts.* 3d ed. Tucson: University of Arizona Press.

Bostick, V. B. (1973). "Vegetation of the McCullough Mountains, Clark County, Nevada." Master's thesis, University of Nevada, Las Vegas.

Bradley, W. G., and Deacon, J. E. (1967). "The biotic communities of southern Nevada." In *Pleistocene studies in southern Nevada.* H. M. Wormington and D. Ellis, eds., pp. 129–200. Nevada State Museum Anthropological Papers 13.

Brown, D. E.; Lowe, C. H.; and Pase, C. P. (1979). "A digitized classification system for the biotic communities of North America, with community (series) and association examples for the Southwest." *Journal of the Arizona-Nevada Academy of Sciences* 14, supp. 1, 1–16.

Brown, D. E., and Lowe, C. H. (1980). *Biotic communities of the Southwest.* U.S. Department of Agriculture, Forest Service, General Technical Report RM-78.

Cole, K. L. (1983). "Late Pleistocene vegetation of Kings Canyon, Sierra Nevada, California." *Quaternary Research* 19, 117–29.

———. (1985). "Past rates of change, species richness, and a model of vegetational inertia in the Grand Canyon, Arizona." *American Naturalist* 125, 289–303.

Cole, K. L., and Webb, R. H. (1985). "Late Holocene vegetation changes in Greenwater Valley, Mojave Desert, California." *Quaternary Research* 23, 227–35.

Cronquist, A.; Holmgren, A. H.; Holmgren, N. H.; and Reveal, J. L. (1972). *Intermountain flora*, vol. 1. New York: Hafner.

Currey, D. R., and James, S. R. (1982). "Paleoenvironments of the northeastern Great Basin and northeastern Basin rim region." In *Man and environment in the Great Basin.* D. B. Madsen and J. F. O'Connel, eds., pp. 27–52. Society for American Archaeology Paper no. 2.

Davis, O. K.; Agenbroad, L.; Martin, P. S.; and Mead, J. I. (1984). "The Pleistocene dung blanket of Bechan Cave, Utah." In *Contributions in*

Quaternary vertebrate paleontology. H. H. Genoways and M. R. Dawson, eds., Carnegie Institute Special Publication 8, 267–82.

Delcourt, P. A., and Delcourt, H. R. (1984). "Late Quaternary paleoclimates and biotic responses in eastern North America and the western North Atlantic Ocean." *Palaeogeography, Palaeoclimatology, Palaeoecology* 48, 263–84.

Dohrenwend, J. C. (1984). "Nivation landforms in the western Great Basin and their paleoclimatic significance." *Quaternary Research* 22, 275–88.

Hansen, R. M. (1978). "Shasta ground sloth food habits. Rampart Cave. Arizona." *Paleobiology* 4, 302–19.

———. (1980). "Late Pleistocene plant fragments in the dungs of herbivores at Cowboy Cave." In *Cowboy Cave.* J. D. Jennings, ed., pp. 179–90. Salt Lake City: University of Utah Anthropological Papers, no. 104.

Harrington, M. R. (1933). *Gypsum Cave, Nevada.* Los Angeles: Southwest Museum Papers no. 8.

Hastings, J. R.; Turner, R. M.; and Warren, D. K. (1972). *An atlas of some plant distributions in the Sonoran Desert.* Technical Report no. 21, Institute of Atmospheric Physics, University of Arizona, Tucson.

King, J. E., and Van Devender, T. R. (1977). "Pollen analysis of fossil packrat middens from the Sonoran Desert." *Quaternary Research* 8, 191–204.

King, T. J., Jr. (1976). "Late Pleistocene—early Holocene history of coniferous woodland in the Lucerne Valley region, Mojave Desert, California." *Great Basin Naturalist* 36, 227–38.

Kutzbach, J. E. (1983). "Modeling of Holocene climates." In *Late-Quaternary environments of the United States.* Volume 2, *The Holocene.* H. E. Wright, Jr., ed., pp. 271–77. Minneapolis: University of Minnesota Press.

Kutzbach, J. E., and Guetter, P. J. (1986). "The influence of changing orbital parameters and surface boundary conditions on climate simulations for the past 18,000 years." *Journal of Atmospheric Sciences* 41, 166–79.

Kutzbach, J. E., and Street-Perrott, F. A. (1985). "Milankovitch forcing of fluctuations in the level of tropical lakes from 18 to 0 kyr B.P." *Nature* 317, 130–34.

La Marche, V. C., Jr. (1973). "Holocene climate variations inferred from treeline fluctuations in the White Mountains, California." *Quaternary Research* 3, 632–60.

Laudermilk, J. D., and Munz, P. A. (1934). *Plants in the dung of Northrotherium from Gypsum Cave, Nevada.* Carnegie Institute Washington Publication 453, pp. 29–37.

Leskinen, P. H. (1975). "Occurrence of oaks in late Pleistocene vegetation in the Mojave Desert of Nevada." *Madroño* 23, 234–35.

Long, A.; Hansen, R. M.; and Martin, P. S. (1974). "Extinction of the Shasta ground sloth." *Geological Society of America Bulletin* 85, 1843–48.

Martin, P. S.; Sabels, B. E.; and Shutler, R., Jr. (1961). "Rampart Cave coprolite and paleoecology of the Shasta ground sloth." *American Journal of Science* 259, 102–27.

Mead, J. I., and Phillips, A. M. III (1981). "The late Pleistocene and Holocene fauna and flora of Vulture Cave, Grand Canyon, Arizona." *Southwestern Naturalist* 26, 257–88.

Mehringer, P. J., Jr. (1965). "Late Pleistocene vegetation in the Mohave Desert of southern Nevada." *Journal of the Arizona Academy of Sciences* 3, 172–88.

———. (1967). "Pollen analysis of the Tule Springs Site, Nevada." In *Pleistocene studies in southern Nevada.* H. M. Wormington and D. Ellis, eds., pp. 129–200. Carson City: Nevada State Museum Anthropological Papers.

Mehringer, P. J., Jr., and Ferguson, C. W. (1969). "Pluvial occurrence of bristlecone pine (*Pinus aristata*) in a Mojave Desert mountain range." *Journal of the Arizona Academy of Sciences* 5, 284–92.

Mehringer, P. J., Jr., and Sheppard, J. C. (1978). "Holocene history of Little Lake, Mojave Desert, California." In *Ancient Californians.* E. L. Davis, ed., pp. 153–66. Natural History Museum of Los Angeles, Science Series 29.

Munz, P. A. (1974). *A flora of southern California.* Berkeley: University of California Press.

Olausson, E., Ed. (1982). *The Pleistocene/Holocene boundary in southwestern Sweden.* Stockholm: Sveriges Geologiska Undersökning, Serie C, NR 74.

Phillips, A. M. III (1977). "Packrats, plants, and the Pleistocene in the Lower Grand Canyon." Ph.D. diss., University of Arizona, Tucson.

———. (1984). "Shasta ground sloth extinction: Fossil packrat midden evidence from the western Grand Canyon." In *Quaternary extinctions: A prehistoric revolution.* P. S. Martin and R. G. Klein, eds., pp. 148–58. Tucson: University of Arizona Press.

Ritchie, J. C.; Eyles, C. H.; and Haynes, C. V. (1985). "Sediment and pollen evidence for an early to mid-Holocene humid period in the eastern Sahara." *Nature* 314, 352–55.

Rowlands, P. G.; Johnson, H.; Ritter, E.; and Endo, A. (1982). "The Mojave Desert." In *Reference handbook on the deserts of North America.* G. L. Bender, ed., pp. 103–45. Westport, Conn.: Greenwood Press.

Ruddiman, W. F., and Duplessy, J.-C. (1985). "Conference on the last deglaciation: Timing and mechanism." *Quaternary Research* 23, 1–17.

Simpson, G. G. (1933). "A Nevada fauna of Pleistocene type and its probable association with man." *American Museum Novitates* 667, 1–10.

Smith, G. I. (1979). *Subsurface stratigraphy and geochemistry of Late Quaternary evaporites, Searles Lake, California.* U.S. Geological Survey Professional Paper 1043.

Smith, G. I., and Street-Perrott, F. A. (1983). "Pluvial lakes of the western United States." In *Late-Quaternary environments of the United States. Volume 1, The late Pleistocene.* S. C. Porter, ed., pp. 190–214. Minneapolis: University of Minnesota Press.

Spaulding, W. G. (1977). "Late Quaternary vegetational change in the Sheep Range, southern Nevada." *Journal of the Arizona Academy of Sciences* 12, 3–8.

———. (1981). "The late Quaternary vegetation of a southern Nevada mountain range." Ph.D. diss., University of Arizona, Tucson.

———. (1983). "Late Wisconsin macrofossil records of desert vegetation in the American Southwest." *Quaternary Research* 19, 256–64.

———. (1985). *Vegetation and climates of the last 45,000 years in the vicinity of the Nevada Test Site, south-central Nevada.* U.S. Geological Survey Professional Paper no. 1329.

———. (1989). *Pluvial climates and paleoenvironments of the southern Great Basin, with special reference to Yucca Mountain.* U.S. Geological Survey Water Resources Investigations Report.

Spaulding, W. G., and Graumlich, L. J. (1986). "The last pluvial climatic episodes in the deserts of southwestern North America." *Nature* 320, 441–44.

Spaulding, W. G.; Leopold, E. B.; and Van Devender, T. R. (1983). "Late Wisconsin paleoecology of the American southwest." In *Late-Quaternary environments of the United States. Volume 1, The Late Pleistocene.* S. C. Porter, ed., pp. 259–93. Minneapolis: University of Minnesota Press.

Spaulding, W. G., and Petersen, K. L. (1980). "Late Pleistocene and early Holocene paleoecology of Cowboy Cave." In *Cowboy Cave.* J. D. Jennings, ed., pp. 163–77. Salt Lake City: University of Utah Anthropological Papers 104.

Spaulding, W. G.; Robinson, S. W.; and Paillet, F. L. (1984). *A preliminary assessment of late glacial vegetation and climate change in the southern Great Basin.* U.S. Geological Survey, Water Resources Research Report 84-4328.

Thompson, R. S. (1984). "Late Pleistocene and Holocene environments in the Great Basin." Ph.D. diss., University of Arizona, Tucson.

Thompson, R. S., and Kautz, R. R. (1983). "Pollen analysis." In *The archaeology of Gatecliff Shelter.* D. H. Thomas, ed., pp. 136–51. New York: American Museum of Natural History Anthropological Paper no. 59.

Thompson, R. S., and Mead, J. I. (1982). "Late Quaternary environments and biogeography in the Great Basin." *Quaternary Research* 17, 39–55.

Thompson, R. S.; Van Devender, T. R.; Martin, P. S.; Foppe, T.; and Long, A. (1980). "Shasta ground sloth (*Nothrotheriops shastense* Hoffstetter) at Shelter Cave, New Mexico: Environment, diet, and extinction." *Quaternary Research* 14, 360–76.

Van Devender, T. R. (1973). "Late Pleistocene plants and animals of the Sonoran Desert: A survey of ancient packrat middens in southwestern Arizona." Ph.D. diss., University of Arizona, Tucson.

Van Devender, T. R. (1977). "Holocene woodlands in the Southwestern deserts." *Science* 198, 189–92.

Van Devender, T. R., and Spaulding, W. G. (1979). "Development of vegetation and climate in the southwestern United States." *Science* 204, 701–10.

Van Devender, T. R.; Spaulding, W. G.; and Phillips, A. M. III (1979). "Late Pleistocene plant communities in the Guadalupe Mountains, Culberson County, Texas." In *Biological investigations in the Guadalupe Mountains National Park, Texas.* H. H. Genoways and R. J. Baker, eds., pp. 13–30. Washington, D.C.: National Park Service Transactions and Proceedings Series no. 4.

Wells, P. V. (1983). "Paleobiogeography of montane islands in the Great Basin since the last glaciopluvial." *Ecological Monographs* 53, 341–82.

Wells, P. V., and Jorgensen, C. D. (1964). "Pleistocene woodrat middens and climatic change in the Mojave Desert: A record of juniper woodlands." *Science* 143, 1171–74.

Wells, P. V., and Berger, R. (1967). "Late Pleistocene history of coniferous woodland in the Mojave Desert." *Science* 155, 1640–47.

Wells, P. V., and Woodcock, D. (1985). "Full-glacial vegetation of Death Valley, California: Juniper woodland opening to *Yucca* semidesert." *Madroño* 32, 11–23.

Chapter 10

Late Quaternary Vegetation and Climate in the Great Basin

Robert S. Thompson

INTRODUCTION

The Great Basin is the largely waterless domain between the Rocky Mountains and the Sierra Nevada (Fig. 10.1). It is a land of broad valleys covered with sagebrush and shadscale interspersed with mountain ranges that harbor woodlands and coniferous forests. Domestic cattle and sheep share the range with wild horses and antelopes. Permanent human settlements are few, and cities are present only on the peripheries of the region.

During the last ice age immense lakes covered the valleys of the Great Basin, and glaciers crept down high montane valleys. Herds of camels, llamas, native horses, and mammoths roamed over the region, and pikas, now restricted to few of the highest mountain ranges, lived in talus near the valley bottoms. Humans arrived soon after the lakes desiccated and as the herds of large Pleistocene animals disappeared. Since that time human societies in the Great Basin have contended with continual climatic and environmental changes.

Packrat middens have been the subject of scientific inquiry in the Great Basin since 1957. In that year Phil Orr published a note in which he described deposits found in caves in western Nevada. The colorful term "amberat," coined (apparently by Orr) following

an observation of a midden in a Montana cave sometime around 1912, denoted the substance's resemblance to amber and its ratlike smell. Although something of a mystery in the 1950s, amberat is now known to be the desiccated urine of packrats (*Neotoma*). It acts as the cementing agent holding together assemblages of plant fragments and other materials collected by packrats, permitting the preservation of paleobotanical assemblages for tens of thousands of years. The Great Basin is the most northerly of the regions where packrat middens have been intensively studied. Due to higher levels of effective moisture and the limited availability of suitable substrates for long-term preservation, this region has not provided the temporal or spatial coverage of middens from regions farther south. On the other hand, packrat middens and stratigraphic palynological data can be found in relative proximity in the Great Basin, providing unusual opportunities to augment and compare vegetational and climatic reconstructions based on either method.

This chapter explores the midden data from the Great Basin and compares the interpretations based on these data with those from palynological and geophysical indicators of

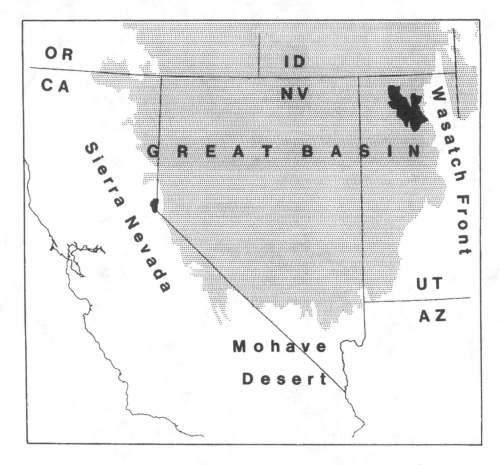

Fig. 10.1. Map illustrating the geographical extent of the Great Basin as the term is used in this chapter.

climatic changes. The comparisons and contrasts among these data sets have implications on the nature of Pleistocene vegetation, the nature and chronology of Pleistocene-Holocene vegetation changes, and the nature, amplitude, and timing of climatic changes over the last forty thousand years.

MODERN ENVIRONMENT

The term "Great Basin" originally referred to the region of internal drainage within the basin-and-range province of the western interior of the United States. In this chapter the term refers to the area within this physio-

graphic province where sagebrush and/or shadscale-steppe dominates the vegetation of the valley bottoms. This is generally a region of internal drainage, although "Great Basin" vegetation grows in portions of adjacent states that have drainage to the oceans. Packrat middens from the Great Basin sections of southern Oregon and northern California are discussed by Mehringer and Wigand in chapter 13.

Over a hundred elongate mountain chains trend north to northeast across the Great Basin (McLane, 1978), with ranges generally 8–24 km wide rising 300–1500 m above the intervening valleys, which are of similar width (Stewart, 1980; Fig. 10.2). In the northern and central portions of the region the mountain heights are commonly 1800–3050 m, and the valley bottoms range from 1200 to 1500 m. South of approximately latitude 38° N, the mountains and valleys are generally 600–900 m lower than their more northerly

counterparts (Stewart, 1980). The Great Basin is bounded on the west by the massive Sierra Nevada, on the east by the Wasatch Front of the Rocky Mountains, and on the southeast by the Colorado Plateau. To the south the high mountains and valleys give way to the low-elevation Colorado River drainage, and on the north the Basin is bounded by the

Fig. 10.2. Map illustrating the physiography of the Great Basin. The solid line is the 1525-m elevational contour. Blackened areas represent elevations above 2135 m.

sagebrush-covered Snake River Plain.

There are strong regional differences in substrate across the Great Basin. Cliff-forming limestone and dolomites are common only in the eastern and southeastern sectors. Igneous and metamorphic rocks dominate the central, northern and western portions, and rock shelters and caves are less common in these areas. These regional differences in rock type have strongly influenced the long-term preservation of packrat middens, and Pleistocene middens are common only in calcareous ranges in the east and in rhyolitic outcrops along the southern periphery of the Great Basin.

Climate

The Great Basin is situated in the rain shadows of the Sierra Nevada to the west and, to a lesser degree, the Rocky Mountains to the east and southeast. Precipitation is highly variable from year to year and is distributed unevenly in space due to local rain shadows and elevational gradients. The region has a continental climate with a dominance of cool-season precipitation (September to May), primarily derived from migratory low-pressure systems following the westerlies off the northern Pacific Ocean. A strong precipitation gradient occurs in June, with the northern sectors receiving abundant moisture while drought prevails to the south.

The eastern and southeastern sections of the Great Basin receive a substantial proportion of their mean annual rainfall in the summer months of July and August from southerly, subtropical "monsoonal" sources. Much of the remainder of the Great Basin is dry during this period because anticyclonic flow prevents moisture-laden air masses from entering the region (Mitchell, 1976; Tang and Reiter, 1984), and the cold California current maintains low sea-surface temperatures and retards the rising of moist air off the ocean (Pyke, 1972).

The expected decrease in temperature with increasing altitude may be complicated by local inversions on lower mountain slopes in the Great Basin. In many temperate areas precipitation increases relatively rapidly with altitude up to a certain elevation (about 1500 m) and then increases at a lower rate above that point (Barry, 1981). In the Great Basin there is apparently a somewhat different pattern in which there is little discernible increase in precipitation over the elevational range between 1000 and 2200 m. This may be an artifact of the sites of climate-recording stations that lie in this elevation range, because most of them are located on valley bottoms or bajadas.

Vegetation

The vegetation of the Great Basin is generally zoned into a series of nearly horizontal altitudinal bands following the elevational gradients in temperature and precipitation (Billings, 1951). Shadscale- and greasewood-dominated steppe (see Table 10.1 for a list of common plant names and their scientific equivalents) occurs in the lower valley elevations. Sagebrush covers the upper bajadas and occurs as an understory element in the wooded communities at higher elevations. In many areas sagebrush species grow from the valley bottoms to the mountain tops, and they dominate at all elevations in the extreme northwestern Great Basin.

Valley-bottom elevations are lower in the Lahontan Basin of western Nevada and in the Bonneville Basin of western Utah than on the intervening "plateau." Shadscale or greasewood desertscrub is dominant in these basins, whereas sagebrush is dominant in the valleys of the central "plateau." Valley-bottom elevations are lower in southern Nevada, and here creosote bush and other Mojave Desert plants are dominant below Great Basin steppe elements that intermix with woodland species on the lower mountain slopes.

Pinyon-juniper woodland occurs on the lower mountain slopes in the eastern and central Great Basin (Fig. 10.3), while in the extreme northern and northwestern portions of the region pinyon is not present and junipers occur in pure stands. In the eastern and southeastern subregions montane forests with ponderosa pine, white fir, and Douglas fir grow above the woodland. This montane assemblage is lacking in the central and northwestern Great Basin, where an upper-sagebrush grassland (Billings, 1951) occurs above the woodland. Subalpine coniferous forests are present on many of the higher Great Basin ranges. Great Basin bristlecone pine, limber pine, and Engelmann spruce are common in this zone in the eastern Great Basin,

Table 10.1. *Plant common names and scientific equivalents.*
Modern ranges of the taxa in vegetation zones in the eastern Great Basin are given on right. Key to vegetation zones: D = shadscale desert; LS = lower sagebrush-grass; WL = woodland; US = upper sagebrush-grass; MF = montane forest; SA = subalpine forest. X = abundant in this zone; O = occurs, but rarely abundant.

Common name	Scientific name	D	LS	WL	US	MF	SA
Bristlecone pine	Pinus longaeva	—	—	—	—	—	X
Budsage	Artemisia spinescens	X	O	—	—	—	—
Cliff-bush	Jamesia americana	—	—	O	O	O	—
Desert peach	Prunus andersonii	O	X	O	—	—	—
Douglas fir	Pseudotsuga menziesii	—	—	—	—	X	—
Engelmann spruce	Picea engelmannii	—	—	—	—	—	X
Fernbush	Chamaebatiaria millifolium	—	—	X	O	O	—
Gambell oak	Quercus gambellii	—	X	—	—	—	—
Greasebush	Forsellesia nevadensis	—	—	X	—	—	—
Greasewood	Sarcobatus vermiculatus	X	O	—	—	—	—
Greenmolly	Kochia americana	X	—	—	—	—	—
Horsebrush	Tetradymia spp.	—	X	O	—	—	—
Joint-fir	Ephedra viridis	—	O	X	—	—	—
Leather-leaf mountain mahogany	Cercocarpus ledifolius	—	—	X	O	O	—
Limber pine	Pinus flexilis	—	—	—	—	O	X
Little-leaf mountain mahogany	Cercocarpus intricatus	—	—	X	—	—	—
Lupine	Lupinus cf. argenteus	—	—	O	X	O	—
Mock-orange	Philadelphus microphyllus	—	—	O	O	O	—
Mountain spray	Holodiscus dumosus	—	—	O	O	O	—
Plains prickly pear	Opuntia cf. polyacantha	—	O	X	—	—	—
Ponderosa pine	Pinus ponderosa	—	—	—	—	X	—
Prickly pear	Opuntia cf. erinacea	—	O	X	—	—	—
Prostrate juniper	Juniperus communis	—	—	—	—	O	X
Quaking aspen	Populus tremuloides	—	—	—	O	X	—
Rock spiraca	Petrophytum caespitosum	—	—	X	—	—	—
Rocky Mountain juniper	Juniperus scopulorum	—	—	O	—	X	—
Rocky Mountain maple	Acer glabrum	—	—	O	O	X	—
Sagebrush (big)	Artemisia tridentata	—	X	X	X	X	O
Service-berry	Amelanchier utahensis	—	—	X	O	O	—
Shadscale	Atriplex confertifolia	X	O	—	—	—	—
Single-needle pinyon pine	Pinus monophylla	—	—	X	—	—	—
Snake-weed	Gutierrezia sarothrae	O	X	O	—	—	—
Snowberry	Symphoricarpos spp.	—	O	O	X	O	—
Sticky rabbitbrush	Chrysothamnus viscidiflorus	O	X	O	O	—	—
Subalpine fir	Abies lasiocarpa	—	—	—	—	O	X
Utah juniper	Juniperus osteosperma	—	—	X	—	—	—
Western juniper	Juniperus occidentalis	—	—	X	—	—	—
White fir	Abies concolor	—	—	O	—	X	—
Whitebark pine	Pinus albicaulis	—	—	—	—	—	X

while spruce is not present in the western subregion. The subalpine forests of the northeastern ranges support whitebark pine and subalpine fir, showing an affinity to the northern Rocky Mountains (Loope, 1969; Critchfield and Allenbaugh, 1969). Pockets of alpine tundra occur on isolated peaks across the Great Basin (Billings, 1978).

A rugged physiography, sharp differences in slope and aspect, rain shadows from nearby

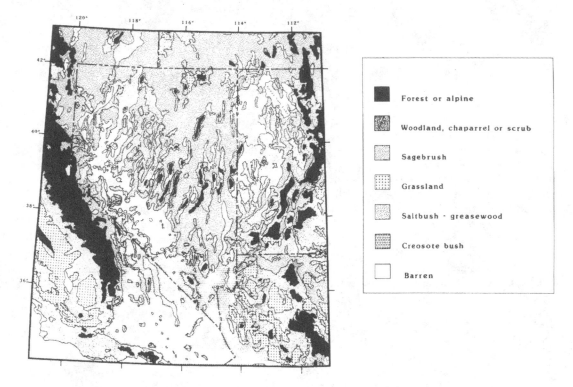

Fig. 10.3. Generalized vegetation map of the Great
Basin region (after Kuchler, 1964).

peaks, intense cold-air drainage, and other terrain features create a range of local microcli-
mates. These, in turn, create microhabitats
that confound the simple elevational array of
plant communities described above. In general, the lower elevational limits of the woodland community are lower in the eastern and
southern Great Basin, where summer precipitation is more abundant (West et al.,
1978). Ponderosa pine, white fir, Douglas fir,
and Rocky Mountain juniper are restricted to
those regions that do not have pronounced
summer drought. The distribution of single-
needle pinyon pine also may be tied to the
pattern of summer precipitation within the
region, although other climatic variables such
as winter and/or summer temperatures may
influence its distribution (West et al., 1978;
Axelrod, 1981; Thompson and Hattori, 1983).

The flora of the Great Basin has much
stronger affinities with that of the Rocky
Mountains than with the Sierra Nevada
(Harper et al., 1978; Loope, 1969; Billings,
1978). As will be discussed in a later section,

this may reflect climatic controls and/or the
availability of access routes into the Great
Basin.

Island Biogeography

Basin-and-range topography creates a situation where montane vegetation (especially
subalpine and alpine vegetation) is discontinuous and separated into habitat islands of
various sizes. These islands are separated
from each other and from Sierran and Rocky
Mountain forests by extensive "seas" of
steppe and desert vegetation. The lower limits of montane vegetation are at higher elevations in the western and northern reaches of
the Great Basin, and montane habitat islands
are correspondingly smaller. Additionally,
some plants may be essentially restricted by
substrate to smaller islands of desirable habitat within these montane islands.

The modern diversity and distribution of
upland small mammals, birds, and vegetation
in the region can be modeled using the methods of island biogeography, as well as histori-

cal factors (MacArthur and Wilson, 1967; Brown, 1971, 1978; Harper et al., 1978; Thompson and Mead, 1982; Wells, 1983; Grayson, 1987). Calculations for pinyon-juniper woodland (West et al., 1978) and total mountain flora (Harper et al., 1978) suggest that the species-area relationships are similar to those on oceanic islands, but that extinction without subsequent replacement has guided the development of these floras. As pointed out by Wells (1983), there is great diversity in the number of montane conifer species present on different mountain ranges in the Great Basin. Some of the largest ranges (> 1000 km^2) have three or fewer species, while some smaller (< 30 km^2) and lower ranges have as many as five species. Wells (1983) also points out a major disparity between the stock of montane conifers present in the eastern versus the western Great Basin. The mountain ranges east of longitude 116° W have nearly the full range of conifers available in the Wasatch Front, and Wells (1983, p. 347) argues that the insular slope ($z = 0.264$) and the good correlation ($r = .78$) between island size and number of species are evidence that these eastern ranges are a "typical set of ecological islands, presumably in dynamic equilibrium between immigration and extinction of montane conifers." Ranges west of 116° W have fewer, or in some cases, no montane conifers. In these ranges extinction appears to have occurred without replacement by subsequent immigrations.

RECONSTRUCTION OF PAST VEGETATION

Plant macrofossil assemblages from packrat middens provide most of the information now available on late Quaternary vegetational changes in the Great Basin. Palynological data from marsh, lake, and dry-cave sediments provide important ancillary data and information on environments not sampled by packrat middens. Additional lines of evidence come from macrofossils preserved in cave deposits, lake sediments, and alpine environments, and from pollen preserved in the matrices of packrat middens.

Packrat Middens

Most packrat midden researchers collect 0.5–1 kg of material from a given strati-

graphic unit of a midden in the field. If further inspection in the laboratory reveals extraneous materials or weathering rinds, these are removed and a subsample is placed in water. After the urine-based matrix dissolves, the residue is screened through soil sieves and washed. The residue is dried and the plant macrofossils, bones, and arthropod remains are sorted by hand. Specimens are selected for radiocarbon dating; frequently, remains of one individual taxon of paleoecological and/or paleoclimatic interest are chosen for dating. These techniques maximize the number of taxa represented but increase the probability of including contaminating materials of different ages. Accelerator/mass spectrometer (AMS) radiocarbon dating techniques now permit the assessment of this potential difficulty (Van Devender et al., 1985).

Using a different technique, Wells (1983) collects small parcels of material that, in his opinion, represent one stratigraphic interval. The plant taxa that are observable on the surface are identified and the entire sample is submitted for radiocarbon dating. Although this technique may reduce the risk of contamination, it also severely reduces the number of identified plant taxa. As discussed in a later section, the differences between Wells's methods and those of other researchers may affect the resolution and nature of the vegetation reconstructions.

Plant macrofossils from packrat middens are identified by comparison with reference materials and herbarium specimens, and through reference to published keys. I believe that each midden macrofossil assemblage represents a time slice of a few years to a few decades (Thompson, 1985).

The available array of packrat middens from the Great Basin are not evenly distributed in time or space. The most intensive coverage is from the eastern and southern portions of the region, where there are outcrops of limestone, dolomite and rhyolite. Ancient middens also have been discovered in granite caves (Thompson, 1984), in tufa caves (Thompson et al., 1986), and in lava tubes (Mehringer and Wigand, 1987; *this volume*, chap. 13). The midden coverage is heavily skewed toward the late glacial and late Holocene (Fig. 10.4; Table 10.2), and coverage is particularly sparse prior to 15 ka and between

Table 10.2. *Packrat midden localities mentioned in text.*

Site	Latitude (N)	Longitude (W)	Age range (ka)	Reference
Eleana Range	37° 07'	116° 14'	10.6 – 49.8	Spaulding, 1985
Westgaard Pass	37° 17'	118° 08'	6.7 – 8.8	Thompson, unpub.
Eureka View	37° 20'	117° 47'	0.5 – 14.7	Spaulding, 1980
Stine	37° 29'	114° 37'	1.8 – 15.0	Madsen, 1973
Fright	37° 31'	114° 16'	0.3 – 4.4	Madsen, 1973
Etna	37° 33'	114° 37'	6.6 – 20.1	Madsen, 1973
Wells #43	37° 34'	114° 32'	12.6	Wells, 1983
Silver Peak	37° 45'	117° 40'	2.7 – 10.4	Thompson, unpub.
Esmeralda	37° 53'	117° 14'	8.7 – 13.1	Thompson, unpub.
Cole's Cave	38° 12'	118° 55'	0.8 – 1.4	Thompson, unpub.
Carlin's Cave	38° 18'	115° 02'	0.3 – 20.4	Thompson, unpub.
Bodie Canyon Cave	38° 19'	118° 56'	1.0 – 6.0	Thompson, unpub.
Wells #32	38° 30'	113° 32'	12.4	Wells, 1983
Wells #31, 35, 36	38° 32'	113° 31'	10.5 – 12.5	Wells, 1983
Wells #30	38° 34'	113° 35'	12.9	Wells, 1983
Wells #26, 27	38° 51'	114° 13'	2.1 – 9.2	Wells, 1983
Garrison	38° 57'	114° 03'	12.2 – 13.5	Thompson, 1984
Wells #12, 14 – 16	38° 58'	114° 08'	11.0 – 11.9	Wells, 1983
Wells #13, 21 – 23, 28	38° 58'	114° 10'	4.6 – 11.8	Wells, 1983
Wells #3, 5, 18	38° 58'	114° 10'	10.6 – >36.3	Wells, 1983
Gatecliff/Mill Canyon	39° 00'	116° 47'	0.0 – 9.5	Thompson and Hattori, 1983
Wells #33	39° 01'	113° 36'	12.2	Wells, 1983
June Canyon	39° 01'	116° 45'	0.4 – 5.2	Thompson, unpub.
Wells #34, 37 – 41	39° 02'	113° 34'	1.7 – 11.9	Wells, 1983
Bloody Arm Cave	39° 08'	114° 25'	10.1	Thompson, 1984
Old Man Cave	39° 10'	114° 08'	10.0	Thompson, 1984
Arch Cave	39° 17'	114° 05'	2.4 – 34.0	Thompson, 1984
Council Hall Cave	39° 20'	114° 06'	4.2 – 6.1	Thompson, 1984
Wells #10, 11, 20, 29	39° 20'	114° 06'	1.7 – 12.1	Wells, 1983
Streamview	39° 20'	114° 06'	6.5 – 17.4	Thompson, 1984
Wells #2, 4, 6 – 9, 17, 19, 24, 25	39° 20'	114° 07'	9.6 – >37.0	Wells, 1983
Smith Creek/Ladder Caves	39° 21'	114° 05'	10.5 – 31.0	Thompson, 1984
Valleyview	39° 30'	114° 43'	6.3 – 6.7	Thompson, 1984
Granite Wash	39° 40'	113° 50'	12.8 – 13.6	Thompson, 1984; Thompson, unpub.
Brown's Hole/Bronco Charlie	40° 09'	115° 31'	0.0 – 3.1	Thompson, 1984
Crypt/Fishbone/Guano Cave	40° 12'	119° 16'	0.4 – 12.4	Thompson et al., 1986; Thompson, unpub.
Kramer Cave/Falcon Hill	40° 16'	119° 20'	3.0 – 12.0	Thompson et al., 1986; Thompson, unpub.
Wells #1	42° 27'	117° 20'	27.3	Wells, 1983

7 and 8 ka. The rarity of older middens may be due to the progressive deterioration of these materials through time and/or to Pleistocene climatic conditions that were unfavorable to long-term preservation. The almost complete lack of samples in the 7–8-ka range remains unexplained.

Fossil Pollen

In the following text palynological data will be discussed in comparison with packrat midden records from the Great Basin. The guidelines used for these palynological works follow standard procedures set forth in textbooks

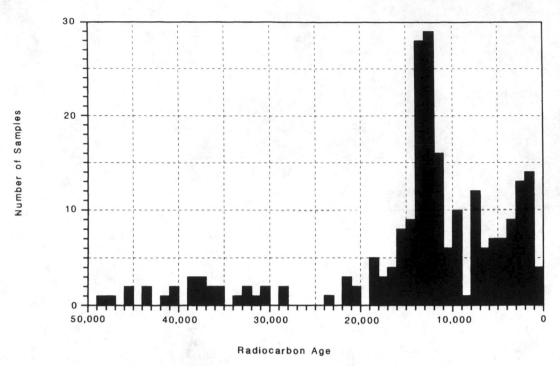

Fig. 10.4. The relationship between radiocarbon age and number of packrat midden samples in the Great Basin.

(e.g., Faegri and Iversen, 1975) and previous reports (e.g., Mehringer, 1967; Thompson and Kautz, 1983), and will not be covered in detail here. It should be pointed out that these two means of vegetation reconstruction differ in their taxonomic and spatial resolution, with packrat middens providing finer-scale information in both cases. They also differ in continuity, with palynological studies of stratigraphic sections and cores generally providing nearly continuous histories of vegetation change.

The pollen contents of packrat midden matrices also have been examined from samples from the Great Basin (Madsen, 1973; Thompson, 1985). Although the pollen preservation is excellent, these analyses have the inherent weaknesses of both data sets—they have the coarse taxonomic resolution of pollen data and the lack of temporal continuity shown by midden data.

LATE PLEISTOCENE VEGETATION

Information on late Pleistocene (around 40–10 ka) vegetation in the Great Basin is available from two different types of environ-

ments. Packrat middens and pollen records from caves provide data on the plants that grew on coarse substrates on the lower mountain slopes and along the margins of the valleys (see Fig. 10.5 for the locations of Pleistocene sites). Pollen data from mid-valley locations and a few packrat middens indicate the character of the vegetation on finer-textured alluvial and colluvial sediments.

The large difference between late Pleistocene and modern climates in the Great Basin is demonstrated by the past occurrence of "pluvial" lakes 100 m or more in depth in many now-arid valleys. Alpine glaciers descended to below 2800 m elevation in many of the larger mountain ranges (Blackwelter, 1931, 1934; Sharp, 1938), while cryogenic features formed at higher elevations across the Great Basin (Dohrenwend, 1984). In this very different climate, plant species were arranged on the landscape in a fashion distinct from today's pattern. Many species that now live in the region had their northern limits in the modern Mojave Desert or in the modern Mojave–Great Basin transition region. Other plants grew at elevations as much as 1000 m

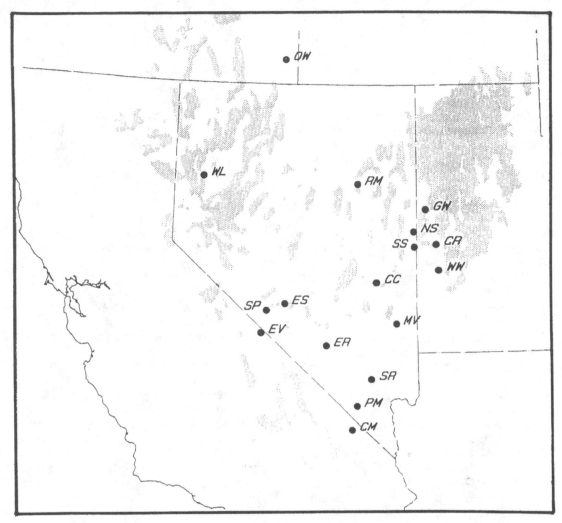

Fig. 10.5. Map of sites of late Pleistocene age discussed in text (gray areas represent Pleistocene lakes at their maximal extents). Key to sites: OW = Owyhee Range, WL = Winnemucca Lake basin, RM = Ruby Marshes, GW = Granite Wash, NS = northern Snake Range, SS = southern Snake Range and Snake Valley, CR = Confusion Range, WW = Wah Wah Mountains, CC = Carlin's Cave, MV = Meadow Valley Wash and Rainbow Canyon, ES = Esmeralda, SP = Silver Peak, EV = Eureka Valley, ER = Eleana Range, SR = Sheep Range, PM = Potosi Mountain, CM = Clark Mountain.

below their common modern lower limits. The Pleistocene records are notable in that there are no known examples of large-scale geographic shifts into the Great Basin of plants that no longer occur in the region.

The precise taxonomic detail available from packrat midden assemblages from the Great Basin provides evidence that, as in other portions of the West (Cole, 1981; Davis, 1981; Spaulding, 1981), plant species responded to Quaternary climatic changes in individual-

istic fashions (Gleason, 1926, 1939). The following text reviews the nature of Pleistocene vegetation changes in the Great Basin. For convenience, I have organized the discussion around the modern groupings of vegetation in this region, keeping in mind that plant species did not respond to climatic changes as communities.

Alpine plants

On the basis of the modern biogeography of

alpine plants in the intermontane West, Billings (1978) suggested that vegetation zones were sufficiently depressed in elevation during the Wisconsin to permit alpine floras to link up across the region. Whatever accounts for the modern alpine biogeography, the packrat midden and pollen data demonstrate the dominance of subalpine trees and steppe species (and the absence of alpine species) on the lower mountain slopes and valley bottoms during the late Pleistocene (see discussion below).

Subalpine Plants

During much of the late Pleistocene, subalpine conifers grew as much as 1000 m below their modern lower limits in the Great Basin. The dominant trees in packrat midden assemblages of this age from the eastern and southern Great Basin include intermountain bristlecone pine, limber pine, prostrate juniper, and, locally, Engelmann spruce. The first three are relatively xerophytic (for subalpine conifers); the Great Basin midden records are notably lacking in immigrant species reflecting high levels of effective moisture. Spruce macrofossils have been recovered from sites on north-facing slopes or near canyon bottoms only in the Snake Range (Thompson, 1979; Thompson and Mead, 1982; Wells, 1983).

During the late Wisconsin in the Snake, Confusion, and Wah Wah ranges of the east-central Great Basin, bristlecone pine and prostrate juniper occurred as low as 1660 m until after 12 ka, approximately 680 m and 840 m below their respective modern limits (Fig. 10.6; Wells, 1983; Thompson, 1984). By Wells's calculations, Engelmann spruce was depressed as much as 1000 m in this region. Bristlecone pine and prostrate juniper potentially could have been depressed by this amount, but the high elevations of valleys in this region may have prevented the expression of this trend.

During the late Wisconsin limber pine grew as low as 1350 m in association with Douglas fir and Utah juniper in Meadow Valley Wash in the southeastern Great Basin (Fig. 10.7). Madsen (1973) originally identified the Pinaceae remains as bristlecone pine and white fir; they have subsequently been revised to limber pine and Douglas fir by Spaulding and Van Devender (1980). Wells (1983) calculated

that in this setting limber pine was > 500 m below its modern occurrence in the Snake Range of eastern Nevada and more than 1000 m below its nearest modern "zone of dominance." However, as pointed out by Spaulding and Van Devender (1980), Meadow Valley Wash is very mesic for its elevation, and Gambel oak and single-needle pinyon pine today grow at elevations 300 m lower than in other nearby settings. Thus Wells's (1983) estimate for the Pleistocene depression of limber pine at this locale may be in error by several hundred meters. Modern limber pine populations in the ranges of southern Nevada are sparse and elevationally restricted compared with the large Pleistocene range of this tree. In the Sheep Range limber pine is uncommon today, while bristlecone pine is comparatively abundant (Spaulding, 1981). The modern rarity of limber pine in the region may reflect some as-yet-unidentified aspect of climatic control on the species, or alternatively may be due to competition with other conifers.

Limber and bristlecone pines are the only subalpine conifers found in nearly all mountain ranges in the southern and central Great Basin. Axelrod (1976) argued that the modern distributions of these pines are relictual from the Tertiary. However, the packrat midden records of these species indicate that these ranges were established or greatly modified during the glacial-interglacial cycles of the Pleistocene (Thompson and Mead, 1982).

Montane Plants

Many of the dominant shrubs and trees of the modern montane communities, including ponderosa pine, Douglas fir, white fir, and, for some of the interval, Rocky Mountain juniper, were apparently absent from the region for much of the late Pleistocene. Some montane and mesic upper-woodland shrubs, such as currant, fernbush, buffalo-berry, and cliffbush, were associated with subalpine conifers in the late Pleistocene vegetational matrix in the eastern Great Basin.

Of the montane species that were absent, many persisted in lower elevational and/or more southerly settings. Among the montane conifers, late Pleistocene records of white fir have been recovered from the Sheep Range (Spaulding, 1981) and Spring Range (Thomp-

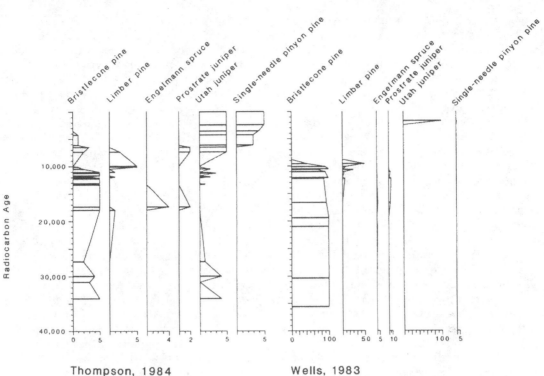

Thompson, 1984   Wells, 1983

Fig. 10.6. Comparison of conifers recovered in the Thompson (1984) and Wells (1983) packrat midden series from the northern Snake Range in the east-central Great Basin. For the Thompson data set, the horizontal axis is a relative abundance scale: 0 = ab-

sent, 1 = rare, 2 = uncommon, 3 = common, 4 = abundant, 5 = very abundant. For the Wells data set, the horizontal axis is ''% total biomass'' (Wells, 1983).

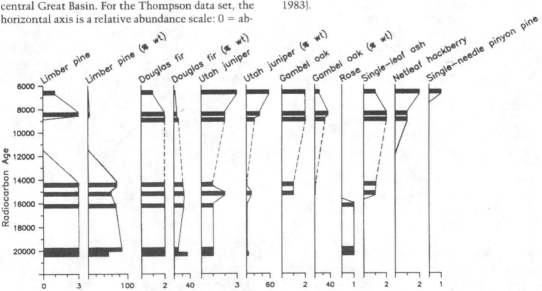

Fig. 10.7. Time series: Plant species recovered from packrat middens from the Etna Site, Meadow Valley Wash (Madsen, 1973). Two scales are used for different portions of the horizontal axis: for those columns with ''% wt'' in the heading, the scale is

percent of the total weight of identified specimens. For the other columns, the scale is a relative abundance ranking: 0 = absent, 1 = rare, 2 = common, 3 = dominant (Madsen, 1973) son and Mead, 1982) of southern Nevada and from Clark Mountain in adjacent southeastern California (Mehringer and Ferguson, 1969). Douglas fir has been recovered from packrat middens in Meadow Valley Wash (Madsen, 1973; Wells, 1983). Pollen of this species was identified from late Pleistocene sediments from Great Salt Lake (Martin and Mehringer, 1965; Mehringer, 1977), indicating that the tree grew somewhere in the drainage of lake Bonneville, either in the easternmost Great Basin or in the adjacent Wasatch Front. Douglas fir macrofossils of Pleistocene age have been found in deposits from sites close to the southeastern and southern limits of the Great Basin. These deposits include packrat middens and cave fill from southeastern Utah (Betancourt, 1984, *this volume*, chap. 12; Spaulding and Van Devender, 1980) and packrat middens from the eastern Grand Canyon of north-central Arizona (Cole, 1982, *this volume*, chap. 11).

Ponderosa pine, lodgepole pine, white fir, and Douglas fir all have distinct Sierran and Rocky Mountain races or subspecies. Critchfield and Allenbaugh (1969) suggested that these differentiated populations were the result of Pleistocene segregation. In support of this hypothesis, Wisconsin packrat midden macrofossil assemblages from the Great Basin lack these taxa (Thompson and Mead, 1982). The fossil midden assemblages document the widespread occurrence of bristlecone pine and limber pine, two species that do not have separate races on either side of the region.

Below the subalpine zone, limber pine today frequently grows in the montane zone wherever more "typical" montane conifers are absent. During portions of the late Pleistocene there may have been a latitudinal zonation of vegetation in the Great Basin, with a bristlecone pine–dominated "subalpine" forest occurring at 38° to about 40° N in the eastern Great Basin. South of this band, this community may have given way to either a mixed bristlecone pine–limber pine community *or* to a limber pine–dominated (especially "montane") community. Wells (1983, p. 370) suggests that "the late-Pleistocene zone dominated by limber pine roughly corresponds to the steep elevational gradient that underlies a steep transition in the existing vegetation gradient from the 'cold' Great

Basin Desert to the 'warm' Mohave Desert." This notion may be confounded by the well-dated occurrences of limber pines in the Deep Creek Range and at the Garrison Site in the Snake Valley (Thompson, 1984). On the other hand, the samples from these two sites date between about 14 and 12 ka and may not be applicable to vegetation reconstructions for the full glacial (around 18 ka), given the great changes in other paleoclimatic indicators (e.g., lake levels and glaciers) between 18 and 14 ka.

Woodland Plants

During the late Pleistocene a few lower-woodland plants such as greasebush, little-leaf mountain mahogany, and rock-spirea grew within their modern elevational and geographic ranges (the apparent occurrence of woodland squawbush [*Rhus trilobata*] in an assemblage dated at 17,960 ± 1100 B.P. [A-2092; Thompson, 1984] has subsequently been disproven—AMS dating indicates the true age of these fossils to be 8810 ± 420 B.P. [AA-776; Van Devender et al., 1985]). Many of these species grow almost exclusively on calcareous substrates, and perhaps this adaptation overrode the effects of climatic change. Utah juniper and Rocky Mountain juniper apparently grew in the Snake Range of the eastern Great Basin during some of the warmer phases of the Wisconsin (Thompson, 1984). Similarly, single-needle pinyon pine grew along the southern fringe of the Great Basin during these warm periods (Van Devender and Spaulding, 1979; Spaulding, 1985).

Plant macrofossils from the Winnemucca Lake basin in the western Great Basin, within the area previously inundated by lake Lahontan, reveal that western juniper grew with sagebrush, desert peach, plains prickly pear, and greasebush at the close of the Pleistocene (12.4–11.2 ka; Thompson et al., 1986; Thompson, unpublished). As this site was covered by lake waters prior to this time, these plants could not have grown there for much of the late Wisconsin. It is not known if woodland plants occurred regionally prior to 12.2 ka, and the Winnemucca middens may have no relevance to reconstructions of full-glacial vegetation.

Throughout the late Pleistocene, Great Basin woodland plants grew in more southerly and lower elevational settings than they occur in today. For example, the Great Basin variety of single-needle pinyon pine grew in the modern Mojave Desert (Van Devender and Spaulding, 1979), the southern Sierra Nevada (Cole, 1982), and the southern Central Valley of California (Mason, 1944).

Steppe Plants

As noted by Billings (1951), sagebrush and other steppe species today are present across nearly the entire elevational gradient in the Great Basin. Indeed, woodland, montane, and even subalpine vegetation can be viewed as overlying a continuous matrix of steppe vegetation. In most situations the steppe taxa are abundant, if not dominant, on fine-textured soils, while the woodland and forest trees grow primarily on the coarser substrates. Packrat middens and pollen data from the Great Basin indicate that this situation pertained during the Pleistocene as well.

Macrofossil evidence from packrat middens from the Snake Range, Deep Creek Range, and Gap Mountain in the eastern Great Basin indicates that steppe plants commonly grew with bristlecone pine and other subalpine and montane plants. The steppe plants recovered from Pleistocene middens in the Snake Range of east-central Nevada include sagebrush, rabbitbrush, horsebrush, snake-weed, and snowberry (Thompson, 1984).

A pollen record from the Ruby Marshes, in a mid-valley location in northeastern Nevada, indicates that sagebrush and other steppe plants were the dominant vegetation through the late Pleistocene (Thompson, 1984). Pollen referable to *Juniperus* and other woodland taxa are at extremely low levels in the Pleistocene portion of this record, suggesting that woodland plants were uncommon or absent. Similar interpretations are supported by palynological data from other sites across the central Great Basin, including Hidden Cave (Wigand and Mehringer, 1985) and Walker Lake (Thompson, unpublished) in the Lahontan Basin. As noted above, many of these steppe species were common associates of subalpine trees on the lower mountain slopes. Six macrofossil assemblages dated from 13.1 to 8.6 ka from a single large packrat midden from the Esmeralda rock shelter in an isolated outcrop in shadscale scrub south of

Tonopah, Nevada, provide evidence for treeless steppe vegetation in the relatively low-elevation corridor between the Lahontan Basin and the Mojave Desert (Thompson, unpublished). Dominant species in the Pleistocene community at this site include sagebrush, shadscale, and snowberry.

The studies discussed above indicate that during the late Pleistocene, sagebrush (and/or closely related species of *Artemisia*) and other steppe shrubs maintained their modern geographic ranges and perhaps even their upper modern limits. This occurred even as their lower limits moved downslope by hundreds of meters and their geographic limits expanded in all directions. These taxa dispersed (or increased in abundance) southward into the Mojave and Sonoran deserts (Van Devender and Spaulding, 1979), westward into the Sierra Nevada, northward into the Pacific Northwest (Barnosky et al., 1987), and perhaps even into the central Colorado Rocky Mountains (Fall, 1985).

Conflicting Reconstructions of Pleistocene Vegetation

Wells and I have each constructed time series of vegetation changes based on packrat midden macrofossil assemblages from the Snake Range and environs in the east-central Great Basin (Thompson, 1979; Thompson and Mead, 1982; Wells, 1983; Thompson, 1984). This provides an opportunity to compare two aspects of packrat midden studies: (1) Do two investigators using different methods produce the same information, or are there systematic differences between the data sets? and (2) Are there major interpretative differences due to disparate methods and/or assumptions?

My series, based on wet-screening, yielded a variety of plant species, including conifers, steppe shrubs, and herbs (Thompson, 1984). In contrast Wells's series, based on dry-sample examination, recorded only conifers. Despite these differences in technique, the conifers reported in the two series of packrat middens generally reflect the same types of changes in presence and abundance (Fig. 10.6). The chronologies of change also are similar. Both midden series suggest that bristlecone pine, limber pine, and prostrate juniper were important elements in the vegetation of the lower mountain slopes in the east-central

Great Basin during the late Pleistocene, or at least were readily accessible to packrats living in the region at the time. Both midden series indicate that these plants persisted into the early Holocene and were essentially absent during the middle and late Holocene.

Differences in the conifer record include my records of Utah juniper from middle Wisconsin and late glacial middens. Wells concluded that this plant was a middle Holocene invader. This disagreement can be clarified by further radiocarbon dating. In a second case, according to my samples Rocky Mountain juniper was in the area through much of the late Wisconsin, while Wells concluded that it arrived at the Pleistocene-Holocene boundary.

There are major differences in our ideas regarding the nature of late Pleistocene vegetation in the east-central Great Basin. On the basis of regional pollen records, the abundance of steppe plants in Pleistocene packrat middens, faunal data, and modern plant-substrate relationships in the region, I have argued that steppe vegetation was dominant in the intermountain valleys during the Wisconsin (Thompson and Mead, 1982; Thompson, 1984). While subalpine conifers descended to relatively low elevations on coarse substrates, coniferous communities probably did not form bridges between the mountain ranges of the eastern Great Basin. Wells (1983) reasoned that since it has been demonstrated that bristlecone pine, limber pine, and prostrate juniper descended to elevations as low as 1800 m, these plants would have formed a coniferous community with "virtual continuity" with the southern Rocky Mountains. In his words: "The great abundance of *Pinus longaeva* in Wisconsinan *Neotoma* deposits at all sites from 1660 to 2680 m . . . clearly indicates a massively continuous single zone of this species at levels now abandoned to a richly diversified zonation of montane coniferous forest, pinyon-juniper woodland, sagebrush shrub-steppe, and shadscale desert" (Wells 1983, p. 358).

I agree that bristlecone pine was a dominant species in the late Pleistocene vegetation when most of the woodland and montane conifers were scarce or absent from the region. However, I believe that the weight of the available evidence indicates that sagebrush steppe was an important part of the as-

semblage, at least on fine-grained substrates, and elements of the shadscale zone were common in the vegetational matrix of the Wisconsin in the east-central Great Basin. Again the matter of sampling technique seems critical. Methods that do not reveal occurrences of shrub species cannot be relied upon to indicate the degree to which trees may have been interspersed with steppe elements.

The lower elevational limits of bristlecone and limber pines during the late Pleistocene provided the potential for continuous forest corridors connecting the Great Basin ranges with each other and with the Wasatch Front. However, I believe that substrate restrictions prevented these conifers from forming extensive valley-bottom stands. In modern subalpine environments these white pines grow primarily on bedrock outcrops and shallow, coarse soils (Loope, 1969; Lepper, 1974; Peet, 1978). In particular, bristlecone pine is generally limited to light-colored basic outcrops (Wright and Mooney, 1965). Sagebrush and other shrubs dominate deeper and finer-textured soils in modern subalpine environments in the Great Basin.

Although forest corridors may not have been continuous across the Great Basin during the late Pleistocene, the lowering of the lower limits of subalpine species would have enlarged the forest habitat islands, as the area available within a given elevational span is much greater at lower elevations than it is today on isolated mountaintops. Additionally, as demonstrated by the packrat middens from the Garrison locality in far-western Utah, isolated stands of subalpine conifers did grow on rock outcrops (but probably not fine-grained substrates) at valley-bottom elevations (Thompson, 1984). The increased size and greater proximity of forest islands, coupled with the presence of small, mid-valley forest isolates, should have promoted the dissemination of subalpine plants across the region by birds and other dispersal methods. Examples of this process may be provided by the Pleistocene occurrence of bristlecone and limber pines at Clark Mountain, California (Mehringer and Ferguson, 1969), and the modern presence of these taxa in the Panamint Range of southeastern California and the Sheep and Spring ranges of southern Nevada. In all of these situations low-elevation divides

(< 1500 m) that were covered with steppe and juniper woodland species during the late Pleistocene (Wells and Berger, 1967) now separate these mountains from each other and from the central Great Basin (Thompson and Mead, 1982).

Geographic Limits

What were the geographic limits of the late Pleistocene subalpine conifer-steppe matrix of the Great Basin? As described below, late Pleistocene vegetation in much of the interior West was dominated by, or strongly influenced by, what today are typically Great Basin plants.

The Great Basin shares many plant species with the Wasatch Front of Utah. Pollen data from Great Salt Lake (Pleistocene lake Bonneville) indicate that spruce, fir, and pine covered the Wasatch Front during the late Pleistocene, with sagebrush being regionally important (Mehringer, 1977; Spencer et al., 1984). Plant macrofossils from lake sediments reveal that subalpine fir (Cottam et al., 1959), spruce, and fir and/or juniper (the latter identified only at this coarse taxonomic level through analysis of fossil wood; Scott et al., 1983) grew along the margin of lake Bonneville. The spruce remains were found at elevations between 1379 and 1494 m and ranged in age from 22.5 to 18.6 ka (Scott et al., 1983). These data suggest that the bristlecone pine–limber pine subalpine vegetation of the central Great Basin was not shared with the Wasatch Front during the Wisconsin. This may reflect the greater amounts of moisture received by the massive Wasatch Front, which lies on the lee side of lake Bonneville.

Pollen data from the Utah-Idaho-Wyoming border region (Beiswenger, 1984) and pollen and macrofossils from the Yellowstone Plateau (Baker, 1976) indicate that sagebrush-dominated steppe or steppe-tundra dominated the north-central Rocky Mountains during the Wisconsin. This vegetation may have been contiguous with similar vegetation on the Snake River plain (Davis, 1981) and in the Pacific Northwest (Barnosky et al., 1987). On the basis of modern floristic isolates in the northern Sierra Nevada, Major and Bamberg (1967) suggested that the northern Great Basin was covered by cold-dry steppe during the late Pleistocene. From floristic relations

among modern Great Basin alpine populations, Billings (1978) inferred that alpine vegetation cover was much expanded during the Wisconsin. No direct evidence yet exists on the past geographic extent of cold steppe and/or tundra on the northern periphery of the Great Basin, or where and how this vegetation (if it existed) graded into the subalpine forest-steppe matrix of the central Great Basin.

To the south Great Basin subalpine elements dispersed into Mojave Desert ranges, including Clark Mountain, California, and the Sheep and Spring ranges of southern Nevada (Mehringer and Ferguson, 1969; Spaulding, 1981; Thompson and Mead, 1982). Bristlecone pine and limber pines grew as much as 570 m and 1100 m, respectively, below their modern elevational limits in the Sheep Range (Spaulding et al., 1983). The occurrence of bristlecone and limber pines on Clark Mountain is particularly notable, in that this southernmost record is on a mountain range that rises only to 2410 m and is separated from the nearest source mountains (the Spring Range) by 50 km (Mehringer and Ferguson, 1969). In the valleys of the Mojave Desert, sagebrush grew in the modern domain of creosote bush (Mehringer, 1967; Quade, 1986). Concurrently, Great Basin woodland plants such as single-needle pinyon pine and Utah juniper spread over much of the modern Mojave Desert. Some of the modern Mojavean elements, such as Joshua tree, grew with these woodland plants (Spaulding, 1981). Single-needle pinyon pine grew with junipers and other woodland species in the far northwestern Mojave Desert (McCarten and Van Devender, 1988).

On the western periphery of the Great Basin, palynological data from drainages on the east slope of the Sierra Nevada indicate expanded coverage of sagebrush and other relatively xerophytic Great Basin taxa in the late Pleistocene (Adam, 1967; Thompson, unpublished). Pollen data from the western slope (Davis et al., 1985) also document increased sagebrush cover, which may have persisted into the early Holocene. Plant macrofossils from packrat middens from Kings Canyon in the southwestern Sierra Nevada (Cole, 1983) indicate that the Great Basin woodland plants were intermixed with relatively xerophytic

Sierran species in the late Pleistocene.

Variations through Time

On global and continental scales, the period from about 40 to 10 ka encompasses a phase of intermediate glacial ice volume (isotope stage 3 or middle Wisconsin), through the interval of maximal glacial development (isotope stage 2 or the late Wisconsin full glacial), to the time of the waning of the continental glaciers and the threshold of modern climatic conditions. Vegetation in the Great Basin underwent discernible changes throughout this period, reflecting environmental variations induced by changing climatic circulations.

Circa 40–20 ka (pre–full glacial). Few records are available and the data available come from only two locales: the Snake Range of eastern Nevada (Figs. 10.6 and 10.8; Thompson, 1984; Wells, 1983) and the Eleana Range of south-central Nevada (Spaulding, 1985). In the Snake Range Utah juniper, which apparently was absent during the full glacial, grew with subalpine conifers at 34, 30, and 27.3 ka (Thompson, 1984). At Ladder Cave shadscale and winterfat were present at 31 and 27.3 ka, but are lacking in full- and late-glacial assemblages. In a pollen profile from Council Hall Cave, also in the northern Snake Range, TCT (Taxaceae-Cupressaceae-Taxodiaceae) pollen is at much higher levels (up to 10%) in samples older than 24 ka than in spectra of apparent full- and late-glacial age (< 5%; Thompson, 1984). While these pollens cannot be confidently assigned to Utah juniper, the co-occurrence of elevated TCT percentages with this juniper in packrat middens of similar ages supports the contention that this woodland species grew in east-central Nevada prior to the full glacial.

Wells (1983) believes that limber pine, which was recorded in trace amounts in two infinite-age midden samples from the Snake Range, disappeared from this region prior to the full glacial. The middle Wisconsin–full-glacial difference is more dramatic at the Eleana Range on the southern periphery of the Great Basin (Spaulding, 1985). A Utah juniper woodland with leather-leaf mountain mahogany and steppe species was present from about 50 to 35 ka. Utah juniper disappeared and was replaced by limber pine at some point prior to 17.1 ka (Spaulding, 1985).

22–14 ka (full glacial). The full-glacial period was characterized by a general dominance of subalpine taxa and the almost complete absence of thermophiles that were present during later periods. Packrat midden assemblages from the Snake Range from this interval are heavily dominated by bristlecone pine and have the highest recorded abundances of Engelmann spruce (Thompson, 1984; Wells, 1983). At Carlin's Cave, at the south end of the Egan Range, bristlecone pine was the only conifer present in assemblages dated between 20 and 16.6 ka (Thompson, unpublished). At this site, as with the Snake Range, bristlecone pines grew with a suite of montane and steppe shrubs, including rabbitbrush, fernbush, snowberry, and horsebrush.

In the record from the Etna site in Meadow Valley Wash (Fig. 10.7), limber pine, Douglas fir, rose, and juniper (either Utah juniper [Madsen, 1973] or Rocky Mountain juniper [Wells, 1983]) were dominant in midden assemblages dated between 20.1 and 16.2 ka (Madsen, 1973). Significant vegetational change apparently occurred between 16.2 and 15.2 ka (see discussion in following section). Spaudling's (1985) midden chronology from the Eleana Range at the southern edge of the Great Basin reveals that limber pine grew with fernbush, leather-leaf mountain mahogany, sagebrush, and other steppe plants, with only minor variations in relative abundance from 17.1 to 13.2 ka.

14–10 ka (late glacial). The late glacial is by far the best-represented Pleistocene interval in the packrat midden record from the Great Basin. At several of the sites where data are available from both the full- and late-glacial periods, significant changes occurred between the two intervals. Unfortunately, many of the midden series go back only as far as the late glacial, and, as a result, our view of Pleistocene vegetation may be skewed.

In the Snake Range of the eastern Great Basin, evidence is available from several sites for the differences between full- and late-glacial vegetation. At Streamview rock shelter, prostrate juniper, Engelmann spruce, cliff-bush, and buffalo-berry, present at 17.3 ka, were absent at 11 ka (Thompson, 1984). At nearby Ladder Cave, limber pine, mock-orange, and buffalo-berry were present at 18 ka but apparently had disappeared by 13.2 ka

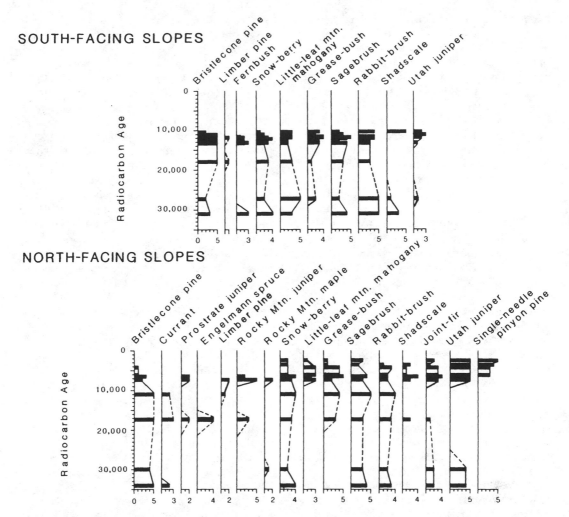

Fig. 10.8. Time series: Plant species from packrat middens from the northern Snake Range according to aspect (Thompson, 1984). The horizontal scale is a relative abundance scale: 0 = absent, 1 = rare, 2 = uncommon, 3 = common, 4 = abundant, 5 = very abundant.

(Thompson, 1984). New plants growing at the site by 13.2 ka include fernbush, snakeweed, and rock-spiraea.

Wells (1983) concluded that limber pine, which was recorded in his pre–30-ka samples, was absent from the region during the full glacial. He believes that limber pine returned sometime after 12 ka, while I have obtained radiocarbon dates on this species from three sites that demonstrate its presence as early as 13–14 ka. Traces of limber pine also occur in some full-glacial assemblages, and perhaps this species persisted regionally throughout the late Wisconsin. I have evidence that lim-

ber pine grew with prostrate juniper, Rocky Mountain juniper, currant, buffalo-berry, and sagebrush at 13.6 ka, derived from a sample at the Granite Wash site in the Deep Creek Range of west-central Utah (Thompson, 1984). As the name implies, this sample was recovered from a granite rock shelter, and perhaps the substrate difference allowed limber pine to outcompete bristlecone pine. If this is true, then limber pine may have been present through the full glacial on noncalcareous substrates.

Quasi-continuous midden chronologies from the northern Snake Range provide evidence of vegetational changes through the late glacial. At Ladder Cave, on a relatively arid south-facing slope, several mesophytic plants present in samples dated at 13.2 and 13.1 ka disappeared over the next two millennia (Fig. 10.8; Thompson, 1979, 1984). Fernbush disappeared between 13.1 and 12.1 ka, cliff-bush and currant between 12.1 and 11.1 ka. Conversely, the relatively thermophilous and drought-tolerant Utah juniper was present only as a trace (or possible contaminant) in the 13.2-ka sample, but was common in the 11.2-ka assemblage. At nearby Smith Creek Cave (on the same slope), Utah juniper was absent in samples dated at 13.3 and 12.2 ka, but was present in 11.9- and 10.5-ka samples. In this midden series shadscale and netleaf hackberry appeared in the 10.5-ka sample, while the abundance of bristlecone pine dropped off sharply.

A plant assemblage associated with a 10,650–10,450-ka archaeological occupation of Smith Creek Cave (Thompson, 1985) provides evidence that relatively mesophytic plants (limber pine, Rocky Mountain juniper, chokecherry) grew with modern dominants (Utah juniper, sagebrush, little-leaf mountain mahogany, greasebush, yucca) at this site at the close of the Pleistocene. This assemblage was not fully modern in that it contained these mesophytes and lacked single-needle pinyon pine.

The packrat middens from the Snake Range studied by Wells (1983) provide corroborating evidence that bristlecone and limber pines survived at relatively low elevations until the end of the Pleistocene. These pines survived until after 10 ka on a south-facing slope, while Engelmann spruce and prostrate juniper persisted until 10.2 ka on a north-facing slope.

Packrat midden assemblages from sites in west-central Utah near the Snake Range provide additional evidence of vegetational change through the late glacial. Two limber pine–dominated samples were collected from the Garrison site on an isolated rock knob in the Snake Valley (Thompson and Mead, 1982; Thompson, 1984). The mesophytes cliff-bush, rose, and buffalo-berry were present in the older (13.5 ka) assemblage but absent in the younger (12.2 ka) one. Concurrently, several relatively thermophilous and/or xerophytic plants (Utah juniper, shadscale, winterfat, and squawbush) appear in the younger sample.

In the Confusion Range Wells (1983) found that bristlecone pine disappeared between 11.9 and 10.3 ka: At the earlier date it was still abundant at the lowest site (1660 m), but it had disappeared from even the highest (2160 m) and north-facing site by 10.3 ka. Although bristlecone pine disappeared, limber pine, Douglas fir, Rocky Mountain juniper, and prostrate juniper persisted at the higher-elevation site. The relatively early loss of bristlecone pine in the Confusion Range, compared with its longer persistence in the Snake Range, has been attributed by Wells to the difference in mass between the two ranges (maximum elevations 2530 m versus 4120 m). New species apparently arrived in western Utah as bristlecone pines declined. The first regional record of Douglas fir is dated at 11.9 ka in the Confusion Range, while the first evidence of Rocky Mountain juniper is dated at 10.5 ka in the Wah Wah Mountains and 10.3 ka in the Confusions.

In Meadow Valley Wash in southeastern Nevada, limber pine, Douglas fir, juniper, and rose were present in full-glacial samples (Madsen, 1973). Rose apparently disappeared by 15.2 ka, as single-leaf ash, squawbush, and Gambel oak appeared (Fig. 10.7). Limber pine and Douglas fir persisted until at least 12.6 ka (Madsen, 1973; Wells, 1983) but were absent at 8.9 ka, when Utah juniper and oak were dominant.

The most pronounced vegetational change during the late glacial occurred at the Eleana Range, site of the southernmost packrat midden series in the Great Basin (Spaulding, 1985). Here, full-glacial vegetation was characterized by limber pine, fernbush, leather-leaf mountain mahogany, sagebrush, lupine, and horsebrush. This assemblage continued with only minor changes in abundance until at least 13.2 ka and then underwent a dramatic turnover between this date and 11.7 ka. Limber pine, fernbush, leather-leaf mountain mahogany, and sagebrush sharply declined in abundance, as *Haplopappus nanus* increased. The new vegetational regime was dominated by new arrivals: Utah juniper, single-needle pinyon pine, purple-sage, and prickly pear (*Opuntia* cf. *erinacea*). This record is particularly striking in that a woodland with few compositional differences from today's was established before 11 ka (Spaulding, 1985).

The discussion above focuses primarily on the modifications in the lower-elevation populations of subalpine and montane conifers. Packrat midden and pollen data also indicate that changes were occurring at the northern geographic limits of montane plants during the waning of the Pleistocene. For example, the northern limit of Douglas fir during the full glacial appears to have been near the latitude of Meadow Valley Wash (37°33′ N; Madsen, 1973). This tree reached the Confusion Range of west-central Utah (39°05 N; Wells, 1983) by 11.9 ka, and the northern limits of the Great Basin and beyond before 10 ka (Bright, 1966; Baker, 1976; Mehringer, 1977, 1986).

HOLOCENE VEGETATION

The modern vegetation of the Great Basin is the result of a progression of events over the last ten thousand years. The processes operating during the Holocene include extinction, isolation, immigration, and responses to continued climatic changes, possible prehistoric human impacts, and historic human influences.

The chronological terms used in this chapter differ somewhat from those used by Van Devender (this volume, chaps. 7 and 8). I have chosen to follow the U.S. Geological Survey's convention of placing the Pleistocene-Holocene boundary at 10 ka, whereas Van Devender detected vegetation changes indicating that 11 ka is a more major paleoclimatic boundary in the southwestern deserts (this volume, chaps. 7 and 9; Van Devender and Spaulding, 1979). Our divisions of the Holocene also differ, and I have chosen to use the informal terms "early" (10–7 ka), "middle" (7–4 ka), and "late" (4 ka to the present) Holocene in a manner that, in my opinion, applies to the vegetation history of the Great Basin.

Chronology of Vegetational Change

The pattern of vegetational change from the latest Pleistocene through the Holocene, as observed from a station within the modern pinyon-juniper woodland, mimicked the modern elevational gradient (see Fig. 10.9 for locations of Holocene sites). This late Pleistocene subalpine vegetation gave way to

a mixture of montane and upper-woodland species, which in turn was replaced by pinyon-juniper woodland. However, compositional changes continued through the Holocene, even after the establishment of the latter vegetational association. The general chronology of vegetation change is described below.

10–7 ka (early Holocene). Few data are available for this period. In the Snake Range of the eastern Great Basin, three midden macrofossil assemblages dated between 9 and 10 ka reveal that limber and bristlecone pines remained well below their modern limits into the early Holocene (Wells, 1983; Thompson, 1984). As the subalpine conifers decreased in abundance, quaking aspen, Rocky Mountain maple, service-berry, and other mesophytic plants joined the vegetational assemblage.

Pollen spectra of apparent early Holocene age from Council Hall Cave, also in the Snake Range, have the lowest relative frequencies of pine pollen in this forty-thousand-year record (Thompson, 1984). *Populus*, Tubuliflorae, *Holodiscus*-type, *Prunus-Rosa*, and *Cercocarpus*-type attain their highest percentages in these levels. Pollen types that are referable to mesophytic herbs today common to the montane upper sagebrush (Caryophyllaceae, Cruciferae, *Galium*, and *Linanthastrum*) are common. TCT pollen increases in the Council Hall Cave record by the end of the early Holocene, and a packrat midden dated at 7.4 ka from a north-facing slope in the northern Snake Range indicates Utah juniper and Rocky Mountain juniper as codominants, with low levels of limber pine, bristlecone pine, and prostrate juniper (Thompson, 1984).

In the nearby Confusion Range of western Utah, Rocky Mountain juniper apparently grew in the absence of other conifers at 8.6 ka at a site where limber pine, bristlecone pine, and prostrate juniper grew at 11.9 ka (Wells, 1983). A similar progression is seen in the midden series from Carlin's Cave, slightly farther south on the Nevada side of the line. Here bristlecone and limber pines were present until at least 12.8 ka but were absent by 8.6 ka, when Rocky Mountain juniper, Utah juniper, and fernbush grew without pines (Thompson, unpublished). At Meadow Valley Wash in southeastern Nevada, limber pine

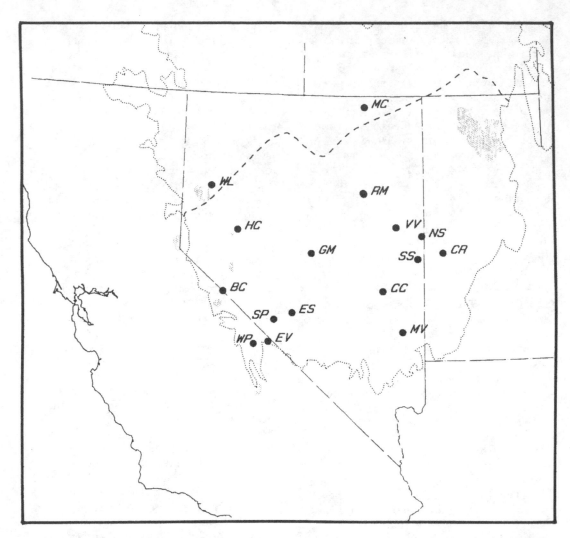

Fig. 10.9. Map of sites of Holocene age discussed in text (gray areas represent modern lakes). Key to sites: MC = Mission Cross Bog, WL = Winnemucca Lake basin, RM = Ruby Marshes, HC = Hidden Cave, GM = Gatecliff Shelter and Mill Canyon, VV = Valley View, NS = northern Snake Range, SS = southern Snake Range and Snake Valley, CR = Confusion Range, CC = Carlin's Cave, MV = Meadow Valley Wash and Rainbow Canyon, BC = Bodie Canyon, ES = Esmeralda, SP = Silver Peak, WP = Westgaard Pass Cave, EV = Eureka Valley.

and Douglas fir become uncommon at the packrat midden sites between 12.6 and 8.9 ka, and were replaced by Utah juniper and Gambel oak (Madsen, 1973).

At Gatecliff Shelter in central Nevada (Fig. 10.10), aspen and rose grew with upper-sagebrush–grassland plants from 9.5 to 9 ka in what is now pinyon-juniper woodland (Thompson and Hattori, 1983). Basal samples from the pollen record in the cave fill from this shelter, dated at about 7 ka, indicate that

sagebrush dominance continued at this site through the early Holocene.

Palynological data from valley settings from northeastern Nevada (Ruby Marshes; Thompson, 1984), western Nevada (Hidden Cave; Wigand and Mehringer, 1985), and the Mojave Desert (Tule Springs; Mehringer, 1967) indicate that sagebrush-dominated steppe persisted at lower-than-modern elevations into the early Holocene. Shadscale steppe (at the Ruby Marshes and Hidden

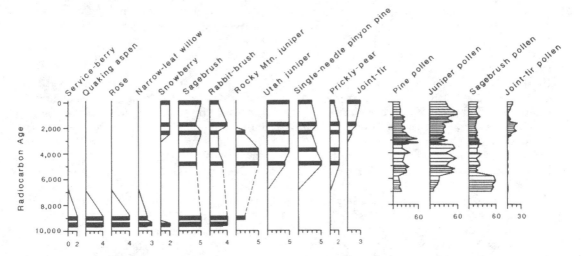

Fig. 10.10. Time series: Packrat midden macro-fossil assemblages from Gatecliff Shelter and Mill Canyon (left) and pollen data from Gatecliff Shelter (right). For the macrofossil data, the horizontal scale is a relative abundance scale: 0 = absent, 1 = rare, 2 = uncommon, 3 = common, 4 = abundant, 5 = very abundant. For the pollen data, the horizontal scale is percent of the terrestrial pollen sum.

Cave) or Mojave Desert scrub (at Tule Springs) replaced sagebrush by 8–6.5 ka.

7–4 ka (middle Holocene). With the exception of the Eleana Range series from the southern periphery (where near-modern vegetation was in place by the close of the Pleistocene; Spaulding, 1985), the most rapid recorded changes in vegetation in the Great Basin occurred between 7 and 6 ka. Essentially modern woodland vegetation was established at the Snake and Schell Creek ranges (Thompson, 1984), Meadow Valley Wash (Madsen, 1973), the Toquima Range (Thompson and Hattori, 1983; Thompson, unpublished), and at Carlin's Cave, the Wassuk Range, and the White Mountains (Thompson, unpublished). With the exception of the Carlin's Cave site, all of these localities are within pinyon-juniper woodland today, and the arrival of single-needle pinyon pine is the hallmark of the early-to-middle Holocene vegetation changes (see below for further discussion of the dispersal of this plant). The first records of ponderosa pine and white fir in the eastern Great Basin also are dated to the middle Holocene (6.1 ka; Thompson, 1984).

Although the middle Holocene woodlands contained most of the modern dominants at the packrat midden sites, they differed from modern assemblages in that they included a number of species that today are absent or rare. In the Schell Creek Range of eastern Nevada, packrat midden assemblages dated at 6.5 and 6.1 ka show the modern dominants at the site (single-needle pinyon pine and Utah juniper) intermixed with montane conifers now absent (limber pine and Rocky Mountain juniper). Similarly, middens dated from 5.3 to 2.4 ka from Gatecliff Shelter and nearby sites in the Toquima Range of central Nevada reflect a pinyon-juniper assemblage with single-needle pinyon pine and Utah juniper sharing dominance with Rocky Mountain juniper, which no longer grows at the site. Additionally, some of the common modern shrubs, such as joint-fir, apparently did not reach the Toquima Range until after the middle Holocene (Fig. 10.10; Thompson and Hattori, 1983; Thompson, unpublished).

Pollen data from valley settings indicate a regional expansion of shadscale steppe at the expense of sagebrush after 7 ka (Byrne et al., 1979; Thompson, 1984; Wigand and Mehringer, 1985), effectively raising the lower elevational limits of sagebrush steppe. Radiocarbon-dated tree and krummholz snags from above the modern treeline (LaMarche and Mooney, 1972; LaMarche, 1973) reveal a concurrent raising of the upper limits of subalpine forests.

4 ka to present (late Holocene). The data available from packrat middens suggest that

few changes in floristic composition have occurred in woodlands in the Great Basin during
the late Holocene. The only notable exception to this statement is, as mentioned above,
the decline of Rocky Mountain juniper and
the arrival of joint-fir in the Toquima Range
(Thompson and Hattori, 1983; Thompson,
unpublished). In the valleys sagebrush expanded its coverage and shadscale steppe contracted (e.g., O'Malley Shelter; Madsen, 1973;
Ruby Marshes; Thompson, 1984), while at
high elevations the treelines generally declined from their middle Holocene heights
(LaMarche and Mooney, 1972; LaMarche,
1973). In a pollen profile from Mission Cross
Bog, a montane setting in the Jarbidge Mountains of northeastern Nevada, the middle-to-
late Holocene transition is marked by significant increases in fir and pine, and reductions
in TCT and Chenopodiineae (Thompson,
1984).

While late Holocene vegetation has been
floristically similar to modern conditions,
there have been notable variations in abundance through the period. Pollen and midden
data from southeastern Oregon indicate an
expansion of western juniper into sagebrush
steppe between about 4 and 2 ka, reflecting
conditions with more effective moisture than
today (Mehringer and Wigand, 1987, *this volume*, chap. 13). Similar cool and/or moist
phases may be reflected in: (1) the fir pollen
maximum dated at 4.2 to 3.7 ka at Mission
Cross Bog, northeastern Nevada (Thompson,
1984); (2) the pine pollen maximum dated at
about 3.0–1.7 ka at Swan Lake, southeastern
Idaho (Bright, 1966); and (3) the pine pollen
maximum dated at about 3.5–2.5 ka at Gatecliff Shelter in central Nevada (Thompson and
Kautz, 1983). The end of the latter period coincides with the demise of Rocky Mountain
juniper at this site. It has not been established
whether these fluctuations are manifestations
of the same phenomenon, in which case the
apparent discrepancies in age reflect impreci-
sion in radiocarbon dating.

Holocene Vegetation Dynamics

The transition from Pleistocene to Holocene
vegetation in the Great Basin was accomplished through the processes of extinction of
low-elevation and southern populations of
subalpine and montane plants, the immigra-

tion of woodland plants, and the interactions
between established species and invaders.
The following text describes the evidence for
the scope and timing of each of these
processes.

Extinctions and isolations. As climatic
conditions changed from the latest Pleistocene into the Holocene, subalpine and montane plant populations growing on the lower
mountain slopes and at southerly sites were
eliminated. The major woody species affected
by this process in the Great Basin include
bristlecone pine, limber pine, Engelmann
spruce, prostrate juniper, mountain currant,
buffalo-berry, and fernbush. In the Snake
Range of east-central Nevada, late Pleistocene
records of fir pollen (Thompson, 1984) and
discoveries of netleaf hackberry seeds in cave
deposits may be evidence of the former existence and subsequent extirpation of subalpine fir and hackberry from this range. As
the lower elevational populations of subalpine and montane plants died off, the ranges
of these species moved upslope. There is less
area available at higher elevations, and the
habitable zone for each species was undoubtedly reduced as its range moved upslope. This
process would lead to greater fragmentation
of the range of the species, greater distances
between populations, and, in accord with the
theory of island biogeography, fewer subal-
pine/montane species could survive over the
long term on these smaller habitat islands.

The Great Basin is host to a large number
of the endemic plants (most of which are
herbaceous; Reveal, 1979; Stutz, 1978), with
approximately 90 and 169 occurring in the
states of Nevada and Utah, respectively
(Gentry, 1986). The changes in vegetation distributions over the last forty thousand years
recorded in packrat middens suggest the
mechanisms responsible for the development
of these unique taxa. These processes include:
(1) the fragmentation of the range of a given
plant, permitting genetic isolation, and rapid
evolution at least at the subspecific level (Reveal, 1979; see Critchfield, 1984, for a discussion of the effects of this process on divergent
evolution in western conifers); and (2) the
opening up of new habitats due to the recessions of Pleistocene lakes and other phenomena related to long-term climatic changes.
Genera such as *Atriplex* may have undergone

"explosive"evolution during these cycles, radiating into these newly created environments (Stutz, 1978).

Immigration of new species. Pleistocene-to-Holocene climatic changes opened up new habitable areas for plant species that apparently were absent from the Great Basin during the full glacial. These late-glacial and Holocene "invaders" include single-needle pinyon pine, ponderosa pine, Douglas fir, white fir, leather-leaf mountain mahogany, and perhaps Utah juniper, Rocky Mountain juniper, and quaking aspen. Given their differing climatic adaptations, these species should have begun their northward dispersals at different times and presumably would have progressed at different rates. As discussed earlier, Douglas fir, Rocky Mountain juniper, and Utah juniper apparently reached the east-central Great Basin prior to 11 ka. These early arrivals may have had a distinct advantage for rapid dispersal: If they moved northward while plant zones were still significantly depressed (relative to today) the distances between habitat islands would have been relatively short and the islands would have been relatively large, providing better targets for propagules.

For species such as single-needle pinyon pine, which apparently dispersed across the central Great Basin in the middle Holocene, the target islands were smaller and the intervening distances greater. Thus the dispersal rates could be expected to be less rapid. Nevertheless, once the climatic/environmental conditions in the central Great Basin were suitable for this plant, it apparently dispersed across much of its modern range in a few thousand years (Fig. 10.11). Corvids cache the nuts of pinyons and other pines (Vander Wall and Balda, 1977) and provide a transport vector that could explain the apparently swift dissemination of single-needle pinyon (Lanner, 1983). An alternative explanation, proposed by Mehringer (1986), is that prehistoric humans took an active interest in the establishment of pinyon nut groves and aided in the dispersal of this tree.

Packrat midden records from various slopes and aspects across the central Great Basin and pollen records from the same region indicate that single-needle pinyon pine did not arrive or was unimportant until after 6.5 ka (Fig.

10.12). By 6–5 ka it was well established within the middle portion of its modern elevational range. No evidence is yet available on the timing of the dispersal of this plant along the relatively low elevation "troughs" on the eastern and western margins of the Great Basin (Fig. 10.2). It is conceivable that pinyons dispersed northward in these low-elevation corridors at the close of the Pleistocene. However, palynological data from Meadow Valley Wash (Madsen, 1973) and the Toquima Range (Thompson and Kautz, 1983) may be interpreted as indicating that this species did not fill in the lower reaches of its elevational range until the late Holocene.

As discussed above, many plant species adjusted their geographic and elevational ranges between the late glacial and the middle Holocene. While many species were apparently within essentially their modern ranges by 6 ka, other species were still undergoing range adjustments in the late Holocene. An example of this is provided by joint-fir, which was present along the southern and eastern flanks of the Great Basin by the latest Pleistocene. A series of packrat middens (Thompson and Hattori, 1983; Thompson, unpublished) and two pollen records (Thompson and Kautz, 1983) indicate that this species did not reach the Toquima Range until after 4 ka. Either inherently slow dispersal rates for this taxon or a climatic cause may be involved. Two points that may favor the latter are: (1) the establishment of joint-fir in the Toquima Range occurred as Rocky Mountain juniper declined, presumably due to climatic change; and (2) a palynological record from Meadow Valley Wash (Madsen, 1973) shows a marked increase in *Ephedra* pollen during the late Holocene, in an area where joint-fir had been established for several thousand years.

Interactions between established and immigrating species. Many of the species dispersing into the Great Basin apparently arrived before the total extirpation of the subalpine and montane species established during the Pleistocene. In the Schell Creek and Snake ranges of eastern Nevada, limber pine and Rocky Mountain juniper formed a community with single-needle pinyon pine and Utah juniper between 7 and 6 ka at sites where the former two taxa are now absent (Thompson, 1984). Was this admixture of gen-

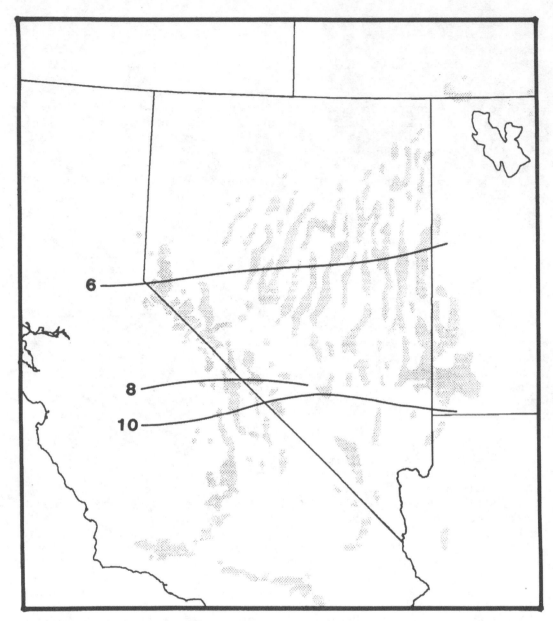

Fig. 10.11. Hypothetical pattern of dispersal of single-needle pinyon pine. Gray stipple represents the modern range of the species. Solid lines repre- sent the hypothesized northern boundaries of the species at 10, 8, and 6 ka.

erally montane species with woodland plants due to climatic conditions similar to the modern montane-woodland interface, or were the montane species able to survive in these settings due to the absence of competition? Limber pine today commonly forms wood- lands with Utah juniper (at elevations as low as 2000 m) in the northern portion of the latter species's range where single-needle pi- nyon pine is absent (Loope, 1969; Thompson, personal observations). Harper et al. (1978) noted that similar "niche expansions" (their term) occur today with subalpine fir, big sage- brush, and fernbush in mountain ranges where

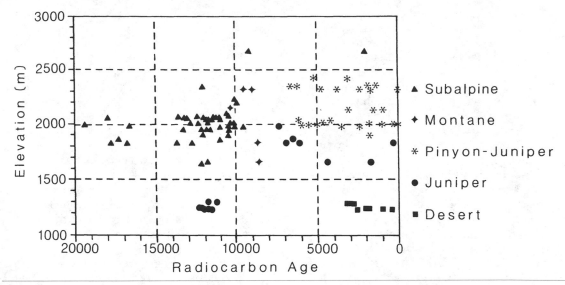

Fig. 10.12. Summary diagram: Time versus elevation for vegetational assemblages (based on packrat
middens) across the central Great Basin (38° N to
40° N) for the last twenty thousand years.

these species grow in the absence of competitor species present in other ranges. Similar competitive interactions could explain the
early Holocene dominance of Rocky Mountain juniper in the eastern Great Basin and its
subsequent replacement by Utah juniper.

Vegetation History: Summary and Discussion

The full-glacial flora of the Great Basin was
apparently impoverished: many of the modern species were absent, and there is no evidence that extralocal plants immigrated into
the region. The modern flora of the Great
Basin mountain ranges contains few Sierran
elements and has strong affinities to the flora
of the Rocky Mountains (Loope, 1969; Harper
et al., 1978; Billings, 1978; Thompson and
Mead, 1982). The packrat midden record indicates that the late Wisconsin coniferous forests of the Great Basin were populated either
by species endemic to this region or by species shared with the Rocky Mountain region.
The record of western juniper from the Winnemucca Caves (Thompson et al., 1986) provides the only known possible occurrence of a
Sierran species east of its current limits during the Wisconsin, although it is not known
whether these fossils represent the Sierran or
northern Great Basin (eastern Oregon) subspecies. Although some Sierran conifers successfully colonized hydrothermally altered
substrates in the western Great Basin (Billings, 1950), these plants have generally been
unable to penetrate farther eastward.

While the Rocky Mountains of Utah may
have served as a biogeographical continent for
small boreal mammals during the late Pleistocene, it may not have done so for Great
Basin subalpine vegetation. Similarly, the
available data indicate that this region was
not the source area for the modern suite of
montane and woodland plants growing in the
Great Basin. These Great Basin floras have
been enriched through time by immigration
from the south, a process that occurred in
parallel with Holocene floristic development
in the western Rocky Mountains (Betancourt,
this volume, chap. 12).

The shift from the vegetational regime of
the late Wisconsin to the modern regime occurred in different fashions at different sites.
At the Eleana Range site (Spaulding, 1985) a
woodland with few compositional differences
from today's was established before 11 ka. In
contrast, most of the sites have histories of
vegetational development that parallel in
many respects the modern elevational zonation of plant communities in the region (Fig.
10.12). At Meadow Valley Wash (Madsen,
1973; Wells, 1983) limber pine and Douglas
fir were replaced by Utah juniper and oak between 12.6 and 8.9 ka. Pinyon pine did not

join the assemblage until about 6.6 ka. The record from the Confusion and Wah Wah ranges of western Utah indicates that bristlecone and limber pines persisted at low elevations until 10.5 ka. Bristlecone pine was absent after this time, while limber pine, Douglas fir, and Rocky Mountain juniper survived until 10.3 ka. By 8.6 ka only Rocky Mountain juniper remained, and by 4.4 ka it had been replaced by the more xerophytic Utah juniper.

The transition from Pleistocene to Holocene vegetation also occurred in stages in the Snake Range (Thompson, 1984; Wells, 1983). Bristlecone pine disappeared from xeric sites by 10.5 ka while it continued to grow on mesic, relatively low-elevation sites with limber pine until at least 9.6 ka. Concurrently, montane plants such as quaking aspen, Rocky Mountain maple, and Rocky Mountain juniper became increasingly important in the vegetational assemblage. By 7.4 ka subalpine and more mesophytic montane plants were rare, and a woodland with Utah and Rocky Mountain junipers was established. Single-needle pinyon pine arrived at the Snake Range and adjoining areas by 6.3 ka, and the remaining montane species disappeared from the lower mountain slopes soon after this time.

CLIMATIC IMPLICATIONS OF PALEOVEGETATION

Pleistocene Climates

The paleoclimatic interpretation of Pleistocene vegetational records is based on two assumptions: (1) climatic conditions were within the range(s) of tolerance of the species present in the packrat midden assemblages; and (2) climatic conditions presumably exceeded in some manner the environmental ranges of the modern species that were apparently absent from these assemblages. Instrumental climatic data from the modern ranges of the various species provide guidance on the past climatic conditions from the areas of the midden collections.

In the eastern Great Basin late Pleistocene packrat midden and cave-sediment vegetational assemblages are dominated by bristlecone pine and other subalpine species, in association with Engelmann spruce in mesic settings and with sagebrush and other steppe species at all packrat midden sites and in the

valleys. Intermountain bristlecone pine grows today in approximately twenty-five mountain ranges scattered from the southwestern edge of the Great Basin to the Wasatch Front. Throughout this range it is generally restricted to calcareous, light-colored substrates at 2900–3500 m elevation. Compared to other subalpine conifers in the western United States, it lives in a generally cold and dry environment. Both intermountain and Rocky Mountain bristlecone pines may be restricted to relatively cold-dry climates and nutrient-poor substrates by competition with other subalpine taxa (Wright and Mooney, 1965; Peet, 1978). At the dry end of the geographic range of intermountain bristlecone pine, where some of largest groves occur (Table 10.3; Crooked Creek station, White Mountains of eastern California), the tree grows under a mean annual precipitation (MAP) level of 318 mm at 3095 m elevation (Wright and Mooney, 1965). In the northeastern portion of its range, in the southern Ruby Mountains of Nevada, scattered bristlecone pines live under a MAP level estimated at 512 mm at 3050 m elevation (Loope, 1969). In a more mesic setting, in the southern Snake Range of east-central Nevada, bristlecone pines grow with Engelmann spruce under a MAP level estimated at over 800 mm (LaMarche and Mooney, 1972). The precipitation regime of intermountain bristlecone pines is skewed toward winter dominance, although bristlecone pines growing in the southeastern portion of the range may receive as much as 30% of their MAP in the summer months.

During the late Pleistocene, bristlecone pines grew as low as 1800–2100 m in the eastern Great Basin and in the Great Basin–Mojave Desert border region, through what is now pinyon-juniper woodland (on appropriate substrates) into the modern lower-sagebrush zone. In the modern woodland, MAP levels are in the range of 220 to 350 mm (Table 10.3), overlapping slightly with the levels on the driest bristlecone pine sites. The presence of Englemann spruce in Pleistocene vegetational assemblages may indicate that past populations of bristlecone pines were not growing under conditions analogous to their driest modern settings. However, spruce was apparently not abundant at the Pleistocene midden

Table 10.3. *Climatic data for plant associations in the Great Basin region.*
Groupings on the left are from Kuchler (1964; see Fig. 10.3). Key to Kuchler associations: CRS = creosote bush desert, SHD = shadscale desert, SGB = sagebrush steppe, PNJ = pinyon-juniper woodland, SRN = Sierran forests, SWF = southwestern forests. Data on right are for Great Basin subalpine (SUB) and alpine (ALP) associations (mean values only, no standard deviations). The climatic variables are: TJAN = mean January temperature, TJUL = mean July temperature, MAT = mean annual temperature, PJAN = mean January precipitation, PJUL = mean July precipitation, MAP = mean annual precipitation. M = mean for N stations, S = standard deviation for N stations, where N = number of stations.

Group N		CRS 12	SHD 29	SGB 61	PNJ 7	SRN 20	SWF 7	SUB WM[b]	SUB SR	SUB RM	ALP WM[b]
TJAN	M	7.4	−0.3	−2.5	−0.9	1.7	−2.1	−7.1	−10.8[d]	−10.7[g]	−9.8
	S	1.6	3.4	2.6	2.6	3.1	3.3				
TJUL	M	30.1	24.8	21.7	21.0	20.1	19.7	11.9	10.2[d]	8.0[g]	7.3
	S	2.4	3.3	2.3	1.6	2.7	4.6				
MAT	M	18.2	11.7	9.0	8.9	10.1	8.0	1.8	−1.4[d]	−2.3[g]	−2.4
	S	2.0	3.2	2.2	2.0	2.7	4.2				
PJAN	M	1.7	1.6	2.7	3.3	19.8	4.6	3.5	—	—	4.2
	S	0.6	0.8	1.3	1.5	8.3	2.5				
PJUL[a]	M	0.9	1.1	1.6	1.5	0.5	5.2	3.9[c]	—	—	4.3
	S	0.6	0.6	1.0	1.0	0.3	1.8				
MAP	M	11.9	17.3	27.0	28.5	103.1	47.2	30.9	81.3[e]	76.2[f]	37.0
	S	2.7	6.9	8.2	6.6	40.8	18.1		>61.0[f]	119.4[h]	
											77.5[i]
											54.6[j]

[a] It should be noted that some of the climatic data are not normally distributed (this is especially notable for July precipitation) and the standard deviations may thus be misleading.

[b] Mooney et al., 1962

[c] This figure gives the somewhat misleading impression of a summer precipitation maximum. Mean June precipitation is only 0.56 cm, and mean August precipitation is only 1.50 cm.

[d] These estimates were made for 3,050 m elevation in the Snake Range of eastern Nevada based on the data from Lehman Cave National Monument (2081 m elev.) and regional lapse rates (see text).

[e] These estimates were made for 3,050 m elevation in the Snake Range of eastern Nevada by Synder and Langbein (1962).

[f] Houghton, 1969, p. 185.

[g] These estimates were made for 3,050 m elevation in the Ruby Mountains of northeastern Nevada based on data from Ruby Marshes Wildlife Refuge (1833 m elev.) and regional lapse rates (see text).

[h] Estimate for 3,050 m elevation in Northern Ruby Mountains from Loope, 1969, p. 45.

[i] Estimate for 3,050 m elevation in Southern Ruby Mountains (Smith Creek) from Loope, 1969, p. 45.

[j] Estimate for 3,050 m elevation in Southern Ruby Mountains (Pearl Peak) from Loope, 1969, p. 45.

sites and was restricted to mesic settings. This may indicate that precipitation levels did not rise into the upper portion of bristlecone pine's range. In temperature terms the modern climatic stations in modern pinyon-juniper woodlands nearest to the Pleistocene midden sites in east-central Nevada have mean July–August temperatures of 19.1°C (Ely), 20.8°C (McGill), and 21.3°C (Lehmann Caves). Summer temperatures in modern bristlecone pine forests are about ten degrees below these values (Table 10.3), suggesting that summer temperatures during the late Pleisto-cene were that much lower than modern conditions.

While bristlecone pines were dominant in the modern pinyon-juniper woodland during the late Pleistocene, modern woodland and montane conifers were absent. Neither Sierran nor Rocky Mountain montane species were able to colonize the central Great Basin during this period. As illustrated in Table 10.3, these plants today grow in climates with July and mean annual temperatures as much as ten degrees warmer than the conditions prevailing in bristlecone pine forests. The apparent absence of tundra indicates that mean summer temperatures did not drop much below 10°C (Houghton et al., 1975).

The climates of Sierran montane forests also differ from that of intermountain bristlecone pines in precipitation amounts. Sierran montane plants may receive as much as 150% of the precipitation levels of bristlecone forests (Table 10.3).

Climatic variations through the late Pleistocene. As noted above, the vegetation of the middle Wisconsin and late glacial differed somewhat from that of the full glacial. These differences, although minor, apparently reflect climatic differences between these periods. In the Snake Range of eastern Nevada, the presence of Utah juniper, joint-fir, and other woodland elements in middle Wisconsin packrat midden assemblages, and of increased juniper and joint-fir pollen in cave-sediment pollen spectra (Thompson, 1984), almost undoubtedly reflect warmer-than-full-glacial temperatures. Similarly, during the late glacial the apparent arrivals of Douglas fir and woodland and montane junipers, coupled with the decline of subalpine plants on xeric sites, suggest that temperatures rose above full-glacial levels, and perhaps summer precipitation increased by a modest amount. On the other hand, for both the middle Wisconsin and late glacial in the eastern Great Basin, the presence of bristlecone pines and the absence of pinyons indicate that temperatures were still much cooler than they are today.

Western juniper grew at the Winnemucca Caves in western Nevada during the late glacial (Thompson et al., 1986). Today this species grows under a wide range of precipitation conditions, living under mean annual precipitation levels of 25–33 cm in the southern portion of its range up to 127–140 cm in the north (Fowells, 1965). In temperature terms it survives under mean annual temperatures ranging from < 6° to > 12°C. Instrument data from Sand Pass, the weather station closest to the Winnemucca Caves, record a MAP of 17.3 cm and a MAT of 10.6°C. While the latter figure is within the range of climatic tolerance of western juniper, a 50% increase in MAP would be required to bring the climate of this site into the lower range of tolerance for this species.

Holocene Climates

As described in a previous section, the Holocene vegetation history of the Great Basin encompasses a dynamic series of extinctions and immigrations. Presumably these phenomena were tied to climatic stimuli, although perhaps with lag times of unknown magnitudes. The possibility of these inherent lags, coupled with the broad climatic tolerances of many of the plant species involved, complicates paleoclimatic interpretations of these data. Nevertheless, a broad-scale picture of climatic change can be reconstructed from the packrat midden and pollen evidence.

Early Holocene. Although the data are sparse for this period, it seems clear that bristlecone pine generally retreated upslope during this period, while limber pine, Rocky Mountain juniper, quaking aspen, and service-berry increased in abundance within the modern woodland zone in the eastern Great Basin. Pollen spectra of apparent early Holocene age from cave sediments in a woodland setting in the region suggest that plants now characteristic of the montane upper-sagebrush–grass zone grew approximately 400–500 m below their modern limits (Thompson, 1984).

The dominance of montane plants in the range of current pinyon-juniper woodlands suggests that cooler-than-modern climatic conditions prevailed during the early Holocene. Although limber pine may be limited in elevational and geographic distributions by competition with other conifers (Lepper, 1974; Peet, 1978), there may still be paleoclimatic implications of its abundance in the eastern Great Basin during the latest Pleistocene and early Holocene. The optimal temperature for photosynthesis in limber pine seedlings is about 15°C (Lepper, 1974). If this is an approximation of the growing-season temperatures optimal for the species, then its abundance could reflect summer temperatures that were about 4°–5°C below those of today. Pollen data from valley-bottom settings indicate that sagebrush steppe remained well below its modern elevational limits until after 8 ka (Mehringer, 1967; Thompson, 1984), providing additional evidence of cooler-than-modern conditions during the early Holocene. These findings are in direct opposition to predictions made by climate modelers who claim that the continental interior was warmer than today during this period (e.g., Kutzbach, 1987).

Middle Holocene. Limber pine, Rocky Mountain juniper, and other montane species are common in packrat middens from the eastern Great Basin dated as young as around 6.5 ka. Younger assemblages either lack these plants or have only very low abundances of them, are dominated by Utah juniper, and contain pinyon pine, ponderosa pine, and/or white fir. In the valleys the lower limits of sagebrush steppe retreated upslope, and shadscale- and greasewood-dominated vegetation expanded. Concurrently, the upper treeline advanced 100 m or more above modern levels (LaMarche and Mooney, 1972; LaMarche, 1973). All of these factors point to a general rise in summer temperatures and perhaps a shift in the seasonality of precipitation. The modern distributions in the Great Basin region of single-needle pinyon pine, ponderosa pine, and white fir appear to correlate with the limits of significant incursions of summer precipitation from subtropical sources (Thompson and Hattori, 1983; Thompson, 1984). Tueller et al. (1979) have suggested that the abundance of Utah juniper in Great Basin woodlands increases as the amount of summer precipitation increases. Thus, the immigration of single-needle pinyon pine and other taxa, coupled with a dramatic increase in Utah juniper, may reflect an increase of summer rainfall over early Holocene levels, and perhaps over modern levels. Greater abundance of summer precipitation may also be the explanation for the persistence of Rocky Mountain juniper in woodlands in the central Great Basin (Thompson and Hattori, 1983).

Late Holocene. Packrat midden assemblages of late Holocene age from the eastern Great Basin differ only slightly from those of middle Holocene age (Thompson, 1984). In the central Great Basin there is evidence of greater change, with Rocky Mountain juniper declining as joint-fir arrived (Thompson and Hattori, 1983). At Mission Cross Bog, in the northeastern portion of the Great Basin, palynological data indicate an increase in fir and other relatively mesic elements after 4 ka, while valley-bottom pollen data indicate an

increase in areal expanse of sagebrush at the expense of shadscale (Thompson, 1984). At higher elevations treelines generally declined through the late Holocene (LaMarche and Mooney, 1972; LaMarche, 1973). In sum, these changes reflect a general trend toward cooler and moister conditions.

Summary and Discussion: Paleoclimatic Implications

As discussed above, the full glacial occurrence of bristlecone pine in the modern pinyon-juniper zone in the eastern Great Basin can be interpreted as indicating climatic conditions with summer temperatures as much as ten degrees below modern levels, and at least a modest increase in precipitation over current values. Spaulding (1985) interpreted his vegetation assemblages from the Great Basin–Mojave Desert boundary as reflecting a full-glacial climate with summer temperatures about seven to eight degrees below modern levels and mean annual precipitation increased by 30%–40% over today's values. Wells's (1979) suggestion that summer rainfall was greatly augmented during the late Pleistocene is not supported by vegetation data from either region.

The changes in vegetation through time in the east-central Great Basin appear to follow the modern elevational gradient. Climatic data from modern vegetation zones suggest that early Holocene temperatures, while considerably warmer than those of the full glacial, were still several degrees Celsius below those of today. These data also indicate a reduction in mean annual precipitation during the transition from glacial to interglacial conditions, although it is difficult to assess the amount of this decrease.

A comparison of vegetation records from across the latitudinal span of the central Great Basin indicates that there may be a geographical gradient in the climatic sensitivity of these sites. A palynological record from the most northern site, the Ruby Marshes (40° N; Thompson, 1984; Thompson, unpublished), contains no evidence of major vegetational changes from around 40 to 8 ka. Packrat midden assemblages from the Snake Range and adjacent ranges (39°–39°20' N; Thompson, 1979, 1984; Wells, 1983) show only modest vegetational changes from the middle

Wisconsin through the late glacial. Farther south, in the Meadow Valley Wash region (37°30'; Madsen, 1973), vegetational fluctuations over this time range occurred earlier and were of larger magnitude. At the southern edge of the Great Basin, the Eleana Range midden series (37°07' N; Spaulding, 1985) displays the largest-magnitude late Pleistocene changes. The discrepancies between these records may be due to the larger number of plant species available in the south, providing a greater number of possible expressions of climatic change. Alternatively, the southern sites may have been closer to a major boundary between air masses, in which case a relatively small degree of climatic change could result in the movement of this boundary over the midden sites, inducing a large vegetation response. Lesser-magnitude changes occurred between 40 and 8 ka farther south in the Sonoran Desert (Van Devender, *this volume*, chap. 8), constraining the region of maximum variability to the Mojavean region.

Comparisons with Other Paleoclimatic Indicators

The Great Basin is rich in paleoclimatic proxy data, and lake-level, glacial, isotopic, and paleontological studies have been conducted across the region. These studies provide opportunities for comparisons and assessments of differing sensitivities and lag times of vegetation and other indicators to climatic change (Fig. 10.13).

Lacustrine records. Recent research suggests that lakes across the central Great Basin were in near synchrony in their changes over the past twenty-five thousand years (Benson and Thompson, 1987a, 1987b; Thompson et al., 1987). Lake levels were at higher-than-modern levels in portions of the middle Wisconsin, generally low in the period immediately prior to 25 ka, and appear to have risen gradually (with some oscillations) until around 16 ka. The Pleistocene lake maxima in the Great Basin apparently did not coincide

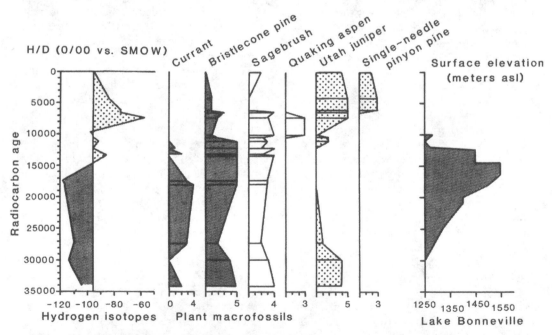

Fig. 10.13. Comparison diagram of paleoclimatic indicators: Changes in lake levels, hydrogen/deuterium ratios, and vegetation in the eastern Great Basin over the last thirty-five thousand years. Gray stipple marks indicators of cooler- and/or moister-than-modern conditions. Horizontal scales are as follows: For the hydrogen isotopic data, the scale is in parts per mil (SMOW = isotopic content of Standard Mean Ocean Water); for the plant macrofossils, the scale is relative abundance: 0 = absent, 1 = rare, 2 = uncommon, 3 = common, 4 = abundant, 5 = very abundant. For lake Bonneville, the scale is lake surface elevation in meters above sea level.

with the continental glacial maximum, although lakes were at least moderately deep during this period. The deep-lake phases of the middle Wisconsin and late-glacial periods apparently occurred under warmer-than-full-glacial climates (as indicated by vegetational data) and conditions that may have had a larger-than-full-glacial southerly component in the precipitation regime. The full glacial, the period of the lowest temperatures in the vegetational record, was represented by moderate-size lakes. Lake-level fluctuations paralleled changes within vegetation but were of much larger scale than those recorded in botanical records of change from the middle Wisconsin through the late glacial.

The maximal recorded lake levels in the Great Basin occurred in the period between 16 and 12.5 ka, after which the lakes desiccated rapidly to very low levels between 12 and 11 ka (Benson and Thompson, 1987a, 1987b; Thompson et al., 1987; Currey and Oviatt, 1985). Botanical data show only minor changes through this period of dynamic lake history, suggesting that these systems have very different climatic sensitivities and response times. In many basins higher-than-historic lake stands occurred in the terminal Wisconsin–early Holocene period (about 11–7 ka). Low lake levels occurred again about 7–4 ka (Antevs, 1948, 1952; Thompson, 1984; Wigand, 1985), following which a series of "neopluvial," higher-than-historic lake stands occurred, separated by relatively arid intervals. In contrast with the terminal Wisconsin events, these lake-level changes have close parallels in the long-term changes in vegetation.

Glacial and periglacial records. Relatively little is known about the chronology of glacial fluctuations in the Great Basin. The fragments of data available suggest that the alpine glacial maxima preceded the maximal lake stands by perhaps several thousand years, and that the glaciers were essentially gone by 13–12.5 ka (Russell, 1885; Antevs, 1952; Mehringer, 1977; Madsen and Currey, 1979; Scott et al., 1983; Wayne, 1984; Benson and Thompson, 1987a). Dohrenwend (1984) identified thousands of nivation basins in the western Great Basin, which he interpreted as indicative of Pleistocene "Marginal periglacial environments" as low as 2100 m at 40°

N, 2100–2400 m at 39° N, and 2400–2600 m at 38° N. On the basis of the regional depression of "full glacial" equilibrium line altitudes (ELAs), Dohrenwend estimated that the mean annual temperature was about seven degrees lower than today. Zielinski and McCoy (1987) also examined the geographical patterning in the amounts of depression of Pleistocene ELAs in the Great Basin. Their results indicate that the differences between Pleistocene and modern climate are not evenly distributed across the region: the northwestern Great Basin either experienced much greater temperature depressions (relative to today) or received a much larger augmentation of precipitation.

The paleoclimatic interpretations of Quaternary vegetational assemblages are in general agreement with the climatic reconstructions based on glacial and periglacial phenomena in terms of temperature depressions and the indications of rising temperatures between the full- and late-glacial periods. However, as with the lake-level data, the glacial records exhibit much larger changes than the vegetational data.

Isotopic Evidence. Siegal (1983) studied deuterium/hydrogen (D/H) ratios of cellulose from plant remains found in packrat middens collected in the Snake Range of eastern Nevada (Fig. 10.13; Thompson, 1984; see also Long et al., *this volume*, chap. 17). Variations in these ratios through time probably reflect several factors, including temperature at the site during the growing season, seasonality and source of precipitation, and sea-surface temperatures. If the most simple case were true, and D/H fluctuations reflect changing temperatures at the midden sites:

1. Temperatures at 34 ka were near modern values.
2. More negative D/H ratios at 29.9 and 27.3 ka suggest temperature depressions of −2.5°C or more below modern values.
3. The most negative D/H values, recorded at 18 ka and 17.4 ka, indicate temperature depression of 4.1°C or more, relative to today.
4. Samples dated from 13.3 to 9.7 ka have D/H ratios near modern values or slightly less negative, reflecting temperature anomalies relative to today of +3.2°

to −1°C. A cooling trend of −0.5°C/1000 yr may have occurred through this period.

5. The D/H ratio at 7.4 ka is the least negative value in the record, reflecting the warmest temperatures during the last thirty-four thousand years. This inferred temperature anomaly is < 7°C higher than today (although it could be considerably less if summer precipitation from subtropical sources was at higher-than-modern levels). Less-negative-than-modern values occurred also at 6.5 and 6.1 ka, reflecting temperatures < 3.6°C and < 4.2°C higher than today, respectively.

6. The D/H value from a 4.2-ka sample continues the trend toward modern values with a temperature estimate of < 2°C higher than today.

In general form, the interpretations of these isotopic data are in agreement with those from the vegetational assemblages that provided the materials for isotopic analyses. In both data sets full-glacial temperatures are significantly colder than those of the middle Wisconsin and late glacial. A major discrepancy exists in that the isotopic data suggest larger-magnitude climatic changes between the full and late glacial than are seen in vegetational data. Additionally, while botanical data indicate that temperatures gradually warmed between 13.3 and 9.7 ka, the D/H ratios can be interpreted as indicating exactly the opposite. For the Holocene, both data sets indicate warmer temperatures and increased summer precipitation during the middle Holocene, and cooler temperatures and (perhaps) decreased summer rainfall in the late Holocene.

Vertebrate paleontology. The remains of horses, camelids, and antelopes in late Pleistocene faunas in the Great Basin (Grayson, 1982, 1987), while not providing a direct climatic signal, do probably reflect the presence of open nonconiferous vegetation. This may be in accord with the pollen-based reconstructions of open steppe vegetation in the valleys during the late Pleistocene, or alternatively could reflect the presence of areas of open vegetation within the bristlecone pine forest. Small mammals provide more direct climatic information on climatic changes because individuals cannot migrate long dis-

tances to escape unfavorable conditions. In this light, the numerous Wisconsin-age finds of boreal small mammals in packrat middens and caves at relatively low-elevation sites in regions where they no longer occur suggest that summer temperatures were relatively cool (Thompson and Mead, 1982; Thompson, 1984). The persistence of pikas in eastern Nevada until 6.8 ka (Thompson, 1984; Van Devender et al., 1985) and in eastern Oregon until < 6.9 ka (Grayson, 1987) suggests that cool summer temperatures persisted through the early Holocene (e.g., Smith, 1974). These data are in accord with the vegetation-based reconstructions of cool summer temperatures during the early Holocene. On the other hand, the survival of the heather vole (*Phenacomys* cf. *intermedius*) until around 5.3 ka in central Nevada (Grayson, 1983) appears to contradict vegetational evidence of warmer-than-modern summer temperatures during the middle Holocene.

Vegetational Inertia? In recent years the potential lag times in vegetational responses to climatic changes have been the focus of debates within the community of paleoecological researchers. In the eastern United States changes in the abundances and distributions of plant species apparently lagged behind major climatic changes by a matter of decades to centuries (e.g., M. B. Davis et al., 1986; Webb, 1986). On the other hand, it has been claimed that in the West lag times may be measured in millennia (Cole, 1985). As far as can be seen from the existing data in the Great Basin, lake levels, glaciers, and hydrogen isotopes all apparently responded to climatic changes during the late Pleistocene more rapidly (and with larger-magnitude changes) than did vegetation. This is especially notable with the apparent regional fall in Pleistocene lakes from their highest to their lowest extents at between 14–13 and 12 ka. Vegetation records show only minor differences through this period, and it is tempting to interpret the absence of major vegetational changes in some subregions during periods of major hydrological changes as evidence of vegetational inertia (Cole, 1985). However, relatively large vegetational changes did occur on the southern periphery of the Great Basin near the time of the lake desiccations (Madsen, 1973; Spaulding, 1985), and the small amounts of vegetational change farther north may reflect the positions of the packrat midden sites in relation to major ecotones and range boundaries (see discussion above). It also is possible that vegetation and lake levels were responding to different aspects of climatic change, and should not be expected to respond to a given stimulus in the same manner. Holocene vegetational records appear to have responded in near synchrony with geophysical changes induced by climatic fluctuations. Whatever the factors responsible for the late Pleistocene differences in response times and mangitudes, they do not seem to have been operative over the last ten thousand years.

Comparison with Reconstructions from Numerical Climate Models

Numerical models of the global climate system provide insights into the dynamics of the atmosphere and into the effects of changes in boundary conditions (such as global ice volume and spatial and seasonal variations in insolation) through time. Although their spatial resolution is limited (4.4° latitude × 7.5° longitude in the model employed by Climates of the Holocene Mapping Program [COHMAP]; Kutzbach and Guetter, 1986; Kutzbach, 1987), model simulations of past climatic states provide descriptions of the major features of circulation that are in essential agreement with geological data at hemispheric to continental scales (e.g., Kutzbach and Wright, 1985).

Full-glacial climates. Simulations of full-glacial climates by COHMAP and Manabe and Broccoli (1985) suggest that the Laurentide ice sheet disrupted westerly flow, splitting the jet stream into two branches. The COHMAP simulations indicate that the mean position of the southerly branch was twenty degrees south of the modern position of the jet stream. This phenomenon placed the Pacific Northwest under easterly anticyclonic flow off the ice sheet, bringing cold-dry environmental conditions to this region (Barnosky et al., 1987). Concurrently, the modern Southwestern deserts received abundant precipitation from storms following the southern branch of the jet stream, and both the COHMAP model runs and paleoenvironmental data (Van Devender and Spaulding, 1979) indicate only moderate depressions below mod-

ern temperatures. The Great Basin lies between these two regions, and both the spatial resolution and the depiction of physiography in the models are too crude to permit a determination of whether this region would have been under desiccating winds from the mid-continent or under westerly moisture-bearing flow off the Pacific during the full glacial. The predominance of cold-tolerant species and the apparent absence of thermophiles in plant assemblages from packrat middens dated to this period may indicate that westerly flow was not great and/or persistent enough to moderate extremely cold winter temperatures. Siegal's (1983; Fig. 10.13) H/D ratios from packrat midden cellulose also indicate that major temperature depressions occurred in the Great Basin during the full glacial.

Late-glacial climates. The COHMAP reconstructions indicate that the climatic conditions of the full glacial persisted until at least 15 ka. Between this time and 12 ka the split jet stream was replaced by a more zonal flow, and the westerlies at 30° N were much weaker than at 18 ka. The exact timing and nature of the transition from full- to late-glacial conditions is difficult to assess in the model given the large uncertainties in estimating key boundary conditions such as ice-sheet height. Paleoenvironmental records from the Pacific Northwest indicate warmer and moister conditions by 14 ka, suggesting diminished easterly flow (Barnosky et al., 1987). In the Great Basin the maximal levels of the late Pleistocene lake systems apparently occurred between around 15 and 13 ka (Benson and Thompson, 1987a), which may reflect a shift of the mean path of the westerlies northward from their full-glacial position in the Southwest.

The demise of the Pleistocene lakes (about 14–13 ka) and the subsequent replacement of subalpine plants by montane and woodland elements (about 12–10 ka) may reflect the continued northward shift of the westerlies as the continental ice sheet retreated. This would place the Great Basin under only intermittent westerly flow and would perhaps allow incursions of southerly air masses. The warmer and drier conditions of this late-glacial regime would have been accentuated by the rising summer insolation anomalies.

Early Holocene climates. The COHMAP reconstructions suggest that the small residual ice sheet present at 9 ka had minimal influence on continental climates. The higher-than-modern insolation values at this time produced large positive temperature anomalies in the interior of North America in the model runs, a situation that would produce increased onshore flow and "monsoonal" precipitation. Given the poor spatial resolution of the model, it is not possible to predict the path of this circulation. Spaulding and Graumlich (1986) believed that they found evidence in the Mojave Desert to support the COHMAP reconstructions, whereas Van Devender (1987) has argued that data from the Sonoran Desert, closer to the source of monsoonal precipitation, negated the model results.

Mitchell et al. (1988) simulated the climate of 9 ka with a model that included interactive soil moisture and ocean-surface temperature components, features lacking in the model available for the COHMAP experiments. They made two sets of model runs for 9 ka: one set with a North American ice sheet absent and the other with it present. Although both are labeled "9 ka simulations," Mitchell et al. (p. 8300) suggest that they could be viewed in chronological order, representing the likely conditions before and after the early Holocene collapse of the last major remnant of the Laurentide ice sheet. In both cases, as with the COHMAP simulations, monsoonal circulation patterns were stronger than at present in the northern hemisphere. However, Mitchell's experiments, which incorporated interactive oceanic and soil-moisture components, produced smaller temperature anomalies for the early Holocene than those predicted by the COHMAP simulations.

With the ice sheet present, the experiments of Mitchell et al. (1988) indicate that winter temperatures in the western United States would be as cold or colder than those of today. According to the model results, July temperatures in the northern hemisphere were around 2.1°C above modern levels. However, the predicted increase in monsoonal circulation is less than in the COHMAP simulations or the Mitchell model runs without an ice sheet. As a result, July precipitation-evaporation levels are simulated to have been

below those of today on a hemispheric basis. With ice sheet absent, Mitchell's simulations indicate a greater difference between early Holocene and modern summer temperatures (about 2.6°C) and increased soil-moisture deficits in the regions of maximum heating. Summertime westerly flow would be greatly diminished in the Southwest, and subtropical flow increased. North of about 45° N winters were warmer than today, whereas south of this latitude they were cooler. This reduced meridional temperature gradient would lessen the strength of the westerlies, decreasing the circulation that brings winter rainfall into much of the western United States. Despite this factor, the model simulations indicate higher-than-modern wintertime soil-moisture levels in the southernmost United States.

If viewed as a chronological sequence, the simulated transition from the climatic state with a residual ice sheet present to one with the ice sheet absent produces conditions that in many regards resemble the environmental sequence for the Southwest proposed by Van Devender and Spaulding (1979). In their construct, cool-moist conditions associated with stronger-than-modern westerlies persisted in the Southwest until around 8000 B.P., after which they were replaced by warmer conditions with more summer rainfall. A similar transition can be seen in the Great Basin, where plant assemblages dated to about 10 ka indicate cooler-than-modern conditions, while assemblages dated to 7.3–6.5 ka reflect warmer conditions and more summer precipitation. Both of the Southwestern and Great Basin data appear to conflict with the COHMAP and Mitchell et al. model results, which predict elevated continental summer temperatures and increased monsoonal circulations by 9 ka. However, as discussed by Mitchell, the boundary conditions of this period can produce a variety of effects not foreseen by analogies to modern patterns. Early Holocene onshore flow in the West could have followed trajectories uncommon today and provided moist air masses from different source areas than those of today. As indicated by Mitchell's reconstructions, a monsoonal regime can produce cooler-than-modern surface temperatures as the increased cloud cover and evaporation offset temperature increases induced by insolation increases.

ACKNOWLEDGMENTS

I thank P. J. Bartlein, J. L. Betancourt, L. V. Benson, J. P. Bradbury, D. K. Grayson, R. M. Forester, P. S. Martin, J. I. Mead, W. G. Spaulding, T. R. Van Devender, and T. Webb III for discussions related to matters covered in this chapter. Betancourt, Bradbury, Martin, and Van Devender also kindly provided critical reviews of the manuscript. Portions of the research discussed in this chapter were funded by National Science Foundation grants to P. S. Martin, T. R. Van Devender, and T. Webb III. Other funding was provided by the Nevada State Historic Preservation Office and the Carson City district office of the U.S. Geological Survey.

REFERENCES

Adam, D. P. (1967). "Late-Pleistocene and Recent palynology in the central Sierra Nevada, California." In *Quaternary paleoecology.* E. J. Cushing and H. E. Wright, Jr., eds., pp. 275–301. New Haven: Yale University Press.

Antevs, E. (1948). "The Great Basin, with emphasis on glacial and post-glacial times: Climatic changes and pre-white man. *Bulletin of the University of Utah* 38, 168–91.

———. (1952). "Cenozoic climates of the Great Basin." *Geologische Rundschau* 40, 94–108.

Axelrod, D. I. (1976). "History of the coniferous forests, California and Nevada." *University of California Publications in Botany* 70, 1–62.

———. (1981). "Holocene climatic changes in relation to vegetation disjunction and speciation." *American Naturalist* 117, 847–70.

Baker, R. G. (1976). *Late Quaternary vegetation history of the Yellowstone Basin, Wyoming.* U.S. Geological Survey Professional Paper 729-E.

Barnosky, C. W.; Anderson, P. M.; and Bartlein, P. J. (1987). "The northwestern U.S. during deglaciation; vegetational history and paleoclimatic implications." In *North America and adjacent oceans during the last deglaciation.* W. F. Ruddiman and H. E. Wright, Jr., eds., pp. 289–321. *Geology of North America,* vol. K-3. Geological Society of America.

Barry, R. G. (1981). *Mountain weather and climate.* New York: Methuen.

Beiswenger, J. M. (1984). "Late Quaternary vegetational history of Grays Lake basin, southeastern Idaho." *Abstracts of the Eighth Biennial Meeting of the American Quaternary Association,* Boulder, p. 8.

Benson, L., and Thompson, R. S. (1987a). "The physical record of lakes in the Great Basin." In *North America and adjacent oceans during the last deglaciation.* W. F. Ruddiman and H. E. Wright, Jr., eds., pp. 241–60. *Geology of North America,* vol. K-3. Geological Society of America.

Benson, L., and Thompson, R. S. (1987b). "Lake-level variation in the Lahontan Basin for the past 50,000 years." *Quaternary Research* 28, 69–85.

Betancourt, J. L. (1984). "Late Quaternary plant zonation and climate in southeastern Utah." *Great Basin Naturalist* 44, 1–35.

Billings, W. D. (1950). "Vegetation and plant growth as affected by chemically altered rocks in the western Great Basin." *Ecology* 31, 62–74.

———. (1951). "Vegetational zonation in the Great Basin of western North America." *Compt. Rend. du Colloque sur les Bases Ecologiques de la Regeneration de la Vegetation des Zones Arides,* pp. 101–22. Paris: Union International Société Biol.

———. (1978). "Alpine phytogeography across the Great Basin." *Great Basin Naturalist Memoirs* 2, 105–17.

Blackwelter, E. (1931). "Pleistocene glaciation in the Sierra Nevada and Basin ranges." *Geological Society of America Bulletin* 42, 865–922.

———. (1934). "Supplementary notes on Pleistocene glaciations in the Great Basin." *Washington Academy of Science Journal* 24, 217–22.

Bright, R. C. (1966). "Pollen and seed stratigraphy of Swan Lake, southeastern Idaho: Its relation to regional vegetational history and to Lake Bonneville history." *Tebiwa* 9, 1–47.

Brown, J. H. (1971). "Mammals on mountaintops: Nonequilibrium insular biogeography." *American Naturalist* 105, 467–78.

———. (1978). "The theory of insular biogeography and the distribution of boreal birds and mammals." *Great Basin Naturalist Memoirs* 2, 209–27.

Byrne, R.; Busby, C.; and Heizer, R. F. (1979). "The altithermal revisited: Pollen evidence from the Leonard Rockshelter." *Journal of California and Great Basin Anthropology* 1, 280–94.

Cole, K. L. (1981). "Late Quaternary environment in the eastern Grand Canyon: Vegetational gradients over the last 25,000 years." Ph.D. diss., University of Arizona, Tucson.

———. (1982). "Late Quaternary zonation of vegetation in the eastern Grand Canyon." *Science* 217, 1142–45.

———. (1983). "Late Pleistocene vegetation of Kings Canyon, Sierra Nevada, California." *Quaternary Research* 19, 117–29.

———. (1985). "Past rates of change, species richness, and a model of vegetational inertia in the Grand Canyon, Arizona." *American Naturalist* 125, 289–303.

Cottam, W. P.; Tucker, J. M.; and Drobnick, R.

(1959). "Some clues to Great Basin postpluvial climates provided by oak distributions." *Ecology* 40, 361–77.

Critchfield, W. B. (1984). "Impact of the Pleistocene on the genetic structure of North American conifers." In *Proceedings of the 8th North American Forest Biology Workshop.* R. M. Lanner, ed.., pp. 70–118.

Critchfield, W. B., and Allenbaugh, G. L. (1969). "The Distribution of Pinaceae in and near northern Nevada." *Madroño* 20, 12–26.

Currey, D. R., and Oviatt, C. G. (1985). "Durations, average rates, and probable cause of Lake Bonneville expansions, stillstands, and contractions during the last deep-lake cycle, 32,000 to 10,000 years ago." In *Problems of and prospects for predicting Great Salt Lake levels.* P. A. Kay and H. F. Díaz, eds., pp. 1–9. Center for Public Affairs and Administration, University of Utah.

Davis, M. B.; Woods, K. D.; Webb, S. L.; and Futyma, R. P. (1986). "Dispersal versus climate: Expansion of *Fagus* and *Tsuga* into the Upper Great Lakes region." *Vegetatio* 67, 93–103.

Davis, O. K. (1981). "Vegetation migration in southern Idaho during the late Quaternary and Holocene." Ph.D. diss., University of Minnesota, Minneapolis.

Davis, O. K.; Anderson, R. S.; Fall, P. L.; O'Rourke, M. K.; and Thompson, R. S. (1985). "Palynological evidence for early Holocene aridity in the southern Sierra Nevada, California." *Quaternary Research* 24, 322–32.

Dohrenwend, J. C. (1984). "Nivation landforms in the western Great Basin and their paleoclimatic significance." *Quaternary Research* 22, 275–88.

Faegri, K., and Iversen, J. (1975). *Textbook of pollen analysis.* New York: Hafner.

Fall, P. L. (1985). "Holocene dynamics of the subalpine forest in central Colorado." In *Late Quaternary vegetation and climates of the American Southwest.* B. L. Fine-Jacobs, P. L. Fall, and O. K. Davis, eds., pp. 31–46. American Association of Stratigraphic Palynologists Contributions Series 16.

Fowells, H. A. (1965). *Silvics of forest trees of the United States.* U.S. Department of Agriculture, Forest Service Agriculture Handbook 271.

Gentry, A. H. (1986). "Endemism in tropical versus temperate plant communities." In *Conservation biology.* M. E. Soule, ed., pp. 153–81. Sunderland, Mass.: Sinauer Associates.

Gleason, H. A. (1926). "The individualistic concept of the plant association." *Bulletin of the Torrey Botanical Club* 53, 7–26.

———. (1939). "The individualistic concept of the plant association." *American Midland Naturalist* 21, 92–110.

Grayson, D. K. (1982). "Toward a history of Great

Basin mammals during the past 15,000 years." In *Man and environment in the Great Basin.* D. B. Madsen and J. F. O'Connell, eds., pp. 82–101. Society of American Archaeology Papers 2.

———. (1983). "Small mammals." In *The archaeology of Monitor Valley.* Volume 2. *Gatecliff Shelter.* D. H. Thomas, ed., pp. 99–126. Anthropological Papers of the American Museum of Natural History 59, pt. 1.

———. (1987). "The biogeographic history of small mammals in the Great Basin: Observations on the last 20,000 years." *Journal of Mammalogy* 68, 359–75.

Harper, K. T.; Freeman, D. C.; Ostler, W. K.; and Klikoff, L. G. (1978). "The flora of Great Basin mountain ranges: Diversity, sources, and dispersal ecology." *Great Basin Naturalist Memoirs* 2, 81–103.

Houghton, J. G.; Sakamoto, C. M.; and Gifford, R. O. (1975). *Nevada's weather and climate.* Nevada Bureau of Mines Special Publication 2.

Kuchler, A. W. (1964). "Potential natural vegetation of the conterminous United States. *American Geographical Society, Special Publication* 36, 155 pp.

Kutzbach, J. E. (1987). "Model simulations of the climatic patterns during the deglaciation of North America." In W. F. Ruddiman and H. E. Wright, Jr., eds., pp. 425–46. *North America and adjacent oceans during the last deglaciation. Geology of North America,* vol. K 3. Geological Society of America.

Kutzbach, J. E., and Guetter, P. J. (1986). "The influence of changing orbital parameters and surface boundary conditions on climate simulations for the past 18,000 years." *Journal of the Atmospheric Sciences* 43, 1726–59.

Kutzbach, J. E., and Wright, H. E., Jr. (1985). "Simulation of the climate of 18,000 yr B.P.: Results for North American/North Atlantic/European sector and comparison with the geologic record." *Quaternary Science Reviews* 4, 147–87.

LaMarche, V. C., Jr. (1973). "Holocene climatic variations inferred from treeline fluctuations in the White Mountains, California." *Quaternary Research* 3, 632–60.

LaMarche, V. C., Jr., and Mooney, H. A. (1972). "Recent climatic change and development of the bristlecone pine, *P. longaeva* Bailey, Krummholz Zone, Mt. Washington, Nevada." *Arctic and Alpine Research* 4, 61–72.

Lanner, R. M. (1983). "The expansion of singleleaf piñon in the Great Basin." In *The archaeology of Monitor Valley.* 2. *Gatecliff Shelter.* D. H. Thomas, ed., Anthropological Papers of the American Museum of Natural History 59, pt. 1.

Lepper, M. G. (1974). "*Pinus flexilis* James, and its environmental relationships." Ph.D. diss. University of California, Davis.

Loope, L. L. (1969). "Subalpine and alpine vegetation of northeastern Nevada." Ph.D. diss., Duke University, Durham, N.C.

MacArthur, R. H., and Wilson, E. O. (1967). *The theory of island biogeography*. Princeton: Princeton University Press.

McCarten, N., and Van Devender, T. R. (1988). "Late Wisconsin vegetation of Robber's Roost in the western Mojave Desert, California." *Madroño* 35, 226–37.

McLane, A. R. (1978). *Silent Cordilleras—the mountain ranges of Nevada*. Reno, Nev.: Camp Nevada Monograph 4.

Madsen, D. B. (1973). "Late Quaternary paleoecology in the southeastern Great Basin." Ph.D. diss., University of Missouri, Columbia.

Madsen, D. B., and Currey, D. R. (1979). "Late Quaternary glacial and vegetation changes, Little Cottonwood Canyon area, Wasatch Mountains, Utah." *Quaternary Research* 12, 254–70.

Major, J., and Bamberg, S. A. (1967). "Some Cordilleran plants disjunct in the Sierra Nevada of California, and their bearing on Pleistocene ecological conditions." In *Arctic and alpine environments*. H. E. Wright, Jr., and W. H. Osborn, eds., pp. 171–88. Indianapolis: Indiana University Press.

Manabe, S., and Broccoli, A. J. (1985). "The influence of continental ice sheets on the climate of an ice age." *Journal of Geophysical Research* 90, 2167–90.

Martin, P. S., and Mehringer, P. J., Jr. (1965). "Pleistocene pollen analysis and biogeography of the Southwest." In *The Quaternary of the United States*. H. E. Wright, Jr., and D. G. Frey, eds., pp. 433–51. Princeton: Princeton University Press.

Mason, H. L. (1944). "A Pleistocene flora from the McKittrick asphalt deposits of California." *Proceedings of the California Academy of Science*, 4th series, 25, 221–23.

Mehringer, P. J., Jr. (1967). "Pollen analysis of the Tule Springs area, Nevada." *Nevada State Museum Anthropological Papers* 13, 129–200.

———. (1977). "Great Basin late Quaternary environments and chronology." *Desert Research Institute Publications in the Social Sciences* 12, 113–67.

———. (1985). "Late-Quaternary pollen records from the interior Pacific Northwest and northern Great Basin of the United States." In *Pollen records of late-Quaternary North American sediments*. V. M. Bryant, Jr., and R. G. Holloway, eds., pp. 167–89. American Association of Stratigraphic Palynologists Foundation.

———. (1986). "Prehistoric environments." In *Handbook of North American Indians*. Volume

11, *Great Basin*. W. L. D'Azevedo, ed., pp. 31–50. Washington, D.C.: Smithsonian Institution.

Mehringer, P. J., Jr., and Ferguson, C. W. (1969). "Pluvial occurrence of bristlecone pine, *Pinus aristata*, in a Mohave Desert mountain range." *Journal of the Arizona Academy of Sciences* 5, 284–92.

Mehringer, P. J., Jr., and Wigand, P. E. (1987). "Western juniper in the Holocene." In *Proceedings of the Pinyon-Juniper Conference*. R. L. Everett, comp. pp. 109–19. U.S. Department of Agriculture Forest Service, Intermountain Research Station, General Technical Report INT-215.

Mitchell, J. F. B.; Grahame, N. S.; and Needham, K. J. (1988). "Climate simulations for 9000 years before present: Seasonal variations and effect of the Laurentide ice sheet." *Journal of Geophysical Research* 93, 8283–8308.

Mitchell, V. L. (1976). "The regionalization of climate in the western United States." *Journal of Applied Meteorology* 15, 920–27.

Orr, P. C. (1957). "On the occurrence and nature of 'amberat.'" *Western Speleological Institute, Observations* 2, 1–3.

Peet, P. K. (1978). "Latitudinal variation in southern Rocky Mountain forests." *Journal of Biogeography* 5, 275–89.

Pyke, C. B. (1972). *Some meterological aspects of the seasonal distribution of precipitation in the western United States and Baja California*. University of California Water Resources Center, no. 139.

Quade, J. (1986). "Late Quaternary environmental changes in the upper Las Vegas Valley, Nevada." *Quaternary Research* 20, 340–57.

Reveal, J. L. (1979). "Biogeography of the intermountain Region." *Mentzelia* 4, 1–92.

Russell, I. C. (1885). *Geological history of Lake Lahontan, a Quaternary lake of northwestern Nevada*. U.S. Geological Survey Monograph 11.

Scott, W. E.; McCoy, W. D.; Shroba, R. R.; and Rubin, M. (1983). "Reinterpretation of the exposed record of the late two cycles of Lake Bonneville, western United States." *Quaternary Research* 20, 261–85.

Sharp, R. P. (1938). "Pleistocene glaciation in the Ruby–East Humboldt Range, northeastern Nevada." *Journal of Geomorphology* 1, 296–321.

Siegal, R. D. (1983). "Paleoclimatic significance of D/H and $^{13}C/^{12}C$ ratios in Pleistocene and Holocene wood." Master's thesis, University of Arizona, Tucson.

Smith, A. T. (1974). "The distribution and dispersal of pikas: Influences of behavior and climate." *Ecology* 55, 1368–76.

Spaulding, W. G. (1981). "The late Quaternary vegetation of a southern Nevada Mountain Range." Ph.D. diss., University of Arizona, Tucson.

———. (1985). *Vegetation and climates of the last 45,000 years in the vicinity of the Nevada Test Site, south-central Nevada*. U.S. Geological Survey Professional Paper 1329.

Spaulding, W. G., and Graumlich, L. J. (1986). "The last pluvial episodes in the deserts of southwestern North America." *Nature* 320, 441–44.

Spaulding, W. G.; Leopold, E. B.; and Van Devender, T. R. (1983). "Late Wisconsin paleoecology of the American Southwest." In *Late-Quaternary enrivonments of the United States*. Volume 1, *The late Pleistocene*. S. C. Porter, ed., pp. 259–93. Minneapolis: University of Minnesota Press.

Spaulding, W. G., and Van Devender, T. R. (1980). "Late Pleistocene montane conifers in southeastern Utah." *University of Utah Anthropological Papers* 104, 159–61.

Spencer, R. J.; Baedecker, M. J.; Eugester, H. P.; Forester, R. M.; Goldhaber, M. B.; Jones, B. F.; Kelts, K.; Mckenzie, J.; Madsen, D. B.; Rettig, S. L.; Rubin, M.; and Bowser, C. J. (1984). "Great Salt Lake and precursors, Utah: The last 30,000 years." *Contributions to Mineralogy and Petrology* 86, 321–34.

Stewart, J. H. (1980). *Geology of Nevada*. Nevada Bureau of Mines and Geology, Special Publication 4.

Stutz, H. C. (1978). "Explosive evolution of perennial *Atriplex* in western America." *Great Basin Naturalist Memoirs* 2, 161–68.

Tang, M., and Reiter, E. R. (1984). "Plateau monsoons of the northern hemisphere; a comparison between North America and Tibet." *Monthly Weather Review* 112, 617–37.

Thompson, R. S. (1979). "Late Pleistocene and Holocene packrat middens from Smith Creek Canyon, White Pine County, Nevada." *Nevada State Museum Anthropological Papers* 17, 362–80.

———. (1984). "Late Pleistocene and Holocene environments in the Great Basin." Ph.D. diss., University of Arizona, Tucson.

———. (1985). "Palynology and *Neotoma* middens." In *Late Quaternary vegetation and climates of the American Southwest*. B. L. Fine-Jacobs, P. L. Fall, and O. K. Davis, eds., pp. 89–112. American Association of Stratigraphic Palynologists Contributions Series 16.

Thompson, R. S.; Benson, L.; and Hattori, E. M. (1986). "A revised chronology for the last Pleistocene lake cycle in the central Lahontan Basin." *Quaternary Research* 25, 1–9.

Thompson, R. S., and Hattori, E. M. (1983). "Packrat (*Neotoma*) middens from Gatecliff Shelter and Holocene migrations of woodland plants." In *The archaeology of Monitor Valley*, 2. *Gatecliff Shelter*. D. H. Thomas, ed., pp. 167–71. Anthropological Papers of the American Museum of Natural History 59, #1.

Thompson, R. S., and Kautz, R. R. (1983). "Pollen analysis." In *The archaeology of Monitor Valley*, 2. *Gatecliff Shelter*. D. H. Thomas, ed., pp. 136–51. Anthropological Papers of the American Museum of Natural History 59, #1.

Thompson, R. S., and Mead, J. I. (1982). "Late Quaternary environments and biogeography in the Great Basin." *Quaternary Research* 17, 39–55.

Thompson, R. S.; Toolin, L. J.; and Spencer, R. J. (1987). "Radiocarbon dating of Pleistocene lake sediments in the Great Basin by accelerator mass spectrometry (AMS). (Abstract). Geological Society of America National Meeting, Phoenix, Ariz.

Tueller, P. T.; Beeson, C. D.; Tausch, R. T.; West, N. E.; and Rea, K. H. (1979). *Pinyon-juniper woodlands of the Great Basin: Distribution, flora, vegetal cover*. U.S. Department of Agriculture Forest Service Research Paper Int-229.

Vander Wall, S. B., and Balda, R. P. (1977). "Coadaptations of the Clark's nutcracker and the piñon pine for efficient seed harvest and dispersal." *Ecological Monographs* 47, 89–111.

Van Devender, T. R. (1977). "Holocene woodlands in the Southwestern deserts." *Science* 198, 189–92.

———. (1987). "Holocene vegetation and climate in the Puerto Blanco Mountains, southwestern Arizona." *Quaternary Research* 27, 51–72.

Van Devender, T. R.; Martin, P. S.; Thompson, R. S.; Cole, K. L.; Jull, A. J. T.; Long, A.; Toolin, L. J.; and Donahue, D. J. (1985). "Fossil packrat middens and the tandem accelerator mass spectrometer." *Nature* 317, 610–13.

Van Devender, T. R., and Spaulding, W. G. (1979). "Development of vegetation and climate in the southwestern United States." *Science* 204, 701–10.

Wayne, W. J. (1984). "Glacial chronology of the Ruby Mountains–East Humboldt Range, Nevada." *Quaternary Research* 21, 286–303.

Webb, T. III (1986). "Is vegetation in equilibrium with climate? How to interpret late-Quaternary pollen data." *Vegetatio* 67, 75–91.

Wells, P. V. (1979). "An equable glaciopluvial in the West: Pleniglacial evidence of increased precipitation on a gradient from the Great Basin to the Sonoran and Chihuahuan deserts." *Quaternary Research* 12, 311–25.

———. (1983). "Paleogeography of montane islands in the Great Basin since the last glaciopluvial." *Ecological Monographs* 53, 341–82.

Wells, P. V., and Berger, R. (1967). "Late Pleistocene history of coniferous woodlands in the Mohave Desert." *Science* 155, 1640–47.

West, N. E.; Tausch, R. J.; Rea, K. H.; and Tueller, P. (1978). "Phytogeographical variation within juniper-pinyon woodlands of the Great Basin." *Great*

Basin Naturalist Memoirs 2, 119–36.

Wigand, P. E. (1985). "Diamond Pond, Harney County, Oregon: Man and marsh in the eastern Oregon desert." Ph.D. diss., Washington State University.

Wigand, P. E., and Mehringer, P. J., Jr. (1985). "Pollen and seed analysis." *Anthropological Papers of the American Museum of Natural History* 61, 108–24.

Wright, R. D., and Mooney, H. A. (1965). "Substrate-oriented distribution of bristlecone pine in the White Mountains of California." *American Midland Naturalist* 73, 257–84.

Zielinski, G. A., and McCoy, W. D. (1987). "Paleoclimatic implications of the relationship between modern snowpack and late Pleistocene equilibrium-line altitudes in the mountains of the Great Basin, western U.S.A." *Arctic and Alpine Research* 19, 127–34.

Chapter 11

Late Quaternary Vegetation Gradients Through the Grand Canyon

Kenneth L. Cole

INTRODUCTION

Objective Perspectives

Paleoecology often is a subjective science. Interpretations of the past are usually expressed in the context of the present. This perspective can be misleading because conceptual units defined for the present (i.e., plant communities, biomes) may have been less meaningful in the past. Some plant communities appear to have developed only recently and perhaps are only transitory associations, still responding to postglacial migrations and anthropogenic alterations. Using a spatial analog for temporal change, this problem can be visualized as classifying the Utah flora using a scheme developed for the Arizona flora. Some communities would be well described, but many would fit poorly within the scheme.

This chapter contains a mixture of objective and subjective analyses of fossil packrat middens from the Grand Canyon. Multivariate statistical techniques of clustering and ordination are used to objectively define past communities, and time-series techniques are used to evaluate past rates of change statistically. However, paleoclimatic estimates ultimately must be based on the assumption that the present is the key to the past.

A Problem of Scale. Because a packrat midden series reflects only the vegetation of a local area, it is unrealistic to compare this sample with broader floristic surveys of entire mountain ranges. The occurrence of a few individual trees in a mountain range is unlikely to ensure that this species would be recorded in a midden assemblage representing only about one hectare. A midden assemblage more accurately represents a small relevé taken at random, usually in a rocky area. As a result, fossil distribution records obtained from the analysis of middens should be compared only to regions where a modern species is common enough that it is reliably recorded within modern packrat middens. The fringes of a plant's distribution will rarely be recorded in a midden. Thus, the fossil record reveals the region of a plant's past dominance rather than its past geographic range.

Terminology and Methods

Time boundaries for past periods are usually time-transgressive, subject to radiocarbon error, and subjectively delimited by movements of the author's favorite species ("indicator species"). As a result, they are merely units described for the sake of convenient communication rather than discrete units with definitive boundaries. They are similar to colors in that they can be named for the sake of communication, but still there are gradational areas between colors.

The following time units are convenient for the midden records used in this chapter.

Time Unit	Radiocarbon Years Before 1950 × 10³
Late Wisconsin	25.0–11.0 ka
Full glacial (Wisconsin)	21.0–15.0 ka
Latest Wisconsin	15.0–11.0 ka
Wisconsin-Holocene transition	12.5–9.0 ka
Early Holocene	11.0–8.5 ka
Middle Holocene	8.5–4.0 ka
Late Holocene	4.0–0.5 ka
Modern	0.5–0 ka

Note that the time units are not always mutually exclusive. The late Wisconsin encompasses both the Wisconsin full glacial and the latest Wisconsin. The Wisconsin-Holocene transition is used to designate that poorly understood time of rapid vegetational change that is neither Wisconsin nor Holocene, but still is a time of special significance meriting a term of its own.

Fossil packrat middens were analyzed by methods outlined in Cole (1981, 1983). Use of cf. and / in presenting fossil data follows Birks and Birks (1980). Taxonomy follows Lehr (1978) unless another authority is noted.

PRESENT AND PAST VEGETATION OF THE GRAND CANYON

The Grand Canyon Desert

The arid vegetation of the Grand Canyon cannot be easily classified into any of the popularly recognized major desert types. It contains plants typical of the Great Basin, Mojave, Sonoran, and Colorado deserts along its 400-km

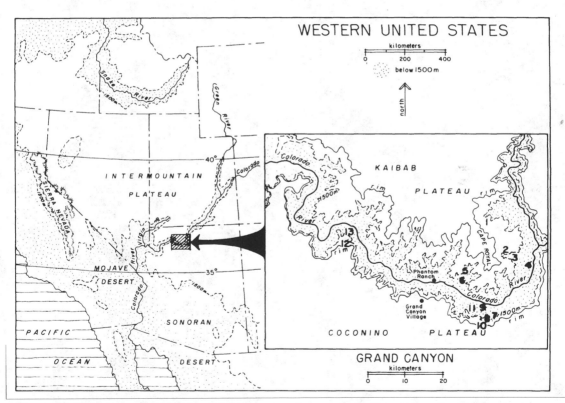

Fig. 11.1. Map of the southwestern United States showing principal geographic features mentioned in text and the 1500-m elevational contour. Midden collection localities from the eastern Grand Canyon are shown in the inset: (1) South Nan- koweap; (2–4) Chuar Valley; (5–6) Clear Creek; (7) Hance Canyon and Bida Cave; (8) Horseshoe Mesa; (9) Cottonwood Canyon; (10) Grandview Point; (11) Grapevine Canyon; (12–13) Bass Canyon.

length, but many of the plant associations are unique to the canyon. It is better understood as a low-elevation corridor extending into the high-elevation Colorado and Intermountain plateaus (Fig. 11.1). Many of the lower-elevation desert plants exist here in close proximity to the steppe and mountain vegetation of the higher plateaus.

The deep canyon creates its own climate as air warmed by the canyon walls recirculates (Malm, 1974), often clearing the clouds from above the canyon. Plants typical of the arid southern deserts either migrated along this corridor during the Holocene or were left as pre-Wisconsin relics in the warm, deep canyon.

Wisconsin Plant Associations

Fifty-two packrat middens collected from the eastern Grand Canyon are listed with their radiocarbon dates and principal macrofossils in Table 11.1. Further information on these middens is available in Cole (1981, 1982). Data from other Grand Canyon middens can be

Table 11.1. *Radiocarbon ages and principal fossil contents of fifty-two fossil packrat middens from the eastern Grand Canyon.*
(Laboratory numbers and dated materials are available in Cole, 1981.)

Radiocarbon Date (B.P.)	Midden	Elevation (m)	Slope Aspect	Principal Macrofossils
950–1220 m				
34,300 ± 3570	CC2	1100	NE	Jumo, Opun, Juos, Artr, Agut, Prfa
29,400 ± 1800	CC3	1100	NE	Jumo, Opun, Juos, Prfa, Rost, Atco
17,400 ± 450	HC4	1100	SE	Juos, Atco, Oper, Symp, Artr
16,400 ± 190	GC1	1100	NE	Juos, Opun, Atco, Rost, Chmi, Pttr
13,800 ± 330	HC3	1100	E	Juos, Opun, Atco, Agut
12,900 ± 200	CC1*	1100	N	Juos, Opun, Fran, Enfr, Rhus, Epne
4802 ± 240	"	1100	N	Enfr
12,380 ± 370	Ch3	970	W	Juos, Oper, Atco, Artr, Symp, Cora
12,030 ± 220	HC8a	1100	W	Juos, Opun, Agut
12,015 ± 365	Ch4	970	W	Juos, Oper, Agut, Artr, Cora
10,760 ± 260	BC1	950	NE	Juos, Pttr, Feru, Fran, Epne, Opun
10,150 ± 120	HC1	1100	W	Juos, Epne, Fran, Oper, Arlu, Prfa
10,110 ± 100	HC2a	1200	W	Juos, Epne, Arlu, Pttr, Rhus, Fran
9400 ± 270	Cl1	1220	SW	Juos, Opun, Agut, Fran, Rhus
8957 ± 96	HM1	1100	W	Juos, Fran, Pttr, Rhus, Oper, Arlu
1345 ± 135	Ch5	970	W	Oper, Prju, Opba, Enfa, Ephd
<1000	HC2b	1200	W	Oper, Epne, Rhus, Agut, Arlu, Yuan
<1000	HM1b	1100	W	Epne, Enfr, Alwr, Oper, Opun, Prfa
1400–1470 m				
20,630 ± 470	HM11	1450	W	Juos, Agut, Ribe, Atco, Rost, Opun
18,630 ± 310	HM6	1450	W	Juos, Rost, Ribe, Psme, Atco, Hodu
16,165 ± 615	Ch2	1450	N	Psme, Abco, Artr, Juos, Symp, Ribe
13,830 ± 790	HM4	1470	W	Juos, Rost, Opun, Ribe, Ephd, Atco
13,780 ± 240	Bi2	1450	W	Opun, Psme, Abco, Rost, Ribe, Pifl
14,170 ± 470	"			
13,540 ± 170	HM7	1450	W	Juos, Opun, Rost, Psme, Abco, Fran
13,470 ± 420	Bi8b	1450	W	Opun, Psme, Rost, Ribe, Abco, Pifl
13,340 ± 150	Bi1c	1450	W	Juos, Rost, Opun, Psme, Oskn, Ribe
12,600 ± 540	"			
11,530 ± 290	Bi8a	1450	W	Opun, Psme, Fran, Juos, Rost, Pifl
10,290 ± 150	Bi3	1450	W	Juos, Oper, Pied, Fran, Agut, Ceoc
8470 ± 100	Bi4	1460	W	Oper, Epne, Juos, Agut, Fran, Cein
6830 ± 175	Ch1	1430	N	Juos, Fran, Opun, Epne, Agut, Pttr
6800 ± 220	Bi6b	1450	W	Oper, Agut, Juos, Epne, Fran, Come
<1000	Ch10	1400	W	Juos, Qutu, Prju, Epne, Agut, Arlu

1600–1900 m

23,350 ± 1110	Ch9	1770	SW	Psme, Abco, Pifl, Juos, Rubu, Hodu
18,800 ± 800	Ch8b	1770	SW	Pifl, Psme, Abco, Artr, Hodu, Juos
18,490 ± 660	Ch8c	1770	SW	Pifl, Psme, Abco, Rubu, Hodu, Juos
15,840 ± 310	Cl2	1600	NE	Psme, Abco, Juos, Pifl, Ribe, Rost
14,050 ± 500	Cl3	1600	NE	Psme, Abco, Juos, Opun, Chmi, Oskn
9070 ± 350	Ch7	1770	NE	Psme, Pied, Juos, Fran, Qutu, Pipo
8900 ± 340	BC3	1900	S	Juos, Psme, Pied, Pipo, Qutu, Pttr
8590 ± 110	BC2	1900	S	Juos, Psme, Pied, Qutu, Pttr, Pipo
8430 ± 400	"			
7110 ± 180	Ch8a*	1770	SW	Agut, Juos, Qutu, Cein, Psme, Pifl
15,915 ± 538[†]	Ch8a*			Pifl
2300 ± 90	Su1	1620	N	Juos, Pied, Fran, Pttr, Come, Cein

2000–2200 m

23,385 ± 772	Na9d*	2020	S	Psme, Pifl, Abco, Pice, Juos, Juco
9520 ± 360[†]	Na9d*			Agut
18,130 ± 350	Na9c	2020	S	Pifl, Psme, Pice, Abco, Juco, Hodu
17,950 ± 600	Na9b	2020	S	Pifl, Pice, Psme, Abco, Juco, Pamy
12,660 ± 230	Na6	2050	NW	Psme, Pifl, Pice, Abco, Juco, Pamy
13,110 ± 240	"			
12,170 ± 210	Na7a	2050	NW	Psme, Abco, Pifl, Pice, Juco, Pamy
7870 ± 140	Na9a	2020	S	Juos, Agut, Qutu, Pied, Opph, Cein
5510 ± 80	GP4	2200	N	Juos, Psme, Fran, Oper, Pied, Pipo
1220 ± 70	Na4	2070	SW	Pied, Juos, Qutu, Agut, Cein, Rhus
1170 ± 80	Na10	2050	NW	Pied, Psme, Pttr, Juos, Oskn, Qutu
<1000	GP6a	2170	W	Pied, Juos, Come, Artr, Cein, Psme
<1000	GP9	2170	SE	Juos, Pied, Ephd, Oper, Agut, Cein

*Midden contains contaminants.
[†]Tandom Accelerator Mass Spectrometer data (Van Devender et al., 1985).

Key to macrofossils: Abco = *Abies concolor*; Acgr = *Acacia greggii*; Agut = *Agave utahensis*; Alwr = *Aloysia wrightii*; Arlu = *Artemisia ludoviciana*; Artr = *Artemisia tridentata*; Atco = *Atriplex confertifolia*; Cein = *Cercocarpus intricatus*; Ceoc = *Cercis occidentalis*; Chmi = *Chamaebatiaria millifolium*; Come = *Cowania mexicana*; Cora = *Coleogyne ramosissima*; Enfa = *Encelia farinosa*; Enfr = *Encelia frutescens*; Ephd = *Ephedra* sp.; Epne = *Ephedra f. nevadensis*; Feru = *Fendlera rupicola*; Fran = *Fraxinus anomala*; Hodu = *Holodiscus dumosus*; Juco = *Juniperus communis*; Jumo = *Juniperus* cf. *monosperma*; Juos = *Juniperus* cf. *osteosperma*; Opba = *Opuntia basilaris*; Oper = *Opuntia erinacea*; Opph = *Opuntia phaeacantha*; Opun = *Opuntia* sp.; Oskn = *Ostrya knowltoni*; Pamy = *Pachystima myrsinites*; Pice = *Picea engelmannii/pungens*; Pied = *Pinus edulis*; Pifl = *Pinus flexilis*; Pipo = *Pinus ponderosa*; Prfa = *Prunus fasciculata*; Prju = *Prosopis juliflora*; Psme = *Pseudotsuga menziesii*; Pttr = *Ptelea trifoliata* var. *pallida*; Qutu = *Quercus turbinella*; Rhus = *Rhus* sp.; Ribe = *Ribes* sp.; Rost = *Rosa* cf. *stellata*; Rubu = *Rubus* sp.; Symp = *Symphoricarpos* sp.; Yuan = *Yucca angustissima*.

Key to localities: BC = Bass Canyon; Bi = Bida Cave; CC = Cottonwood Canyon; Ch = Chuar Valley; Cl = Clear Creek; GC = Grapevine Canyon; GP = Grandview Point; HC = Hance Canyon; HM = Horseshoe Mesa; Na = Nankoweap (Novinger Butte); Su = Supai Formation above Horseshoe Mesa.

found in Phillips (1977, 1984), Mead and Phillips (1981), and Van Devender and Mead (1976). Other regional studies of middens have been conducted in the Sheep Range of Nevada (Spaulding, 1981) and the central and eastern Colorado Plateau (Betancourt, *this volume*, chap. 12).

Reconstruction of the vegetation zones of the Grand Canyon region during the last twenty-four thousand years (Fig. 11.2) shows that Holocene zones are 800–1000 m higher in elevation than the most similar late Wis-consin zones. However, rather than being simply elevated, the Holocene vegetation zones also differ in composition from the late Wisconsin zones (Fig. 11.3). Two major association types can be defined from the late Wisconsin records: juniper-desertscrub and mixed-conifer forest.

Late Wisconsin Juniper-Desertscrub (450–1450 m). The juniper-desertscrub middens contain large amounts of juniper (*Juniperus californica* cf. var. *osteosperma* Benson) 800–1000 m below its modern lower eleva-

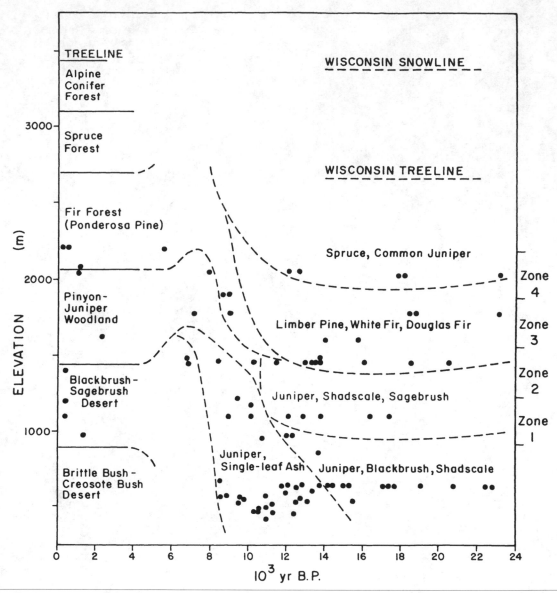

Fig. 11.2. Zonation of major vegetation units in the Grand Canyon during the last twenty-four thousand years (figure from Cole, 1985).

tional limit (Fig. 11.4). Other species abundant in these juniper-desertscrub associations include Mojave prickly pear (*Opuntia* cf. *erinacea*), desert rose (*Rosa stellata/woodsii* Lindl.), sagebrush (*Artemisia* sec. tridentatae), and Utah agave (*Agave utahensis*), all growing well below their modern lower limits (Cole, 1981). Shadscale (*Atriplex confertifolia*) was especially abundant in middens below 1200 m, while blackbrush (*Coleogyne ramosissima*) and snowberry (*Symphoricarpos* cf. *longiflorus*) were abundant below 700 m in the western Grand Canyon (Phillips, 1977; Mead and Phillips, 1981). Some mesophytic species such as Knowlton hop hornbeam (*Ostrya knowltoni*) and gooseberry (*Ribes* sp.)

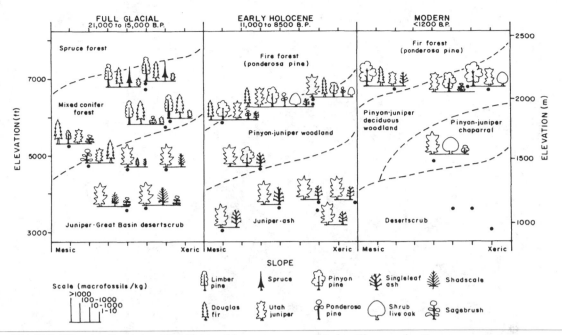

Fig. 11.3. Direct ordinations of fossil assemblages during the full glacial, early Holocene, and Recent in the eastern Grand Canyon. Sizes of symbols are varied to illustrate fossil concentrations as shown on the scale. Key to midden localities is in Table 11.1. Figure modified from Cole (1982).

were less frequently recorded in these middens, suggesting that abundant moisture was available in some microhabitats, perhaps at seeps in shaded locations at the bases of cliffs.

Late Wisconsin Mixed-Conifer Forest (1450–2200 m). In order of increasing elevation, the late Wisconsin mixed-conifer forest assemblages are dominated by fossils of Douglas fir (*Pseudotsuga menziesii*), limber pine (*Pinus flexilis*), and Engelmann spruce (*Picea engelmannii*) (Figs. 11.4, 11.5, and 11.6). White fir (*Abies concolor*) also is an important component in these assemblages. Spruce is abundant in the highest middens (2200 m), indicating proximity to the lower boundary of a spruce forest. These highest middens also contain abundant mountain lover (*Pachystima myrsinites*) and common juniper (*Juniperus communis*), common understory plants in modern spruce-fir forests.

Only two of these late Wisconsin dominants, white fir and Douglas fir, now occur in sufficient numbers within the canyon to ensure a reasonable chance of being sampled in a modern packrat midden. They now grow 720–800 m higher in elevation than during the late Wisconsin. The other late Wisconsin dominants are either entirely absent from the modern canyon or are very rare. These mixed-conifer communities are generally similar to those now found farther upriver or farther north in Utah (Cole, 1981, 1982).

Missing Species. As would be expected from an ice-age record, many species now abundant at low elevations, such as creosote bush (*Larrea divaricata* Cav.), brittlebush (*Encelia farinosa*), ocotillo (*Fouquieria splendens*), and white bursage (*Ambrosia dumosa*), were either not recorded or are represented by only a few trace fossils (possibly contaminants) from the late Wisconsin. Surprisingly, some middle-elevation species also were unrecorded. In this category are many species of the interior chaparral (Pase and Brown, 1982) such as shrub live oak (*Quercus turbinella / undulata*), little-leaf mountain mahogany (*Cercocarpus intricatus*), and quinine bush (*Cowania mexicana*). These species are now so abundant that they are found routinely in modern packrat middens. They were either absent from the late Wisconsin Grand Canyon or were present in such small numbers as to preclude their occurrence in the record.

Despite its wide distribution at middle ele-

Fig. 11.4. View from the Tonto Platform (1100 m), on top of the Tapeats Sandstone, toward Woton's Throne (2300 m). Plant species in the foreground include blackbrush, prickly pear, Datil yucca, and Nevada ephedra. Pleistocene middens from similar sites contain juniper, prickly pear, shadscale, and sagebrush.

vations today, pinyon pine (*Pinus edulis/ monophylla*) is present in only one full-glacial midden at Peach Springs Wash, near the western Grand Canyon (Van Devender and Spaulding, 1979), and a latest Wisconsin midden from the western Grand Canyon (Phillips, 1977). Also, several middens collected from Wupatki National Monument (65 km south of the Grand Canyon) demonstrate the presence of pinyon by at least 11.5 ka and possibly sooner (Cinnamon and Hevly, 1988). These data suggest that the full-glacial distribution of pinyon was at middle elevations in the western to middle Grand Canyon, although quite restricted even in these areas. The Grand Canyon was at or near the fringe of the late Wisconsin pinyon distribution. Additional sampling may eventually turn up isolated pockets of late Wisconsin pinyon in the eastern Grand Canyon.

Ecotonal Stability. The physiography of the Grand Canyon determines the limits of past and present plant associations. The late Wisconsin ecotone between spruce forest and mixed-conifer forest was located at about the same elevation as the present ecotone between pinyon-juniper and fir forest (Fig. 11.6).

The late Wisconsin ecotone between mixed-conifer forest and juniper-desertscrub was located at about the same elevation as the modern ecotone between pinyon-juniper woodland and desertscrub.

This stability of ecotonal areas throughout different climatic regimes has been noted in other studies (Spaulding, 1981) and should be expected. An ecotone represents a sharp transition between plant associations and often is located at a break in physiography or along a geologic contact. For example, the modern ecotone between pinyon-juniper and desertscrub in much of the eastern Grand Canyon occurs near 1450 m at the top of the Redwall Limestone. Below 1450 m lie steep limestone cliffs with xerophytic vegetation, while above, more gentle slopes, underlain by shales of the Supai Group, are far more hospitable to moisture-loving plants such as those that occur in the modern pinyon-juniper woodland and the late Wisconsin mixed-conifer forest.

Major biotic boundaries also were controlled by physiography during the Pleistocene. For example, the high base level of the Colorado Plateau precluded the northward extension of many plants that had an upper

Fig. 11.5. View of middens in an alcove in the Redwall Limestone in Chuar Valley. Midden ages are as follows: Chuar 9, 23,350 yr B.P.; Chuar 8b, 18,800 yr B.P.; and Chuar 8d, > 27,700 yr B.P. Chuar 8a (7110 yr B.P.) is farther inside the alcove at right. Older middens are located near entrance of shelter, suggesting that rates of cliff retreat influence the age distribution of middens (Cole and Mayer, 1983).

Fig. 11.6. View in Nankoweap Basin. Tapeats Sandstone is in the foreground (1580 m) and Redwall Limestone forms cliff in background (1860 to 2070 m) below Alsap Butte (2280 m) at top right. Modern vegetation is pinyon pine and juniper with some Douglas fir at the top of the Redwall Limestone. Pleistocene middens from similar sites contain mostly Douglas fir and white fir on Tapeats sandstone with limber pine dominating on the Redwall Limestone. Sites at the top of the Redwall Limestone contain spruce.

limit of about 1500 m (Betancourt, 1987), except within the Grand Canyon corridor, just as it does today.

Early Holocene Associations

The early Holocene is represented by unique plant assemblages unlike modern or late Wisconsin ones (Fig. 11.7). Below 700 m in the western Grand Canyon shadscale and blackbrush have become rare during the Wisconsin-Holocene transition, while single-leaved ash (*Fraxinus anomala*) has become codominant with juniper (Figs. 11.2 and 11.3). Between 900 and 1450 m in the eastern Grand Canyon shadscale and sagebrush have decreased during the Wisconsin-Holocene transition, while single-leaved ash and hop tree (*Ptelea pallida*) have become codominant with juniper. Above 1450 m, in the absence of limber pine and spruce, ponderosa pine (*Pinus ponderosa*), pinyon pine, and shrub live oak merge with Douglas fir.

FOSSIL HISTORIES OF SELECTED SPECIES

Limber Pine. The predominant fossil in the higher-elevation late Wisconsin records from the Grand Canyon is limber pine. Needles of this tree are abundant in all Pleistocene middens above 1500 m. Surprisingly, this tree, which often grows on rocky outcrops (Lepper, 1974), is now absent from the Grand Canyon, where it is apparently limited by high temperatures, low precipitation, or competition from ponderosa pine. Concentrations of limber pine needles in middens decrease between the full glacial and the latest Wisconsin, and the few needles in middens postdating 12 ka are probably contaminants collected by packrats from older middens nearby. One such needle was accelerator-dated at 16.3 ka (AA-533) from a midden (Chuar 8a) containing a middle Holocene-type assemblage and a 7.1-ka agave leaf (Table 11.1). The needle was evidently incorporated into the Holocene midden from a nearby Pleistocene midden, perhaps Chuar 8b or 8c, which are less than a meter away and of similar ages (Fig. 11.5).

Pinyon Pine. The eastern Grand Canyon assemblages are slightly north and east of late Wisconsin records of pinyon (Betancourt, 1987, *this volume*, chap. 12). Fossil records confirm the late Wisconsin presence of pinyon on and south of the Mogollon Rim (Van Devender et al., 1985; Jacobs, 1985) and in the central and western Grand Canyon (Van Devender and Spaulding, 1979; Phillips, 1977). It also was present during the Wisconsin-Holocene transition south of the Grand Canyon at Wupatki (Cinnamon and Hevly, 1988). During the Pleistocene-Holocene transition, pinyon pine either migrated into the eastern Grand Canyon from the west and/or south, or simply expanded from undetected local refugia within the eastern Grand Canyon.

Early Holocene pinyon fossils are predominantly two-needled (*Pinus* cf. *edulis*). However, one- and even three-needled fascicles also are preserved in small percentages. The early Holocene pinyons were evidently similar to the modern low-elevation pinyons of the Grand Canyon, which often contain single-needled fascicles on predominantly two-needled trees. These pinyons with variable needle number have been considered either hybrids (*P. edulis* × *monophylla*) (Lanner, 1974) or a separate variety (*P. edulis* var. *fallax*) (Little, 1968). Regardless of the taxonomic rank preferred, the fossils are similar to many modern local specimens.

Ponderosa Pine. Despite its large modern range across the Colorado Plateau, ponderosa pine was either absent or too infrequent to be recorded in the Grand Canyon during the late Wisconsin. The absence of this species suggests summer-dry conditions during the late Wisconsin because it can tolerate cold but not dry conditions, as indicated by its absence today from the summer-dry interior of the Great Basin (Thompson, 1984).

Ponderosa pine is first recorded around 9 ka in three middens from steep slopes within the canyon. Today this pine rarely descends the canyon's steep slopes and these sites now support pinyon-juniper woodland. The abundant ponderosa pine in these early Holocene middens probably is a result of either high summer precipitation during the early Holocene (Spaulding and Graumlich, 1986; Cole, 1982), deep relictual soils left from the Wisconsin (Cole, 1981, 1985), or a combination of the two.

Chaparral Species. Like pinyon pine, other southerly species characteristic of the interior chaparral (Pase and Brown, 1982), such as shrub live oak, single-leaved ash, hop tree, quinine bush, little-leaf mountain mahogany, New Mexico locust (*Robinia neomexicana*), and hackberry (*Celtis reticulata*), are common in early Holocene middens. These species evidently either migrated north and up the canyon or expanded from undetected refugia. On the whole, the abundance of these species in the early Holocene middens suggests that precipitation was high, perhaps exceeding modern values, and that climate was more equable than during the late Wisconsin.

LATITUDINAL VERSUS ELEVATIONAL SHIFTS

Most plant species moved 600–1000 m upward in elevation during the Wisconsin-Holocene transition. However, this elevational shift can be more accurately viewed as a latitudinal shift producing an apparent upward movement because of the correlation between elevation and latitude (Cole, 1982). The latitudinal nature of the shift is demonstrated by the dominance of northerly species during the late Wisconsin and southerly species during the Holocene.

Northerly species such as limber pine, spruce, common juniper, mountain lover, and buffalo-berry (*Shepherdia canadensis*), are now so rare in and south of the Grand Canyon that the prospect of finding them recorded in a randomly selected modern midden is remote. But these species are often dominants in the modern vegetation far to the north of the Grand Canyon. Southerly species, such as pinyon pine and shrub live oak, became common enough to be represented in the midden record only after the end of the Wisconsin.

The flora of the highest-elevation habitats in the late Wisconsin Grand Canyon is recorded in six middens from two sites representing approximately two hectares. It is likely that other species typical of even higher latitudes were not common enough to have been present in this small-area sample.

MULTIVARIATE ORDINATION OF MIDDEN ASSEMBLAGES

Plant assemblages may be classified into groups or along gradients based on mathematical analysis of their contents (Gauch, 1982; Pielou, 1984). These analyses can reveal striking similarities and differences between assemblages that may escape notice in more subjective analyses. In order to study the relationships between fossil assemblages from the eastern Grand Canyon, I have undertaken two types of multivariate analysis, cluster analysis and an ordination using Detrended Correspondence Analysis (DCA or DECORANA).

Cluster Analysis. Figures 11.7 and 11.8 show the results of two types of cluster analysis performed on fossil associations from packrat middens. The first analysis (Fig. 11.7) clusters associations based on the presence or absence of macrofossils in forty-nine middens from the eastern Grand Canyon. Similarity between middens was measured using Jaccard's similarity index (Mueller-Dombois and Ellenberg, 1974), and clustering follows techniques specified in McCammon and Wenninger (1970).

The great variety of plant associations occurring over a 1250-m (4070-foot) elevational gradient over a thirty-four-thousand-year period is responsible for the poor formation of clusters in Figure 11.7. However, despite the poor clustering, the segregation of some late Wisconsin and Holocene associations is very clear. The Pleistocene mixed-conifer samples are clustered in subgroups with site elevation increasing to the right. The three early Holocene mixed-conifer forest samples cluster together (Ch7, BC3, and BC2), as do the early Holocene juniper-ash–hop tree samples (BC1, HC2, HC1, and HM1a). These results emphasize that the late Wisconsin and early Holocene associations were in fact unique rather than simply elevationally displaced modern communities. Future work should include modern midden samples from proposed analog localities upriver and farther north.

A second cluster analysis (Fig. 11.8) using only late Wisconsin middens older than 12 ka produced segregated groups correlating well with elevation. For this analysis the macrofossil data were transformed into eight concentration classes to consider macrofossil concentration rather than simply presence or absence. A similarity matrix was then computed and the clustering program BMDP-2M was applied using sum-of-squares distances with no standardization of variance (Dixon et al., 1985).

Figure 11.8 is arranged with midden site elevation increasing to the right. With the exception of the two interstadial assemblages

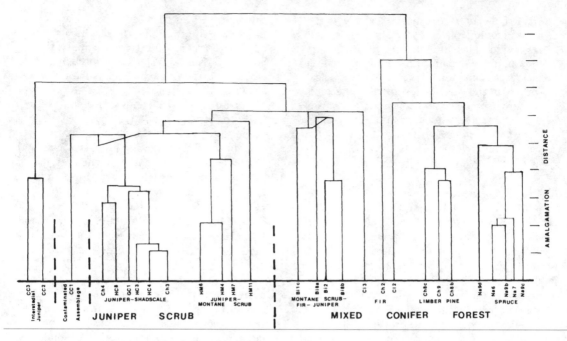

Fig. 11.7. Cluster analysis of forty-nine Wisconsin and Holocene middens from the eastern Grand Canyon using a similarity matrix based on presence and absence of perennial species. Midden localities are identified in Table 11.1.

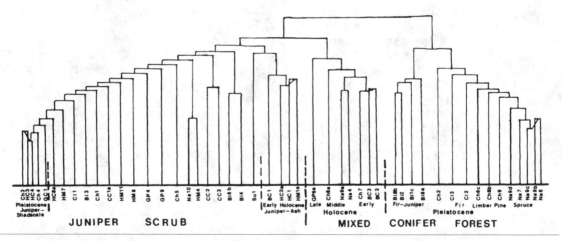

Fig. 11.8. Cluster analysis of twenty-eight late Wisconsin middens using a similarity matrix based upon macrofossil concentration of perennial species. Midden groups representing Pleistocene communities are arranged with elevation increasing from left to right. Midden localities are identified in Table 11.1.

(CC2 and CC3) and a midden containing contaminants (CC1a), middens at similar elevations but thousands of years apart in age clustered together. This result suggests very stable conditions during the late Wisconsin.

The results of the second analysis (Fig. 11.8) correlate well with clustering of these data based on presence-absence data alone, suggesting that packrat fossil assemblages that are similar in presence-absence also will be similar when macrofossil abundances are considered.

Analyses based on Holocene data alone produce very poorly defined clusters. Although the numbers of Holocene and Pleistocene middens are similar, the Holocene middens are not as easily classified into groups. The amount of vegetational change during the Holocene is too great for the small number of samples (21). Most Holocene middens are simply outliers; that is, they are not very similar to any other assemblage. This result suggests that during the last eleven thousand years plant associations were far less stable than they were between 24 and 12 ka. The late Wisconsin seems to have been a time of exceptional stability compared to the Holocene (Cole, 1985).

Detrended Correspondence Analysis.

Figure 11.3 shows three ordinations of the vegetation of the full glacial, early Holocene, and latest Holocene–modern middens. In this figure the plant associations are *directly* arranged according to the modern elevation and insolation of the midden sites. In contrast, an *indirect* multivariate ordination of the late Wisconsin midden data using DCA is shown in Figure 11.9, in which the plant associations are indirectly arranged with respect to one another using only proximity in the multidimensional space computed from their fossil contents. Concentration values were transformed into concentration classes, and rare species were downweighted (Hill, 1979).

The results of the DCA analysis are very similar to the direct ordination for the full glacial shown in Figure 11.3. The first axis of the ordination shows a very high correlation with elevation of the fossil deposits ($r^2 = .996$), indicating that despite the twelve-thousand-year time span represented (24–12 ka), elevational zonation was very stable. In other words, associations within elevational zones did not change much during this long period. The second ordination axis represents some aspects of insolation and soil development.

PAST RATES OF VEGETATIONAL CHANGE

One method of analyzing vegetational change involves comparing midden assemblages against each other or with the modern vegetation at the site using a similarity index (Cole,

1985). Values for Sørensen's index (Mueller-Dombois and Ellenberg, 1974) comparing eastern Grand Canyon middens with the modern vegetation at the collection localities show that late Wisconsin and modern assemblages were not very similar and that the similarity increases as the middens get younger (Fig. 11.10). Trends can be extracted from the data by averaging the values into two-thousand-year intervals in order to filter out noise and equally distribute the data points in time. Past rates of change in Sørensen's index (the slope of the curve in Fig. 11.10) is plotted in Figure 11.11A.

Figure 11.11 shows high rates of change in similarity occurring between 10 and 6 ka. Because these measurements use the perspective of the modern vegetation (by basing 100% similarity on the modern associations), they measure the development of the modern plant associations. Thus the modern associations mostly developed between 10 and 6 ka. However, unless the late Wisconsin to modern changes were unidirectional, proceeding as an orderly replacement of late Wisconsin species with modern species, this statistic will not measure the true rates of change through time. Rates of change could be better calculated as the similarity between successive middens at a single locality (if enough middens could be collected from one locality) or as the amount of change within similar habitats or elevational zones.

Because the eastern Grand Canyon middens were collected from many localities, I divided the data into the four elevational zones shown in Figure 11.11. Times of arrivals and departures of species from each of these zones were then calculated and classified into two-thousand-year intervals and plotted in Figure 11.11B. Each unit value in Figure 11.11B represents the first or last record of a species in an elevational zone. Since this statistic measures the number of species moving in or out of a zone in a unit of time, I have termed it "species flux." Species flux reaches its highest values between 12 and 8 ka.

Species flux shows high rates of change two thousand years before the Sørensen's index calculations. The principal reason for this is that many late Wisconsin species were last recorded well before Holocene species arrived. This becomes evident if the two components

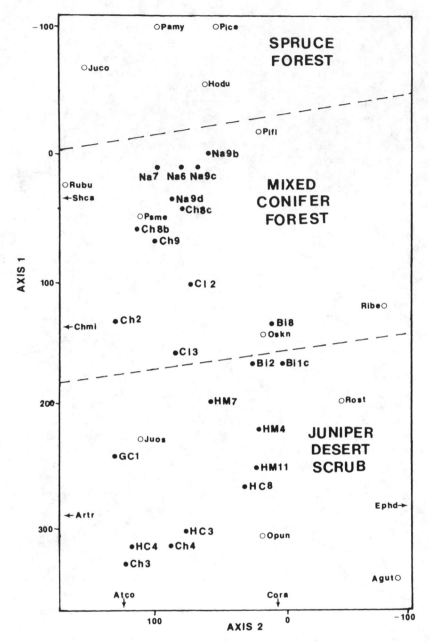

Fig. 11.9. Detrended Correspondence Analysis (DCA) ordination of late Wisconsin midden data from the eastern Grand Canyon. Axis 1 has a high correlation (r^2 = .996) with site elevation. Abbreviations for midden assemblages (dots) and plant taxa (open circles) are defined in Table 11.1.

of species flux—arrivals and departures—are separated (Fig. 11.12). If late Wisconsin species departed before the arrival of Holocene species, then the species richness of individual assemblages was probably lower during the interim. A moving average of species number (Fig. 11.12B) shows that this is the case.

Late Wisconsin Stability. Records from southwestern packrat middens demonstrate

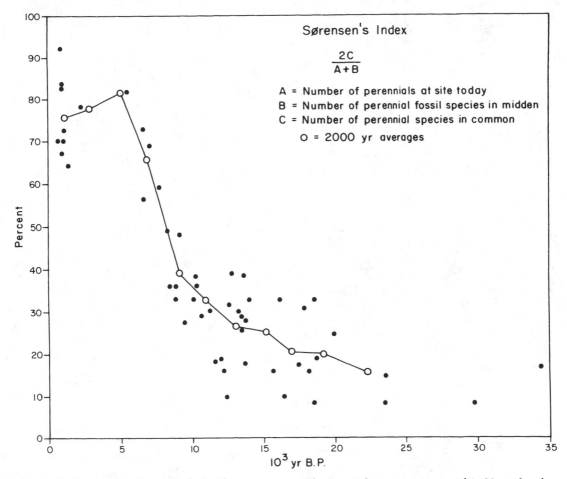

Fig. 11.10. Sørensen's index of similarity comparing perennial species present in midden deposits with perennial species present within 30 m of each fossil site (from Cole, 1985).

that late Wisconsin vegetation was fairly stable. This is supported by other authors (Spaulding et al., 1983; Spaulding, *this volume*, chap. 9; Van Devender, *this volume*, chap. 8), as well as by the statistical analyses reported in this chapter. Although some changes between the full glacial and late Wisconsin (Cole, 1981; Phillips, 1977) are evident, they are relatively minor. The stability of the late Wisconsin stands in sharp contrast to the drastic changes recorded for the last twelve thousand years in every ecosystem with an adequate paleoecological record. Because the late Wisconsin plant communities were relatively stable, they were likely in close equilibrium with the existing climate. That is, major migrations or competitive al-

terations were not occurring during this time.

Wisconsin-Holocene transition. The Wisconsin-Holocene transition encompassed the most radical shifts recorded in the packrat midden record (Cole, 1985). Rates of change in plant communities were high, perhaps lowering the species richness during the early portions of the period as some Pleistocene species were eliminated before their Holocene replacements arrived (Fig. 11.12). In other words, the Holocene species appear to have been delayed en route to their eventual community dominance by migrational or competitive lag.

The plant communities during the early Holocene were unlike any observed either today or in the fossil record of other times. I be-

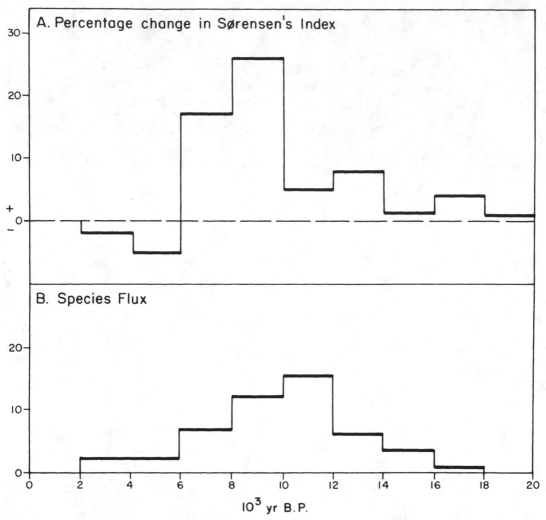

Fig. 11.11. (A) Percentage change in Sørensen's index between two-thousand-year periods. Line = slope of the line in Figure 11.10. (B) Species flux: First and last occurrences of dominant taxa within elevational zones (from Cole, 1985).

lieve this is because the vegetation was in extreme disequilibrium with the climate during this transition. The immigration of new species into their modern zones, rather than into lower zones, suggests that the Wisconsin-Holocene climatic change had already occurred *before* their arrival. Although most plants seemed to have reached their modern limits by the middle Holocene, some, such as ocotillo, still may be adjusting to past climatic fluctuations (Cole, 1986).

Many paleoclimatologists prefer to attribute the strange early Holocene plant asso-

ciations to a distinct climate (Markgraf, 1986; Webb and Bryson, 1971; Webb, 1986; Spaulding and Graumlich, 1986; Van Devender, 1977) and consider any migrational or competitive dynamics of the vegetation to be insignificant. This viewpoint has support in the probable dynamics of the early Holocene monsoon (Kutzbach, 1981) and the seasonal progression of thermal maxima (Davis, 1984).

However, the unique early Holocene vegetation also resulted in part from the massive, continental alterations of ecosystems accompanying the end of the Wisconsin. Evidence

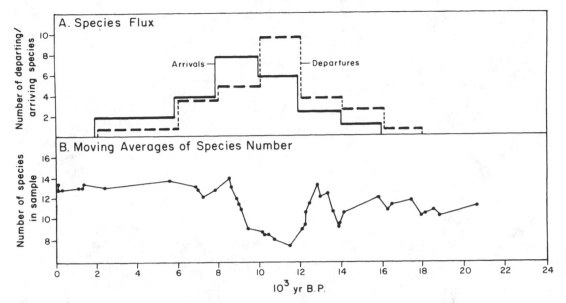

Fig. 11.12. Species flux divided into taxa making first appearance (arrivals = solid line) and last appearance (departures = dashed line) within the four elevational zones shown on the right of Figure 11.2.

(B) Five-sample moving average of species number from eastern Grand Canyon middens (from Cole, 1985).

for this includes: (1) the slow migration of species northward following the late Wisconsin compared to the rapid elevational migration of species (Cole, 1987); which resulted in (2) the invasion of early Holocene dominants into their modern zones rather than into lower zones (Cole, 1985); and (3) the "lag" in response between physical proxy data (glaciers, lakes) and biological proxy data. As the complexity of plant population dynamics becomes better understood, these biologically oriented theories will gain ground.

CLIMATIC SYNOPSIS

Climatic Reconstructions. The task of climatic reconstruction is complicated by the lack of modern midden data from analog areas comparable to modern pollen samples in the eastern United States (Webb and Bryson, 1971; Delcourt et al., 1983). Given the paucity of such data, certain assumptions are unavoidable: (1) the distribution of a taxon is primarily dependent on climate rather than other factors (i.e., substrate, history); and (2) the investigator understands the modern range of the taxon well enough to judge where within its overall range it is common enough to be sampled regularly in packrat middens.

Although these assumptions need to be addressed more fully in the future, for the sake of this discussion I will assume that these obstacles have been overcome.

Full-Glacial Climates. A comparison of modern and full-glacial assemblages from the Grand Canyon demonstrates that individual taxa and comparable communities shifted upward about 800 m at the close of the Wisconsin. Using the modern elevation and precipitation values from Phantom Ranch (783 m, 21 cm) and the South Rim (2120 m, 35.6 cm), a precipitation lapse rate of 10.9 cm per 1000 m is calculated. Using this lapse rate and the 800-m elevational shift, Wisconsin mean annual precipitation should have been 8.7 cm greater than modern values, representing a 24% increase for the South Rim and a 41% increase for Phantom Ranch.

A similar analysis using the mean annual temperature of Phantom Ranch (20.6°C) and the South Rim (9.4°C) yields a lapse rate of 8.4°C per 1000 m. Using this lapse rate and the 800-m elevation shift, Wisconsin mean annual temperatures are calculated to have been 6.7°C less than modern temperatures.

These values serve only as general guides. More precise estimates require additional detailed study of the modern climate of the Grand Canyon and areas of more analogous vegetation now found hundreds of miles up the Colorado and Green rivers and in central and northern Utah. The Wisconsin plant assemblages suggest a winter precipitation regime consistent with other studies (Van Devender, *this volume*, chaps. 7 and 8). However, because these areas of analogous vegetation are generally much cooler (Cole, 1981), and Wisconsin winter circulation patterns probably increased winter and spring cloud cover, some of the apparent differences in precipitation between the Wisconsin and modern values probably result from lower rates of evaporation. The marked exclusion of chaparral species and predominance of Great Basin species, which typically dominate in areas of greater seasonality, suggest that the climate was more continental (Cole, 1981). This result is in contradiction to the equable, warm winter climates that may have occurred south of the Colorado Plateau (Van Devender, *this volume*, chap. 8).

Early Holocene Climates. The Early Holocene fossil assemblages are similar to associations now present along the Mogollon Rim, south of the Canyon. The presence of chaparral elements during the early Holocene and records of ponderosa pine from steep slopes of Coconino Sandstone (BC2 and BC3) and Redwall Limestone (CH7) are reminiscent of communities now found near Sedona, Arizona. Because the modern absence of ponderosa pine on steep slopes within the Grand Canyon seems to be caused by edaphic conditions as much as climate, its growth there in the early Holocene may have been encouraged by the retention of deeper Pleistocene soils (Cole, 1981).

Reconstructing climates during the early Holocene is also complicated by the complex dynamics of vegetational change, but, using the Mogollon Rim area as a guide, Grand Canyon climates must have had a summer precipitation greater than today with temperatures similar to or as much as one degree greater than today.

Middle Holocene Climates. The midden record definitely shows a middle Holocene with greater evaporative stress than is the case today. The lower limits of juniper and pinyon were at higher elevations. The middle

Holocene climates were probably about one degree warmer with slightly less mean annual precipitation than today.

CONCLUSIONS

In this chapter I have used several statistical techniques to reinforce the classification and estimation of rates of change present within fossil midden assemblages from the Grand Canyon. In addition, the view of a midden as a plant relevé from the past rather than an absolute record of the limits of species occurrences has shaped my conclusions. Climatic reconstructions can differ markedly if the modern analog area for a past extralocal species is taken as the nearest region where it is a frequent member of the flora rather than the nearest isolated occurrence of the species.

The conclusions inferred from these analyses are: (1) the late Wisconsin was a time of relatively stable conditions followed by rapid changes during the Wisconsin-to-Holocene transition; (2) vegetation was most likely in disequilibrium with climate during the Wisconsin-to-Holocene transition; and (3) past elevational shifts in vegetation and climate are more accurately viewed as latitudinal shifts.

ACKNOWLEDGMENTS

The work reported in this chapter was supported by grants and fellowships from the National Science Foundation, the University of Arizona, the Cave Research Foundation, and the Museum of Northern Arizona. Other necessary support was received from Ray Turner of the U.S. Geological Survey, the National Park Service, and the Laboratory of Isotope Geochemistry at the University of Arizona. This manuscript was critically reviewed by Julio Betancourt, Geoffrey Spaulding, Thomas Van Devender, and Paul S. Martin.

REFERENCES

Betancourt, J. L. (1987). "Paleoecology of pinyon-juniper woodlands: Summary." *Proceedings of the Pinyon-Juniper Conference*, pp. 129–39. U.S. Department of Agriculture, Forest Service, General Technical Report INT 215.

Birks, H. J. B., and Birks, H. H. (1980). *Quaternary paleoecology*. Baltimore: University Park Press.

Cinnamon, S. K., and Hevly, R. H. (1988). "Late Wisconsin macroscopic remains of pinyon pine on the southern Colorado Plateau, Arizona." *Current Research in the Pleistocene* 5, 47–48.

Cole, K. L. (1981). "Late Quaternary vegetation of the Grand Canyon: Vegetational gradients over the last 25,000 years." Ph.D. diss., University of Arizona, Tucson.

———. (1982). "Late Quaternary zonation of vegetation in the Grand Canyon." *Science* 217, 1142–45.

———. (1983). "Late Quaternary vegetation of Kings Canyon, Sierra Nevada, California." *Quaternary Research* 19, 117–29.

———. (1985). "Past rates of change, species richness, and a model of vegetational inertia in the Grand Canyon, Arizona." *American Naturalist* 125, 289–303.

———. (1986). "The lower Colorado River valley: A Pleistocene desert." *Quaternary Research* 25, 392–400.

———. (1987). "Differential vegetational response along elevational and latitudinal gradients: A test of vegetational inertia." (Abstract). Twelfth Annual Meeting of the International Union for Quaternary Research, p. 145. Ottawa: National Research Council of Canada.

Cole, K. L., and Mayer, L. (1982). "Use of packrat middens to determine rates of cliff retreat in the eastern Grand Canyon." *Geology* 10, 597–99.

Davis, O. K. (1986). "Multiple thermal maxima during the Holocene." *Science* 225, 617–19.

Delcourt, P. A.; Delcourt, H. R.; and Davidson, J. L. (1983). "Mapping and calibration of modern pollen-vegetation relationships in the southeastern United States." *Review of Palaeobotany and Palynology* 39, 1–45.

Dixon, W. J.; Brown, M. B.; Engelman, L.; Frane, J. W.; Hill, M. A.; Jennrich, R. I.; and Teporek, J. D. (1985). *BMDP statistical software.* Berkeley: University of California Press.

Gauch, H. G. (1982). *Multivariate analysis in community ecology.* Cambridge: Cambridge University Press.

Hill, M. O. (1979). *DECORANA: A Fortran program for detrended correspondence analysis and reciprocal averaging.* Ithaca: Cornell University Press.

Jacobs, B. F. (1985). "A middle Wisconsin pollen record from Hay Lake, Arizona." *Quaternary Research* 24, 121–30.

Kutzbach, J. E. (1981). "Monsoon climate of the early Holocene: Climate experiment with the Earth's orbital parameters from 9000 years ago." *Science* 214, 59–61.

Lanner, R. M. (1974). "Natural hybridization between *Pinus edulis* and *Pinus monophylla* in the American Southwest." *Silvae Genetica* 23, 108–16.

Lehr, J. H. (1978). *A catalogue of the flora of Arizona.* Phoenix: Desert Botanical Garden.

Lepper, M. G. (1974). "*Pinus flexilis* James and its

environmental relationships." Ph.D. diss., University of California, Davis.

Little, E. L., Jr. (1968). "Two new pinyon varieties from Arizona." *Phytologia* 17, 329–41.

McCammon, R. B., and Wenninger, G. (1970). *The dendrograph.* Kansas Geological Survey Computer Contributions, no. 48.

Malm, W. C. (1974). "Air movement in the Grand Canyon." *Plateau* 46, 125–32.

Markgraf, V. (1986). "Plant inertia reassessed." *American Naturalist* 127, 725–26.

Mead, J., and Phillips, A. M. III. (1981). "The late Pleistocene and Holocene fauna of Vulture Cave, Grand Canyon, Arizona." *Southwestern Naturalist* 26, 257–88.

Mueller-Dombois, D., and Ellenberg, H. (1974). *Aims and methods of vegetation ecology.* New York: Wiley and Sons.

Pase, C. P., and Brown, D. E. (1982). "Interior chaparral." In *Desert plants.* D. E. Brown, ed., 4:95–99. Tucson: University of Arizona.

Phillips, A. M. III. (1977). "Packrats, plants, and the Pleistocene in the lower Grand Canyon." Ph.D. diss., University of Arizona, Tucson.

———. (1984). "Shasta ground sloth extinction: Fossil packrat midden evidence from the western Grand Canyon." In *Quaternary extinctions.* P. S. Martin and R. G. Klein, eds., pp. 148–58. Tucson: University of Arizona Press.

Pielou, E. C. (1984). *The interpretation of ecological data: A primer on classification and ordination.* New York: Wiley and Sons.

Spaulding, W. G. (1981). "The late Quaternary vegetation of a southern Nevada Mountain Range." Ph.D. diss., University of Arizona, Tucson.

Spaulding, W. G., and Graumlich, L. J. (1986). "The last pluvial climatic episodes in the deserts of southwestern North America." *Nature* 320, 441–44.

Spaulding, W. G.; Leopold, E. B.; and Van Devender, T. R. (1983). "Late Wisconsin paleoecology of the American southwest." In *Late Wisconsin of the United States.* S. C. Porter, ed., pp. 259–93. Minneapolis: University of Minnesota Press.

Thompson, R. S. (1984). "Late Pleistocene and Holocene environments in the Great Basin." Ph.D. diss., University of Arizona, Tucson.

Van Devender, T. R. (1977). "Holocene woodlands in the southwestern deserts." *Science* 198, 189–92.

Van Devender, T. R.; Martin, P. S.; Thompson, R. S.; Cole, K. L.; Jull, A. J. T.; Long, A.; Toolin, L. J.; and Donahue, D. J. (1985). "Fossil packrat middens and the tandem accelerator mass spectrometer." *Nature* 317, 610–13.

Van Devender, T. R., and Mead, J. I. (1976). "Late Pleistocene and modern plant communities of Shinumo Creek and Peach Springs Wash, lower

Grand Canyon, Arizona." *Journal of the Arizona Academy of Sciences* 11, 16–22.

Van Devender, T. R., and Spaulding, W. G. (1979). "The development of climate and vegetation in the southwestern United States." *Science* 204, 701–10.

Webb, T. III. (1986). "Is vegetation in equilibrium with climate? How to interpret late-Quaternary pollen data." *Vegetatio* 67, 75–91.

Webb, T. III, and Bryson, R. (1971). "Late and post-glacial climatic change in the northern midwest, USA: Quantitative estimates derived from fossil spectra by multivariate statistical analysis." *Quaternary Research* 2, 70–115.

Chapter 12

Late Quaternary Biogeography of the Colorado Plateau

Julio L. Betancourt

In Utah it may be miles from the association
of desertshrubs on the lower elevation across a
mixture of shrubs and junipers before the pure
stands of junipers are reached at higher altitudes.   *Gleason, 1926*

INTRODUCTION

The individualistic concept of the plant association holds that species are distributed in
space and time more or less independently of
one another (Shreve, 1914; Gleason, 1926)
rather than as discrete communities (Clements, 1916). As such, plant communities can
be thought of as arbitrary positions along a
gradient of continuous compositional change.
In time, then, these positions must represent
only temporary aggregations determined by
chance, climate, disturbance history, and,
to some degree, the changing balance of species interactions. A historical perspective
seems essential for understanding such aggregations, specifically for predicting how
modern plant communities and vegetation
patterns are apt to change in both ecological
and geological time.

As Gleason's example from Utah implies,
the individualistic concept is embodied in the
distribution and density of plants according to
elevation. Among others, Whittaker and Niering (1965) demonstrated the gradual appearance and disappearance of species along elevational gradients without abrupt replacements.

More recent analyses indicate that, though
most elevational distributions appear to be individualistic, abrupt replacements by competing species do indeed occur. Some species
achieve their maximum density near their
upper limits, where they are apparently replaced by other species with overlapping habitat requirements (e.g., Yeaton et al., 1980).

In the western United States, a region
where local relief can exceed 4000 m, similarities in plant distributions and climatic
gradients over broad areas have inspired various schemes to classify vegetation as elevational zones (Merriam, 1890; Daubenmire,
1943). Unfortunately, the zonal concept has
been overextended in matters of historical
biogeography. In pondering the past, biogeographers have raised or lowered modern vegetation zones to the tempo of major climatic
changes. Opportunities for major reorganiza-
tion of community structure and composition
have been virtually ignored. This tendency
does not necessarily reflect an endorsement of
Clementsian ecology. Rather it stems from
insufficient knowledge of modern vegetation

dynamics and, until recently, an inadequate fossil record.

Packrat middens offer a unique way to document reshuffling of plant distributions along elevational and other physical gradients in response to climatic change. The rat's foraging range, normally less than one hectare, limits the source area of well-preserved plant remains identifiable to species. Middens yield information not only about geographic ranges through time, but also a plant's preference for elevation and slope aspect as well as its local association with other plant and animal species.

Here, midden data are used to infer the chronology and dynamics of late Quaternary vegetation history on the Colorado Plateau. An attempt is made to determine whether elevational displacements through time and resulting vegetation patterns are consistent with modern ecological observations. Key questions are addressed about regional biogeography, including the origin of disjunct taxa in low-elevation alcoves, the effect of rock type in modulating past and modern plant distributions, and the role of historical processes in explaining patterns of species richness on isolated mountains. The following stratigraphic terms were adopted for convenience: middle Wisconsin (65–35 ka); late Wisconsin (35–11 ka) (full glacial [22–15 ka], late glacial [15–11 ka]); early Holocene (11–8 ka); middle Holocene (8–4 ka); and late Holocene (4–0 ka). Note that the Wisconsin-Holocene boundary at 11 ka does not conform to convention (10 ka). Slope aspect and elevation are given in parentheses after first mention of a midden site (e.g., W, 1980 m).

Previous Work

Until quite recently, paleoenvironmental research in the Colorado Plateau focused primarily on pre-Quaternary deposits. We are more sure of facies changes in Cretaceous strata than in canyon fills of the Holocene. Wisconsin glaciation is documented but poorly dated for the White Mountains (Merrill and Pewe, 1977) and San Francisco Peaks (Sharp, 1942) in Arizona, the High Plateaus of Utah (Flint and Denny, 1958), and the La Sal Mountains on the Utah-Colorado border (Richmond, 1962). These alpine glaciers apparently were concurrent with the San Juan

Mountain ice sheet, which was about 5000 km^2 in area and up to 1 km thick (Carrara et al., 1984). Protalus ramparts suggesting minimum elevation for orographic snowline occur down to 2600 m on Navajo Mountain (Blagbrough and Breed, 1969). The timing of deglaciation is poorly known except in the San Juan Mountains, where the ice cap had melted by 15 ka (Carrara et al., 1984). South-facing cirques were ice-free by 11 ka and north-facing ones by 9 ka (Andrews et al., 1975).

Unlike the massive Rockies, isolated mountains of the Colorado Plateau harbor few lacustrine settings amenable to continuous pollen deposition and preservation. The vegetation of the lowlands has been inferred from a few pollen stratigraphies on the Mogollon Rim in Arizona (Whiteside, 1964; Hevly, 1964; Batchelder and Merrill, 1976; Jacobs, 1983), the San Agustin Plains (Clisby and Sears, 1956; Markgraf et al., 1984) and Chuska Mountains (Wright et al., 1973) in New Mexico, the San Juan (Maher, 1961; Andrews et al., 1975) and La Plata mountains (Petersen and Mehringer, 1976) in Colorado, and the Wasatch Mountains in Utah (Madsen and Currey, 1979). These extrapolations were untested until recently.

Quaternary research in the lowlands has focused primarily on the late Holocene. This period gave rise to the explosive cultural evolution from Basketmaker to Anasazi and the spectacular prehistoric architecture of places like Chaco Canyon, Mesa Verde, and Canyon de Chelly. Both alluvial stratigraphic techniques and dendrochronology were pioneered on the Colorado Plateau in support of archaeological studies.

Although middens were discovered initially during early exploration surveys in the Four Corners (Simpson, 1850; Yarrow, 1875), their paleoecological use on the Colorado Plateau was delayed until the late 1970s. Here middens were first studied to evaluate long-term stability of vegetation on an ungrazed, isolated butte in Canyon de Chelly (Schmutz et al., 1976). Midden residues also were identified as the source for unusual algal mat growth on submerged cliff faces of Lake Powell (Potter and Pattison, 1976).

A salient feature of the plateau is the abundance of large rock shelters and caves in cliffs of massive sandstone and limestone from 2430 to 1000 m elevation. Fossil middens are most common in caves located away from the head of an alcove and with relatively small aquifer areas feeding seeps along bedding planes and contacts. Pleistocene middens are commonly found in steep talus just inside the dripline and wherever Holocene sedimentation, particularly roof collapse, has been negligible. Locations meeting the above criteria can yield an impressive suite of middens spanning twenty thousand years or more. A series of five to ten middens from a single cave, selected to represent variability in macrofossil contents, can be thought of as a long-term census of a vegetation plot of less than one hectare. This eliminates the often unavoidable ambiguity that arises by grouping middens of different ages from different sites to build a single chronology.

PHYSICAL SETTING

The Colorado Plateau is a well-marked physiographic province that covers almost 400,000 km² in northwestern New Mexico, northeastern Arizona, western Colorado and southeastern Utah (Fig. 12.1). Elevations range from 4600 to 1000 m, with about 45% below 1850 m. In the region's interior great sedimentary systems of alternating resistant and easily erodable rocks dip gently to the northeast, opposite the predominant drainage direction. Intrusive, igneous rocks occur in the interior as laccolithic ranges, including the Navajo, Henry, La Sal, Abajo, and Carrizo mountains. Volcanic rocks are more common along the periphery in the southern Rockies, the Datil-Mogollon volcanic plateau, the San Francisco Mountains, and the lava-capped High Plateaus. Physiographically, the major landforms are high plateaus on the upfolds, hogbacks on their flanks, lower plateaus between the upfolds, laccolithic complexes, and an intricate set of canyons, which are most spectacular in the Canyon Lands and Grand Canyon sections (Hunt, 1956). Areas of rimrock, slickrock, and canyon walls are more common than in other parts of the West. Thus the midden records from the Colorado Plateau may give a better representation of regional vegetation patterns than middens from other regions with less rocky terrain.

The present climate of the Colorado Plateau is strongly affected by topography, with the interior occupying a major rain shadow in the lee of highlands to the south, east, and west. Mean annual precipitation is less than 220 mm at most stations below 1800 m. By comparison, precipitation is commonly doubled at stations of equivalent altitude on or south of the Mogollon Rim. Annual snowfall amounts range from zero in the bottom of the Grand Canyon to a few meters in the uplands. Precipitation amounts during the warm season (April–September) range from 35% to 65% of the annual total. The proportion of winter rainfall increases generally to the northwest and locally toward higher elevations. Mean annual temperature exceeds 10°C at altitudes below 2100 m, except in some valley-bottom stations affected by nocturnal cold-air drainage.

Modern Vegetation

Altitudinal patterns of plant zonation on the Colorado Plateau were first pointed out by Merriam (1890). An alpine zone occurs above 3480 m in restricted areas, specifically the San Francisco Peaks, the La Sal Mountains, and the Tushar Plateau. The uppermost tree zone (3480–2900 m) is dominated by *Picea engelmannii* and *Abies lasiocarpa*, with only sparse representation of *Pinus longaeva, P. aristata*, and *P. flexilis*. Both the alpine and spruce-fir communities occur predominantly on igneous rocks, the most common substrate on plateau summits and mountain peaks. At the elevation of the spruce-fir forest, bunchgrass (*Festuca, Stipa, Poa, Muhlenbergia, Andropogon*) communities, commonly referred to as "tussock grasslands," create a false timberline on dry, steep slopes dominated by shale and porphyry talus.

Below the subalpine zone (2900–2600 m) lies a mixed-conifer forest of *Pseudotsuga menziesii, Abies concolor*, and *Picea pungens*. As in the Great Basin and the Rockies, *Populus tremuloides* stands attain their best development at elevations normally occupied by mixed-conifer forest, particularly on shale substrates. *Pinus ponderosa* forest is best developed between 2500 and 2100 m, most impressively on the Mogollon Rim. Pinyon-juniper woodlands with *Pinus edulis* occupy the extensive lower plateaus from 2100 to 1600 m. *Juniperus monosperma* is the domi-

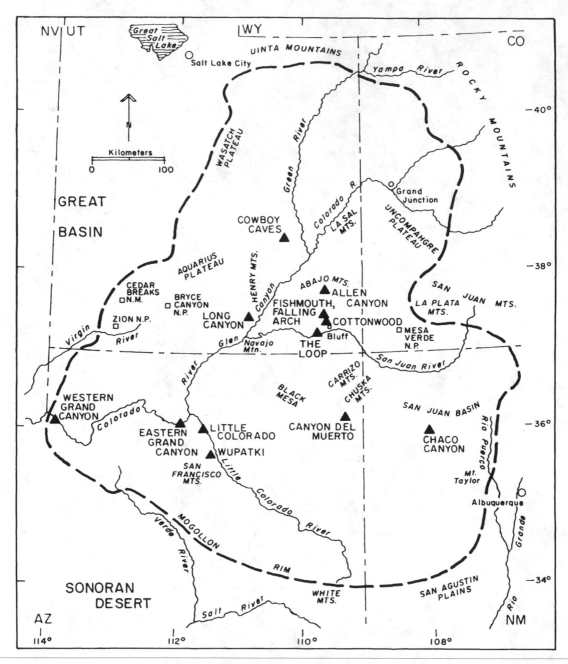

Fig. 12.1. Map of Colorado Plateau (area within dashed line) and midden sites discussed in text (solid triangles).

nant juniper in the southern and eastern parts of the Colorado Plateau, with *J. osteosperma* occupying the broader areas to the west. The lower limits of pinyon-juniper woodland are depressed from the hypersaline Mancos Shale (2130 m) to porphyry (1830 m) to sandstone (1500 m) (Everitt, 1970). *Artemisia tridentata* dominates wherever soils are relatively deep and noncalcareous throughout the *P. ponderosa* and pinyon-juniper zones.

The desertscrub zone below 1600 m is variable. The Dixie Corridor area in southwesternmost Utah and the Grand Canyon in Arizona defines the eastern range of thermophiles such as *Larrea divaricata, Ambrosia dumosa,* and *Krameria parviflora,* mainly below 1200 m. Desertscrub and grassland communities below 1600 m elsewhere on the plateau include those characterized by *Atriplex confertifolia, A. corrugata, Coleogyne ramossisima,* and *Hilaria-Aristida* grassland. Moist alcoves at desert elevations or glens (the source of the name Glen Canyon) include prairie grasses, deciduous trees, and shrubs and herbs of boreal affinities (Welsh and Toft, 1981). At issue is whether these alcove elements are relicts of a formerly widespread flora or achieved their present distribution through long-distance dispersal.

MIDDEN SEQUENCES

Midden sequences have been developed from Chaco Canyon in the San Juan Basin (Betancourt and Van Devender, 1981; Betancourt et al., 1983), Canyons de Chelly and del Muerto west of the Chuska Mountains (Schmutz et al., 1976; Betancourt and Davis, 1984), Allen Canyon Cave on the south slope of the Abajo Mountains (Betancourt, 1984), the San Juan country (White Canyon: Mead et al., 1987; Fishmouth Cave: Betancourt, 1984; Falling Arch, Cottonwood Cave, and the Loop: Van Devender et al., 1987), Glen Canyon (Long Canyon Cave: Van Devender et al., 1987), the lower basin of the Little Colorado River (Van Devender and Spaulding, 1979; Cinnamon, 1988), and the eastern Grand Canyon (Cole, 1981, 1982, 1985; Mead, 1983; Fig. 12.1). Plant macrofossil assemblages comparable to packrat middens also have been recovered from sediment in Cowboy (Spaulding and Petersen, 1980) and Bechan caves (O. K. Davis et al., 1984). A total of 180 middens have been dated and analyzed from the Colorado Plateau.

Chaco Canyon, New Mexico

Fifty-two middens have been analyzed from Chaco Canyon, the locus of a spectacular array of prehistoric ruins constructed mostly between 1 and 0.8 ka (Fig. 12.2). Modern vegetation in the vicinity grades from pinyon-juniper woodland in the upper canyon (Chacra Mesa), to shrub grassland in the lower end of the canyon, the latter containing most of the ruins. The mostly treeless landscape contrasts with over 200,000 conifers preserved as constructional timber in the ruins. Based on alluvial pollen, Hall (1977) inferred local conditions little different from today over most of the Holocene.

All middens were collected from the Upper Cretaceous Cliff House Sandstone at elevations between 1860 and 2020 m, and at various sites from Chacra Mesa to the mouth of the canyon. Near the mouth, a series of seven middens from Atlatl Cave (W, 1910 m) provide sketches of local vegetation history during the Holocene, supported by several shorter sequences elsewhere in the canyon. Atlatl Cave middens dated at 10.6–10.0 ka and 10.5–9.5 ka are dominated by *Pseudotsuga menziesii, Juniperus scopulorum,* and *Pinus flexilis,* with traces of *Picea pungens.* The presence of *Artemisia tridentata, Atriplex canescens, Yucca angustissima, Opuntia polyacantha, O. whipplei, Echinocereus,* and the Colorado Plateau endemics *Berberis fendleri* and *Sclerocactus mesaverdii* suggests an open woodland of *P. flexilis* and *J. scopulorum. Pseudotsuga menziesii* and *Picea* were perhaps restricted to shady alcoves and canyon walls. The relative importance of open steppe or woodland on the mesa tops and broad valleys cannot be ascertained from the plant macrofossil record, but it can be inferred from other evidence at Sheep Camp Canyon Cave, some 15 km to the west.

At the Sheep Camp locality (N, 1980 m) cave fill containing needles of *P. flexilis* and *Picea* is associated with over 90% combined *Picea* and *Pinus* pollen. *Juniperus* pollen is a minor component compared to the early Holocene middens from Atlatl Cave, where it comprised up to 60% of the total pollen (probably from *J. scopulorum*) (Hall, 1985). The same undated fill yielded remains of *Equus* and cf. *Platygonus,* indicating a Pleistocene age. Although *Artemisia* was unimportant in the pollen assemblage, remains of animals now restricted to sagebrush communities of the Great Basin and Columbia Plateau were recovered, along with remains from montane, forested habitats in the south-

Fig. 12.2. Photograph of Pueblo Bonito (bottom) and north slope of Chaco Canyon. Just prior to 10 ka, cliffsides supported *Pinus flexilis, Pseudotsuga menziesii, Juniperus scopulorum,* and *Picea pungens.* By 8.3 ka *P. flexilis* and *P. pungens* were replaced by *Pinus ponderosa, P. edulis,* and *J. monosperma. Pseudotsuga menziesii* and *P. pon-* *derosa* persisted as isolated trees along the upper cliffsides until about 2.4 ka. Stands of *P. edulis* and *J. monosperma* were greatly reduced between 1.2 and 0.5 ka, most likely due to intensive fuel harvesting by the Anasazi. The present vegetation is desertscrub-grassland. (Photo courtesy of Paul Logsdon, Santa Fe.)

ern and central Rockies (Gillespie, 1985). Evidence from the Sheep Camp locality suggests a mosaic of sagebrush steppe, most likely on the mesa tops and valleys, with forest and woodland on the canyon slopes. This pattern may have continued into the early Holocene, as recorded by evidence from Atlatl Cave.

A midden at Atlatl Cave dated at 8.3 ka records the departure of *Picea* and *Pinus flexilis*, the persistence of *J. scopulorum* and *Pseudotsuga menziesii*, and the appearance of *Pinus ponderosa, P. edulis*, and *J. monosperma*. This midden also provides the only record of *Rosa* sp., a shrub no longer present in Chaco Canyon. A midden dated at 7.9 ka from Werito's Rincon (NE, 2025 m), about 7 km southeast of Atlatl Cave, also contains *Pseudotsuga menziesii, Pinus ponderosa, P. edulis*, and *Juniperus monosperma*. The middens at 8.3 and 7.9 ka represent the youngest records of *J. scopulorum*. A few isolated individuals persist today on Chacra Mesa. Extralocal *Pseudotsuga menziesii* and *Pinus ponderosa* were recorded in eight middens dated between 6.7 and 2.4 ka. These trees disappeared locally between 2.8 and 2.3 ka at Atlatl Cave, sometime after 2.8 ka at Sheep Camp Canyon Cave, after 2.4 ka at the head of Mockingbird Canyon (N, SW, 1940 m), between 4.8 and 2.2 ka at the Indian Ruins locality (NW, 1930), and after 3.7 ka at Chacra Mesa (NW, 1930, *Pseudotsuga menziesii* only) (Betancourt et al., 1983). Together these sequences suggest that *Pinus ponderosa* and *Pseudotsuga menziesii* persisted—most likely as isolated trees along the upper cliffs—until between 2.4 and 2.2 ka.

The reduction of *P. edulis* between 1.2 and 0.5 ka is one of the most dramatic vegetation changes recorded during the late Holocene of the Southwest (Betancourt and Van Devender, 1981; Betancourt et al., 1983). At Chaco Canyon over three hundred middens containing *P. edulis* were noted in places where it no longer grows, primarily on the south-facing side of the canyon. Thirty-two middens with *P. edulis* and twelve without were dated to determine if its reduction (and in many cases that of *J. monosperma*) was synchronous with the peak of Anasazi activity. The early Holocene mixed-conifer records aside, all middens older than 1.2 ka had abundant macrofossils of *P. edulis*, and middens lacking this species were

no older than 0.5 ka (Fig. 12.3).

The large amounts of *Pinus ponderosa* used in construction between 1 and 0.8 ka were not harvested locally. Beams identified as *Picea* and *Abies* suggest that the Anasazi imported timber from mountains over 75 km away (Betancourt, Dean, and Hull, 1986). However, middle and late Holocene middens demonstrate that stands of *Pinus edulis* and *Juniperus monosperma* were available to the Anasazi for fuel. There are two possible explanations for reduction of these stands between 1.2 and 0.5 ka. The first alternative is a climatic event stressful enough to produce a shift from woodland to desertscrub. If this was the case, it was the only time this occurred in over eight thousand years at Chaco Canyon, and it has no parallel elsewhere in the Southwest. The second alternative relates woodland depletion to the postulated fuel needs of the Anasazi. Samuels and Betancourt (1982) showed by computer simulation that Anasazi fuel needs over a two-hundred-year period could easily account for decimation of a woodland of 13,000 ha, assuming a maximum density of 15 cords/ha.

Canyons de Chelly and del Muerto, Arizona

Four middens dating to 11.9, 6.2, 3.1, and 1.9 ka are reported from Canyons de Chelly and del Muerto on the west side of the Chuska Mountains. These sparse data allow an interesting comparison with a pollen profile from Dead Man Lake on the crest of the Chuskas, at 2780 m elevation (Wright et al., 1973). This pollen record is poorly dated, may embrace the middle and late Wisconsin, and has been analyzed with particular care for identification of pine pollen types. *Artemisia* sp. pollen percentages are high (30%–60%) throughout the Pleistocene sediments, dropping to below 20% in the Holocene, while pollen assigned to *Pinus ponderosa* comprises 60%–80% of the pine pollen in middle Wisconsin and Holocene sediments. Pollen attributed to *Pinus flexilis* peaks in the full-glacial middens. *Pinus edulis* pollen, a common component throughout the Pleistocene, is least abundant in the Holocene samples. For the full- and late-glacial vegetation, Wright et al. (1973) inferred a "telescoping" of zones, with *Pinus ponderosa* not expanding below the piedmont of the Chuska Mountains. This interpretation

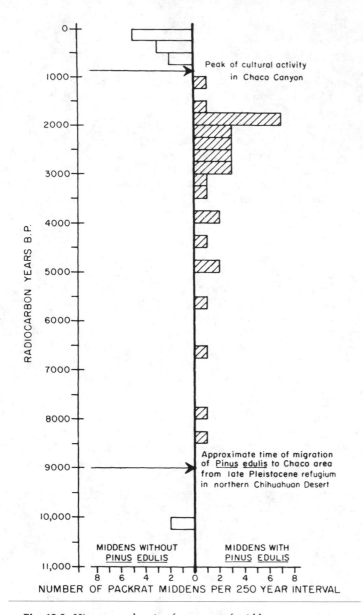

Fig. 12.3. Histogram showing frequency of middens with (hatched bars) and without (open bars) *Pinus edulis*, from sites where it no longer grows in Chaco Canyon, New Mexico. *P. edulis* immigrated to Chaco Canyon in the early Holocene but was greatly reduced between 1.2 and 0.5 ka, coincident with the peak in cultural activity.

contrasted with Martin and Mehringer's (1965) map of Pleistocene vegetation of the Southwest, which shows *Pinus ponderosa* forest in these lowlands. Instead, Wright et al. (1973) suggested dominance of *Artemisia* steppe and pinyon-juniper woodlands in the lowlands, a reconstruction also advocated by Hall (1985).

The midden record from the Plateau supports neither Wright nor Martin and Mehringer. The nearest evidence is a midden dated at 11.9 ka from Canyon del Muerto, 1000 m lower and only 40 km west of Dead Man Lake (Betancourt and Davis, 1984). The midden was collected from desertscrub hatitat at 1770 m elevation along a steep, north-facing cliff of Permian de Chelly Sandstone. This late-glacial midden contains *Pseudotsuga menziesii, Pinus flexilis, Picea pungens, Juniperus communis, J. scopulorum, Cornus stolonifera, Acer glabrum, Rosa* sp., *Rubus* sp., and *Sambucus* sp. The assemblage is duplicated today in riparian settings as low as 2250 m in the Chuskas, although *P. flexilis* no longer occurs in the range. The most abundant pollen types in the midden are *Artemisia* (38.3%) and *Pinus* (18.0%). A nearby midden dated at 3.1 ka contains mostly herbs and shrubs now growing at the site, traces of *P. edulis* and *J. osteosperma*, and a single needle of *Pseudotsuga menziesii*, a probable contaminant. *Quercus* (16.6%), *Cleome* (15.1%), and *Pinus* (10.8%) are the most abundant pollen types. Pollen of *Zea mays* and elevated values for *Cleome*, a disturbance plant on tilled soils, suggest an early date for farming in the area.

The displacement of *Picea, P. flexilis,* and *J. communis* well below the base of the Chuskas, even in this riparian setting, and the absence of *P. ponderosa* or *P. edulis*, do not support a model of "telescoping" of vegetation zones as inferred by Wright et al. (1973). Pleistocene middens from 2200–1000-m localities elsewhere on the Colorado Plateau, from xeric as well as mesic sites, have yet to produce a single macrofossil of *P. ponderosa*. These middens record an impressive suite of vegetation types, from spruce-fir forest to juniper woodland. On the Colorado Plateau the oldest records of *P. ponderosa* are dated at 10.3 ka. *Pinus edulis* appears in a few isolated records from the western Grand Canyon as early as 16.6 ka (Cave of the Early Morning Light,

1300 m elev.) and at Wupatki National Monument, just north of Flagstaff, Arizona, by 13.7 ka (see later discussion). However, pinyons were not an important component of the late Wisconsin vegetation of the Colorado Plateau.

Wright et al. (1973) did allow for the possibility that *Pinus flexilis* may have replaced *P. ponderosa* in the full glacial as it now does in northwestern Wyoming. The midden record suggests that this also may have been the case in the late glacial. *Pinus edulis* and *P. ponderosa* pollen in full- and late-glacial sediments from Dead Man Lake can be accounted for only by long-distance dispersal from expanded woodlands several hundred kilometers to the south. The middle Wisconsin rise in *P. ponderosa* pollen, contemporaneous with elevated percentages of *P. edulis* in lake sediments from the White Mountains of Arizona (Jacobs, 1985), could indicate expansion of pinyon-juniper woodland and *P. ponderosa* forest during a relatively warm interstadial (Spaulding et al., 1983; Betancourt, 1987).

Abajo Mountains, Utah

Allen Canyon Cave (Fig. 12.4), overlooking Cliff Dwellers Pasture in the southern piedmont of the Abajo Mountains, faces a lush, sandstone-rimmed valley bordered by pinyon-juniper woodland and *Quercus gambelii–Cercocarpus* spp. chaparral. In Cliff Dwellers Pasture clumps of *Betula occidentalis* grow in rich meadows. *Q. gambelii* and *Acer grandidentatum* cover the rest of the valley floor. The steep talus in front of the shelter supports a dense stand of *Acer grandidentatum* and *Populus tremuloides*, with several towering *Pseudotsuga menziesii* and *P. ponderosa* growing just beyond the dripline. *Pinus edulis* and *Juniperus osteosperma* occur on the ridgetop above the shelter.

Nine middens from Allen Canyon Cave record vegetation changes between 11.3 and 1.8 ka (Fig. 12.5; Betancourt, 1984). A sample dated at 11.3 ka is the only Wisconsin midden from the western United States with an assemblage typical of modern spruce-fir forest. The four most common plants in this midden are *Picea engelmannii, Abies lasiocarpa, Pinus flexilis,* and *Juniperus communis. Pseudotsuga menziesii* is present but contributes less than 0.2% of the plant matrix (in-

Fig. 12.4. Photograph of Allen Canyon Cave
(2200 m) on the slopes of the Abajo Mountains,
southeastern Utah.

cluding unidentifiable material), compared to
1.0%–5.6% for the other conifers. In the understory *J. communis* was accompanied by
Shepherdia canadensis, Sambucus sp., *Rosa*
sp., *Pachystima myrsinites, Cornus stolonifera, Acer,* and *Amelanchier.* A rich grass
flora (seven taxa) suggests meadows on the
valley floor.

Middens dated at 10.4, 10.1, and 10.0 ka
record a sharp reduction in *Picea engelmannii* and *Abies lasiocarpa,* the arrival of *Picea
pungens, Pinus ponderosa, Quercus gambelii,* and *Cercocarpus intricatus,* the persistence of *Pinus flexilis* as a dominant, and a
twentyfold increase in the abundance of
Pseudotsuga menziesii, averaged over the
three middens. *Picea, Abies, Pinus flexilis,
Shepherdia canadensis,* and *Sambucus* sp. are
missing from a 7.2-ka midden, replaced by *J.
osteosperma, J. scopulorum, Rhus trilobata,*
and sixty times more *Pinus ponderosa* than

any other sample in the cave. *P. edulis* is first
recorded at 3.4 ka, which also is the youngest
record of *Juniperus communis* at the site.
The late Holocene samples dated at 3.4, 3.0,
and 1.8 ka yielded the only records of *Opuntia polyacantha, Yucca angustissima,* and
Oryzopsis hymenoides in the series. *Juniperus scopulorum* and *Pachystima myrsinites* are not in the 1.8-ka sample, though
they still occur elsewhere in Cliff Dwellers
Pasture. It is interesting that several understory elements remain constant throughout
the series, including *Amelanchier* sp., *Cornus
stolonifera, Pachystima myrsinites,* and
Rubus sp. The mosses *Hypnum cuppressi-
forme* and *Thuidium abietenum,* which are
present in the late-glacial as well as several of
the Holocene middens, still grow at the site.

The middle Holocene dominance of *Pinus
ponderosa* and absence of *Pinus edulis* suggest an interesting hypothesis. Throughout

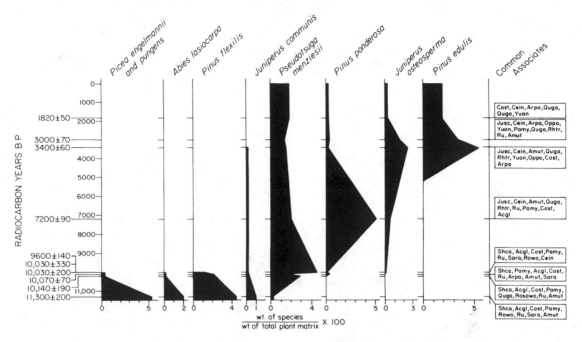

Fig. 12.5. Chronosequence of select taxa and their macrofossil abundance in middens from Allen Canyon Cave. The succession of midden floras from the late Wisconsin to the present parallels present zonation from the lower range of spruce-fir to the upper range of pinyon-juniper woodland. Abbreviations: Acgl, *Acer glabrum*; Amut, *Amelanchier utahensis*; Arpa, *Arctostaphylos patula*; Cein, *Cercocarpus intricatus*; Cost, *Cornus stolonifera*; Jusc, *Juniperus scopulorum*; Pamy, *Pachystima myrsinites*; Oppo, *Opuntia polyacantha*; Rowo, *Rosa woodsii*; Quga, *Quercus gambelii*; Ru, *Rubus*; Sara, *Sambucus racemosa*; Shca, *Shepherdia canadensis*; Yuan, *Yucca angustissima*.

the southern Rockies *P. edulis* attains its highest density at its upper limits before it is replaced by *P. ponderosa*. Yeaton et al. (1980) invoke interspecific competition to explain a similar replacement of *P. sabiana* by *P. ponderosa* on the western slopes of the Sierra Nevada. In the arid White Mountains of California, dominant trees of the pinyon-juniper belt are subject to little competition from elements of higher vegetation zones. There *Pinus monophylla* achieves its greatest density in the middle of its elevational range. *Pinus monophylla* also has a conspicuously high upper limit in the White Mountains (West et al., 1978).

For the southern Rockies one might postulate an initial expansion of *Pinus ponderosa* below its present lower limits during the early Holocene and early part of the middle Holocene, as suggested by its dominance at 7.2 ka in Allen Canyon and its greater abundance from 8.3 to 6.7 ka at Chaco Canyon. *P.*

edulis was already established in the middle and lower part of its modern range on the Colorado Plateau by 10–8 ka. *Pinus edulis* failed to extend its range upward to the higher elevations it now occupies, possibly because of competition from *Pinus ponderosa* and *Quercus gambelii*, which already occupied the middle elevations. Competition was relaxed when these species retreated upward in the middle to late Holocene.

The San Juan Country, Utah

This area, which includes White Canyon (Natural Bridges National Monument), Comb Ridge, lower Cottonwood Wash, and the upper end of San Juan River Canyon, is the core of Gregory's (1938) San Juan country. Comb Ridge, a conspicuous hogback of eastward-dipping rocks, begins opposite Allen Canyon and continues south across the San Juan River to Kayenta, Arizona, a total distance of 160 km. Cottonwood Wash, the main drainage of the Allen Canyon country, parallels Comb Ridge 15 km to the east and empties into the San Juan River near Bluff, Utah. From Bluff the river flows through a low, wide gap at Comb Ridge, then swings northwest through a deep, tortuous canyon of spectacular meanders, typified by the Goosenecks near Mexican Hat. Six midden localities have been investigated in the San Juan country: White Canyon (S, 1820 m) in Natural Bridges National Monument; Fishmouth Cave (NE, 1585 m), Falling Arch (SE, 1460 M) and North Fork (N, 1800 m) on Comb Ridge north of the river; Cottonwood Cave (S, 1390 m) in lower Cottonwood Wash; and the Loop (SW and SE, 1525 m), an abandoned meander in San Juan River Canyon.

In a shelter (S, 1820 m) perched high above the floor of White Canyon, Natural Bridges National Monument, Mead et al. (1987) recovered both loose and indurated middens interspersed in cave sediments, which also contained dung of the extinct mountain goat *Oreamnus harringtoni*. The canyon habitat is relatively moist, with pinyon-juniper woodland interrupted by restricted stands of *Pinus ponderosa* and *Pseudotsuga menziesii* on north-facing slopes.

A one-meter-deep unit was excavated by Mead et al. (1987). The basal layer (90–99 cm depth) contains *Oreamnus* dung dated to > 39.8 ka. A loose packrat midden (23–90 cm depth) has abundant *Picea, Pseudotsuga menziesii, Pinus flexilis,* and *Rosa* cf. *woodsii*. Above it is another layer of *Oreamnus* dung dated to 23.4 ka, bracketing the age of the lowest midden between > 39.8 ka and 23.4 ka. At 26–53 cm depth is another loose midden, dated at 21.3 ka. This deposit contains the same taxa as the lower midden, with the addition of *Juniperus communis* and *J. scopulorum*. An overbank mud flow (16–26 cm depth), originating from a flood when the canyon floor was perhaps much higher, contained a *Betula* twig dated to 26.5 ka. Above it (0–16 cm depth) is an indurated midden dated to 9.7 ka, which is dominated by *Pseudotsuga menziesii* and contains only traces of *Picea, Pinus flexilis,* and *Rosa woodsii*. Also present are *J. communis* and *J. scopulorum*, along with taxa now at the site, including *Acer negundo* and *Celtis reticulata*. Microhistological remains suggest that *Oreamnus* browsed

predominantly on *Pseudotsuga menziesii* and *Pinus flexilis*.

Fishmouth Cave is probably the largest cave in the area (Fig. 12.6). Eight middens were sampled spanning the period 13.8–2.3 ka (Betancourt, 1984; Van Devender et al., 1987). Common plants now growing within 100 m are *Juniperus osteosperma, Amelanchier utahensis, Fraxinus anomala, Rhus trilobata, Shepherdia rotundifolia, Celtis reticulata, Cercocarpus intricatus, Atriplex canescens, Yucca angustissima, Opuntia polyacantha,* and *O. erinacea*. Two late-glacial samples dated at 13.8 and 12.8 ka are dominated by *Pinus flexilis, Pseudotsuga menziesii, Juniperus scopulorum, J. communis,* and *Cornus stolonifera*, with lesser amounts of *Picea pungens, Rosa* cf. *woodsii,* and *Cercocarpus montanus* (Fig. 12.7). Both middens contain *Ephedra* sp., *Yucca angustissima,* and *Opuntia polyacantha*. A midden at 10.4 ka records the loss of *Picea pungens* and *Juniperus communis*, the persistence of *Pinus flexilis, Pseudotsuga menziesii,* and *Juniperus scopulorum*, and the first records of *Quercus gambelii* from the site. By 9.7 ka *P. flexilis, J. scopulorum,* and *Cercocarpus montanus* were replaced by *J. osteosperma* and *C. intricatus*. Middens dated at 9.7 and 6.1 ka contain modest amounts of *Pseudotsuga menziesii* and *Quercus gambelii*. *Quercus* pollen makes up 20%–40% of the total in the early Holocene middens but does not exceed 10% in the other middens from the cave (O. K. Davis, 1987). Middens dated at 3.7 and 3.6 ka contain twice the weight of *Juniperus osteosperma* macrofossils than that found in other Holocene middens. These middens contain unusually high percentages of *Juniperus* pollen (40%–60%). The sample at 3.7 ka provides the only record of *Pinus edulis* from the site; a few isolated individuals still occur within 1 km. *Fraxinus anomala* and *Shepherdia rotundifolia* appear for the first time in the late Holocene samples. There is a progressive increase in the number of grass taxa, from one (the ever-present *Oryzopsis hymenoides*) in the late glacial to eight in a midden dated at 2.3 ka.

A similar sequence was obtained from six middens at Falling Arch, a small rock shelter on Comb Ridge, about 5 km south of Fish-

Fig. 12.6. Aerial photograph of Fishmouth Cave (1585 m) and Comb Ridge, southeastern Utah. Late Wisconsin middens contain *Pinus flexilis, Pseudotsuga menziesii, Juniperus scopulorum, J. communis,* and *Picea pungens. Pinus flexilis* and *J. scopulorum* persisted until 10 ka, but *Pseudotsuga* was common until at least 6.1 ka. *J. osteosperma* has been the dominant conifer at the site for the past ten thousand years.

mouth Cave. This locality (SE, 1460 m) is located high on a ridge of Jurassic Navajo Sandstone, with slickrock as the primary substrate within 50 m. Common woody perennials growing in cracks and pockets of shallow soils include *Coleogyne ramossisima, Amelanchier utahensis, Juniperus osteosperma, Ephedra viridis, Fraxinus anomala, Cowania mexicana,* and *Shepherdia rotundifolia.* The canyon bottom 30 m below the shelter supports *Populus fremontii, Quercus gambelii, Juniperus osteosperma,* and *Artemisia nova.* There is a single *Pinus ponderosa* growing in the canyon bottom at 1500 m elevation, some 250 m upstream of the midden site. The nearest *P. ponderosa* stand is at the source of Butler Wash (1700 m), about 20 km to the north.

The Falling Arch sequence includes middens dated at 19.7, 13.1, 9.2, 5.8, 3.8, and 2.4 ka. The full-glacial sample (19.7 ka) is dominated by *Pinus flexilis,* with only modest amounts of *Pseudotsuga menziesii* and *Juniperus scopulorum* and the abundant shredded cambium of *Artemisia tridentata*-type. Mesophytic shrubs common in this assemblage are *Rosa* cf. *woodsii* and *Cornus stolonifera.* The midden also contains several xerophytes such as *Atriplex canescens, Cercocarpus intricatus,* and *Opuntia polyacantha.* The late-glacial (13.1 ka) midden is quite similar in composition to the full-glacial assemblage. Significant differences are a twentyfold increase in *J. scopulorum,* only traces of *Cercocarpus intricatus,* and abundant seeds of *Sambucus* sp., a shrub that was missing from the full-glacial sample. None of the montane plants recorded in the older middens is represented in a sample dated at 9.2 ka, which is dominated by *Juniperus osteosperma.* This midden lacks *Cercocarpus intricatus, Fraxinus anomala,* and *Coleogyne*

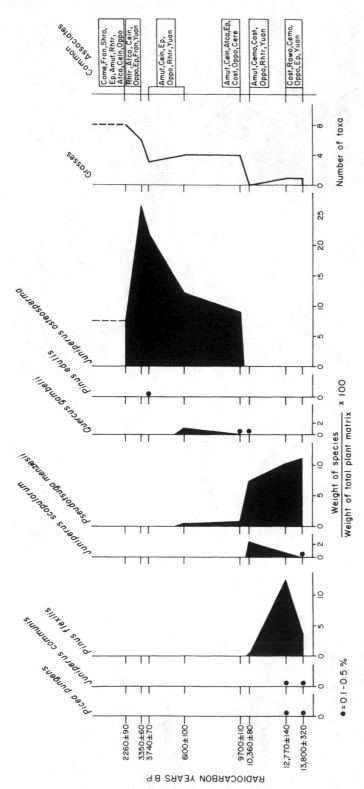

Fig. 12.7. Chronosequence of select taxa and their macrofossil abundance in middens from Fishmouth Cave. Abbreviations: Amut, *Amelanchier utahensis*; Atca, *Atriplex canescens*; Cein, *Cercocarpus intricatus*; Cemo, *Cercocarpus montanus*; Cere, *Celtis reticulata*; Come, *Cowania mexicana*; Cost, *Cornus stolonifera*; Ep, *Ephedra* sp.; Fran, *Fraxinus anomala*; Oppo, *Opuntia polyacantha*; Rhtr, *Rhus trilobata*; Rowo, *Rosa woodsii*; Shro, *Shepherdia rotundifolia*; Yuan, *Yucca angustissima*.

ramossisima, which are common elements in samples dated at 5.8, 3.8, and 2.4 ka. A single *Pinus edulis* nut in the youngest midden (2.4 ka) was probably introduced by man. This midden also contains charcoal and twine, attesting to human occupation of the shelter.

Middens also were collected from the North Fork locality in a shallow canyon at the north end of Comb Ridge. Pinyon-juniper woodland surrounds this small shelter in Navajo Sandstone. Several *Pseudotsuga menziesii* grow in a deep alcove at 1950 m elevation about 1.5 km west of the site, and a grove of *Populus tremuloides* occurs in a similar alcove at 1790 m, only 0.5 km to the east. The alcove at the North Fork locality also includes several mesophytes commonly found at higher elevations, including *Acer grandidentatum, Pachystima myrsinites, Berberis repens*, and *Ribes montigenum*. A midden dated at 3.7 ka contains abundant *Pseudotsuga menziesii* and *J. scopulorum*, plants not recorded in younger middens or presently at the site.

Cottonwood Cave (S, 1390 m) is around 15 km east of Falling Arch and some 3 km upstream of the confluence of Cottonwood Wash and the San Juan River at Bluff, Utah (Fig. 13.1). The shelter is in the Jurassic Bluff Sandstone member of the Morrison Formation and occupies the north wall of a large alcove. Principal woody perennials growing at the site are *Quercus gambelii, Celtis reticulata, Forestiera neomexicana, Shepherdia rotundifolia, Rhus trilobata, Artemisia nova, Atriplex canescens, A. confertifolia, Lycium pallidum*, and *Sarcobatus vermiculatus*. A perennial spring nearby supports dense stands of *Populus fremontii, Salix exigua, Phragmites australis*, and *Juncus* sp. *Juniperus osteosperma* does not occur within 1 km.

The Cottonwood Cave series includes five middens dated at 15.7, 12.7, 12.6, 12.1, and 6.0 ka. The full-glacial sample (15.7 ka) is dominated by *Picea pungens, Rosa* cf. *woodsii, Cornus stolonifera*, and *Betula occidentalis*. It contains several *Pinus flexilis* nuts (no needles) and a single needle of *Pseudotsuga menziesii*. The middens dated at 12.7 and 12.6 ka are nearly identical; they are dominated by *Pinus flexilis*, with lesser amounts of *Pseudotsuga menziesii, Juniperus scopulorum*, and *Picea pungens. Rosa* cf.

woodsii, Cornus stolonifera, Rubus sp., and *Betula occidentalis* also occur with *Opuntia polyacantha* and *Yucca angustissima* in the middens. The 12.1-ka midden records an increase in *Juniperus scopulorum*, the loss of *Picea pungens*, a reduction in *Rosa* cf. *woodsii*, and the presence of *Atriplex canescens*. This midden also provides the only fossil record of *Berberis fendleri* outside of Chaco Canyon and southwestern Colorado (Betancourt, Van Devender, and Rose, 1986). In Utah it is presently known only from Island in the Sky (1800 m) in Canyon Lands National Park (Welsh and Toft, 1981). The middle Holocene midden (6 ka) lacks conifers and contains *Quercus gambelii, Opuntia polyacantha, Celtis reticulata, Atriplex canescens, Ephedra* sp., *Oryzopsis hymenoides, Acer negundo*, and *Berberis fremontii*. The latter two plants no longer grow at the site.

A midden sequence also was collected from a cliff of alternating clayey sandstone and limestone overlooking the Loop (Fig. 12.8), an abandoned cutoff meander of the San Juan River about 5 km west of Comb Ridge. Leopold and Bull (1979) noted twigs and seeds of *J. osteosperma*, a tree that no longer grows in the vicinity, in what they thought were late-glacial middens along the upper cliffs. Four middens collected from the north rim (E, W, 1500 m) of the Loop were dated at 9.5, 9.4, 9.4, and 1.2 ka. The three early Holocene middens are dominated by *Juniperus osteosperma* in association with *Symphoricarpos* cf. *oreophilus, Berberis fremontii, Artemisia tridentata*-type, *Ephedra, Coleogyne ramossisima, Oryzopsis hymenoides, Opuntia*, and *Atriplex confertifolia*. Only the latter five plants still grow at the site. The midden dated at 1.2 ka is very similar to the modern vegetation, reflecting the dominance of *Ephedra* sp., *Coleogyne ramossisima, Oryzopsis hymenoides, Atriplex confertifolia*, and *Gutierrezia microcephala*.

Midden records from the San Juan country track vegetation history over the past twenty thousand years, though they are somewhat biased toward mesic alcove sites. Full-glacial assemblages from Falling Arch (19.7 ka) and Cottonwood Cave (15.7 ka) are quite different. If they are assumed to be contemporaneous, vegetation patchiness could be inferred for the full glacial, with *Picea* extending below

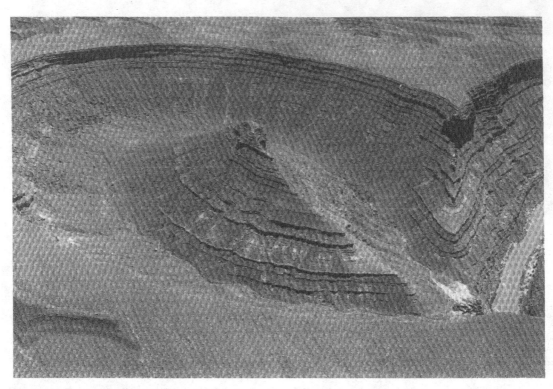

Fig. 12.8. Aerial photograph of the Loop, an abandoned cut-off meander of the San Juan River (lower right). Middens from the upper cliffs, dated at 9.5 ka, contain extralocal *Juniperus osteosperma*, *Berberis fremontii*, *Symphoricarpos* sp., and *Artemisia tri-* *dentata*-type, along with some of the modern dominants on the calcareous slopes—*Coleogyne ramossisima*, *Atriplex confertifolia*, *Ephedra*, and *Oryzopsis hymenoides*.

the lower limits of *Pinus flexilis* in more riparian settings. However, the middens are separated by four thousand years and may reflect temporal changes in vegetation. At both sites dissimilarities between full- and late-glacial floras do not have a clear climatic interpretation. At Falling Arch the full-glacial subdominance of *Atriplex canescens* and *Cercocarpus intricatus*, a xerophytic segregant of *Cercocarpus ledifolius*, contrasts with the late-glacial abundance of *Sambucus* sp., a streamside and forest understory plant. At Cottonwood Cave the decrease in *Picea pungens* and the increase in *Pinus flexilis*, *Pseudotsuga menziesii*, and *Juniperus scopulorum* indicate a marked change from full to late glacial, though the nature of possible climatic change remains unclear.

Late-glacial middens from Fishmouth Cave (13.8 and 12.8 ka), Falling Arch (13.1 ka), and Cottonwood Cave (12.7, 12.6, and 12.1 ka)

can be regarded as contemporaneous. These middens suggest that: (1) on sandstone a woodland of *Pinus flexilis* and *Juniperus scopulorum* mixed with *Pseudotsuga menziesii* extended from about 2000 m at the base of the Abajo Mountains down to 1300 m on the San Juan River; (2) *Picea pungens* and other mesophytes such as *Betula occidentalis*, *Juniperus communis*, *Cornus stolonifera*, *Rubus* sp., and *Sambucus* sp. were common in shady alcoves and riparian settings, gaining importance with increasing elevation; and (3) the presence of desertscrub or grassland on fine-grained soils is suggested from records of *Atriplex canescens*, *Ceratoides lanata*, *Yucca angustissima*, *Opuntia polyacantha*, and *Ephedra* sp., in late-glacial middens at all three sites.

Glen Canyon

The area of Glen Canyon forms one of the

largest near-continuous exposures of bedrock in North America, with vegetation hardly noticeable in satellite imagery. Seven middens were collected from a large rock shelter in Long Canyon (S, 1390 m, Fig. 12.1), which drains the extreme southwestern end of the Waterpocket Fold. Colluvial slopes at the site are vegetated by sparse desertscrub, including *Coleogyne ramossisima*, *Ephedra* spp., *Atriplex canescens*, *A. confertifolia*, *Gutierrezia* spp., and *Chrysothamnus* spp. Above 1300 m deep sands against canyon walls and alcove sites support sparse stands of *Quercus gambelii*, *Q. turbinella*, *Juniperus osteosperma*, *Fraxinus anomala*, *Amelanchier utahensis*, and *Shepherdia rotundifolia*. A few individuals of *Pinus edulis* occur at Long Canyon Cave.

The Long Canyon Cave sequence includes middens dated at 17.4, 14.9, 9.5, 7.6, 6.6, 5.2, and 2.4 ka. The oldest sample is the only full-glacial midden above 1000 m in the entire Southwest lacking conifers. Two twigs of *Juniperus osteosperma*, which lack the calcium carbonate encrustations typical of other macrofossils from this midden, are probably younger contaminants. A *Quercus* sp. acorn is in the same poor condition as other remains from the sample and may actually be of full-glacial age. The most common remains, comprising about half of the plant matrix, are shredded cambium of *Artemisia tridentata*-type. Other abundant macrofossils are *Rosa* cf. *woodsii*, *Opuntia polyacantha*, *Cornus stolonifera*, *Symphoricarpos oreophilus*, *Rhus trilobata*, *Oryzopsis hymenoides*, and *Cirsium* sp. The 14.8-ka midden contains similar taxa but also has abundant *Picea pungens* and *Betula occidentalis*. By 9.5 ka mesophytes drop out of the sequence and are replaced by *Juniperus osteosperma* and *Quercus gambelii*. *Artemisia tridentata*-type, which is missing from the middle and late Holocene samples, is common at 9.5 ka. Middens dated at 7.6, 6.6, and 5.2 ka record reduced *Juniperus osteosperma*, common *Quercus gambelii*, and the oldest records of *Fraxinus anomala*, *Celtis reticulata*, and *Atriplex canescens* from the site. The 2.5-ka sample records an apparent return to dominance of *J. osteosperma*. Like other sequences from the central and eastern Colorado Plateau, *Oryzopsis hymenoides* and *Opuntia*

polyacantha are present from the full glacial to the late Holocene. *Pinus edulis* was not found in any of the Holocene middens, even though its needles litter the cave floor today.

The 14.8-ka assemblage is replicated by macrofossils identified from a late-glacial mammoth-dung deposit at Bechan Cave (SW, 1310 m), about 5 km southeast of the Long Canyon locality. This unusual deposit, dated to 12.9–11.7 ka, contains abundant *Picea pungens*, *Betula occidentalis*, *Sambucus* sp., *Cornus stolonifera*, *Rosa* sp., *Opuntia polyacantha*, *Oryzopsis hymenoides*, and *Carex* spp. A *Quercus* acorn was found in a dung bolus, and *Quercus* wood from the site was dated to 16.5 ka (O. K. Davis et al., 1984). The absence of Pinaceae other than *Picea pungens* may mean that *Pseudotsuga menziesii* and *Pinus flexilis* reached their lower limits at around 1300–1400 m on sandstone, with *Picea pungens* extending to lower elevations along permanent streams and springs. Cowboy Cave (S, 1710 m), located in a tributary canyon east of Dirty Devil River (Fig. 12.1), yielded a similar dung deposit with abundant *Pseudotsuga menziesii*, *Picea pungens*, and *Picea engelmannii*, but no *Pinus flexilis* (Spaulding and Petersen, 1980). *Pinus edulis* needles in the dung deposit were dated by tandem accelerator mass spectrometer (TAMS) as modern (R. S. Thompson, personal communication). The only other strata to yield this pine at Cowboy Cave are of late Holocene age.

The Lower Little Colorado River

Five middens have been analyzed from two localities in the lower Little Colorado River (Van Devender and Spaulding, 1979), some 50 km upstream of where it empties into the eastern Grand Canyon opposite Cole's (this volume, chap. 11) Chuar Valley locality. An additional twenty-five middens have been collected from Permian Kaibab Limestone on the Koney Mountain Monocline at Wupatki National Monument (Cinnamon, 1988).

Three middens (Tappan Wash: 1310–1370 m) were collected on the Coconino Rim south of the canyon of the Little Colorado. Present vegetation grades from pinyon-juniper woodland above 1800 m down to desertscrub at lower elevations. A midden dated at 11.6 ka (1310 m) contains abundant *Juniperus osteosperma*, *Opuntia whipplei*, *Coleogyne* ramossisima, *Forsellesia*, and *Symphoricar-pos longiflorus*. It also has a single *Pinus edulis* nut fragment. Two other middens, dated at 1.5 and 1.0 ka (1370 m), contain abundant *J. osteosperma* even though the nearest trees are several kilometers away. Two samples were analyzed from Lee Canyon (1890 m), 10 km downstream of the Tappan Wash locality. *P. edulis* and *Artemisia tridentata* are the modern dominants, with a few *J. osteosperma* beyond 50 m of the midden site. Samples dated at 8.8–8.4 ka are dominated by *J. osteosperma*, *Yucca* sp., and *Opuntia* sp. The lack of *P. edulis* is curious considering its occurrence in early Holocene middens from Chuar Valley (1770 m, 9.1 ka), Bass Canyon (1900 m, 8.9–8.6 ka), and Bida Cave (1430 m, 10.3 ka), all within 40 km of Lee Canyon (Cole, 1981).

New data from Wupatki National Monument, also in the Little Colorado valley, bear directly on the northern limits of *Pinus edulis* during the late Wisconsin. Four middens collected from 1490–1525-m sites on Kaibab Limestone were dated at 13.7, 11.5, 11.2, and 10.8 ka. The dominant plants in these samples are *Pinus edulis* and *Juniperus scopulorum*; no other conifers are represented. An additional twenty-one middens were dated from 1.9 ka to the present, with an eight-hundred-year hiatus during intensive human occupation at Wupatki. Cinnamon (1988) suggests that this hiatus might relate to human predation on packrats, a notion also suggested to explain a critical gap in midden data from Chaco Canyon (Betancourt et al., 1983).

The midden records from the eastern Grand Canyon and the lower Little Colorado differ significantly from those of the central and eastern Colorado Plateau, primarily because of the contrast in predominant rock types (limestone versus sandstone) and the increase in the average lowest elevation of the land surface (800 m at Phantom Ranch in the Grand Canyon versus 1300 m at Bluff on the San Juan River). The Pleistocene occurrence of *Abies concolor* from 2050 to 1450 m elevation in the eastern Grand Canyon (Cole, *this volume*, chap. 11) is not duplicated from similar elevations on sandstone one hundred or more kilometers to the east; neither is the Wisconsin prevalence of *Juniperus osteosperma* between 970–2020 m in the eastern

Grand Canyon. As elsewhere on the Co Plateau, *Pinus flexilis* and *Pseudotsuga ziesii* extended down to at least 1450 m *perus scopulorum*, present only in trace amounts at the Nankoweap site, and *Pi pungens*, absent from the fossil record, ently played a lesser role in the Pleistoc vegetation of the eastern Grand Canyon elsewhere on the plateau.

Wisconsin records for *Pinus edulis* an now known for two localities in the we Grand Canyon and another site near W in the Little Colorado River valley. The patki records of *P. edulis* from limeston at 1490–1525 m, dated from 13.7 to 10 could be interpreted in two ways. The l Little Colorado valley might have been northernmost outpost of this tree durin the full and late glacials. This is suppor the full-glacial record of *P.* cf. *edulis* at of the Early Morning Light (Van Deven Spaulding, 1979). Alternatively, this ea record represents only an isolated stand *P. edulis* began migrating northward fro south of the Mogollon Rim at the end o full glacial, arriving near Wupatki by at 13.7 ka. In either case there remains co able contrast between *Pinus edulis* at 1 1525 m elevation at Wupatki and assem dominated by *Pinus flexilis*, *Abies con Pseudotsuga menziesii*, and *Picea pung* down to 1390 m only 100–200 km to t north and northwest. These low-elevat mixed-conifer communities persisted u about 10 ka. *Pinus edulis* is not recorde the eastern Grand Canyon until 10.3 k *this volume*, chap. 11). Apparently, the tion of pinyon northward from the Wu area was relatively slow. Its absence fr early Holocene middens at Lee Canyon (1890 m), where it now grows, suggests the expansion of *P. edulis* northward w not uniform.

DISCUSSION

The midden record from the Colorado allows a few generalizations and a bit o speculation concerning various aspects Quaternary biogeography. The influenc rock type on past and modern vegetatio either a local or regional scale, can har ignored. Tentative climatic inferences drawn from vertical displacements of p the conspicuous absence of modern taxa in the late glacial, and the sequence of departures and arrivals depicted in midden chronologies. The midden record also is discussed as evidence for the individualistic nature of plant associations. Special reference is made to the relative size and habitat heterogeneity of geographic ranges in producing associations and disassociations incongruent with modern vegetation (disequilibrial assemblages). Finally, I discuss how historical factors might affect behavior of isolated mountains as ecological islands.

Effect of Rock Type

The edaphic diversity of the Colorado Plateau may account for the high degree of plant endemism (Welsh, 1978; Reveal, 1979) and helps explain many of the systematic departures from classic Rocky Mountain plant zonation (Daubenmire, 1943). Restriction of Pleistocene middens to certain rock types (sandstone and limestone) limits inferences to those substrates. Three aspects of geological control on vegetation merit consideration: the vertical limits of plants, disruption or total exclusion of vegetation zones from unfavorable substrates, and the persistence of edaphically specialized plants despite dramatic climatic change. Both lower and upper limits tend to be higher on limestone than on volcanic or metamorphic rocks. On the Colorado Plateau pinyon-juniper woodland reaches its lowest elevation on mesic sandstone substrates.

In general, the Pleistocene and modern vegetation of the plateau appear "wetter" than similar elevations farther west. In the Wisconsin *Picea* spp. occurred at relatively low elevations on the plateau but are missing from limestone sites west of longitude 112° W (Fig. 12.9). Pleistocene conifer diversity decreased to the west, just as it does today. In addition, the Wisconsin upper limits of *Juniperus osteosperma* were much higher to the west. To what extent can these dichotomies be attributed to the increasing westward influence of the Sierra Nevada rain shadow and decreasing influence of moist, tropical air, or simply to the transition from predominant sandstone on the Colorado Plateau to carbonate and igneous rocks of the Great Basin?

Today spruce-fir forests occur above 2900 m, mostly on igneous rocks on the summits of laccolithic ranges, volcanic peaks, basalt-capped plateaus, or on limestone (e.g., the Kaibab Plateau). Where sandstone occurs at high elevations (e.g., Navajo Mountain), these forests are depressed another 200–300 m. The lower elevational range of spruce-fir forest during the Wisconsin would have placed the forest predominantly on substrates of sedimentary rock, with the lowest limits achieved on sandstone. Because conifers are generally intolerant of heavy clay soils, spruce-fir expansion was probably forestalled on shale.

At high elevations where other substrates normally support mixed-conifer and *Pinus ponderosa* forests, shales commonly support tussock grasslands, *Artemisia* shrublands (particularly *Artemisia arbuscula* var. *longiloba*), extensive clonal stands of *Populus tremuloides*, and *Quercus gambelii* scrub. It is an interesting coincidence that the dominant plants on high-elevation shales are rhyzomatous grasses and deciduous trees capable of asexual reproduction. There are several selective advantages to asexual reproduction on an unfavorable substrate such as shale. For example, a single plant may colonize a relatively large area, and temperature and moisture requirements for seed production, germination, and seedling establishment need not be met. Regeneration from mature root systems—and the fast initial growth that follows—is an effective way of avoiding drought and catastrophic freezes at the seedling stage. Ramet populations have a low mortality rate and a greater life expectancy (Deevey type II curve). They can expand and contract depending on current climatic conditions while maintaining certain gene combinations (Abrahamson, 1980). Extensive clonal stands of *Populus tremuloides* and *Quercus gambelii*, perhaps several millennia old, are common throughout Utah (Cottam, 1954) and are widespread on shale at elevations commonly occupied by conifers.

The zonational anomalies created by shaly substrates are best illustrated in the La Sal Mountains (Fig. 12.10). This range, with altitudes up to 3880 m, includes three Tertiary stock and laccolithic complexes intruded into Mesozoic sedimentary rocks. The outward-dipping host rocks are the clay-rich Jurassic Morrison Formation and Cretaceous Mancos

12.9A

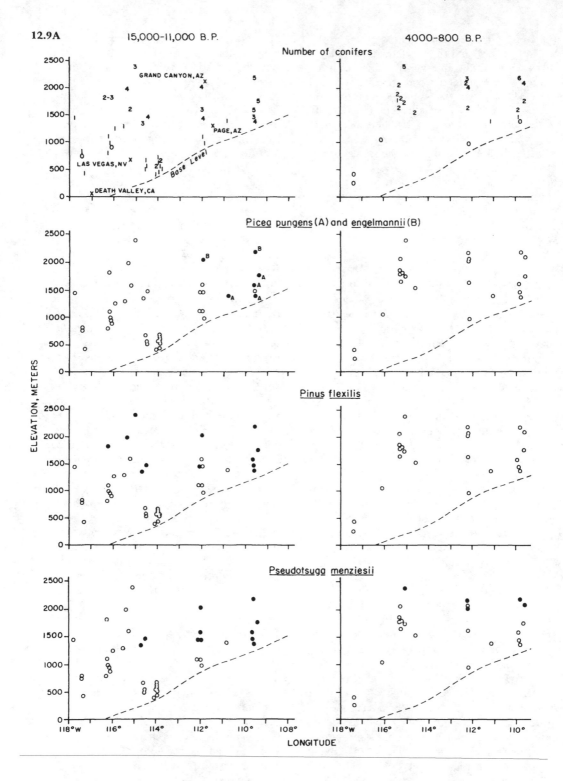

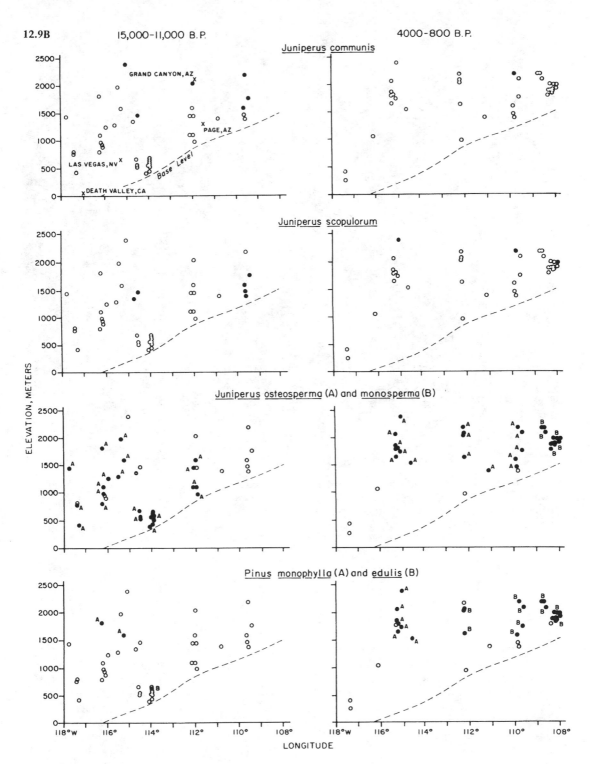

Fig. 12.9. Elevational and longitudinal distribution of select conifers in late-glacial (A) and late Holocene (B) middens between 35° and 38° N. All middens of relevant age are plotted to note both presence (closed circles) and absence (open circles) of particular taxa.

Fig. 12.10. Aerial photograph showing distribution of vegetation zones on the west slope of the La Sal Mountains. AT, alpine tundra; SF, spruce-fir; AM, *Populus tremuloides* meadows; PP, *Pinus ponderosa*; OM, *Quercus gambelii–Cercocarpus*; PJ, *Pinus edulis–Juniperus osteosperma*; SB, *Arte-* *misia tridentata*. The broad bands occupied by AM and OM coincide with continuous outcrops of Mancos Shale, while at the same elevations PP is restricted to outcrops of Dakota Sandstone (from Richmond, 1962).

Shale, which outcrop between 3000–2200 m. Mixed-conifer and *Pinus ponderosa* forests are poorly devclopcd on thcse shales, replaced instead by an upper belt of *Populus tremuloides* and meadows and a lower zone of *Quercus gambelii–Cercocarpus* spp., just above pinyon-juniper woodland. Stands of *Pinus ponderosa* are restricted to outcrops of Navajo Sandstone to the north and Cretaceous Dakota Sandstone to the east and south. Its absence on the west slope, attributed to a rain shadow effect by Richmond (1962), is more likely due to the preponderance of shale and lack of sandstone at critical elevations. In this region rain shadows are more pronounced on the east than on the west side of highlands. The east slope of the nearby San Juan Mountains receives about one-third the annual rainfall of the west slope. During the Wisconsin the upper range of spruce-fir was probably truncated at the sandstone-shale boundary, where it was replaced by expanded alpine tundra, *Artemisia*, and bunchgrass communities. Most of the La Sal Mountains were probably treeless. Present clonal stands of *Populus tremuloides* and *Quercus gambelii* on shales in this and other similar ranges were probably established by seed perhaps as early as 10 ka. An interesting hypothesis is that *P. tremuloides* and *Q. gambelii* populations should exhibit less genetic variability on shale than other substrates on the plateau.

Shales also may have played an important role in the development of a regional flora rich in endemics and in the presence of halophytic desertscrub during the Wisconsin. Welsh (1978) tallied 340 endemics from the plateau and segregated them according to elevational range and substrate. Endemics restricted to shale or clayey soils constitute about 25% of the total, and endemics that grow above 1980 m contribute 26%. If we take into account relative area, shales and higher elevations contribute a higher incidence of endemics than other substrates and lower elevations (Welsh, 1978). Shales occur throughout a broad elevational range and are not restricted geographically. Endemics with shale specialization may have adapted to changing climates by simply shifting their elevational ranges.

The absence of *Coleogyne ramossisima* on the western foothills of the Henry Mountains may be due to the predominance of shale at critical elevations (Everitt, 1970). Near Thompson, Utah, an abrupt boundary between communities dominated by *Coleogyne* and *Ephedra* on Jurassic sandstones and *Atriplex corrugata* on Mancos Shale stretches over 100 km, defined by the geologic contact (Cannon, 1960). Common desertscrub taxa on saline shales (salt concentrations of as much as 30,000 ppm are common) and other alkaline soils include *Atriplex corrugata, A. confertifolia, Ceratoides lanata, Grayia spinosa, Sarcobatus vermiculatus, Suaeda torreyana,* and *Allenrolfea occidentalis.* Most of these halophytes occur as far north as Canada and obviously can tolerate extreme cold. Salts accumulated in these plants act as an "antifreeze" in colder climates. The prevalence of *Sarcobatus* and other Chenopodiaceae pollen in Wisconsin profiles throughout the region (e.g., Maher, 1961) could mean that these plants were widespread, particularly on shales and greatly expanded alkaline flats enhanced by higher and seasonally fluctuating water tables. As yet, we cannot rule out Wisconsin desertscrub communities on the plateau, particularly those dominated by halophytic shrubs.

Late Glacial Biogeographic Gradients

Midden data from 2200–1000-m sites allow some generalizations about glacial-age plant zonation on the Colorado Plateau (Fig. 12.11). Given the nature of the midden record, the reconstructions are most pertinent to cliffsides and coarse, shallow soils. They are least relevant to open topographic settings with deeper soils; that is, sites where middens are not preserved. Upper treeline elevations are imprecisely known for the period between 15 and 11 ka. Wright et al. (1973) suggested that the upper treeline was at around 2600 m in the Chuska Mountains, implying a 900-m lowering below present timberline. However, at Alkali Basin (2800 m) in the Crested Butte area, *Pinus* and *Picea* pollen percentages of 50% and 30%, respectively, indicate a much higher treeline by 13 ka (Markgraf and Scott, 1981). For the purpose of this discussion I have adopted a late-glacial (15–11 ka) upper treeline of 2900 ± 100 m.

The relative importance of Rocky Mountain alpine plants in late-glacial shrub and herb communities above treeline cannot be ascertained. Alpine plants on the Colorado Plateau now occur on the only three mountaintops with significant areas above 3660 m: the Tushar Plateau, the San Francisco Peaks, and the La Sal Mountains. A vicariant explanation for this distribution seems doubtful unless these plants migrated as understory elements in continuous boreal forests or during a previous glacial period colder than the Wisconsin. Many of the plants in question are poor dispersers, so it is unlikely that they ever achieved a uniform distribution across treeless communities on the plateau. This must have created an open niche for *Artemisia tridentata* and its close relatives, as well as tussock grasses. High *Artemisia* pollen percentages ubiquitous in Wisconsin sediments throughout the western United States may originate not only from an expanded *Artemisia* steppe at lower elevations but also from its dominance in extensive areas above treeline.

Late-glacial boreal forests dominated by *Picea engelmannii* and *Abies lasiocarpa* extended down to 2000 m (about 900 m lower than today) in mesic canyon habitats. No macrofossil records of *Pinus longaeva, P. aristata,* or *P. contorta* are known for the Colorado Plateau. Their pollen occurs, however, in pollen sequences from the Chuska Mountains (Wright et al., 1973) and the White Mountains of Arizona (Jacobs, 1985). The

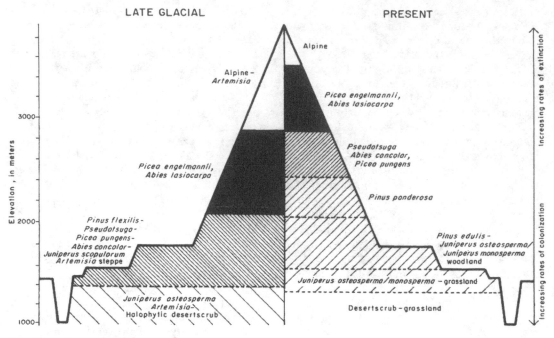

Fig. 12.11. Generalized late-glacial and present plant zonation on the Colorado Plateau.

nearest macrofossil records that allow some estimation of the Wisconsin lower limits of bristlecone pine are from the Confusion Range of west-central Utah, where they occurred as low as 1660 m (Wells, 1983). *Pinus longaeva* and *P. aristata* presently have spotty distributions on the plateau, with *P. longaeva* growing as low as 2400 m on an isolated patch of sandstone in the Mammoth Creek area east of Cedar Breaks, Utah. This tree was probably more common in the late glacial but, unlike in the Great Basin, assumed only a secondary role in subalpine forests and woodlands.

On rocky sites conifer woodlands and forests, dominated by *Pinus flexilis, Pseudotsuga menziesii, Juniperus communis, J. scopulorum,* and *Picea pungens,* extended from about 2000 to 1300 m in the San Juan country and Glen Canyon. These species probably had expanded distributions in this elevational range on sandstone outcrops throughout the plateau. The lower limit of these communities appears to have been 150–200 m higher on limestone in the eastern Grand Canyon, where *Abies concolor* was an important component of late-glacial vege-

tation. Boreal and montane elements occurred down to the lowest elevations and were integrated regionwide by riparian corridors in tributaries of the San Juan and Colorado rivers. Thus, Welsh and Toft's (1981) hypothesis that boreal elements present in low-elevation alcoves are the result of long-distance dispersal seems inconsistent with the midden data.

The absence of *Pinus edulis* and *P. ponderosa* in Wisconsin middens, even in habitat types where they are now common, is a curious pattern unanticipated prior to the advent of midden analysis (Martin and Mehringer, 1965; Wright et al., 1973). The absence of *P. edulis* may be due to a Wisconsin upper limit that coincided with relatively high minimum elevations in the region. Middens from south-facing limestone ridges in southern New Mexico suggest an upper limit between 1550 and 1705 m (Van Devender et al., 1984; Betancourt, 1987). The higher latitude of the plateau may have excluded *Pinus edulis,* though it was present by at least 16.6 ka in the western Grand Canyon and 13.7 ka in the Wupatki area north of Flagstaff. The more xerophytic *Juniperus osteosperma* survived this effect because of its greater cold tolerance.

This juniper presently extends several hundred kilometers beyond the northern limits of *P. edulis* or *P. monophylla*. Its seeds germinate in the spring rather than mid-to-late summer as those of pinyons do. This allows *J. osteosperma* seedlings to become well established by the end of the growing season and prior to autumn frosts (Meagher, 1943).

The absence of *Pinus ponderosa*, presently the most widespread pine in North America, is not easily explained. *P. ponderosa* var. *scopulorum* is the most likely variety to have had a late-glacial distribution on the Colorado Plateau. Whatever limits its success over most of Nevada, southern Idaho, and western Wyoming—a vast and conspicuous hiatus in its distribution—may also apply to an explanation of its late-glacial absence on the plateau. Competitive release could explain the expansion of *Pinus flexilis* at elevations where *P. ponderosa* might be expected from an equal and orderly displacement of modern vegetation zones. In Idaho, Montana, and Wyoming *P. flexilis* today is important at middle-to-low elevations wherever *P. ponderosa* is missing (Peet, 1978). In the Wisconsin, areas presently in *P. edulis–J. osteosperma* (or *J. monosperma*) woodlands may have supported a physiognomic equivalent dominated instead by *P. flexilis* and *J. scopulorum*. Although the players were quite different, the vegetation may have looked much the same. A suitable analog may be the woodland now confined to prairie margins in Montana (Fig. 12.12). In the Wisconsin *P. flexilis* may have expanded as far south as the submogollon highlands.

Because of the relatively high minimum elevations in the region, the zonal sequence from the eastern Colorado Plateau "bottomed out" near the lower limits for *Pseudotsuga menziesii*, *Pinus flexilis*, and *Juniperus scopulorum*. Xerophytic woodlands dominated by *Juniperus osteosperma*, documented between 1500 and 900 m in the eastern Grand Canyon (Cole, *this volume*, chap. 11), had a restricted distribution elsewhere on the plateau. Perhaps they occurred no farther east than the limestone cliffs in the Cataract and San Juan River canyons.

Holocene Vegetation Dynamics

Site-specific vegetation dynamics can be conveniently summarized in terms of departures and arrivals at midden sites (e.g., Cole, 1985). The relative success of this approach lies in repetition and comparable temporal resolution between midden sequences. Both the nature and timing of turnovers may be unique to a given site. For example, the timing of extinctions for Pleistocene relicts may be a function of elevation and other site characteristics. *Pseudotsuga menziesii* became locally extinct between 13 and 9.2 ka at Falling Arch, while it persisted until after 6.1 ka at Fishmouth Cave, only 100 m higher and 5 km to the north.

Cole's (1985) claim that departures and arrivals of individual species were not simultaneous, in support of vegetational inertia, applies in some cases but not in others. Species richness, as recorded in the middens, appears to have been lowest during the Pleistocene-Holocene transition, most likely because colonizations lagged behind extirpations. In some cases, however, extirpations were delayed until arrival of a potential competitor. For example, *Pinus edulis* vacated desert lowlands to the south and presumably began migrating upward and northward by 11 ka (Betancourt, 1987). *Pinus flexilis* persisted at Chaco Canyon (1910 m) and Fishmouth Cave (1585 m) until 10 ka, when *P. edulis* is first recorded north of Wupatki. While hardly conclusive, the data suggest that migrational lag may have delayed competitive interactions and thus forestalled some extirpations for several centuries, if not millennia. However, migrational lag cannot explain the failure of *Juniperus osteosperma* to replace montane conifers at Fishmouth Cave and other sites until after 10 ka. Late-glacial stands of *J. osteosperma* were certainly within 100 km and potentially within 10 km of these sites. The early presence of *Pinus edulis* at Wupatki on the southern edge of the Colorado Plateau also implies unfavorable climates during the late-glacial transition rather than migrational lag as the reason for its delayed arrival at sites only 100–200 km to the north and northwest.

Arrival times are dependent on migration rates and distances for the various plants now dominant in the vegetation. Postglacial migrations could have been rapid, like the historic spread of *Bromus tectorum* in western

Fig. 12.12. During the Wisconsin, woodlands of *Pinus flexilis–Juniperus scopulorum*, such as the one shown here near Choteau, Montana, probably covered areas similar to modern *P. edulis–J. osteo-* *sperma* woodlands on the Colorado Plateau. (Photo from *Plant geography: with special reference to North America* [Academic Press, 1978] with permission of the author, R. F. Daubenmire.)

North America (Mack, 1981), or sluggish, as with *Castanea* in the Holocene of New England (M. B. Davis, 1986). Neither the present density nor temporal coverage of midden sites on the plateau is sufficient to resolve questions about the relative pace of immigration. However, Holocene migrations apparently occurred both by expansion along continuous corridors and through long-distance dispersal. For example, *Coleogyne ramossisima, Pinus edulis, Juniperus osteosperma,* and *J. mono-sperma* could easily have migrated unimpeded by barriers to fill their present distributions. *Abies concolor* and *Pinus ponderosa,* on the other hand, probably island-hopped across moderate distances to reach isolated ranges (Wells, 1983). The probable hit-and-miss quality of such Holocene dispersals may explain anomalous distributional patterns such as the conspicuous absence of *Abies*

concolor from the Chuska Mountains and Mount Taylor in northwestern New Mexico.

While some plants only recently achieved their maximum distributions, the ranges of others may have expanded and contracted during the Holocene. Evidence from Allen Canyon Cave, Chaco Canyon, and the eastern Grand Canyon suggests that *Pinus ponderosa* and *Quercus gambelii* were more widespread at low elevations during the early to middle Holocene, only to contract their ranges in the late Holocene. In western Utah hybrids of *Q. gambelii × turbinella* presently occur up to 400 km north of the nearest stands of *Q. turbinella.* Cottam et al. (1959) argued that middle Holocene warmth allowed expansion of *Q. turbinella* to northern Utah, a connection severed by cooler temperatures during the late Holocene. Harper et al. (1985) counter that the hybrids may have resulted from long-range pollination. Resolution of this debate ultimately rests on collection of middens on a transect along the presumed path of dispersal.

Climatic Inferences

Most paleoclimatic reconstructions from ecological data in the southwest are based on the amount of elevational displacement from modern plant ranges figured into regressions of current temperature and precipitation with elevation. Such gradients, however, should not be treated as constant. Many microclimatic factors, such as orographic effects, are modified by seasonality of rainfall and character of the prevailing storm types. These were undoubtedly inconstant across the Pleistocene-Holocene boundary.

The lower limits of plants are presumably set by deficiencies in effective moisture. Depression of lower limits can be accomplished by cooler growing-season temperatures and by greater rainfall at critical times in the plant's annual cycle. Except at upper treeline, competitive interactions seem to play an important role in determining upper limits of key taxa. Many plants achieve their maximum densities at their upper limits, where they come into contact with other species better adapted to climates at higher elevations (e.g., Yeaton et al., 1980). In spite of these complications, tentative climatic inferences can be drawn from midden sequences on the plateau. Full- and late-glacial midden floras from the Colorado Plateau show significant differences coincident with deglaciation of Rocky Mountain ice fields by 14 ka. This deglaciation could have been accomplished by a rise of 250–500 m in equilibrium-line altitude, only 25%–50% of the rise that occurred at the end of the late glacial (Porter et al., 1983). The contrast between full- and late-glacial midden floras varies from site to site. In Long Canyon the full-glacial assemblage (17.4 ka) lacks conifers, while the late-glacial (14.9 ka) assemblage is dominated by *Picea pungens*. At Cottonwood Cave dominance shifts from *Picea pungens* and *Betula occidentalis* (15.7 ka) to *Pinus flexilis*, *Pseudotsuga menziesii*, *Juniperus scopulorum*, and *Opuntia polyacantha* (12.7 ka). At Falling Arch a relatively xerophytic assemblage dominated by *P. flexilis*, *Cercocarpus intricatus*, and *Atriplex canescens* (19.7 ka) is re-placed by one in which *Pseudotsuga menziesii*, *Juniperus scopulorum*, and *Sambucus* sp. (13.1 ka) are dominant. Depending on the site, either an increase or a decrease in effective moisture could be inferred. The only safe conclusion at this time is that local communities were unstable from the full glacial to the late glacial due to an unspecified change in climate. Other lines of evidence are equally ambiguous. The rise in equilibrium-line altitude suggests warmer ablation seasons in the late glacial, but could also result from less precipitation (Porter et al., 1983).

Two inferences can be made about full- and late-glacial climates. Wisconsin summers must have been much cooler and drier than those of today. During the full- and late-glacial periods, *Picea pungens* grew down to at least 1390 m (based on midden evidence) and to 1310 m if macrofossils from the Bechan Cave dung deposit originated locally. In the plateau its average lower limit along permanent streams is now about 2380 m. Because they are watered by stream flow, riparian stands of *Picea pungens* are curtailed at their lower limit not by precipitation but by warm temperatures during the growing season. This is demonstrated by the elevational range in which horticultural plantings of *P. pungens* can be maintained by irrigation—well below its present limits but above warm, desert elevations. During the Wisconsin this tree occurred at least 1070 m lower than it does today. Its glacial-age lower limit presumably lay along permanent streams and spring-fed alcoves now ephemeral or intermittent. The displacement can be expressed in climatic terms by multiplying the elevational difference between present and inferred past lower limits by the lapse rate for mean growing-season temperatures. The vertical gradient in mean temperatures from April to September, calculated from eighteen stations in southeastern Utah, is 0.59° C/100 m (Betancourt, 1984). Thus, warm-season temperatures at least 6.3°C cooler than today are indicated.

The second inference about Wisconsin climates, based on the absence of *Pinus ponderosa* and the unimportance in midden floras of summer-flowering herbs and grasses, is that summers were relatively dry. *Pinus ponderosa* var. *scopulorum* is now absent from areas with dry summers like the Great

Basin. It reaches its westernmost stations in the calcareous ranges of eastern and southern Nevada, just east of stations where summer precipitation (July–August) contributes less than 10% of mean annual rainfall. By comparison, summer rainfall over most of the plateau accounts for 20%–30% of the total. The absence of *Pinus ponderosa* from the Colorado Plateau might imply a reduction in summer rainfall from 20%–30% of the present annual average to less than 10% during the Wisconsin. Coincident with the westward decrease in summer rainfall is a lesser frequency of cloud-to-ground lightning strikes (Reap, 1986). Reports of presettlement fire intervals indicate a range from almost yearly in northern Arizona (Dieterich, 1980) to eighteen years in the Sierra Nevada (Kilgore and Taylor, 1979). The ecology of *Pinus ponderosa* has been linked to fire by many authors. For example, the lack of regeneration in the Southwest over the past fifty years or more has been blamed on fire suppression. The winter-wet–summer-dry region of the late Wisconsin, much like the present climate of the western Great Basin, was conducive to less-frequent fires through a reduction of thunderstorm activity and lightning strikes. It may be that changes in fire regimes best explain the comings and goings of *Pinus ponderosa* on the Colorado Plateau.

Cooler summers and a shifting of maximum warmth to earlier in the annual life cycle also may help to explain the glacial-age absence of *Pinus ponderosa* on the plateau. In the San Francisco Mountains its period of active growth lasts from May 15 to September 20, as much as a month earlier, and several weeks later, than other montane conifers monitored by Pearson (1931). The seeds of *P. ponderosa* also mature later, around October 1–20. A shift in the march of temperatures across the growing season could have a critical effect on successful completion of its life cycle.

It is known from orbital geometry, specifically the twenty-two-thousand-year cycle of precession, that the monthly insolation pattern shifts on a millennial scale (Berger et al., 1980). Many times during the Quaternary an insolation maximum in May shifted progressively toward July–August as an insolation minimum in February moved toward April.

The early-summer (May–June) maximum occurred during the full- to late-glacial periods, the midsummer (July–August) maximum in the early Holocene, and the late-summer–early-fall (September–October) maximum in the middle Holocene (Berger et al., 1980). The effect of this shift on the summer march of temperatures for a particular latitude has yet to be modeled, but conceivably it could shift temperature regimes by several days to a few weeks. During the late Wisconsin relatively cool temperatures associated with lower insolation in late summer and fall could have prohibited successful completion of the annual cycle for *Pinus ponderosa*. In particular, temperatures at the time of seed maturation in October might have been too cold for successful germination.

The relative importance of vegetational inertia versus climatic change in explaining the pattern of local extirpations and immigration across the Pleistocene-Holocene transition is uncertain. Extinction forced by competition, if it occurred, could have been forestalled by slow migration of woodland dominants (*Pinus ponderosa* and *P. edulis*) from Pleistocene refugia far to the south. However, *Juniperus osteosperma*, which occurred on the plateau, failed to expand into its present elevational range until after 10 ka. The turnover at about 10 ka may reflect abrupt warming, which is supported by *Picea* wood recovered from a site above the present timberline in the San Juan Mountains, with the oldest fragments dated to 9.6 ka (Carrara et al., 1984).

Between 10 and 6 ka *Pinus ponderosa* and *Quercus gambelii* expanded beyond their present distributions, while some montane conifers persisted at sites where they no longer occur. Even during the latter part of this period, commonly associated with hot-dry conditions (the altithermal, hypsithermal, or climatic optimum), it appears to have been wetter than today on the Colorado Plateau. Lowered elevational ranges for montane and transitional forest elements on the plateau between 10 and 6 ka agree with evidence for spruce-fir forest 100 m lower than today in central Colorado (Markgraf and Scott, 1981; Fall, 1985) and northern Utah (Madsen and Currey, 1979); now-riparian trees growing on exposed, rocky slopes in the Sonoran Desert (Van Devender, *this volume*, chap. 8); and relatively high lake stands in the San Agustin Plains and Estancia Basin of New Mexico (Markgraf et al., 1984). The most likely explanation is more copious subtropical moisture than today, either from an enhanced Bermuda High in the Atlantic or more frequent tropical storms in the Pacific Ocean.

The Individualistic Nature of Plant Associations

Because of their local origin and ease of identification, midden floras offer a unique opportunity to demonstrate the individualistic responses of plants to a changing climate. Midden records from various regions demonstrate unequal displacement of species along elevation gradients during the Wisconsin and reveal the ephemeral nature of current zonational patterns (Spaulding, 1981; Cole, 1981; Wells, 1983; Van Devender, 1986). Midden floras provide several examples of late-glacial combinations of plants that presently are allopatric and at opposite extremes of environmental gradients. Such is the case for *Pinus longaeva* and *Atriplex confertifolia* in the Sheep Range of southern Nevada (Spaulding, *this volume*, chap. 9). Examples from the Colorado Plateau include *Picea pungens* and *Juniperus communis* with *Ephedra* spp., *Yucca* spp., *Opuntia* spp., and *Atriplex canescens*. Some of the anomalous midden floras have withstood tests of contemporaneity through direct dating; others have not (Van Devender et al., 1985).

Disequilibrial assemblages could have their origins in contracting and expanding geographic ranges across the Pleistocene-Holocene boundary. For a particular taxon, an expanding geographic range should mean an increase in the number of co-occurring species due to species-area effects. A contracting range should have the opposite effect. During the late Wisconsin, *Pinus remota*, presently relictual in west Texas and northeastern Mexico, was the widespread pinyon at low elevations in the Chihuahuan Desert. *Pinus edulis* had a rather limited range in the northern Chihuahuan Desert and on the southern edge of the Colorado Plateau. *Pinus remota* and *Quercus hinckleyi*, common associates in Wisconsin woodlands of the Chihuahuan Desert, now occupy quite separate relictual ranges (Van Devender, 1987; Wells, 1987).

Pinus edulis obviously did not associate with endemic floras of the Colorado Plateau or with many of the northern species now common in pinyon-juniper woodlands.

The increase in area with lowering of vegetation zones also could explain many of the anomalous associations found in midden floras. For instance, *Pinus flexilis* once covered far more extensive areas than it does now in the Great Basin and Colorado Plateau. Potentially, it could have co-occurred with a greater number of species than it does today. On the Colorado Plateau the increase in environmental heterogeneity encompassed by this plant's range was accentuated by expansion to the basal elevations, in this case the canyon country. During the last glacial age montane plants commonly grew in topographic settings they now rarely encounter. Climbing dunes against a massive, sandstone cliff probably supported an aggregate of species that would seem anomalous by today's standards.

Mountains as Islands: Historical Considerations

A dominant theme in the biogeography of the western United States involves the relative contributions of the Sierra Nevada and Rocky Mountain "mainlands" in populating montane "islands" of the intermountain region. More recently, the notion of isolated mountaintops as ecological islands has encouraged various applications of the theory of island biogeography (MacArthur and Wilson, 1967). In its simplest form the theory states that the number of species on an island represents a dynamic equilibrium between contemporary rates of extinction (governed by island area) and colonization (as affected by distance from the mainland).

The theory's relevance to explaining differential stocking of isolated mountaintops in the Great Basin has been examined for birds (Johnson, 1975; Behle, 1978), flightless mammals (Brown, 1971, 1978), and plants (Harper et al., 1978; Wells, 1983). These exercises have been complicated by: the interactions of area, habitat heterogeneity, and maximum elevation of the islands; whether or not the assumed mainland is the actual source of the colonizations; the lack of environmental uniformity along the transects defining mainland-island distances; and the role of climatic change in determining the nature and rates of extinction and colonization.

Wells (1983) and Thompson and Mead (1982) have offered examples of how paleo-ecological knowledge contributes to island biogeography in the Great Basin. Observations here are limited to two aspects of the problem that merit attention: the lack of contemporaneity between rates of extinction and colonization, and misinterpretation of the Rocky Mountain cordillera as the main source of propagules. Extinctions and colonizations on insular mountains occur on time scales similar to those of the climatic changes that, in part, determine composition of the biota. This influence of time should be evident in differing rates of extinction and colonization with elevation. Since the end of the Pleistocene, rates of extinction should have been greatest at the highest elevations, where habitat areas are smallest and the biota is relictual. Because of their great isolation these relictual populations can no longer be rescued by immigration from afar.

For flightless mammals Brown (1971, 1978) found that the steepest species-area slopes were obtained by restricting the sample to progressively higher elevations. He concluded that insular mammals were derived from a common set of Pleistocene colonists by subsequent extinction in the absence of colonization. It should follow that rates of colonization are greatest for low elevations at the leading edge of Holocene immigrations, where rates of extinction also are dampened by larger habitat areas. Historical factors rather than stochastic and contemporary rates of extinction and colonization may largely explain the species-area slopes for biota of insular mountains. It is unlikely that pulses of colonization have kept exact pace with extinction to maintain dynamic equilibrium throughout the Holocene.

The second qualification for mountains as ecological islands involves the history of the assumed source or mainland as it affects the interaction between diversity and isolation. During the late Wisconsin spruce-fir forest extended from the upper treeline at about 2900 ± 100 m to the base of the Rocky Mountains (35°–39° N) at about 2000 m. If we perceive the Rockies as a steep-sided, high plateau, spruce-fir forest probably covered less area in the late Wisconsin than today and was restricted to steep slopes and deep valleys leading down to the basal plains. On the west slope of the Rockies between Crested Butte and Durango, Colorado, shale is the predominant rock type that outcrops between 3000–2000 m, and the boreal forest would have thrived only in the most favorable sites. Therefore, during the late Wisconsin the Rocky Mountain mainland was biologically impoverished, save for greatly expanded treeless communities above the boreal forest and below the ice. The greatest pool of montane plants and animals existed in the lowlands and isolated highlands of the Colorado Plateau (including the Wasatch Mountains) and southern Colorado Rockies. The montane biota of the Rockies is presently more diverse than that of the isolated highlands, in part because it has suffered fewer extinctions at high elevations while receiving the full complement of Holocene immigrants at low-to-intermediate elevations.

SUMMARY

More than 180 middens have been dated and analyzed from outcrops of Paleozoic limestone and Mesozoic sandstone on the Colorado Plateau. The sites' elevational range (2200–900 m) allows reconstruction of late Wisconsin and Holocene plant zonation on rocky sites. Aside from greatly lowered vertical distributions for individual species, the contrast between past and modern zonation is due to southerly displacement of *Pinus ponderosa* and *P. edulis* during the late Wisconsin. In their absence *Pinus flexilis* expanded from its current discontinuous subalpine distribution to dominance in communities in roughly the same elevational range as modern pinyon-juniper woodlands. During the late Wisconsin montane communities with *Pinus flexilis, Pseudotsuga menziesii, Picea pungens, Abies concolor, Juniperus communis,* and *J. scopulorum* grew directly above *J. osteosperma* woodland and probably desert-scrub. This pattern repeated itself westward across the southern Great Basin (Thompson, *this volume,* chap. 10) and northern Mojave Desert (Spaulding, *this volume,* chap. 9).

Upstream of Glen Canyon perennial surface flow and cool summer temperatures supported riparian and montane plants down to the lowest elevations. Current disjunctions in well-watered alcoves resulted not from long-distance dispersal but from disintegration of a montane flora that was formerly more widespread.

The number of plants co-occurring with montane taxa during the Wisconsin probably was greater owing to their expanded geographic range. This may have produced associations incongruent with modern experience. Anomalous assemblages also could be explained by expansion of boreal elements from mountaintops and plateau summits, where igneous rocks are most common, to different substrates and the more varied topography of the canyon country.

Historical factors may explain many of the insular patterns of the intermountain region, including the Colorado Plateau. Ecological extinctions and colonizations on montane islands happened on similar time scales to major climatic changes. Rates of Holocene extinctions have been greatest on mountaintops, which serve as refugia for Pleistocene relicts. The highest rates of colonization are in the piedmonts, at the leading edge of immigrations. Except for truly alpine species, the main cordillera of the Rocky Mountains probably played only a minor role as the mainland source for colonizations by long-distance dispersal. Plant distribution patterns on the insular mountains are probably independent of distance to the Rockies. Although there is a perceived relation between species richness and distance to the Rockies, this pattern may not have developed in the manner specified by island biogeography theory.

ACKNOWLEDGMENTS

Much of the work summarized in this chapter was supported by the National Geographic Society, the National Park Service, the Department of Energy, and the U.S. Geological Survey. I especially wish to thank Austin Long, Laboratory of Isotope Geochemistry, University of Arizona, for providing radiocarbon support during lulls in funding. Cathy Barnoski, K. L. Cole, P. S. Martin, W. G. Spaulding, R. S. Thompson, R. M. Turner, and T. R. Van Devender gave valuable insights and comments on the manuscript. W. B. Gillespie accompanied me on most trips to the Colorado Plateau; it has become difficult to tell where his ideas end and mine begin.

REFERENCES

Abrahamson, W. G. (1980). "Demography and vegetative reproduction." In *Demography and evolution in plant populations*. O. T. Solbrig, ed., pp. 89–106. University of California Botanical Monographs 15.

Andrews, J. T.; Carrara, P. E.; King, F. B.; and Stuckenrath, R. (1975). "Holocene environmental changes in the alpine zone, northern San Juan Mountains, Colorado: Evidence from bog stratigraphy and palynology." *Quaternary Research* 5, 173–97.

Batchelder, G. L., and Merrill, R. K. (1976). "Late Quaternary environmental interpretations from palynological data, White Mountains, Arizona." *American Quaternary Association Abstracts*, p. 125.

Behle, W. H. (1978). "Avian biogeography of the Great Basin and intermountain region." *Great Basin Naturalist Memoirs* 2, 55–80.

Berger, A.; Guiot, J.; Kukla, G.; and Pestiaux, P. (1980). "Long-term variations of monthly insolation as related to climatic changes." *Internationales, Alfred-Wegener Symposium*. Berliner Gewissen shaftliche Abhandlunge, Reihe A, Band 19, pp. 24–25.

Betancourt, J. L. (1984). "Late Quaternary plant zonation and climate in southeastern Utah." *Great Basin Naturalist* 44, 1–35.

———. (1987). "Paleoecology of pinyon-juniper woodlands: Summary." *Proceedings of the Pinyon-Juniper Conference*, pp. 129–39. U.S. Department of Agriculture, Forest Service, General Technical Report INT 215.

Betancourt, J. L., and Davis, O. K. (1984). "Packrat middens from Canyon de Chelly, northeastern Arizona: Paleoecological and archeological considerations." *Quaternary Research* 21, 56–64.

Betancourt, J. L.; Dean, J. S.; and Hull, H. M. (1986). "Prehistoric long-distance transport of construction beams, Chaco Canyon, New Mexico." *American Antiquity* 51, 370–75.

Betancourt, J. L.; Martin, P. S.; and Van Devender, T. R. (1983). "Fossil packrat middens from Chaco Canyon, New Mexico: Cultural and ecological significance." In *Chaco Canyon country: A field guide to the geomorphology, Quaternary geology, paleoecology, and environmental geology of northwestern New Mexico*. S. G. Wells, D. W. Love, and T. W. Gardner, eds., pp. 207–217. American Geomorphological Field Group. Field Trip Guidebook.

Betancourt, J. L., and Van Devender, T. R. (1981). "Holocene vegetation in Chaco Canyon, New Mexico." *Science* 214, 656–58.

Betancourt, J. L.; Van Devender, T. R.; and Rose, M.

(1986). "Comparison of plant macrofossils in woodrat (*Neotoma* sp.) and porcupine (*Erethizon dorsatum*) middens from the western U.S." *Journal of Mammology* 67, 266–73.

Blagbrough, J. W., and Breed, W. J. (1969). "Protalus ramparts on Navajo Mountain, southern Utah." *American Journal of Science* 265, 759–72.

Brown, J. H. (1971). "Mammals on mountaintops: Nonequilibrium insular biogeography." *American Naturalist* 105, 467–78.

———. (1978). "The theory of insular biogeography and the distribution of boreal birds and mammals." *Great Basin Naturalist Memoirs* 2, 209–27.

Cannon, H. L. (1960). *The development of botanical methods of prospecting for uranium on the Colorado Plateau.* U.S. Geological Survey Bulletin 1085-A.

Carrara, P. E.; Mode, W. N.; Meyer, R.; and Robinson, S. W. (1984). "Deglaciation and postglacial timberline in the San Juan Mountains, Colorado." *Quaternary Research* 21, 42–55.

Cinnamon, S. K. (1988). "The vegetation community of Cedar Canyon, Wupatki National Monument, as influenced by prehistoric and historical environmental change." Master's thesis, Northern Arizona University, Flagstaff.

Clements, F. E. (1916). *Plant succession.* Carnegie Institute of Washington Publication no. 242.

Clisby, K. H., and Sears, P. B. (1956). "San Agustin Plains: Pleistocene climatic changes." *Science* 124, 537–39.

Cole, K. L. (1981). "Late Quaternary environments in the eastern Grand Canyon: Vegetational gradients over the last 25,000 years." Ph.D. diss., University of Arizona, Tucson.

———. (1982). "Late Quaternary zonation of vegetation in the eastern Grand Canyon." *Science* 217, 1142–45.

———. (1985). "Past rates of change, species richness, and a model of vegetational inertia in the Grand Canyon, Arizona." *American Naturalist* 125, 289–303.

Cottam, W. P. (1954). "Prevernal leafing of aspen in Utah Mountains." *Journal of the Arnold Arboretum* 35, 239–50.

Cottam, W. P.; Tucker, J. M.; and Drobnick, R. (1959). "Some clues to Great Basin postpluvial climatic distributions provided by oak distributions." *Ecology* 40, 361–77.

Daubenmire, R. F. (1943). "Vegetational zonation in the Rocky Mountains." *Botanical Review* 9, 325–93.

Davis, M. B. (1986). "Climatic instability, time lags, and community disequilibrium." In *Community ecology.* J. Diamond and T. J. Case, eds., pp. 269–84. New York: Harper and Row.

Davis, O. K. (1987). "Palynological evidence for historic juniper invasion in central Arizona." *Proceedings of the Pinyon-Juniper Conference,* pp. 120–24. U.S. Department of Agriculture, Forest Service, General Technical Report INT 215.

Davis, O. K.; Agenbroad, L.; Martin, P. S.; and Mead, J. I. (1984). "The Pleistocene dung blanket of Bechan Cave, Utah." *Carnegie Museum of Natural History Special Publications* 8, 267–82.

Dieterich, J. (1980). *Chimney Spring forest fire history.* U.S. Department of Agriculture, Forest Service Research Paper, RM-220.

Everitt, B. L. (1970). *A survey of the desert vegetation of the northern Henry Mountains region, Hanksville, Utah.* Baltimore: Johns Hopkins University Press.

Fall, P. L. (1985). "Holocene dynamics of the subalpine forest in central Colorado." *American Association of Stratigraphic Palynologists Contribution Series* 16, 31–46.

Flint, R. F., and Denny, C. S. (1958). *Quaternary geology of Boulder Mountain, Aquarius Plateau, Utah.* U.S. Geological Survey Bulletin 1061-D.

Gillespie, W. B. (1985). "Holocene climate and environment in Chaco Canyon." In *Environment and subsistence in Chaco Canyon.* F. J. Mathien, ed., pp. 13–46. Publications in Archeology 18E, Chaco Canyon Studies, National Park Service.

Gleason, H. A. (1926). "The individualistic concept of the plant association." *Torrey Botanical Club Bulletin* 53, 7–26.

Gregory, H. E. (1938). *The San Juan country: A geographic and geologic reconnaissance of southeastern Utah.* U.S. Geological Survey Professional Paper 188.

Hall, S. A. (1977). "Late Quaternary sedimentation and paleoecologic history of Chaco Canyon, New Mexico." *Geological Society of America Bulletin* 88, 1593–1618.

Hall, S. A. (1985). "Quaternary pollen analysis and vegetational history of the Southwest." In *Pollen records of late-Quaternary North American sediments.* V. M. Bryant and R. G. Holloway, eds., pp. 95–123. Dallas: American Association of Stratigraphic Palynologists Foundation.

Harper, K. T.; Freeman, D. C.; Ostler, W. K.; and Klikoff, L. C. (1978). "The flora of Great Basin mountain ranges: Diversity, sources, and dispersal ecology." *Great Basin Naturalist Memoirs* 2, 81–103.

Harper, K. T.; Wagstaff, F. J.; and Kunzler, L. M. (1985). *Biology and management of the Gambel oak vegetative type: A literature review.* U.S. Department of Agriculture, Forest Service, General Technical Report INT-179.

Hevly, R. H. (1964). "Paleoecology of Laguna Salada." *Fieldiana, Anthropology* 55, 171–87.

Hunt, C. B. (1956). *Cenozoic geology of the Colorado Plateau.* U.S. Geological Survey Professional Paper 279.

Jacobs, B. (1983). "Past vegetation and climate of the Mogollon Rim area, Arizona." Ph.D. diss., University of Arizona, Tucson.

———. (1985). "A middle Wisconsin pollen record from Hay Lake, Arizona." *Quaternary Research* 24, 121–30.

Johnson, N. K. (1975). "Controls of numbers of bird species on montane islands in the Great Basin." *Evolution* 29, 545–67.

Kilgore, B. M., and Taylor, D. (1979). "Fire history of a sequoia–mixed conifer forest." *Ecology* 60, 129–42.

Lanner, R. M., and Van Devender, T. R. (1974). "Morphology of pinyon pine needles from fossil packrat middens in Arizona." *Forest Science* 20, 207–11.

Lanner, R. M., and Van Devender, T. R. (1981). "Late Pleistocene pinyon pines in the Chihuahuan Desert." *Quaternary Research* 15, 278–90.

Leopold, L. B., and Bull, W. B. (1979). "Base level, aggradation, and grade." *Proceedings of the American Philosophical Society* 123, 168–202.

Loope, W. L. (1977). "Relationships of vegetation to environment in Canyonlands National Park." Ph.D. diss., Utah State University, Logan.

MacArthur, R. H., and Wilson, E. O. (1967). *The theory of island biogeography.* Princeton: Princeton University Press.

Mack, R. N. (1981). "Invasion of *Bromus tectorum* L. into western North America, an ecological chronicle." *Agro Ecosystem* 7, 145–65.

Madsen, D. B., and Currey, D. R. (1979). "Late Quaternary glacial vegetation changes, Little Cottonwood Canyon area, Wasatch Mountains, Utah." *Quaternary Research* 12, 254–70.

Maher, L. J. (1961). "Pollen analysis and postglacial vegetation history in the Animas Valley region, southern San Juan Mountains, Colorado." Ph.D. diss., University of Minnesota, Twin Cities.

———. (1964). "*Ephedra* pollen in sediments of the Great Lakes region." *Ecology* 45, 391–95.

Markgraf, V.; Bradbury, J. P.; Forester, R. M.; Singh, G.; and Sternberg, R. S. (1984). "San Agustin Plains, New Mexico: Age and paleoenvironmental potential reassessed." *Quaternary Research* 22, 336–43.

Markgraf, V., and Scott, L. (1981). "Lower timberline in central Colorado during the past 15,000 yr." *Geology* 9, 231–34.

Martin, P. S., and Mehringer, P. J., Jr. (1965). "Pleistocene pollen analysis and biogeography of the Southwest." In *The Quaternary of the United States.* H. E. Wright, Jr., and D. G. Frey, eds., pp. 423–35. Princeton: Princeton University Press.

Mead, J. I. (1983). "Harrington's extinct mountain goat (*Oreamnus harringtoni*) and its environ-

ment in the Grand Canyon." Ph.D. diss., University of Arizona, Tucson.

Mead, J. I.; Agenbroad, L. D.; Phillips, A. M. III; and Middleton, L. T. (1987) "Extinct mountain goat (*Oreamnus harringtoni*) in southeastern Utah." *Quaternary Research* 27, 323–31.

Meagher, G. S. (1943). "Reaction of pinyon and juniper seedlings to artificial shade and supplemental watering." *Journal of Forestry* 41, 480–82.

Mehringer, P. J., Jr., and Ferguson, C. W. (1969). "Pluvial occurrence of bristlecone pine (*Pinus aristata*) in a Mohave Desert mountain range." *Journal of the Arizona Academy of Sciences* 5, 284–92.

Merriam, L. H. (1890). "Results of a biological survey of the San Francisco Mountain region and desert of the Little Colorado River in Arizona." *North American Fauna* 3, 1–37. U.S. Department of Agriculture, Division of Ornithology and Mammalogy, Washington, D.C.

Merrill, R. K., and Pewe, T. L. (1977). *Late Cenozoic geology of the White Mountains, Arizona.* Arizona Bureau of Geology and Mineral Technology Special Paper 1.

Pearson, G. A. (1931). *Forest types in the Southwest as determined by climate and soil.* U.S. Department of Agriculture, Forest Service, Technical Bulletin 247.

Peet, R. K. (1978). "Latitudinal variation in southern Rocky Mountain forests." *Journal of Biogeography* 5, 275–89.

Petersen, K. L., and Mehringer, P. J., Jr. (1976). "Postglacial timberline fluctuations, La Plata Mountains, southwestern Colorado." *Arctic and Alpine Research* 8, 275–88.

Phillips, A. M. III (1977). "Packrats, plants, and the Pleistocene in the lower Grand Canyon." Ph.D. diss., University of Arizona, Tucson.

Porter, S. C.; Pierce, K. L.; and Hamilton, T. D. (1983). "Late Wisconsin mountain glaciation in the western United States." In *Late-Quaternary environments of the United States.* Volume 1, *The late Pleistocene.* S. C. Porter, ed., pp. 71–114. Minneapolis: University of Minnesota Press.

Potter, L. D., and Pattison, N. B. (1976). *Shoreline ecology of Lake Powell.* Washington, D.C.: National Science Foundation, Lake Powell Research Project Bulletin 29.

Reap, R. M. (1986). "Evaluation of cloud-to-ground lightning data from the western United States for 1983–84 summer seasons." *Journal of Climatology and Applied Meteorology* 25, 785–99.

Reveal, J. L. (1979). "Biogeography of the intermountain region: A speculative appraisal." *Mentzelia* 4, 1–92.

Richmond, G. M. (1962). *Quaternary stratigraphy of the La Sal Mountains, Utah.* U.S. Geological

Survey Professional Paper 324.

Samuels, M. L., and Betancourt, J. L. (1982). "Modeling the long-term effects of fuelwood harvests on pinyon-juniper woodlands." *Environmental Management* 6, 505–15.

Schmutz, E. M.; Dennis, A. E.; Harlan, A.; Hendricks, D.; and Zauderer, J. (1976). "An ecological survey of Wide Rock Butte in Canyon de Chelly National Monument, Arizona." *Journal of the Arizona Academy of Sciences* 11, 114–25.

Sharp, R. P. (1942). "Multiple Pleistocene glaciation on San Francisco Mountain, Arizona." *Journal of Geology* 50, 481–503.

Shreve, F. (1914). *A montane rainforest: A contribution to the physiological plant geography of Jamaica.* Carnegie Institute of Washington Publication no. 199.

Simpson, J. H. (1850). *Journal of a military reconnaissance from Santa Fe, New Mexico to the Navajo country . . . in 1849.* Thirty-first Congress, 1st session, Senate Executive Document 64, pp. 55–168. Washington, D.C.: Government Printing Office.

Spaulding, W. G. (1981). "The late Quaternary vegetation of a southern Nevada mountain range." Ph.D. diss., University of Arizona, Tucson.

Spaulding, W. G.; Leopold, E. B.; and Van Devender, T. R. (1983). "Late Wisconsin paleoecology of the American Southwest." In *Late-Quaternary environments of the United States.* Volume 1, *The late Pleistocene.* S. C. Porter, ed., pp. 259–93. Minneapolis: University of Minnesota Press.

Spaulding, W. G., and Petersen, K. L. (1980). "Late Pleistocene and early Holocene paleoecology of Cowboy Cave." *University of Utah Anthropological Papers* 104, 163–77.

Thompson, R. S., and Mead, J. I. (1982). "Late Quaternary environments and biogeography in the Great Basin." *Quaternary Research* 17, 39–55.

Van Devender, T. R. (1986a). "Climatic cadences and the composition of Chihuahuan Desert communities. The late Pleistocene packrat midden record." In *Community ecology.* J. Diamond and T. J. Case, eds., pp. 285–99. New York: Harper and Row.

———. (1987). "Late Quaternary history of pinyon-juniper-oak woodlands dominated by *Pinus remota* and *P. edulis.*" *Proceedings of the Pinyon-Juniper Conference,* pp. 99–103. U.S. Department of Agriculture, Forest Service, General Technical Report INT-215.

Van Devender, T. R.; Betancourt, J. L.; and Wimberly, M. (1984). "Biogeographic implications of a packrat midden sequence from the Sacramento Mountains, south-central New Mexico." *Quaternary Research* 22, 344–60.

Van Devender, T. R.; Martin, P. S.; Thompson, R. S.;

Cole, K. L.; Jull, A. J. T.; Long, A.; Toolin, L. J.; and Donahue, D. J. (1985). "Fossil packrat middens and the tandem accelerator mass spectrometer." *Nature* 317, 610–13.

Van Devender, T. R., and Spaulding, W. G. (1979). "Development of vegetation and climate in the southwestern United States." *Science* 204, 701–10.

Van Devender, T. R.; Thompson, R. S.; and Betancourt, J. L. (1987). "Vegetation history of the deserts of southwest North America, the nature and timing of the late Wisconsin–Holocene transition." In *North America and adjacent oceans during the last deglaciation.* W. F. Ruddiman and H. E. Wright, Jr., eds., pp. 323–52. *The Geology of North America,* vol. K-3. Boulder: Geological Society of America.

Wells, P. V. (1983). "Paleobiogeography of montane islands in the Great Basin since the last glaciopluvial." *Ecological Monographs* 53, 341–82.

———. (1987). "Systematics and distribution of pinyons in the late Quaternary." *Proceedings of the Pinyon-Juniper Conference.* U.S. Department of Agriculture, Forest Service, General Technical Report INT-215.

Welsh, S. L. (1978). "Problems in plant endemism on the Colorado Plateau." *Great Basin Naturalist Memoirs* 2, 191–95.

Welsh, S. L., and Toft, K. A. (1981). "Biotic communities of hanging gardens in southeastern Utah." *National Geographic Research Report* 13, 663–81.

West, N. E.; Tausch, R. J.; Rea, K. H.; and Tueller, P. T. (1978). "Phytogeographical variation within juniper-pinyon woodlands of the Great Basin." *Great Basin Naturalist Memoirs* 2, 119–36.

Whiteside, M. C. (1964). "Paleoecological studies of Potato Lake and its environs." Master's thesis, Arizona State University, Tempe.

Whittaker, R. H., and Niering, (1965). "Vegetation of the Santa Catalina Mountains, Arizona, a gradient analysis of the south slope." *Ecology* 46, 429–52.

Wright, H. E., Jr.; Bent, A. M.; Hansen, B. S.; and Maher, L. J., Jr. (1973). "Present and past vegetation of the Chuska Mountains, northwestern New Mexico." *Geological Society of America Bulletin* 84, 1155–80.

Yarrow, H. C. (1875). "Collections of batrachians and reptiles." In *Exploration and Surveys, Zoology,* vol. 5. G. M. Wheeler, ed., pp. 559–62. Washington, D.C.: U. S. Army Department of Engineering.

Yeaton, R. I.; Yeaton, R. W.; and Horenstein, J. E. (1980). "The altitudinal replacement of digger pine by ponderosa on the western slopes of the Sierra Nevada." *Bulletin of the Torrey Botanical Club* 107, 487–95.

PART III
Special Studies

Chapter 13

Comparison of Late Holocene Environments from Woodrat Middens and Pollen: Diamond Craters, Oregon

Peter J. Mehringer, Jr.   Peter E. Wigand

INTRODUCTION

The continuing investigations of eastern Oregon's late Quaternary environments began in the mid-1970s as a part of the Steens Mountain Prehistory Project. Paleoecological and chronological data have since been derived from radiocarbon, tephra, and paleomagnetic dating; studies of fossil pollen, algae, and plant macrofossils; textural analyses of lake sediments; and the stratigraphy and chronology of desert dunes (Aikens et al., 1982; Mehringer, 1985, 1986, 1987; Mehringer and Wigand, 1986, 1987; Verosub and Mehringer, 1984; Verosub et al., 1986; Wigand, 1987).

According to these studies, the steppe of southeastern Oregon has undergone punctuated differences in Holocene importance of juniper, grasses, sagebrush, and fire. For example, radiocarbon-dated pollen records from Diamond Pond, supported by macrofossils from nearby woodrat middens, disclose juniper's apparently rapid response to climatic events (Mehringer and Wigand, 1987). At Fish and Wildhorse lakes on Steens Mountain (Fig. 13.1), pollen ratios indicate changes in the abundance of grass and sagebrush (Mehringer, 1985, Fig. 12). Counts of microscopic charcoal fragments (Patterson et al., 1987) attest to climatic regimes producing many to few fires with occasional conflagra-tions (Mehringer, 1987).

Although these findings seemed evident when averaged over decades and centuries, we wished to relate them more clearly to questions of range and cultural resource management. We were particularly interested in determining, first, if the rapidity of western juniper's (*Juniperus occidentalis*) historic incursions into steppe represented a unique event during the Holocene and, second, the relative significance of this recent "invasion" (Burkhardt and Tisdale, 1976). It was necessary, therefore, to improve temporal resolution and compare fossil sequences with short-lived but ecologically significant events such as the droughts of the 1920s and 1930s. Thus the time separating samples would ideally represent less than one generation of the organism(s) being studied.

Though western juniper may produce cones by twenty-five years of age, it does not approach its potential for pollen production for another fifty years or more (L. E. Eddleman, personal communication). Nonetheless, the history of western juniper would be best resolved by a fossil pollen sequence with samples separated by no more than twenty-five years. Perhaps less than twenty years, the span of a single generation, should separate samples used to evaluate potential human re-

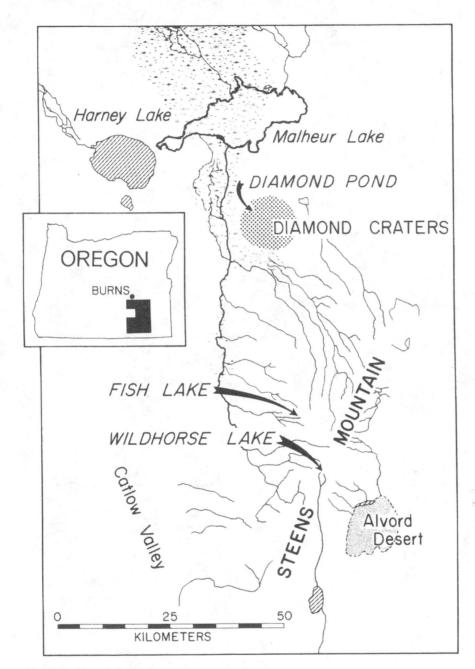

Fig. 13.1. Map of the study area.

sponses to varying environments in the northern Great Basin. When considering people, however, these relationships are not so simple. Human "resource managers" of the prehistoric period have affected the state of steppe and woodland on this continent for at least twelve thousand years. They diverted water, accidentally and intentionally spread propagules, and otherwise modified local habitats to their needs (Downs, 1966; Lawton et al., 1976). Because burning was their most important means of manipulating the environment (Barrett and Arno, 1982; Gruell, 1985), prehistoric fire regimes must have varied with population densities and lifestyles. In any case, details that could be revealed only by close-interval sampling were prime objectives of these studies.

In this chapter we describe and compare: (1) plant macrofossils recovered from twenty-four late Holocene woodrat middens from the Bureau of Land Management Outstanding Natural Area, Diamond Craters, Oregon; (2) pollen content of twelve of these middens and another two from the Catlow Rim; and (3) pollen, algae, and microscopic charcoal fragments from close-interval analyses of the last two thousand years, using cores from Diamond Pond (Fig. 13.2). We evaluate the paleoclimatic and chronological significance of western juniper remains from the middens by comparing their age distribution with continuous juniper pollen records from nearby Diamond Pond (Malheur Maar). New close-interval analyses of pollen samples spanning the last two thousand years provide a basis for understanding the significance of macrofossil assemblages from woodrat middens as compared with pollen records of the past five thousand years.

STUDY AREA

Diamond Craters Volcanic Complex (Fig. 13.2) is situated in the southern Harney Basin, a 13,727-km² area of large-scale post-Miocene subsidence lying at the eastern end of the Brothers Fault Zone (Walker, 1979). The craters consist of six major (and one minor) structural domes that range from 50 to 120 m high and cover an area over 9 km wide (Peterson and Groh, 1964). On the western-most dome at 1265 m elevation lies Diamond Pond (Fig. 13.3), a 2-m-deep, 46-m-wide perennial pond that fills the 92-m-wide Malheur Maar—a volcanic explosion crater—to within 9 m of the rim.

Sagebrush and its associates dominate most of Diamond Craters, but western juniper is important on low-lying lava flows north of North Dome (Fig. 13.4). Shadscale—salt-bushes (*Atriplex*) and greasewood (*Sarcobatus*)—occupies thin, stony soils on the south and west sides of West Dome and on the valley floor around the craters. Within Malheur Maar, south-facing slopes are dominated by

shadscale and north-facing slopes by sagebrush communities. A variety of emergent and littoral plant species form a fringe around the pond, and thick growths of pondweed and hornwort clog its waters.

Drainage from surrounding uplands collects in the north-central part of Harney Basin in two large, shallow, marsh-surrounded lakes that coalesce in times of unusually deep water (Fig. 13.1). Since early in 1826, when Europeans first entered the Harney Basin, variations in rainfall are known to have dramatically influenced the volumes of these lakes (Piper et al., 1939; Walker and Swanson 1968; Hubbard, 1975).

METHODS

Multiple, overlapping ten-centimeter-diameter cores of lake sediment were collected in plastic core barrels from near the center and deepest area of Diamond Pond (2 m) in 1978 (Fig. 13.5). These were placed in cold storage at Washington State University. We were able to remove more samples accurately because core barrels were measured and marked, and the sediments were also marked when initially logged and sampled. One hundred new analyses were completed covering the last two thousand years at Diamond Pond.

For these studies it was necessary to extend close-interval sampling through the historic period. Because sediments near the mud-water interface are easily disturbed by coring, 30–40-cm-deep sections were frozen in place with a "box-freezer" that utilizes dry ice and alcohol (Huttunen and Merilainen, 1978; O'Sullivan, 1983). These sections, collected on November 7th, after sagebrush had flowered, extended the pollen, charcoal, and algae records through 1985. The frozen sections of loose lake bottom were correlated with the uppermost cores by distinctive stratigraphic marker beds and verified by overlapping pollen spectra. As a result, the 12 cm of loose lake bottom deposited during the last fourteen years were added to the top of the previous Diamond Pond section.

To conform with the height of all other samples, we used a band saw to cut the frozen slabs horizontally into 1-cm strips whose volume was determined (after the samples melted) by centrifuging them in graduated 10-ml tubes for one minute at 1,000 RPM. This

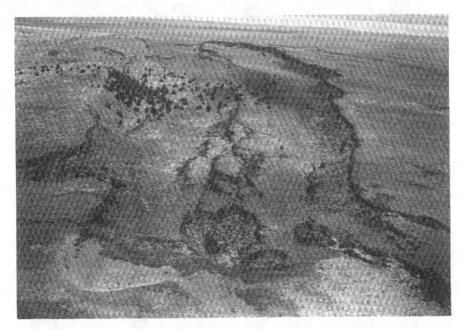

Fig. 13.2. Oblique air photo of Graben Dome, Diamond Craters, Oregon, shows sagebrush vegetation with scattered junipers and wooded patches (photo by E. Benedict, 1982).

Fig. 13.3. Fringed by cat-tail and bulrush, Diamond Pond straddles the sagebrush-shadscale ecotone east of Malheur Marshes and north of Steens Mountain (photo by W. Bright, September 1978).

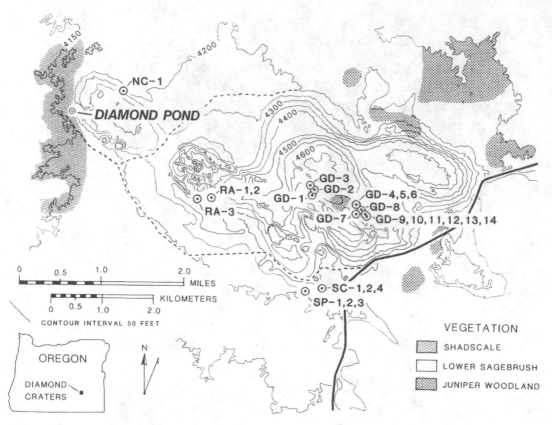

Fig. 13.4. Location of radiocarbon-dated woodrat middens, Diamond Craters, Oregon.

compacted sediment volume more closely approximates down-core sediment samples used for pollen analyses and weight loss on ignition (Dean, 1974). Volume and dry weight were measured for all samples. *Lycopodium* spore tracers, introduced as the first step in pollen extraction, facilitated population estimates of pollen, algae, and microscopic charcoal fragments.

Ages of lake cores were determined from thirty-eight radiocarbon dates from three sites (Diamond Pond, and Fish and Wildhorse lakes on Steens Mountain) that are correlated by tephra. At Diamond Pond and Fish Lake cesium-137 abundance indicated approximate positions of samples dating from A.D. 1945 and 1964; though preliminary, these estimates are probably accurate to within a decade. Four time scales—as befits the topic—are used in presenting ages of pollen spectra and woodrat middens.

We use depth and radiocarbon years B.P. where years between events are not of primary importance or for comparing other data so treated. Tree-ring-corrected radiocarbon ages B.P. and tree-ring-corrected years A.D./ B.C. enhance comparisons with tree-ring records and estimates of years within or between samples; for example, in estimating pollen or charcoal accumulation rates. Calendar corrections for core ages were determined by converting radiocarbon dates to tree-ring ages (Klein et al., 1982). Core age for any particular level comes from deposition rate derived by curve fitting (fourth-order polynomials) of radiometric dates or their tree-ring corrections. S. W. Robinson's (1986) computer program converted radiocarbon dates of woodrat middens to calendar years.

We prepared woodrat middens by first removing uneven and weathered surfaces with a hammer and chisel. Except for the very

Fig. 13.5. Washington State University students coring at Diamond Pond (October 1978). One of the many explosion craters in this volcanic terrane (Malheur Maar) holds a 46-m-diameter pond that has accumulated pollen, algae, seeds, and molluscs for at least six thousand years.

smallest samples middens were then split along bedding planes to identify and separate distinct stratigraphic units; larger middens were sawed, vertical to their apparent planes of deposition, to better reveal depositional sequences. The stratigraphically discrete pieces of urine-cemented dung and plant remains were then washed thoroughly in hot water and soaked for ten to twenty-four hours. The undissolved midden and freed plant remains, dung and sediments were then poured onto nested, stainless steel sieves (8, 10, 24, and 42 mesh) and washed thoroughly. This process was repeated until midden contents no longer adhered and were free of amberat (usually three to five days).

Pollen samples were taken from middens during the soaking and sieving process. After a sample of consolidated midden was reduced by at least one-half and washed, it was placed in distilled water for several hours. The sample was again poured onto the nested screens and washed with distilled water; everything passing the 42-mesh screen was retained and mixed, and about 200 ml were poured through a 100-mesh screen into a 250-ml beaker. *Lycopodium* spore tracers were added as a check on extraction. The sample was then concentrated by centrifuging in 50-ml tubes, extracted using routine hydrofluoric acid and acetylation methods, stained, and mounted in silicon oil.

After drying and sorting, we identified plant remains by comparison with our collections and those of the Marion Ownbey Herbarium, Washington State University. They were counted, dried at 40°C, cooled in a desiccator, and weighed to five places and rounded to three; thus .001 indicates a trace that actually may weigh less (Table 13.1). Abundance, state of preservation, and usefulness of the parts preserved for distinguishing particular taxa

Table 13.1A. *Macrofossils (weight, grams) from woodrat middens, BLM outstanding natural area, Diamond Craters, Oregon*

	Spattercone Cave		
	SP-1 2350 ± 50	SP-3 2670 ± 70	SP-2 3045 ± 120
TREES AND SHRUBS			
Juniperus occidentalis			
twigs	12.817	8.206	4.392
seeds and berries	3.035(98W/6F)	1.423(55W/5F)	.412(19W/2F)
bark	.126	.121	
galls	.012	.170	.014
Ribes aureum (seeds)			
Ribes cereum			
seeds			
leaves			
Artemisia (sec. Tridentatae)			
twigs	.259	.018	.017
buds and bracts			
bark	.378	.637	.208
leaves	.263	.044	.012
Chrysothamnus nauseosus			
bracts	.032	.001	.002
pappi and seeds	.001	.001	.001
Tetradymia glabrata			
twigs			.006
bracts and pappi			
Atriplex confertifolia			
fruit			
leaves			
Atriplex spinosa (seeds)			
FORBS			
Amsinckia tessellata (nutlets)		.003(1W)	
Astragalus sp.			
Brickellia microphylla			
twigs			
bracts			
leaves			
Castilleja chromosa (seeds)	.005(13W)		
Cicuta douglasii (seeds)			
Chenopodium sp.			
Collinsia parviflora (seeds)			
Cryptantha ambigua (seeds)			
Cryptantha circumscissa (seeds)			
Cryptantha pterocarya (seeds)			
Cryptantha torreyana (seeds)			
Cryptantha cf. *torreyana* (flowers and leaves)	.002	.001	
Cryptantha watsonii (seeds)			
Cryptantha sp. (leaves)			
Cryptantha sp. (bracts)			
Descurainia pinnata (siliques)			
Epilobium paniculatum (seeds)			
Eatonella nivea (seeds)	.001(2W)		

Table 13.1A. *(Continued)*

	Surprise Cave			Rat Amber Ridge			Nolf Crater
	SC-4 2680 ± 90	SC-2 2700 ± 80	SC-1 2800 ± 110	RA-2 2340 ± 90	RA-3 2455 ± 90	RA-1 3000 ± 75	NC-1 4075 ± 90
	6.909	.977	4.976	2.362	1.831	5.934	.043
	3.187(105W/8F)	.805(10W)	2.228(78W/45F)	.965(33W/4F)	3.845(126W/4F)	1.380(50W/6F)	.058(3W/18F)
	.052		.049	.320	.028	.850	
	.017						
							.013(17W)
	.210			.019	.033	.022	
				.003		.001	
	.552		.264	.344	1.577	.075	.018
	.016		.001	.036	.128	.003	.002
	.066		.013	.006	.010	.006	
	.001				.001		.001
					.002(1W)		
					.006(7W)		
				.001(1W)	.059(21W/4F)		.015(15W)
					.004(1W)		
				.001			
					.002(1W)		
			.001(1W)	.003(5W)	.018(17W)	.001(1W)	
				.001(2W)			.001(3W)
	.002		.001			.002	
					.005		.001
				.001		.001	
				.001(2W)			.001(1W)
				.001(3W)	.001(1W)	.001(3W)	

Fabaceae (seeds)			
Galium aparine var. echino-spermum (seeds)	.004(2W)		
Gilia inconspicua (seeds)			
Lactuca cf. serriola (seeds)			
Leptodactylon pungens (bracts)		.002	
Lupinus sp. (seeds)			
Mentzelia albicaulis (seeds)			
Mentzelia laevicaulis (seeds)			
Montia perfoliata (seeds)			
Penstemon deustus (capsules)	.055	.009	.009
Phacelia linearis (seeds)			
Phacelia lutea var. lutea (seeds)		.001(2W)	
Phacelia rattanii (seeds)			
Plagiobothrys kingii var. harknessii (nutlets)	.023(11W)		
Plectritis macrocera (fruits)	.001(2W)	.001(2W)	.001(1W)
Polemoniaceae (capsules)	.003	.001	.002
Suaeda (seeds)			
GRASSES			
Agropyron spicatum (spikelets)			
Distichlis stricta var. dentata			
Elymus cinereus (spikelets)			
Oryzopsis hymenoides (caryopses)	.027(22W)	.079(11W)	
Poa secunda var. sandbergii			
Sitanion hystrix (spikelets)			
Stipa comata (spikelets)	.003(4W)		.007(1W)
Stipa occidentalis (spikelets)			
Stipa thurberiana (spikelets)	.002(3W)		.004(1W)
Vulpia octoflora var. hirtella (spikelets)			
MOSSES			
Grimmia spp.	.088	.054	.026

W = whole seeds, fruits, nutlets or spikelets; F = seed fragments

Table 13.1B. *Macrofossils (weight, grams) from woodrat middens, BLM outstanding natural area, Diamond Craters, Oregon*

	Graben Dome					
	GD-2 835 ± 55	GD-11 1100 ± 90	GD-7 1540 ± 130	GD-9 1560 ± 60	GD-3 1625 ± 150	GD-13 1700 ± 90
TREES AND SHRUBS						
Juniperus occidentalis						
twigs	7.189	.948	15.229	5.074	2.229	14.197
seeds and berries	2.450(94W/3F)	.612(27W/1F)	3.319(121W/65F)	.191(6W/1F)	.666(21W/6F)	1.334(38W/9F)
bark	.030		.998	.221		.205
galls		.001	.003	.017		.004
Ribes aureum (seeds)						

							.002(2W)
	1.775	.157	1.080	.001	.146	.632	.003
.009	.030	.010	.042	.001	.030	.037	.003
.264	4.413	.385	1.199	.008	1.442	3.231	.056
.023	.175	.020	.358	.001	.190	.084	.004
.004	.033	.050	.071		.014	.072	.005
.001	.010	.014	.029		.025	.003	.001
			.075				
					.013		
.001(1W)	.007(5W)	.001(3W)				.003(5W)	.001(1W)
	.055		.001		.002		
.001	.001	.001			.004		
	.010	.001	.026		.006	.006	.001
	.001(1W)	.001(3W)					
						.003(5W)	
				.001(1W)			
.003(11W)		.002(6W)				.009(8W)	
	.001(1W)	.001(3W)	.001(1W)		.001(2W)	.001(3W)	.001(1W)
						.001(1W)	
					.003(11W)	.001(2W)	
	.002		.001		.001	.001	
	.003		.001		.001	.001	.001
.001(1W)		.001(1W)	.001(14W)		.001(5W)	.002(5W)	
					.001(4W)	.001(2W)	
	.002(1W)		.006(4W)		.001(1W)	.002(1W)	
.001	.001	.002	.002		.031	.078	
						.003(1W)	
.002(7W)	.001(1W)	.001(2W)	.002(8W)	.001(1W)	.003(11W)	.001(3W)	.001(8W)
						.039	

Phacelia linearis (seeds)			.013(48W)			
Phacelia lutea var. lutea (seeds)						
Phacelia rattanii (seeds)						
Plagiobothrys kingii var. harknessii (nutlets)			.025(28W)	.001(2W)		.007(8W)
Plectritis macrocera (fruits)				.001(1W)		
Polemoniaceae (capsules)	.001		.005	.001	.004	.002
Suaeda (seeds)						
GRASSES						
Agropyron spicatum (spikelets)						
Distichlis stricta var. dentata						.001(1W)
Elymus cinereus (spikelets)			.001(2W)			
Oryzopsis hymenoides (caryopses)	.047(29W)	.021(20W)	.303(215W)	.007(15W)	.101(35W)	.082(70W)
Poa secunda var. sandbergii			.003(11W)	.001(3W)		
Sitanion hystrix (spikelets)			.038(3W)	.004(4W)	.001(1W)	.001(2W)
Stipa comata (spikelets)	.002(1W)	.017(5W)	.020(11W)	.007(7W)	.004(2W)	.023(17W)
Stipa occidentalis (spikelets)	.004(2W)					
Stipa thurberiana (spikelets)			.002(4W)			.002(5W)
Vulpia octoflora var. hirtella (spikelets)			.001(4W)			
MOSSES						
Grimmia spp.	.014		.029			

W = whole seeds, fruits, nutlets or spikelets; F = seed fragments

determined levels of identification. Identification of some grasses and composites required special expertise beyond our competence that was provided by Paul M. Peterson and Joy Mastrogiuseppe of the Marion Ownbey Herbarium.

THE WOODRAT MIDDENS

Following our initial report on the contents of twelve middens (Mehringer and Wigand, 1987), we collected twelve more middens from Diamond Craters and continued with the identification of undetermined macrofossil remains. Here we report on all twenty-four middens, with twenty-five additional taxa. Middens were collected from Diamond Craters without regard to their apparent age or content. The few located but not taken were left only because they occupied inaccessible deep crevices or high ledges. Following clues from topographic maps, air photos, and informants, we estimate that we investigated at least 70% of the Diamond Craters terrane most likely to hold ancient middens (Fig. 13.4). Most potential sites are too moist for preservation of middens. In fact, this midden series includes fist-sized midden remnants and lumps of amberat with no exposed fossil plants. The ten (unstudied) middens collected from Catlow Valley represent only a small sample of those located, and they are large by comparison. Midden contents are arranged in Table 13.1 by site, area, and age. Radiocarbon dates are presented in Table 13.2.

Dung and dust trapped within woodrat middens hold abundant pollen that could include fossil taxa not represented by macrofossils. Also, pollen from middens may give clues to the regional vegetation beyond the small home range of woodrats, and it may prove important in evaluating assumptions that woodrat midden macrofossils represent only very local assemblages tied to rimrock and, therefore, less influenced by effects of fire. Table 13.3 gives pollen and charcoal counts from fourteen middens.

Macrofossils from Woodrat Middens

Nomenclature and identifications follow descriptions and illustrations in Cronquist et al.

						.001(1W)	
.001(2W)	.006(4W)		.007(5W)		.001(2W)	.003(3W)	
						.001(1W)	
	.001	.001	.001		.002	.001	
	.005(2W)						
.040(73W)	.034(55W)	.006(16W)	.005(20W)	.006(4W)	.012(18W)	.050(51W)	.015(7W)
.002(3W)	.003(6W)	.006(5W)	.002(2W)		.001(3W)	.002(3W)	
.004(3W)	.014(9W)	.005(5W)	.006(6W)	.001(1W)	.001(1W)		
						.001(2W)	
	.006(6W)				.004(3W)		
			.001(1W)			.001(2W)	
		.001	.001				

(1977, 1984) or Hitchcock and Cronquist (1973). Several taxa deserve special comment, and some taxonomic revisions may not correspond to common usage. For example, we followed Cronquist et al. (1984) in listing Sandberg's bluegrass as *Poa secunda* var. *sandbergii* (Table 13.1) rather than the more familiar *Poa sandbergii*.

Rabbitbrush comprises many species, subspecies, and hybrids (Anderson, 1986a, 1986b). Because its abundant bracts possessed strongly keeled, generally acute, dark-tipped phyllaries, we attributed all rabbitbrush in these middens to *Chrysothamnus nauseosus* without reference to subspecies. Perhaps expert examination of these fossils would reveal more complexity (Anderson, 1980). Likewise, *Artemisia spinescens* was not observed and, because of taxonomic complexity, all sagebrush remains were referred only to subgenus Tridentatae (McArthur et al., 1979; Shultz, 1986). By contrast, size, surface texture, and shape of *Phacelia* seeds permitted their specific identification.

At least four genera of the Boraginaceae and five species of *Cryptantha* were represented by distinctive seeds (Table 13.1). The nutlet of *Cryptantha pterocarya* contains three-winged seeds surrounding a central seed lacking wings; we recovered both types of seeds. As in *Brickellia microphylla*, characteristic twigs, bracts, seeds, or leaves often occurred together and in abundance. Other taxa were determined with assurance only to genera or family, and among the rare macrofossils another dozen or so genera remain unidentified.

All the middens examined in this study occur between 2 km and 100 m from the nearest patches of juniper woodland—outside the foraging range of woodrats—and as much as 1 km from the nearest single tree. At Graben Dome some middens (GD-4, GD-5, GD-7, GD-12, GD-13, and GD-14) are within 50 m of a standing snag, stump, or young tree (Fig. 13.5, 13.6). Thus, as a group, the Graben Dome middens require the least change in vegetation to account for western juniper's presence in fossil middens. The age distribution of all middens suggests a chronological pattern to juniper abundance at sites

Table 13.2. *Radiocarbon dates and tree-ring-corrected radiocarbon ages from woodrat middens from Diamond Craters and the Catlow Rim, southeastern Oregon.*

Laboratory Number	Radiocarbon Date (B.P.)	Tree-ring-corrected Age (A.D./B.C.)	Site	Material Dated	Elevation (meters)	Slope Direction
Diamond Craters						
WSU-2725	835 ± 55	1200 ± 60	GD-2	juniper	1410	ENE
*WSU-3479	1055 ± 130	975 ± 130	GD-11	dung	1400	WSW
*WSU-3584	1140 ± 120	890 ± 125	GD-11	juniper	1400	WSW
WSU-3484	1540 ± 130	510 ± 130	GD-7	juniper	1395	NNE
WSU-3478	1560 ± 60	500 ± 75	GD-9	juniper	1400	WSW
WSU-2727	1625 ± 150	420 ± 160	GD-3	juniper	1410	ENE
WSU-3480	1700 ± 90	335 ± 110	GD-13	juniper	1400	WSW
WSU-3476	1960 ± 110	50 ± 125	GD-4	dung	1400	WSW
WSU-3485	2210 ± 110	265 ± 145	GD-12	juniper	1400	WSW
WSU-3481	2260 ± 80	330 ± 95	GD-14	juniper	1400	WSW
WSU-2575	2340 ± 90	420 ± 125	RA-2	juniper	1330	WNW
WSU-2447	2350 ± 80	425 ± 115	SP-1	juniper	1280	WSW
WSU-3477	2360 ± 80	440 ± 120	GD-8	juniper	1400	WSW
WSU-2726	2390 ± 260	495 ± 330	GD-1	juniper	1435	WSW
WSU-2572	2455 ± 90	590 ± 170	RA-3	juniper	1330	ENE
WSU-2581	2670 ± 70	845 ± 50	SP-3	juniper	1280	WSW
WSU-2574	2680 ± 90	855 ± 75	SC-4	juniper	1290	WSW
WSU-2448	2700 ± 80	870 ± 65	SC-2	dung	1290	WSW
WSU-2441	2800 ± 110	980 ± 135	SC-1	juniper	1290	WSW
WSU-2573	3000 ± 75	1250 ± 110	RA-1	juniper	1330	WNW
WSU-3570	3045 ± 120	1305 ± 180	SP-2	juniper	1280	WSW
WSU-3482	3270 ± 100	1565 ± 115	GD 5	juniper	1400	WSW
WSU-3483	3380 ± 80	1690 ± 100	GD-6	juniper	1400	WSW
WSU-3475	4075 ± 90	2645 ± 135	NC-1	dung	1295	ESE
Catlow Rim						
WSU-3501	890 ± 80	1135 ± 90	DC-2	juniper	1460	SSW
WSU-3500	6100 ± 200	5040 ± 225	TM-1	dung	1460	WSW

*Two dates from GD-11 averaged as WSU-3484, 1100 ± 90 B.P. (A.D. 935 ± 90).

where juniper is absent today (Fig. 13.7).

Determining a universally applicable method of quantifying plant remains from woodrat middens would be difficult at best. The relative weights of middens from southeastern Oregon often are influenced by their content of cobble-to silt-sized weathered basalt and wind-borne dust. Relative abundances of plant remains, dung, and amberat influence weight as well. Midden size is subject to the same constraints. The volume (by displacement of water) of middens examined in this study varied from 110 cc (GD-14) to 1000 cc (GD-7). Larger samples tend to have more plant remains and variety. However, GD-14 is only one-half the size of GD-10 but

has greater variety because GD-10 is composed primarily of amberat with few plant remains. In short, to quantify remains based on original sample weight or volume, one must first determine the amount of stone, amberat, dust, dung, branches, twigs, etc.—not just a midden's or subsample's weight or volume. To allow comparison among these samples we weighed all identified plant remains and counted the seeds and fruits (Table 13.1).

Western juniper (*Juniperus occidentalis*) dominates all Diamond Craters middens. Sagebrush is the most abundant shrub, whereas Indian ricegrass (*Oryzopsis hymenoides*) and squirrel tail (*Sitanion hystrix*) are paramount among the grasses. Excepting abundance of juniper, the fossil plant assemblages suggest little vegetational change over the last four thousand radiocarbon years. For example, *Penstemon deustus* is characteristically scattered on Surprise Flow, and its remains are most common in Surprise and nearby Spattercone caves. It is otherwise important only in the GD-6 midden, which is adjacent to a basalt flat with patches of thin soil. By contrast, Indian ricegrass is abundant in most Graben Dome middens, where it is currently favored by deeper pumice soils. *Ribes aureum* was found only in the 4075 B.P. midden from Nolf Crater. Today it is common in moist rock rubble on the shaded crater slopes. Seep-weed (*Suaeda* sp.) also was recognized only in the Nolf Crater midden, but we did not observe it there in 1986.

Fossil Pollen from Woodrat Middens

Cheat grass (*Bromus tectorum*) is a recent introduction; still, its awns are found deep within cracks in ancient woodrat middens, adhering to sticky amberat and completely embedded in hard-surface amberat, which becomes viscous when moist. Despite removal of outer surfaces and special attention to presence of cheat grass, it was discovered in two samples. If cheat grass can be incorporated in amberat, then it is likely that pollen postdating occupation of a midden has been incorporated in amberat-filled cracks and surface amberat over centuries or millennia. Either *Neotoma* dung or the interiors of samples being prepared for macrofossil study might provide samples of pollen for comparison with other midden remains. These would best reflect regional and local wind-borne pollen, and pollen that adhered to woodrats and to plants they ate or collected to build the den and nest.

Pollen from the interiors of twelve middens from Diamond Craters corresponds to the macrofossils in their monotonous predominance of juniper and sagebrush pollen (Table 13.3). Distant conifers, oak, and alder contributed to the minimal expected background of nonlocal pollen. We anticipated sedge and cattail pollen from marshes of the Blitzen Valley, 10 km downwind; the presence of the algae *Botryococcus* suggests recirculation in dust blown from floors of ephemeral ponds and seasonally dry marsh margins (Thompson,

1985). Paucity of *Sarcobatus* and other chenopod pollen, however, is surprising considering their abundance in Diamond Pond cores of the same age, but in agreement with their rarity as midden macrofossils. Charcoal values of 9.2%–61% of total pollen reflect differences expected in samples representing short enough periods to distinguish nearby fires. They confirm the prehistoric place of fire in steppe.

Other taxa represented by macrofossils are rare or absent as pollen. The 10,941 pollens counted from the Diamond Crater middens included not a single member of the Boraginaceae, so common among the seeds in most samples (Table 13.1). Likewise, pollens of *Leptodactylon, Phacelia, Astragalus, Galium,* and *Mentzelia* were present but rare. Potential local taxa not recognized as macrofossils include *Navarretia, Physalis, Delphinium,* and three genera of Umbelliferae. These also could have been wind-borne long distances along with pollen of *Pinus, Cercocarpus,* and *Ceanothus.* By comparison with the macrofossils, pollen from Diamond Craters middens was relatively uninformative. It suggests, however, that in the past sagebrush and juniper were locally dominant and probably regionally abundant.

Pollen from two middens from the lower-sagebrush zone below Catlow Rim (eastern margin of Catlow Valley), about 60 km to the south, provides a comparison (Table 13.3). Though not studied, cursory examination reveals that one of these contains western juniper (Dry Creek-2; WSU-3501, 890 ± 80 B.P.). The other lacks juniper, although western juniper trees are now common there (Three Mile-1; WSU-3500, 6100 ± 200 B.P.; Fig. 13.8).

These two sites reflect contrasting local and regional pollen pictures. Like the Diamond Craters middens, DC-1 is dominated by juniper with sagebrush pollen, whereas TM-1 shows abundant sagebrush pollen and little juniper. The TM-1 midden suggests a sagebrush-dominated landscape; pollen of willow, sedge, and monkey-flower (*Mimulus guttatus*) and spores of horsetail (*Equisetum*) reveal nearby sources along Three Mile Creek.

THE DIAMOND POND CORES

As with the woodrat middens, our pollen

Table 13.3. Percentage (total pollen) of microfossils and charcoal from woodrat middens (arranged chronologically).

	Diamond Craters												Catlow Rim	
	GD-11	GD-7	GD-9	GD-13	GD-4	GD-12	GD-14	GD-8	GD-5	GD-6	GD-10	NC-1	DC-2	TM-1
Trees														
Juniperus	64.5	34.3	49.3	57.5	57.4	47.7	59.5	51.5	78.7	60.2	39.7	27.5	92.0	1.2
Pinus	5.1	15.9	12.4	3.3	8.7	12.2	4.4	5.2	4.2	9.6	6.9	15.3	0.4	9.1
Abies			0.1			0.2						0.3		0.1
Tsuga												0.1		
Pseudotsuga-Larix									0.1					
Arceuthobium												0.1		
Betula-Corylus					0.1	0.2		0.1						
Cercocarpus-type		0.1	0.3	0.4										0.3
Quercus				0.1				0.2						0.2
Fraxinus											0.2			
Shrubs														
Ribes				0.1										
Ceanothus							0.2	0.1						0.1
Holodiscus-type					0.1		0.2							0.4
Prunus-type														
Artemisia	20.9	40.0	29.6	31.3	27.0	29.8	25.2	36.9	14.4	23.8	41.9	45.7	5.2	57.0
Ambrosia-type	0.3	0.2	0.1	0.3	0.1	0.2		0.1			0.2	0.1	0.1	0.8
Liguliflorae					0.1									0.6
Tubuliflorae	4.3	1.5	3.0	1.9	2.0	2.1	4.4	2.5	1.2	1.8	2.9	3.0	1.2	3.7
Sarcobatus	0.3	1.0	1.2	0.1	0.8	1.2	0.3	0.8	0.5	0.5	0.9	1.0	0.9	4.3
Other Chenopods	1.4	2.4	1.5	1.6	1.2	3.0	2.6	1.2	0.4	1.3	4.4	1.4	0.3	4.7
Grasses and Forbs														
Gramineae	0.3	2.6	0.9	0.9	1.0	0.5	0.9	0.5	0.1	1.5	0.6	2.1	0.3	5.7
Eriogonum				0.1							0.2	0.2	0.1	0.4
Malvaceae												0.1		
Phacelia	0.8	0.1	0.1	0.3							0.3	0.8	0.2	1.0
Galium	0.2			0.3							0.2	0.6		0.2
Montia				0.1										
Claytonia														0.3
Amsinckia													0.1	0.6
Mentzelia	0.2		0.1	0.1							0.2			0.3
Ranunculus			0.1	0.1		0.2								0.1

Aconitum-type														0.1
Delphinium-type		0.1												0.1
Umbelliferae			0.3					0.1		0.1		0.1		1.0
Phlox						0.5	0.2							1.0
Gilia	0.5		0.1	0.1		0.2	0.2			0.1	0.2		0.1	
Leptodactylon		0.1		0.1					0.1					
Navarretia			0.1	0.3					0.1					
Saxifragaceae						0.2	0.2							0.8
Primulaceae			0.1									0.1		0.2
Physalis							0.3							
Heliotropium										0.1				0.1
Collinsia														
Cruciferae		0.3					0.2	0.1						
Euphorbia														0.1
Astragalus	0.2			0.1		0.9								0.5
Other legumes													0.1	0.4
Undetermined	0.3	0.4	0.5	0.1		0.2	0.7	0.1			0.2	0.5	0.1	1.3
Riparian and Marsh														
Salix		0.1					0.3	0.3	0.2	0.3				1.6
Alnus				0.3							0.2			0.2
Mimulus guttatus														0.1
Urtica														
Cyperaceae	0.5	1.2	0.5	0.1		0.4		0.1	0.1	0.5	0.7		0.1	1.3
Typha (tetrad)	0.2			0.1		0.4		0.1	0.1		0.2		0.1	0.1
Equisetum														0.5
Spores														
Trilete Spores														
Monolete Spores						0.2	0.2			0.1				
Sporormiella		0.1										0.1		
Bryophytes			0.1											0.1
Total	623	1129	758	799	737	574	587	1478	1905	779	682	890	1957	931
Algae														
Botryococcus							0.2						0.1	
Charcoal > 25 μm	20.4	29.3	29.0	12.6	13.2	30.9	18.5	9.2	9.7	16.8	9.9	61.1	1.2	27.0

Fig. 13.6. The basalt rim in the distance held Graben Dome middens (GD-9–GD-14); note the charred western juniper snag and two young trees on the lower slope (March 14, 1986).

analyses of lake cores amount to twice the number reported previously. The sum of terrestrial pollen—which averaged 757 grains for each of the 197 Diamond Pond analyses—excludes plants that occur primarily in or around aquatic habitats (e.g., sedges, cat-tails, pondweeds), spores (e.g., ferns and quillwort), and acid-resistant algae (e.g., *Pediastrum, Tetraedron,* and *Botryococcus*), which are many times more abundant than pollen. On the average, 638 *Lycopodium* tracers and 1453 charcoal fragments were recorded while counting the 757 terrestrial pollen grains. Because relative abundance of grass and juniper pollen could be an especially important ecological indicator in the lower-sagebrush zone, the Diamond Pond pollen-counting plan called for at least 30 grass pollens (mean = 48, range 30–122) and 20 juniper pollens (mean = 42, range 18–92) in every sample.

A summary pollen diagram for the last two thousand years of the Diamond Pond core is presented in Figure 13.9; details for 2000–

5000 B.P. are in Wigand (1987, Fig. 14). Percentages of terrestrial pollen were calculated after excluding pollen of distant conifers. Most pine pollen, with occasional grains of spruce, fir, Douglas fir, and hemlock, originates in the mountains of western Oregon and northern California. Pine pollen accounts for about 14% of all terrestrial pollen at Diamond Pond.

Figure 13.9 diagrams changing values of pollen abundance produced by local sagebrush-shadscale shrubs, grasses, juniper, and pond-fringe species (*Rumex,* Cyperaceae, and *Typha*). Water depth at Diamond Pond results from regional recharge and evaporation, and fossils of aquatic and pond-edge species record local water conditions. Likewise, relative abundance of *Sarcobatus* and other chenopod pollen reflects the importance of local shadscale species and downwind areas of drowned or dry lake bed. Records of *Ruppia, Sarcobatus,* sagebrush, and juniper provide especially interesting examples of past environ-

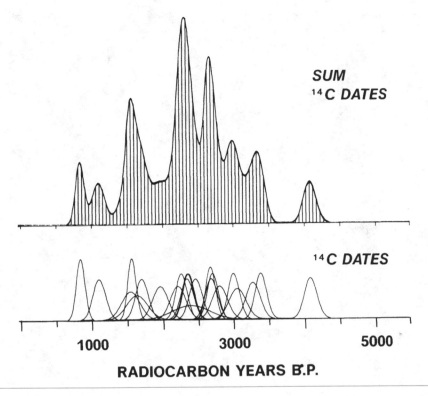

SUM
^{14}C DATES

^{14}C DATES

1000   3000   5000

RADIOCARBON YEARS B.P.

Fig. 13.7. Plots of radiocarbon dates from twenty-three woodrat middens from Diamond Craters, Oregon. Each date is plotted as a normal distribution with a probability of one; width at the base is three standard deviations. Because the areas under all curves are equal, a date with a smaller standard deviation gives a taller peak than one with a larger standard deviation. Areas under the curves for all dates are summed above in ten-year increments (Geyh, 1980; Geyh and de Maret, 1982).

ments at Diamond Craters.

The introduction, rise, and demise of ditch-grass (*Ruppia maritima*) and coontail (*Ceratophyllum demersum*), with varying values of pondweed (*Potamogeton* spp.), indicate changes in pond size, water depth, and water quality. *Ruppia* thrives in brackish water and its propagules must have reached Diamond Pond thousands of times. Yet, according to fossil pollen and seeds (Fig. 13.9; Wigand, 1987), and excepting single samples, *Ruppia* was successfully established only once in over six thousand years. It persisted for nearly four hundred years (A.D. 1400s to the late 1700s) then perished, to be replaced by *Potamogeton* and *Ceratophyllum* (represented by seeds and leaf spines rather than pollen). *Ruppia*'s arrival and success coincides with large *Sarcobatus* pollen values and probably an episode of shallow brackish water at Diamond Pond.

The curve for *Sarcobatus* pollen may give the clearest clues to changing environments near Diamond Craters, particularly lake lev-

Fig. 13.8. One of the woodrat middens from this site on the Catlow Rim at Three Mile Creek dates to 6100 B.P. Though juniper is common here today, the midden lacks juniper macrofossils and contains only 1.2% juniper pollen (March 10, 1987).

els. *Sarcobatus* would have flourished with less effective moisture and shrinking lakes in the Harney Basin. Expanding lakes would drown *Sarcobatus* habitat and that of other chenopods, among them shadscale (*Atriplex confertifolia*). Because the chenopods are prolific pollen producers, changes in their numbers greatly influence the percentages of all other pollen types.

Figure 13.10 shows the most important aspects of the Diamond Pond core analyses relative to the woodrat midden sequence—juniper pollen percentages and changing relations of grass to sagebrush and charcoal to pollen. Closely spaced samples from the last two thousand years clarify the continued corresponding fluctuations in percentage of juniper pollen and charcoal/pollen ratios. They also confirm that grass/sagebrush ratios do not follow the earlier (3800–2000 B.P.) pattern of abundant grass pollen. Rather, increasing sagebrush pollen (in relation to grass) charac-

terizes the last two thousand years at Diamond Pond. In sum, the pollen record suggests that, although steppe persisted through the Holocene, both terrestrial and aquatic vegetation responded to natural environmental perturbations driven by varying climate.

Juniper from Woodrat Middens and Lake Cores

Juniper pollen from a closely sampled core and radiocarbon-dated juniper macrofossils from nearby woodrat middens give a notion of western juniper's history and the relative significance of its recent success—despite chaining, bulldozing, cutting, poisoning, and burning. Detailed pollen analyses were necessary to evaluate apparent correspondence between the radiocarbon ages of juniper-dominated middens and distinct episodes of abundant charcoal and juniper pollen (Figs. 13.7, 13.9; Mehringer and Wigand, 1987, Fig. 4). Also, using the pollen and seeds of aquatic plants from Diamond Pond cores, Wigand (1987, Fig. 24) recognized these same general periods as times of relatively deeper water. The apparent clustering of radiocarbon ages of middens became as intriguing as their macrofossils and pollen.

Though different from the present in their abundant juniper, Diamond Craters middens from rimrock and lava tubes did not record some events apparent in the fossil sequence from Diamond Pond's mud. Nonetheless, ages of middens corresponded closely to other events marked by juniper pollen and indicators of deeper water. Because these apparent agreements came from two independently dated series (juniper or woodrat fecal pellets, and lake sediments), their comparison required that we consider: (1) potential for contamination in different environments; (2) short-term perturbations in ^{14}C production; and (3) and progressive disparity between radiocarbon and sidereal calendar years resulting in a thousand-year discrepancy from two thousand to seven thousand years ago (Stuiver et al., 1986). We evaluated contamination potential via selection and pretreatment of samples before they were dated. The latter two considerations led us to convert radiocarbon ages to calendar years (Robinson 1986) in presenting the woodrat midden–Diamond Pond core comparisons (Fig. 13.11).

Because of natural variations in atmospheric ^{14}C, a single radiocarbon age may have more than one calendar date and the standard deviation may be considerably larger than indicated by counting statistics alone. It was conceivable that the pattern of dates of juniper-dominated woodrat middens (Fig. 13.7) was in part an artifact of differences in rates of radiocarbon production. The Diamond Pond core samples, on the other hand, are in stratigraphic order. Changing patterns of juniper values are in proper sequence regardless of how accurately they are dated. Also, in this case, close-interval analyses showed that juniper pollen peaks gradually rose and fell over several samples representing a decade, a century, or longer.

To confirm that clusters or single radiocarbon dates from woodrat middens matched periods of larger juniper pollen values, both data sets were calibrated to calendar years. Of special interest was the apparent double peak of

midden dates around 2360 B.P. and 2700 B.P. that corresponded to distinctive events in tree-ring-correction curves of radiocarbon ages (Pearson and Stuiver, 1986; Robinson, 1986). Comparing Figures 13.7 and 13.11, it is clear that tree-ring corrections do not alter the pattern of dates but further distinguish the peaks (tree-ring-corrected to about 440 B.C. and 870 B.C.) that match the independently dated largest juniper pollen percentages. Thus, distribution of woodrat midden dates is not merely an artifact of perturbations in atmospheric ^{14}C.

Tree Rings, Pollen, and Charcoal

Having established the chronological correspondence between individual juniper-containing middens and the continuous record of juniper pollen from Diamond Pond, it is appropriate to ask what the juniper pollen percentages represent (e.g., numbers of trees, stages of woodland development, pollen production, responses to fire or drought). Figures 13.12 and 13.13 illustrate the distribution of samples analyzed over 12 m and 2.5 m of the Diamond Pond core. All samples are 1 cm tall; however, because deposition rates varied through the core, a sample at 7.4 m (four thousand years ago) represents about 10.4 years, at 2 m (four hundred years ago) about 2.6 years, and at 0.30 m (forty years ago) about 1.4 years. Greater variation among adjacent samples would be expected as years per sample decreased; the factors being measured also would change. Thus, the apparent response of junipers, as discerned from their fossil pollen, is partly a matter of scale.

A series of pollen analyses of samples averaging ten years' duration might record a wave of juniper expansion or decline. Another averaging a year or two in duration could reveal shorter events along the wave's crest such as differences in vigor of pollen-producing trees or the immediate effects of fire or disease. Agreement between broad swells in the pollen and macrofossil records clearly represents the former (Fig. 13.11). The latter is apparent, though not fully understood, in the details revealed by comparisons of tree rings with juniper pollen and charcoal (Fig. 13.14).

Though potentially giving the best estimates of juniper pollen abundance, pollen accumulation rates are suspect at Diamond

DIAMOND POND
(Summary Diagram to 2000 B.P.)

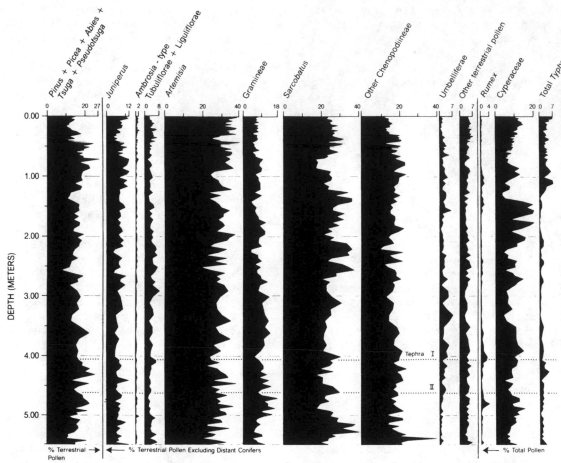

Fig. 13.9. Summary percentage pollen diagram for 126 samples dated to 2000 B.P. from Diamond Pond, Oregon. Note the changing values of shrubs, pond-fringe species, aquatic plants, and algae that all show variation not apparent from the woodrat middens alone. Exclusion of pollen of distant conifers

Pond. Because the small basin has steep slopes and sediments have varied (Wigand, 1987), pollen and charcoal accumulation rates were influenced by the intensity of storms and the vegetation cover's ability to impede slope wash. Changing water depth, surface area, aquatic vegetation, and the filtering fringe of bulrushes and cat-tails all have influenced depositional environments. Recirculation in dry desert dust also might operate to increase annual influx of pollen during droughts.

Pollen percentages are influenced by constraints that may act in concordance with en-

vironmental factors to emphasize small juniper pollen values. In droughty periods such as the 1930s, juniper pollen production could have faltered with tree vigor. Also, Harney and Malheur lakes were small or completely dry (Piper et al., 1939); prolific pollen producers such as *Atriplex* and *Sarcobatus* invaded their empty beds. Thus, percentage of juniper pollen might have declined even more as Malheur Lake shrank. Perhaps it is this combination of factors that affords such good agreement with the chronologies for juniper pollen, juniper from woodrat middens, and water-depth indicators. We explore

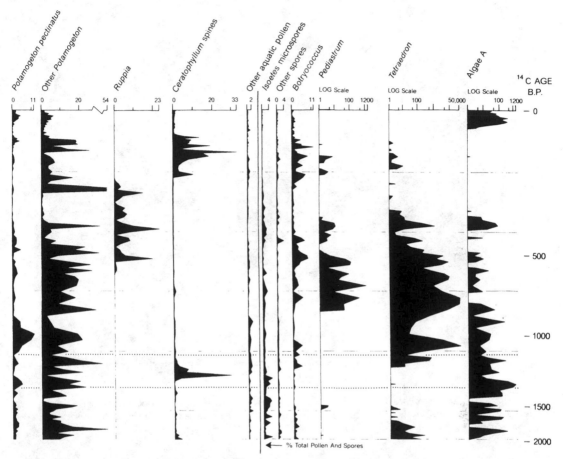

from the terrestrial pollen sum as shown here is followed in calculating percentage of juniper pollen in Figures 13.10 and 13.14. Algae are given as per- centage of pollen and spores; Algae A is an undeter- mined genus of dinoflagellate.

this possibility by reference to details of the past five hundred years and by comparing pollen and charcoal values with a tree-ring series from western juniper (Holmes et al., 1986).

Over fifty years ago Jessup (1935) recognized a correspondence between wide annual rings in eastern Oregon's western junipers and plentiful precipitation. According to recent studies of climate-response functions from tree rings and monthly weather records, mild winters with ample precipitation and cool springs favor wide annual rings in western juniper (Earle and Fritts, 1986; Fritts and Xiangding, 1986). The same factors would

lead to larger lakes with loss of lake-bed and playa-edge pollen producers and *relatively* more juniper in the regional pollen rain— even if juniper pollen production were un- affected. The smallest juniper values (juniper $cm^{-2} yr^{-1}$ and percentage) in the last five hun- dred years coincide with a distinct tree-ring- indicated drought in the early 1500s and with abundant charcoal (Fig. 13.14). Over the next two hundred years juniper values built gradu- ally while charcoal values declined. Together with the charcoal, this sequence suggests that junipers were relatively abundant by the late 1700s but waned with multiple burns through

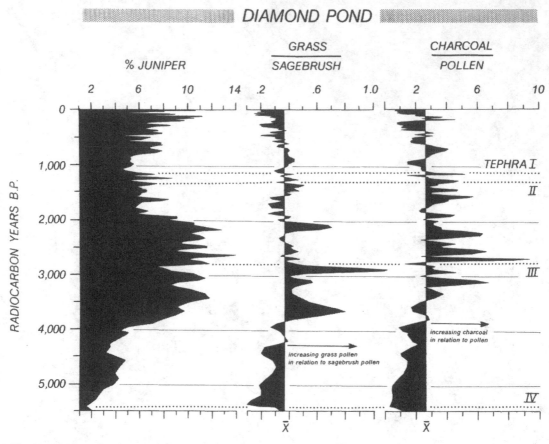

Fig. 13.10. Three-level smoothed (normal curve-smoothing function, Holloway, 1958) juniper pollen percentages and ratios from Diamond Pond are plotted about the mean of the average values for each five-hundred-year interval since 5500 B.P.

the 1860s. Further decline in the 1890s may reflect drought and cutting by early settlers.

Juniper expansion over the last hundred years is historically documented; yet Figure 13.14 shows a complicated sequence. Juniper pollen production must have varied with vigor wrought by vagaries of short, sharp events in annual distribution of precipitation and temperature, and by barely perceptible maturation of pollen-producing trees. As a matter of percentage constraint alone, percentage of juniper pollen also rose and fell with expanding and shrinking lakes in the Harney Basin. These effects are seen especially in the increases in the early 1900s, the precipitous fall through the drought of the 1930s, and the sharp rise thereafter. Accumulation rates show relatively small amounts of juniper pollen from about 1850 to 1940—with distinct fluctuations matching the 1890s and 1930s percentage declines—then a fivefold increase to the 1980s.

Relatively little charcoal and rising juniper values chronicle fire suppression and expanding woodlands over the last fifty years (Steele et al., 1986). Perhaps trees from the first waves of historic expansion are just now reaching their potential for pollen production, while the post-1930s trees are beginning to contribute as well. The lag between establishment of juniper stands and significant contributions to the regional pollen rain is not obvious over centuries or millennia because samples containing sediments deposited over many years smooth events that even then are considerably shorter than the uncertainties in radiocarbon dates. Thus, chronological agreement between juniper pollen values from

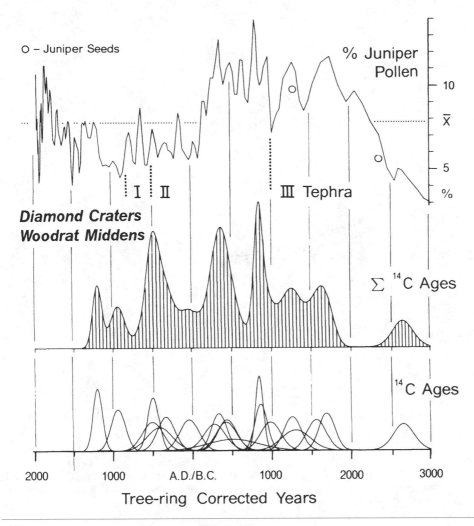

Fig. 13.11. Smoothed Diamond Pond juniper pollen percentages (Fig. 13.10) and tree-ring-corrected woodrat midden dates plotted as a normal distribution with a probability of one; width at the base is three standard deviations.

cores, and macrofossils and pollen from woodrat middens is all the more remarkable.

Despite limitations, relationships between tree-ring widths and pollen and charcoal values since A.D. 1500 demonstrate the applicability of fossil records in explaining natural variation in eastern Oregon's steppe and woodland. Further understanding will require a more precise chronology of the last two hundred years through close-interval cesium-137 and lead-210 dating of the Diamond Pond cores. Comparisons with western juniper fire-scar chronologies and stand characteristics (Eddleman, 1987; Young and Evans,

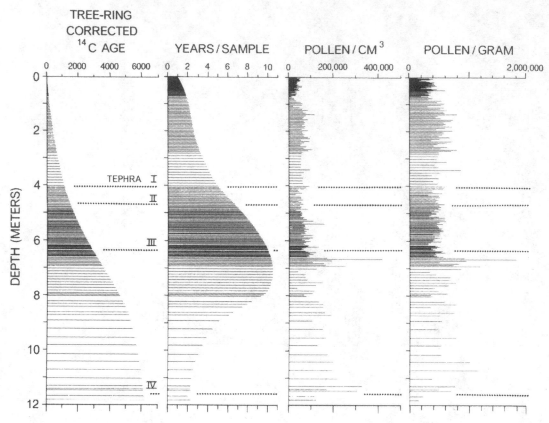

Fig. 13.12. Distribution of 195 pollen samples spanning the past 6300 tree-ring-corrected radiocarbon years B.P. (about 4300 B.C.) from 12 m of Diamond Pond core. All samples are 1 cm tall, but, because deposition rates vary, individual samples may contain fragments covering one to eleven years. Though some mixing in the bottom sediments no doubt tends to smooth variation in pollen deposited annually, it cannot explain similarity in pollen influx over the last 3000 years (6.2 m), despite changing deposition rates. In short, pollen accumulation rates from Diamond Pond are suspect; they are used here only for the last 500 years in Fig. 13.14.

1981) at Diamond Craters and on Steens Mountain could help to evaluate our findings of the last few hundred years and calibrate them for interpreting longer records. For example, would fire-scar studies indicate that the charcoal differences (e.g., between A.D. 1650–1750 and A.D. 1760–1860) resulted from changes in regional fire frequency, intensity, or both?

For whatever reasons, charcoal and juniper pollen values seem related on a scale of decades (Fig. 13.14) or centuries (Fig. 13.10). Over the last five hundred years juniper values generally fell with fire and rose with declining charcoal. The period of corresponding greatest juniper pollen and chronological clusters of juniper-containing woodrat middens (2200–3600 B.P.) also shows the largest charcoal/pollen and grass/sagebrush ratios. On a scale of centuries we see a time when abundant juniper fueled the fires that favored grass and discouraged sagebrush (Fig. 13.10). Charcoal from the midden pollen samples confirms its general abundance during this period of juniper pollen importance at Diamond Pond (Table 13.3).

DISCUSSION AND CONCLUSIONS

We have emphasized the relationships of three kinds of information bearing on the vegetation history of southeastern Oregon—macrofossils from woodrat middens, pollen

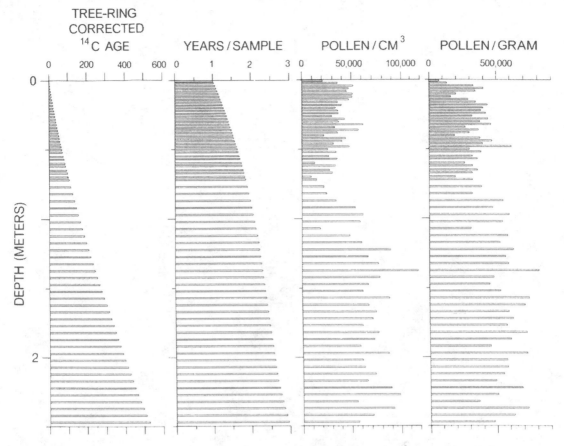

Fig. 13.13. Distribution of sixty-seven pollen samples spanning the last five hundred tree-ring- corrected radiocarbon years. Each sample is 1 cm tall and covers, on the average, one to three years.

from woodrat middens, and pollen from cores. Macrofossils from Diamond Pond cores were described by Wigand (1987). In addition we have attempted to decipher the message left by charcoal fragments in the pollen samples. We conclude by briefly considering what might be said about past vegetation using only one of these sources of informa- tion as compared with all three.

The most detailed information for a single time at a single locality comes from the woodrat midden macrofossils. This series contains fifty-four identified taxa, with as many as thirty-four in a single midden. Taken individually or as a group they indicate a landscape with juniper more widespread but otherwise little different from now. Though apparent clusters of dates might be intriguing, the Diamond Craters midden macrofossils,

taken by themselves, suggest neither varying environments nor a potential significance to the pattern in their ages.

Pollen analyses of Graben Dome and Nolf Crater middens confirmed preeminence of ju- niper and sagebrush, and might suggest their regional importance as well. However, uncer- tainty of their near or distant origin and the collecting proclivities of woodrats limit util- ity of these data, which give few additional clues to the nature or richness of the local flora. Charcoal fragments in the pollen samples suggest nearby and distant fires. Pol- len also revealed local absence and apparent regional unimportance of western juniper, while a midden from the Catlow Rim, at Three Mile Creek, accumulated about 6100 B.P.

Because most macrofossils in the middens

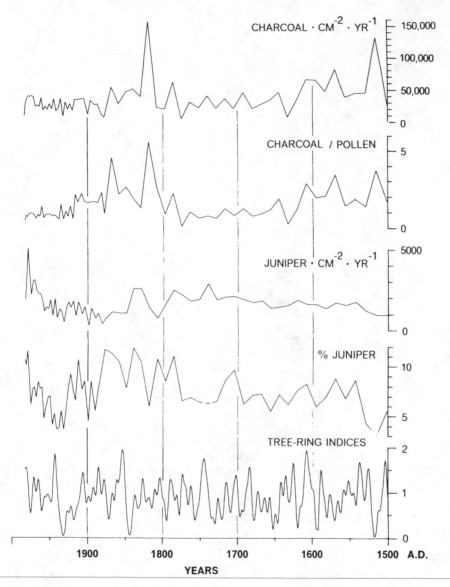

Fig. 13.14. A comparison of thirteen-level smoothed tree-ring indices (low-pass filters, Fritts, 1976, p. 270) of western juniper (*Juniperus occidentalis*) from Little Juniper Mountain, Oregon (Holmes, et al., 1986, p. 90), with percentage of juniper pollen (Fig. 11.9), the charcoal/pollen ratio, and accumulation rates of juniper pollen and charcoal for sixty-seven Diamond Pond samples dating from A.D. 1500 through 1985. Twenty samples fall within the last fifty years and thirty-four in the last one hundred years.

must have come from within the 50 m or less that woodrats usually cover when foraging, these midden records indicate actual movement of juniper populations rather than just increased density. We suspected this fact but could not have proven it with the pollen data alone. The woodrat middens gave local vege-tational details, and woodrat midden pollen spectra suggested regional, though similar, in-terpretations of juniper and sagebrush domi-nance. But only pollen, seeds, and algae from the Diamond Pond cores revealed the pace and regional consequences of environmental change. Water depth at Diamond Pond results from regional recharge and evaporation; fossils of aquatic and pond-edge species record local water conditions. Likewise, relative abundance of *Sarcobatus* and other chenopod pollen reflects the importance of local shadscale species and downwind areas of drowned or dry lake bed.

Together, pollen of aquatic and terrestrial species from Diamond Pond afford a measure and a sequence of past climatic fluctuations lacking from the midden records. Without the midden fossils, however, one could only surmise that *Oryzopsis*, *Sitanion*, and *Stipa* were important grasses or that genera producing little or indistinct pollen but sharing importance with western juniper in middens elsewhere (e.g., *Purshia* and *Cercocarpus*; Mehringer and Wigand, 1987, Table 2) were probably absent from the Diamond Craters flora over the last five thousand years.

Though different from the present in their abundant juniper, Diamond Craters middens apparently obscured events along the hard rimrock that left clear tracks in Diamond Pond's mud. Nonetheless, ages of middens corresponded closely to some of these events. Complacency of the woodrat midden assemblages is not well explained by unchanging Holocene vegetation on basalt flows and rimrock, nor by midden records that send only a very local signal different from that of core pollens. In fact, the midden macrofossils and Diamond Pond microfossils come together comfortably in a scenario of waves of juniper crossing a sagebrush sea in times of high water. It seems that Diamond Craters middens mostly date to climatic episodes that favored western juniper near its lower elevational limits and thereby, perhaps, woodrats as well.

Without Diamond Pond's continuous sequence, the regional supremacy of sagebrush, punctuated by episodes of juniper, grass, shadscale, and the charcoal that chronicles fire, would remain unknown. Without woodrat middens the details of the paleoflora and vegetation history would remain unclear. Together the two complementary methods give greater resolution that either could give alone.

ACKNOWLEDGMENTS

These studies, initiated by National Science Foundation grants BNS-77-12556 and BNS-80-06277, were supported primarily by Bureau of Land Management Contrast no. YA551-CT5-340075. We appreciate cooperation of personnel of the Bureau of Land Management, especially Y. A. Witherspoon, and thank S. A. Andersen, E. Benedict, J. F. Knobbs, J. C. Mehringer, S. K. Urstad, and P. E. Van de Water for field or laboratory assistance; J. Mastrogiuseppe and P. M. Peterson for identifying plants; A. J. Gilmartin for extending the hospitality of the Marion Ownbey Herbarium, J. C. Sheppard for radiocarbon dating, F. H. Whicker for cesium dating, and J. L. Betancourt for helpful suggestions on the manuscript.

REFERENCES

Aikens, C. M.; Grayson, D. G.; and Mehringer, P. J., Jr., (1982). *Final project report to the National Science Foundation on the Steens Mountain Prehistory Project*. Part III, *Technical description of projects and results*. Department of Anthropology, University of Oregon, Eugene.

Anderson, L. C. (1980). "A new species of fossil *Chrysothamnus* (Asteraceae) from New Mexico." *Great Basin Naturalist* 40(4), 351–52.

———. (1986a). "An overview of the genus *Chrysothamnus* (Asteraceae)." In *Proceedings of the Symposium on the Biology of* Artemisia *and* Chrysothamnus, *July 9–13, 1984, Provo, Utah*. E. D. McArthur and B. L. Welch, comp., pp. 29–54. Ogden, Utah: U.S. Forest Service, Intermountain Research Station, General Technical Report INT-200.

———. (1986b). "Sympatric subspecies in *Chrysothamnus nauseosus*." In *Proceedings of the Symposium on the Biology of* Artemisia *and* Chrysothamnus, *July 9–13, 1984, Provo, Utah*. E. D. McArthur and B. L. Welch, comp., pp. 98–103. Ogden, Utah: U.S. Forest Service, Intermountain Research Station, General Technical Report INT-200.

Barrett, S. W., and Arno, S. F. (1982). "Indian fires as an ecological influence in the northern Rockies." *Journal of Forestry* 80(10), 647–51.

Burkhardt, J. W., and Tisdale, E. W. (1976). "Causes of juniper invasion in southwestern Idaho." *Ecology* 57(3), 472–84.

Cronquist, A.; Holmgren, A. H.; Holmgren, N. H.; Reveal, J. L.; and Holmgren, P. K. (1977). *Intermountain flora, vascular plants of the intermountain West, U.S.A.* Volume 6, *The monocotyledons*. New York: Columbia University Press.

Cronquist, A.; Holmgren, A. H.; Holmgren, N. H.; Reveal, J. L.; and Holmgren, P. K. (1984). *Intermountain flora, vascular plants of the inter-*

mountain West, U.S.A. Volume 4, *The subclass Asteridae (except Asteraceas).* New York: Columbia University Press.

Dean, W. E., Jr. (1974). "Determination of carbonate and organic matter in calcareous sediments and sedimentary rocks by loss on ignition: Comparison with other methods." *Journal of Sedimentary Petrology* 44, 242–48.

Downs, J. F. (1966). "The significance of environmental manipulation in Great Basin cultural development." In *The current status of anthropological research in the Great Basin: 1964.* L. d'Azevedo et al., eds., pp. 39–56. Reno: University of Nevada, Desert Research Institute Social Sciences and Humanities Publications 1.

Earle, C. J., and Fritts, H. C. (1986). *Reconstructing river flow in the Sacramento Basin since 1560.* Report for California Department of Resources, Agreement no. DWR B-55395. Laboratory of Tree-ring Research, University of Arizona, Tucson.

Eddleman, L. E. (1987). "Establishment and stand development of western juniper in central Oregon." In *Proceedings of the Pinyon-Juniper Conference, January 13–16, 1986, Reno, Nevada,* pp. 255–59. Ogden, Utah: U.S. Forest Service, Intermountain Research Station, General Technical Report INT-215.

Fritts, H. C. (1976). *Tree rings and climate.* New York: Academic Press.

Fritts, H. C., and Xiangding, W. (1986). "A comparison between response-function analysis and other regression techniques." *Tree-Ring Bulletin* 46, 31–46.

Geyh, M. A. (1980). "Holocene sea-level history: Case study of the statistical evaluation of ^{14}C dates." *Radiocarbon* 22(3), 695–704.

Geyh, M. A., and de Maret, P. (1982). "Histogram evaluation of ^{14}C dates applied to the first complete Iron Age sequence from west central Africa." *Archaeometry* 24(2), 158–63.

Gruell, G. E. (1985). "Fire on the early Western landscape: An annotated record of wildland fires 1776–1900." *Northwest Science* 59(2), 97–107.

Hitchcock, C. L., and Cronquist, A. (1973). *Flora of the Pacific Northwest, an illustrated manual.* Seattle: University of Washington Press.

Holloway, J. L., Jr. (1958). "Smoothing and filtering of time series and space fields." In *Advances in geophysics.* H. E. Landsberg and J. Van Mieghem, eds., pp. 351–89. New York: Academic Press.

Holmes, R. L.; Adams, R. K.; and Fritts, H. C. (1986). *Tree-ring chronologies of western North America: California, eastern Oregon and northern Great Basin with procedures used in the chronology development work including users' manuals for computer programs COFECHA and ARSTAN.* Chronology Series 6, Laboratory of Tree-Ring Research, University of Arizona, Tucson.

Hubbard, L. L. (1975). *Hydrology of Malheur Lake, Harney County, southeastern Oregon.* Portland, Ore.: U.S. Geological Survey, Water Resources Investigations 21-75.

Huttunen, P., and Merilainen, J. (1978). "New freezing device providing large unmixed sediment samples from lakes." *Annales Botanici Fennici* 15(2), 128–30.

Jessup, L. T. (1935). "Precipitation and tree growth in the Harney Basin, Oregon." *Geographical Review* 25(2), 310–12.

Klein, J.; Lerman, P. C.; Damon, P. E.; and Ralph, E. K. (1982). "Calibration of radiocarbon dates." *Radiocarbon* 24, 103–50.

Lawton, H. W.; Wilke, P. J.; DeDecker, M.; and Mason, W. M. (1976). "Agriculture among the Paiute of Owens Valley." *Journal of California Anthropology* 3(1), 13–50.

McArthur, E. D.; Blauer, A. C.; Plummer, A. P.; and Stevens, R. (1979). *Characteristics and hybridization of important intermountain shrubs. III. Sunflower family.* Ogden, Utah: U.S. Forest Service, Intermountain Research Station, Research Paper INT-220.

Mehringer, P. J., Jr. (1985). "Late-Quaternary pollen records from the interior Pacific Northwest and northern Great Basin of the United States." In *Pollen records of late-Quaternary North American sediments.* V. M. Bryant, Jr., and R. G. Holloway, eds., pp. 167–87. Dallas: American Association of Stratigraphic Palynologists.

———. (1986). "Prehistoric environment." In *Handbook of North American Indians.* Volume 11, *The Great Basin.* W. L. d'Azevedo, ed., pp. 31–50. Washington, D.C.: Smithsonian Institution.

———. (1987). *Late Holocene environments on the northern periphery of the Great Basin.* Final Report to Bureau of Land Management, Oregon State Office, Portland, for contract YA551-CT5-340075.

Mehringer, P. J., Jr., and Wigand, P. E. (1986). "Holocene history of Skull Creek Dunes, Catlow Valley, Oregon, U.S.A." *Journal of Arid Environments* 11, 117–38.

Mehringer, P. J., Jr., and Wigand, P. E. (1987). "Western juniper in the Holocene." In *Proceedings of the Pinyon-Juniper Conference, January 13–16, 1986, Reno, Nevada.* R. L. Everett, comp., pp. 109–19. Ogden, Utah: U.S. Forest Service, Intermountain Research Station, General Technical Report INT-215.

O'Sullivan, P. E. (1983). "Annually-laminated sediments and the study of Quaternary environmental changes—A review." *Quaternary Science Reviews* 1, 245–313.

Patterson, W. A. III; Edwards, K. J.; and Maguire,

D. J. (1987). "Microscopic charcoal as a fossil indicator of fire." *Quaternary Science Reviews* 6, 3–23.

Pearson, G. W., and Stuiver, M. (1986). "High-precision calibration of the radiocarbon time scale, 500–2500 B.C." *Radiocarbon* 28(2B), 839–62.

Peterson, N. V., and Groh, E. H. (1964). "Diamond Craters, Oregon." *Ore Bin* 26, 17–33.

Piper, A. M.; Robinson, T. W.; and Park, C. F., Jr. (1939). *Geology and ground-water resources of the Harney Basin, Oregon.* U.S. Geological Survey, Water-Supply Paper 841.

Robinson, S. W. (1986). *A computational procedure for utilization of high-precision radiocarbon curves.* Open-file Report, U.S. Geological Survey, Menlo Park.

Shultz, L. M. (1986). "Taxonomic and geographic limits of *Artemisia* subgenus Tridentatae (Beetle) McArthur (Asteraceae: Anthemideae)." In *Proceedings of the Symposium on the Biology of Artemisia and Chrysothamnus, July 9–13, 1984, Provo, Utah.* E. D. McArthur and B. L. Welch, comp., pp. 20–28. Ogden, Utah: U.S. Forest Service, Intermountain Research Station, General Technical Report INT-200.

Steele, R.; Arno, S. F.; and Geier-Hayes, K. (1986). "Wildfire patterns change in central Idaho's ponderosa pine–Douglas-fir forest." *Western Journal of Applied Forestry* 1(1), 16–18.

Stuiver, M.; Kromer, B.; Becker, B.; and Ferguson, C. W. (1986). "Radiocarbon age calibration back to 13,300 years B.P. and the ^{14}C age matching of the German oak and U.S. bristlecone pine chronologies. *Radiocarbon* 28(2B), 969–79.

Thompson, R. S. (1985). "Palynology and *Neotoma* middens." In *Late Quaternary vegetation and climates of the American Southwest.* B. F. Jacobs, P. F. Fall, and O. K. Davis, eds., pp. 89–112. Dallas: American Association of Stratigraphic Palynologists Foundation.

Verosub, K. L., and Mehringer, P. J., Jr. (1984). "Congruent paleomagnetic and archeomagnetic records from the western United States: A.D. 750 to 1450." *Science* 224, 387–89.

Verosub, K. L.; Mehringer, P. J., Jr.; and Waterstraat, K. L. (1986). "Holocene secular variation in western North America: The paleomagnetic record from Fish Lake, Harney County, Oregon." *Journal of Geophysical Research* 91(B3), 3609–23.

Walker, G. W. (1979). *Revisions of the Cenozoic stratigraphy of Harney Basin, southeast Oregon.* U.S. Geological Survey Bulletin 1475.

Walker, G. W., and Swanson, D. A. (1968). *Summary report on the geology and mineral resources of the Harney Lake and Malheur Lake areas of the Malheur National Wildlife Refuge, north-central Harney County, Oregon and Poker Jim Ridge and Fort Warner areas of the Hart Mountain National Antelope Refuge, Lake County, Oregon.* U.S. Geological Survey Bulletin 1260-L.

Wigand, P. E. (1987). "Diamond Pond, Harney County, Oregon: Vegetation history and water table in the eastern Oregon desert." *Great Basin Naturalist* 47(3), 427–58.

Young, J. A., and Evans, R. A. (1981). "Demography and fire history of a western juniper stand." *Journal of Range Management* 34(6), 501–6.

Chapter 14

The Ecology and Paleoecology of Grasses in Selected Sonoran Desert Plant Communities

Thomas R. Van Devender   Laurence J. Toolin   Tony L. Burgess

Well-preserved plant macrofossils from fossil packrat (*Neotoma* spp.) middens in North American deserts have allowed excellent reconstructions of vegetation and climate histories for rocky slope habitats over the last thirty thousand years (Van Devender and Spaulding, 1979; Van Devender et al., 1987). In the late Wisconsin glacial period pinyon-juniper or juniper woodlands were widespread in the modern Sonoran and Mojave deserts of Arizona and California (Spaulding, *this volume* chap. 9; Van Devender, 1987, *this volume* chap. 8). Characteristic Mojave Desert plants such as *Yucca brevifolia* (Joshua tree) and *Salvia mohavensis* (Mojave sage) were once widespread in Arizona in areas that now are occupied by characteristic Sonoran Desert dominants such as *Carnegiea gigantea* (saguaro) and *Cercidium microphyllum* (foothills palo verde).

The ecology of the grasses in the Sonoran Desert has received little study, and their fossil record is just emerging. Soreng and Van Devender (1989) discussed the remains of *Poa fendleriana* (muttongrass), a mesic C3 perennial of montane forests, in late Wisconsin samples from Pontatoc Ridge in the Santa Catalina Mountains and Montezuma's Head in the Ajo Mountains, Arizona. Grasses were reported in middens from the Puerto Blanco Mountains of southwestern Arizona (Van De-

vender, 1987) and the Picacho Peak area of southeastern California (Cole, 1986) but were not discussed in detail. Here we discuss adaptations and ecology of grasses found in selected Sonoran Desert plant communities and review their paleoecology over the last eleven thousand years based on remains from packrat midden sequences in four study areas.

METHODS

Fossil packrat middens are hard, dark, organic deposits preserved in dry rock shelters. They contain abundant, well-preserved plant remains that were collected by the packrats within about 30–50 m of the sites. Middens often yield more species of plants than the modern flora because fifty or more years may be involved in the formation of an indurated midden. Unlike many other plants, grass remains are typically uncommon in midden assemblages, establishing their presence but not their abundance near the site. Radiocarbon dating allows the plant macrofossil assemblages to be placed in a time sequence. Radiocarbon ages cited in this chapter are in ka (thousands of years before the present), with full dates and associated data listed in Table 14.1.

Sixty-two fossil packrat middens provide chronological sequences of local rocky-slope vegetation from four Sonoran Desert study areas: the Puerto Blanco, Tinajas Altas, and Butler mountains in southwestern Arizona and the Whipple Mountains in southeastern California (Table 14.1, Fig. 14.1; see also Van Devender, *this volume*, chap. 8). Midden grass fossils were identified through comparison with reference collections in the University of Arizona Herbarium and the Laboratory of Paleoenvironmental Studies (Fig. 14.2). Detailed floristic notes and field collections were made near each site and in the general area. Voucher specimens of modern plant collections are in herbaria at the University of Arizona and Organ Pipe Cactus National Monument. A. C. Sanders's specimens from the Whipple Mountains are in the University of California at Riverside Herbarium. Modern grass distributions were determined by surveying the University of Arizona Herbarium. Nomenclature follows Lehr (1978), with authorities cited for exceptions. Precipitation at the midden sites was estimated using elevational lapse rates derived from regional climatic data compiled in the U.S. Geological Survey Ecohydrology Project Office.

ENVIRONMENTAL SETTING

The Sonoran Desert is the arid to semiarid subtropical area centered around the head of the Gulf of California in western Sonora, southwestern Arizona, southeastern California, and much of the Baja California peninsula (Shreve, 1964; Turner and Brown, 1982; Fig. 14.1). It features many distinctive life forms, including trees, shrubs, columnar cacti, rosette succulents, and so on. Elevations range from below sea level in the Imperial Valley of California and 278 m at Needles on the Colorado River to about 1000 m at the northeastern margin (Turner and Brown, 1982). Most parts of the Sonoran Desert occasionally experience freezing temperatures, but the duration of an episode is rarely more than one night. The northern boundary of the Sonoran Desert in Arizona roughly coincides with the range limit of *Carnegiea gigantea*, which can be killed by a freeze of more than thirty-two hours (Shreve, 1911; Turnage and Hinckley, 1938).

Subdivisions of the Sonoran Desert were proposed by Shreve (1964) and refined by Turner and Brown (1982). Our study areas are in the Arizona Upland and the lower Colorado River valley (Fig. 14.1). The former receives 200–245 mm of precipitation per year, with 30%–60% in the summer (June–August; Turner and Brown, 1982), while the latter receives less than 100 mm/yr (Ezcurra and Rodrigues, 1986). The Arizona Upland desertscrub communities are dominated by *Carnegiea gigantea*, *Cercidium microphyllum*, and *Ambrosia deltoidea* (triangle-leaf bursage) in association with many shrubs and succulents.

The vegetation of the lower Colorado River valley is mostly an open desertscrub dominated by *Larrea divaricata* Cov. (creosote bush) and *Ambrosia dumosa* (white bursage), with *Hilaria rigida* (big galleta) on sandy flats and these three mixed with other shrubs on rocky slopes. Arizona Upland dominants such as *Carnegiea gigantea* and *Cercidium microphyllum* are locally absent or restricted to washes. The lower Colorado River valley subdivision can be viewed as the central subtropical unit in a series of desertscrub communities along a winter temperature gradient. Colder winters characterize the Mojave Desert, which adjoins this subdivision to the north. To the south it merges with the more tropical Central Gulf Coast and Plains of Sonora subdivisions, the latter also distinguished by a greater amount of summer rain.

DESERT ADAPTATIONS OF SONORAN GRASSES

The grasses in the study areas have quite an array of morphological, physiological, and phenological adaptations that allow them to survive in arid environments. Several unrelated species show similar suites of characters corresponding to convergent patterns of soil-moisture exploitation and demography. Two important climatic features for desert grasses are temperatures—when soil moisture is seasonally available (winter-spring and middle summer–early fall)—and the duration of drought, especially the length of the hot, dry foresummer in May and June. While freezes may halt the growing season for some perennials, or even kill small ephemeral grasses, they are less limiting for grasses than for subtropical shrubs and succulents. Adaptations to high temperatures and drought are evident in the uses of different photosynthetic pathways and the life histories of individual grass

Table 14.1. *Radiocarbon dates from packrat middens from Puerto Blanco (Ajo Loop = AL, Cholla Pass = CP, Twin Peaks = TP), Tinajas Altas (TA), and Butler (BM) mountains, Pima and Yuma counties, Arizona; and Whipple Mountains (Falling Arches = FA, Redtail Peaks = RP, Tunnel Ridge = TR, Whipple Mountains = WM), San Bernardino County, California. Asterisk indicates tandem accelerator mass spectrometer date.*

Sample	Elevation (m)	Age (B.P.)	Lab. No.	Material Dated
Puerto Blanco Mountains				
TP#1	565	14,120 ± 260	A-3986	*Juniperus californica* twigs
CP#1A	605	10,540 ± 250	A-4276	*Neotoma* fecal pellets
TP#5	560	9720 ± 460	*AA-625	*Encelia farinosa* twigs
CP#1B	605	9070 ± 170	A-4278	*Neotoma* fecal pellets
CP#1C	605	8790 ± 210	A-4277	*Neotoma* fecal pellets
AL#1D	550	7970 ± 130	A-3981	*Neotoma* fecal pellets
TP#6	580	7580 ± 100	A-3980	*Neotoma* fecal pellets
AL#1G	550	7560 ± 170	A-3985	*Neotoma* fecal pellets
AL#1A	550	5240 ± 90	A-4213	*Neotoma* fecal pellets
AL#5B	535	3480 ± 80	A-4233	*Neotoma* fecal pellets
CP#2	590	3440 ± 90	A-4279	*Neotoma* fecal pellets
AL#6C	550	3400 ± 100	A-4235	*Neotoma* fecal pellets
AL#6B	550	3220 ± 100	A-4234	*Neotoma* fecal pellets
TP#2	580	2340 ± 70	A-3973	*Neotoma* fecal pellets
TP#3	550	2160 ± 60	A-3982	midden debris
TP#4	550	1910 ± 50	A-3998	*Neotoma* fecal pellets
AL#3A	550	990 ± 50	A-3972	*Neotoma* fecal pellets
AL#4	550	980 ± 70	A-4214	*Neotoma* fecal pellets
AL#2C	550	130 ± 50	A-3971	*Neotoma* fecal pellets
AL#2B	550	30 ± 50	A-3970	*Neotoma* fecal pellets
Tinajas Altas Mountains				
TA#15A	550	15,680 ± 700	*AA-452	*Pinus monophylla* needles
TA#2B	365	10,950 ± 90	USGS-958	*Neotoma* fecal pellets
TA#3B	580	10,300 ± 100	USGS-959	*Neotoma* fecal pellets
TA#12A	565	10,070 ± 110	A-3730	*Neotoma* fecal pellets
TA#14	550	9900 ± 200	A-3570	*Juniperus californica* twigs
TA#3A	580	9700 ± 100	USGS-957	*Neotoma* fecal pellets
TA#2A	365	8970 ± 75	USGS-956	*Neotoma* fecal pellets
TA#16B	380	8700 ± 110	A-3647	*Neotoma* fecal pellets
TA#16A	380	7860 ± 100	A-3521	*Neotoma* fecal pellets
TA#13B	580	5940 ± 70	A-3507	*Neotoma* fecal pellets
TA#13A	580	5860 + 60	A-3506	*Neotoma* fecal pellets
TA#20C	460	5080 ± 80	A-3565	*Neotoma* fecal pellets
Butler Mountains				
BM#3A	240	11,250 ± 410	*AA-769	*Larrea divaricata* twigs
BM#3B	240	11,060 ± 360	*AA-770	*Juniperus californica* twigs
BM#3B	240	10,170 ± 610	*AA-768	*Larrea divaricata* twigs
BM#3B	240	10,615 ± 355	average of *AA-768 & *AA-770	
BM#5	250	10,360 ± 430	*AA-537	*Larrea divaricata* twigs
BM#1	245	8160 ± 250	A-3652	*Neotoma* fecal pellets
BM#6	245	6490 ± 90	A-3389	*Neotoma* fecal pellets
BM#4	250	3820 ± 70	A-3517	*Neotoma* fecal pellets
BM#3C	255	750 ± 50	A-3911	*Neotoma* fecal pellets
BM#2A	255	610 ± 50	A-3790	*Neotoma* fecal pellets
BM#2B	255	220 ± 70	A-3910	*Neotoma* fecal pellets

Whipple Mountains

RP#11	520	13,810 ± 270	A-1650	*Juniperus californica* twigs
FA#1	320	11,650 ± 190	A-1548	*Juniperus californica* twigs
RP#10B	490	10,490 ± 110	A-3944	*Juniperus californica* twigs
TR#2	365	10,330 ± 300	A-1470	*Juniperus californica* twigs
WM#2	520	9980 ± 180	A-1538	*Juniperus californica* twigs
RP#4B	510	9330 ± 110	A-3734	*Neotoma* sp. fecal pellets
RP#2A	510	9310 ± 150	A-3731	*Neotoma* sp. fecal pellets
RP#1	520	8910 ± 380	A-1580	*Juniperus californica* twigs
WM#6	520	8540 ± 100	A-3943	*Neotoma* sp. fecal pellets
WM#7	525	8180 ± 130	A-3732	*Neotoma* sp. fecal pellets
RP#12	510	8040 ± 120	A-3650	*Neotoma* sp. fecal pellets
WM#7L	525	7500 ± 80	A-3733	*Neotoma* sp. fecal pellets
WM#5S2L	520	6710 ± 160	A-3942	*Neotoma* sp. fecal pellets
WM#4	525	5020 ± 240	A-2160	*Larrea divaricata* twigs
RP#9B	510	4240 ± 70	A-3649	*Neotoma* sp. fecal pellets
WM#5S2	520	940 ± 70	A-3941	*Neotoma* sp. fecal pellets
WM#5S1	520	470 ± 60	A-3940	*Neotoma* sp. fecal pellets
TR#3	370	200 ± 70	A-3939	*Neotoma* sp. fecal pellets
TR#6	370	Modern	A-2159	*Neotoma* sp. fecal pellets

species that occur in the Sonoran Desert.

Photosynthetic Pathways

Because most metabolic activities proceed more rapidly at higher temperatures, photosynthesis should be greater under warm conditions. But for most plants with normal Calvin cycle, or C3, photosynthesis, the net gain of carbon decreases at higher temperatures due to photorespiration, a metabolic artifact of photosynthetic enzymes (Lorimer, 1981; Smith, 1976). Photorespiration can be circumvented by means of a syndrome of biochemical and anatomical features known as C4 metabolism. C4 plants generally have higher rates of carbon fixation under warm conditions, higher levels of light saturation, and lower carbon dioxide compensation points than C3 plants (Hatch and Osmond, 1976; Laetsch, 1974). Available evidence suggests that C4 metabolism is usually less efficient than C3 photosynthesis under cool temperatures or low light levels (Ehleringer, 1978; Ehleringer and Bjorkman, 1977). Fossil grasses with C4 anatomy have been reported from the late Miocene of California and Kansas (Thomasson et al., 1986), and the C4 syndrome appears to have evolved independently in different grass lineages (W. V. Brown, 1977; Gutierrez et al., 1974; Hattersley and Watson, 1976).

In the Chihuahuan (Kemp, 1983), Sonoran, and Mojave (Mulroy and Rundell, 1977) des-

erts ephemerals are predominantly segregated into C3 species that grow in response to cool winter-spring rains and C4 species exploiting warm summer rains. In areas with biseasonal rainfall a few C3 ephemeral dicots flower in the summer, but this is not true for C3 ephemeral grasses that are winter-spring obligates. Teeri and Stowe (1976) found the relative abundance of C4 grasses in a regional flora to be positively correlated with the minimum (i.e., nighttime) temperature during the growing season. In the mountains of Arizona, where temperature decreases and precipitation increases with elevation, C3 perennial grasses (*Agrostis, Bromus, Elymus, Festuca, Koeleria, Panicum, Poa*) predominate in cool mountaintop forests, often in association with montane C4 perennials (*Bouteloua, Muhlenbergia, Panicum*). C4 perennials are more common in warm, lowland grasslands and deserts (Table 14.2). More rainfall in both seasons at high elevations and more in winter in northern latitudes favor C3 perennial grasses. The subtropical summer rainfall of lower latitudes favors C4 perennial grasses. In the Sonoran Desert there is also an east-west shift in the predominant season of rainfall from summer in the Arizona Upland to winter in the western lower Colorado River valley. The dry, warm summers of the Mojave Desert inhibit C4 perennial grasses, whereas cool-season C3 perennials such as *Stipa speciosa* (desert needlegrass) are com-

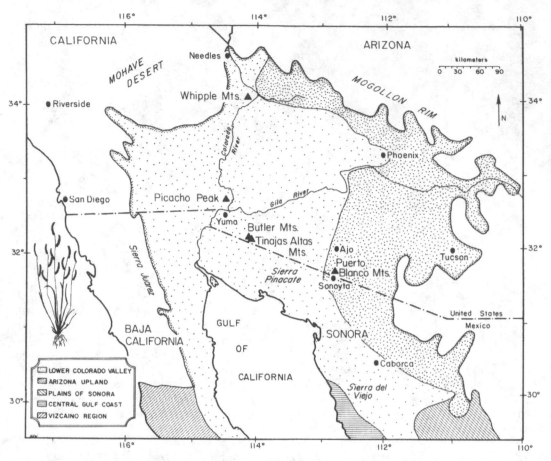

Fig. 14.1. Map of areas mentioned in text. Boundaries of Sonoran Desert and subdivisions follow Shreve (1964). Solid triangles are packrat midden sites.

mon. Farther south there is less contrast between winter and summer temperatures. This allows C4 grasses, which predominate at lower latitudes, to have a longer growing season.

In summarizing the available literature, Osmond et al. (1982) could find few ecological trends in C3/C4 plants that could be differentiated from phylogenetic trends. For example, are C4 grasses better adapted to aridity or is the pattern an artifact of the worldwide dominance of Eragrosteae and Aristideae in arid regions and of Chlorideae in the North American deserts? North of about latitude 30° N on the mainland, ephemeral grasses show separation in growing seasons: C3 winter, C4 summer. In warmer areas farther south, C4 ephemerals often extend their survival into winter and spring, although the pattern is obscured in Baja California. Almost all perennial grasses in the Sonoran Desert are C4 types. The most prominent exceptions, *Oryzopsis hymenoides* (Indian ricegrass) and *Stipa speciosa*, are ecologically important only in the northern and western margins of the Sonoran Desert where winter rainfall is much greater and more dependable than summer rain.

Size and Longevity

Survival of grasses in desert environments depends on a delicate balance between the availability of soil moisture and the leaf area conducting photosynthesis and transpiration. Plant size can indicate roughly how a grass utilizes moisture and survives drought. In deserts rainfall and hence production are usually low, and a medium-to-large perennial life-

Fig. 14.2. Fossil grasses from Tinajas Altas (TA) and Butler (BM) mountains packrat middens. From left: *Aristida* cf. *glauca* floret (BM#6, 6–5 ka); *Muhlenbergia microsperma* cleistogene (BM#6, 6–5 ka); *Hilaria rigida* floret (TA#12A, 10.1 ka); *Bouteloua barbata* floret (BM#1, 8.2 ka); *B. aristidoides* floret (BM#1, 8.2 ka); *Stipa speciosa* floret (TA#3A, 9.7 ka).

form can result only from a long life span, which is achieved by an ability to survive long droughts. Furthermore, individual long-term survival often depends on utilizing transient soil moisture from brief showers, which requires a rapid response in root absorption and photosynthesis (Nobel, 1980; Sala and Lauenroth, 1982). For example, *Hilaria rigida* is a large, essentially chamaephytic grass that is locally abundant in very hot, dry climates of the lower Colorado River valley (Table 14.2). Although its growth is limited by moisture availability, this grass can use low soil-moisture levels very efficiently and can extract water from depths up to 42 cm, which is probably deeper than most rainfall penetrates (Nobel, 1980, 1981). Rainfall triggers rapid shoot and leaf growth from older stems, and the rate of CO_2 uptake is the highest mea-

sured for any vascular plant (Nobel, 1980, 1981). Longevity of established clumps could be hundreds of years (Nobel, 1981), and perhaps only major catastrophic events cause unusual mortality. For example, a large decline in *H. rigida* numbers evident in matched photographs from 1907 and 1959 in the Pinacate region of northwestern Sonora (Martin and Turner, 1977) was attributed to a severe drought. Partial resurgence of the population was apparent by 1970.

On the other end of the spectrum, ephemeral grasses exploit wetter soil provided by sporadic heavy rains or sites where runoff is concentrated. Seeds may remain dormant in the soil until conditions are favorable (Shreve, 1964; Went, 1949). Individual size and fecundity are strongly dependent on the amount and duration of usable soil moisture. We use

Table 14.2. *Characteristics of selected Sonoran Desert grasses.*
Photosynthetic pathway taken from sources in footnotes. Question marks indicate that the
pathway is unknown for this species, but other members of the genus are of this type. Flowering
times in the Sonoran Desert are based on specimens in the University of Arizona Herbarium and
field observations. Seasons in parentheses are secondary or infrequent, opportunistic flowering
periods. Summer records for C3 species usually indicate a June extension of the spring rainy
period (i.e., *Bromus rubens, Schismus barbatus, Stipa speciosa, Vulpia microstachys,* and
V. octoflora). Winter records for C4 species may indicate extension of warm, moist fall
conditions into December (i.e., *Bothriochloa barbinodis, Digitaria californica, Echinochloa
colona, Leptochloa filiformis,* and *Panicum hirticaule*).

Size	Photosynthetic Pathway	Flowering Season
PERENNIAL		
Large		
Bothriochloa barbinodis	C4 (5)	Su-F (W)
Hilaria rigida	C4 (3)	W-Sp-Su(F)
Muhlenbergia porteri	C4 (4)	Su-F(W-Sp)
* *Sorghum halapense*	C4 (5)	Su-F(W-Sp)
Medium-Large		
Aristida ternipes	C4 (5)	Su-F(Sp)
Heteropogon contortus	C4 (5)	Su-F(W-Sp)
Sporobolus cryptandrus	C4 (5)	Su-F
Medium		
Oryzopsis hymenoides	C3 (5)	Sp-Su(F)
Stipa speciosa	?C3	W-Sp(Su)
Aristida glauca	C4 (5)	Su-F(W-Sp)
Aristida hamulosa	C4 (4)	Su-F(Sp)
Bouteloua eriopoda	C4 (4)	Su-F
Bouteloua rothrockii	?C4	Su-F(W)
Digitaria californica	C4 (5)	Su-F(W)
* *Echinochloa colona*	C4 (5)	Su-F(W)
* *Eragrostis lehmanniana*	C4 (5)	Su-F
Pappophorum mucronulatum	?C4	Su-F(Sp)
Setaria macrostachya		
(incl. *S. leucopila*)	C4 (4)	Su-F(W-Sp)
Small-Medium-Large		
Aristida calfornica	C4 (5)	Sp(Su-F-W)
Small-Medium		
Bouteloua chondrosioides	?C4	Su-F
Bouteloua repens	C4 (5)	Su-F(W-Sp)
Panicum hallii	C4 (5)	Su-F(Sp)
Tridens muticus	C4 (1)	Sp-Su-F(W)
Small		
Bouteloua trifida	?C4	All year
Cathestecum erectum	?C4	Su-F
PERENNIAL/EPHEMERAL		
Small		
Enneapogon desvauxii	C4 (1)	All year
Erioneuron pulchellum	C4 (4)	Sp-Su-F(W)
EPHEMERAL		
Medium		
Bromus arizonicus	C3 (5)	Sp
Leptochloa filiformis	?C4	Su-F(W)
Small-Medium		
Bromus rubens	C3 (5)	Sp(W-Su)

Hordeum murinum	C3 (5)	Sp(W-Su)
Hordeum pusillum	C3 (5)	Sp(W)
Phalaris minor	C3 (5)	Sp
†Cottea pappophoroides	?C4	Su-F
Eragrostis cilianensis	C4 (5)	Su-F(W-Sp)
Eriochloa acuminata	C4 (5)	Su-F
Muhlenbergia microsperma	?C4	W-Sp(Su-F)
Panicum hirticaule	C4 (2)	Su-F(W)
Small		
Poa bigelovii	?C3	W-Sp
*Schismus barbatus		
(incl. *S. arabicus*)	C3 (5)	W-Sp(Su)
Trisetum interruptum	?C3	Sp
Vulpia microstachys	?C3	Sp(Su)
Vulpia octoflora	C3 (1)	W-Sp(F)
Aristida adscensionis	C4 (5)	All year
Bouteloua annua	?C4	W
Bouteloua aristidoides	C4 (2)	Su-F(Sp)
Bouteloua barbata	C4 (2)	Su-F(W-Sp)
Brachiaria arizonica	?C4	Su-F
Muhlenbergia brandegei	?C4	W-Sp

(1) Kemp (1983)
(2) Mulroy and Rundel (1977)
(3) Nobel (1980)
(4) Syvertsen et al. (1976)
(5) Waller and Lewis (1979)

* Introduced species.
† Medium perennial over most of its range; some populations in the Sonoran Desert appear to be ephemeral or perennial/ephemeral.

the term "ephemeral" in the same sense as Shreve (1964) to describe a plant whose life cycle may be completed within one or two months in response to a single rainy episode. Shreve thought this term more appropriate than "annual," which includes plants whose life cycles encompass most of the frost-defined growing season. Ephemerals are usually less restricted to specific habitats than perennials. In dry years they may be absent from slopes and nearly so from flats, yet relatively common along desert washes. In years of high rainfall plants of the same species often form extensive stands on interfluves and upland sites. When conditions are favorable, reproduction by large numbers of individuals ensures that the seed bank is replenished. *Bouteloua aristidoides* (six-weeks needle grama) and *B. barbata* (six-weeks grama) pass years as seed in the summer-dry Mojave and western Sonoran deserts, depending on rare incursions of tropical storms to complete their life cycles (Huning, 1978; Tevis, 1958). In contrast, virtually all viable seeds of *Aristida contorta*, an Australian desert ephemeral or short-lived perennial, germinate whenever rain occurs (Mott, 1974). This

suggests that while seeds readily survive droughts, in some grass species they may not persist in the soil for long periods. In the case of *A. contorta*, seeds disperse widely from favorable microsites where plants can reproduce even in dry years, allowing replenishment of short-lived seed banks on drier sites.

Most medium-sized perennial grasses have demographies and water utilization strategies between these extremes. *Bouteloua gracilis* (blue grama), a widespread dominant of grasslands in the Great Plains and higher parts of the Southwest, can respond to a rainfall of less than 10 mm within twelve hours (Sala and Lauenroth, 1982), yet it is uncommon in more arid areas, probably because it is unable to reproduce effectively where drought is frequent. On the Jornada del Muerto of south-central New Mexico maximum life spans of seven medium-sized perennial grasses ranged from seven years for *Hilaria mutica* (tobosa) to twenty-eight years for *Bouteloua eriopoda* (black grama; Wright and Van Dyne, 1976). The probability of year-to-year survival increased with an individual's age, and these larger plants can produce large numbers of seeds. Significant population fluctuations probably occur in response to climatic fluctuations on the order of five to twenty years because recruitment may hinge on unusually favorable conditions. Neilson (1986) has shown episodic seedling establishment of *B. eriopoda* on the Jornada del Muerto.

In the Sonoran Desert the sizes, life histories, and water-utilization strategies of grass species change along an east-to-west gradient characterized by the following climatic trends: (1) total annual rainfall decreases; (2) droughts are longer; (3) rainfall distribution shifts from biseasonal to mostly winter. Along this gradient the grasses shift from predominantly medium-to-large perennials to mostly ephemerals and small perennials. *Hilaria rigida*, a large, long-lived perennial common at the arid end of the gradient, is an exception. However, on other continents similar large chamaephytic grasses can be found in deserts (Shmida and Burgess, 1988). Although the gradient is correlated with elevation, the elevational distributions of grasses at latitude 32° N in Arizona diverge from those at 28°–29° N in Sonora. For example, *Aristida ternipes* (spider grass), *Bouteloua repens* (slender grama), and *Digitaria californica* Henr. (cottontop) have ranges from about 1830 to 550 m in the Arizona Upland, yet they reach sea level around Guaymas, near the southern margin of the Sonoran Desert.

Individual and population sizes also shift from east to west. As rainfall decreases, the average size of individuals of most perennial grasses decreases, corresponding to a shorter life expectancy. Average population size also decreases, and distribution is patchier, a result of more episodic recruitment and higher rates of drought-induced mortality. As a species approaches its arid limits, older individuals become more restricted to relatively mesic microhabitats, and the erratic rainfall selects for quicker maturation on less-favorable sites. *Erioneuron pulchellum* (fluff grass) shifts from a predominantly summer-flowering perennial in the Chihuahuan Desert to a predominantly spring-flowering ephemeral that rarely perennates in the northern Mojave Desert (Beatley, 1970). In the western populations the exceptional individuals that live more than one or two years may ensure population persistence through drought. *Enneapogon desvauxii* (feather pappusgrass) is a tufted

perennial of the Chihuahuan and eastern Sonoran deserts that often flowers in its first year. This annual-perennial flexibility can be viewed as an intermediate step in the evolution of ephemerals, but its widespread occurrence indicates that the "A-P" strategy may be well-suited to some types of arid climates (Shmida and Burgess, 1988). The most arid climates select for ephemeral species that depend on seeds to survive drought (Beatley, 1967).

Phenology and Reproduction

The timing of flowering in desert grasses is related to soil-moisture availability, temperature, carbon pathway, size, and longevity. *Aristida ternipes*, *Heteropogon contortus* (tanglehead), and *Muhlenbergia porteri* (bush muhly) are common, large C4 perennials that mostly flower in summer in the Arizona Upland (Table 14.2). *Pappophorum mucronulatum* Nees (pappus grass) and *Sporobolus cryptandrus* (sand dropseed) are medium-sized perennials that enter the Arizona Upland only in the Tuscon and Waterman mountains and a few more western localities. They flower mostly in the summer and fall.

The robust *Hilaria rigida* of the lower Colorado River valley flowers mostly from February to June in the cool season (Munz, 1974) and, less often, in response to occasional summer rains (Nobel, 1980). *Aristida californica* (Mohave three awn), a small-to-large C4 perennial common on sandy areas around the head of the Gulf of California, flowers in the spring and probably other seasons as well (Felger, 1980). *Stipa speciosa* is a winter-spring-flowering C3 perennial that is rare in most of the Sonoran Desert but common along its western margin and in the Mojave Desert.

Medium and small C4 perennials that flower mostly in the summer include *Aristida glauca* (*A. purpurea*; Reverchon three awn), *Bouteloua repens*, *Digitaria californica*, *Erioneuron pulchellum*, and *Setaria macrostachya* (*sensu lato* including *S. leucopila*; plains bristlegrass). These plants are opportunistic and also may flower in the fall if there is sufficient rainfall, or in the spring if it is warm and moist. *Bouteloua repens* is an example of a species that flowers rarely in the spring in Arizona but more regularly in Sonora. A few flowers were seen on *Bouteloua repens* and *Heteropogon contortus* on 25 February 1988 and on *Hilaria belangeri* and *Tridens muticus* on 7 March 1988 at 975 m in the Tucson Mountains. *Bouteloua chondrosoides* (sprucetop grama), a common species in higher grasslands, flowered only in August and September.

Droughts are survived in dormancy by these perennials, although the modest carbohydrate reserves and shallow root systems leave the plants vulnerable to more than a few drought years. On the other hand, large numbers of seeds are produced in good years. Plants can disperse rapidly along washes and roads into drier areas at lower elevations during a few consecutive years of good summer rainfall. A number of vigorous young *Aristida ternipes* probably dispersed into the Puerto Blanco Mountains study area during the wet summers of 1983 and 1984. In the same period, at slightly higher elevations on the Papago Indian Reservation, there were remarkable increases in *Aristida glauca, A. ternipes, Bouteloua repens, B. rothrockii* (*B. barbata* var. *rothrockii*; Rothrock grama), *Digitaria californica, Erioneuron pulchellum,* and *Heteropogon contortus. Bouteloua repens* and *B. rothrockii* can dominate open areas in the Arizona Upland after a few consecutive wet summers.

Erioneuron pulchellum, the smallest of the perennial grasses discussed here, can be locally abundant in rocky habitants or on roadsides. Although individuals in protected spots may reach ages comparable to larger grasses, populations probably fluctuate in tandem with relatively short climatic events. Individual plants flower at smaller sizes during spring activity than in the summer when plants can grow more vigorously. In the Mojave Desert portion of its range, where summer rainfall is unimportant, flowering is predominantly from February to May (Munz, 1974), although plants respond well to occasional tropical storms in late summer or early fall (Beatley, 1970; A. Sanders, personal communication, 1987).

Obligate native winter-spring C3 ephemerals include *Bromus arizonicus* (Arizona brome), *Poa bigelovii* (Bigelow bluegrass), *Trisetum interruptum* (prairie trisetum), and *Vulpia octoflora* (six-weeks fescue). *Bromus*

arizonicus occurs below 1525 m down to about 540 m in sandy washes in the Arizona Upland and down to 305 m in irrigated agricultural areas. It does not occur in undisturbed desertscrub in the lower Colorado River valley. *Trisetum interruptum* is uncommon in the Tucson Mountains near Tucson and the Waterman Mountains near Silver Bell. *Vulpia octoflora* reaches lower, drier areas than *Poa bigelovii* but is rare or absent below 300 m elevation on the Yuma Mesa and the Gran Desierto (Felger, 1980). The introduced C3 spring ephemeral *Schismus barbatus* (including *S. arabicus*; Mediterranean grass) is very common at all elevations in the Sonoran Desert. *Bromus rubens* (red brome) and *Hordeum murinum* L. (mouse barley) are common European C3 spring ephemerals that have invaded the Arizona Upland. *Bromus rubens* also is common in some areas in northwestern Sonora (R. S. Felger, personal communication, 1987) and Baja California Norte.

The C4 ephemeral grasses of the Sonoran Desert, including *Aristida adscensionis* (six-weeks three awn), *Bouteloua aristidoides, B. barbata, Brachiaria arizonica* (Scribn. & Merr.) S. T. Blake (Arizona panic grass), *Leptochloa filiformis,* and *Muhlenbergia microsperma* (littleseed muhly), are probably derived from perennial relations that respond to relatively dependable warm-season rainfall. *Brachiaria arizonica* and *Leptochloa filiformis* flower mostly in summer (August–September) but can linger as late as December in a warm, wet fall. *Cottea pappophoroides* (Cott pappus grass) has been collected in the Tucson Mountains. These three grasses do not flower in the spring and do not enter the lower, drier portions of the Sonoran Desert. *Bouteloua aristidoides* and *B. barbata* also are mostly summer-fall plants that occasionally flower in the spring. Both *B. aristidoides* and *B. barbata* flower more often in winter and spring in Sonora and Baja California, where winters are warmer. *Aristida adscensionis* is quite opportunistic and can flower throughout the year if moisture is available. At cooler, higher elevations, flowering is mostly in the summer. In the Mojave Desert *A. adscensionis* flowers mostly from February to June (Munz, 1974); warm-season flowering occurs occasionally in years with late sum-

mer–early fall tropical storms (A. Sanders,
personal communication, 1987). *Muhlenbergia microsperma* is predominantly a
winter-spring plant that occasionally flowers
in the summer. Several endemic Baja California C4 ephemerals (*Bouteloua annua* Swallen
and *Muhlenbergia brandegei* C. G. Reeder)
are virtual winter-spring obligates (Gould and
Moran, 1981). When large tropical storms
bring heavy rains into the area, flowering of
summer ephemerals may extend into late November or early December, and a few new
plants may germinate. To the south in Sonora, *Bouteloua barbata* may linger through
the fall and have a second seed crop on short,
new branches if late-fall or winter storms
reach the area, and finally die in January or
February at the beginning of the spring-
summer drought.

An interesting observation is that seedlings
of some perennials may behave as ephemerals. *Aristida californica* and *A. ternipes* can
flower in their first year. On 9 September
1987 *A. ternipes* at the Arizona-Sonora Desert Museum in the Tucson Mountains were
flowering five weeks after germination at 330
mm height in response to 47 mm rainfall.
Aristida glauca seedlings, which germinate in
late summer, can flower in the spring and die
in late April or May. These plants are similar
in size to large *A. adscensionis*, a sympatric
ephemeral, and much smaller than established *A. glauca*. Seedlings often appear in exposed sites where long-term establishment is
unlikely, but this may be a mechanism to
help dispersal into new areas. The occasional
A. adscensionis that flowers in the spring has
to grow vegetatively before flowering at an
even smaller size than seedling *A. glauca*. In
Nacapulis and La Pintada canyons, near
Guaymas, Sonora, *Bouteloua repens* flowers
occasionally in December or January of its
first year. The flowers on these young plants
may or may not be fertile. In the Sonoran
Desert seedling *Erioneuron pulchellum* can
germinate in either fall or spring, flower at
very small sizes, and not survive the arid foresummer into a second year. Flowering seedlings of *Aristida glauca* and *Erioneuron
pulchellum* were seen on 8 September 1984
24 km east of Why, and 16 April 1987 in the
Waterman Mountains of Arizona. In the Mojave Desert of eastern California, first-year

flowering is apparently common, although established perennials are usually present (A.
Sanders, personal communication, 1987).

The following observations illustrate the
adaptive advantage of perennials when
moisture is present for short periods. In 1987
the early summer drought was intense and
prolonged in the eastern Sonoran Desert with
the advent of the monsoons delayed until the
beginning of August. On 5 August in the
limestone Waterman Mountains (730–790 m
elev.), vigorous vegetative growth was observed in *Bouteloua trifida* (red grama),
Erioneuron pulchellum, *Heteropogon contortus*, and *Tridens muticus* (slim tridens). A
single plant of *Bouteloua curtipendula* on a
shady, north-facing cliff base had new flowers.
By 13 August *B. trifida* and *Aristida ternipes*
had begun to flower. By 19 August *Aristida
glauca*, *Digitaria californica*, *Erioneuron
pulchellum*, *Muhlenbergia porteri*, and *Tridens muticus* were also in flower and *Bouteloua trifida* was setting seed. The only
ephemeral grasses observed were tiny seedlings of *Panicum hirticaule* (witchgrass). By 4
September most of the perennials were continuing to flower and setting seed. *Heteropogon contortus* and *Panicum hirticaule* had
begun to flower.

SONORAN DESERT PACKRAT MIDDENS
Puerto Blanco Mountains

Modern vegetation. The rhyolitic Puerto
Blanco Mountains surround the Visitor Center
in Organ Pipe Cactus National Monument,
Pima County, Arizona. Estimated precipitation for 550–605 m elevation in the study
area is 205–215 mm/yr with 40% in summer.
The vegetation on the south-facing slopes is a
sparse Arizona Upland desertscrub dominated
by *Encelia farinosa* (brittle bush) in association with *Cercidium microphyllum*, *Stenocereus thurberi* (Engelm.) Buxb. (organ pipe
cactus), *Sapium biloculare* (Mexican jumping
bean), and *Hyptis emoryi* (desert lavender).
Carnegiea gigantea, *Ferocactus covillei* (Coville barrel cactus), and *Opuntia bigelovii*
(teddy bear cholla) are uncommon on the
slopes. The nearby Ajo Mountains rise to an
elevation of 1465 m and support desert grassland and a xeric chaparral-woodland (Bowers,
1980).

Today grasses are a relatively minor component of the open desertscrub vegetation on the rocky slopes near the midden sites. Most common are C4 ephemerals such as *Bouteloua aristidoides*, *B. barbata*, *Aristida adscensionis*, and *Muhlenbergia microsperma*. *Leptochloa filiformis* is a C4 summer ephemeral occasionally found in moist microhabitats on rocky slopes and along washes. *Vulpia octoflora* is the most common C3 spring ephemeral on the slopes, and *Poa bigelovii* is occasional. *Schismus barbatus* is an abundant introduced C3 spring ephemeral on flats and also may be common on slopes in some years.

Erioneuron pulchellum can be common, widely scattered, or absent on rocky slopes. *Aristida glauca* is a locally common, medium-sized perennial restricted to more mesic microsites on rocky slopes. *Aristida ternipes*, *Digitaria californica*, *Muhlenbergia porteri*, *Setaria macrostachya*, and *Tridens muticus* are occasional to rare in relatively mesic microhabitats. A few *Hilaria rigida* were found on a northeast slope but were not observed on the surrounding flats. A single *B. rothrockii* was found on the flats below a site.

At 550 m in the Puerto Blanco Mountains several perennial grasses, including *Aristida ternipes*, *Digitaria californica*, *Muhlenbergia porteri*, and *Setaria macrostachya*, approach their lower elevational limits in Arizona. (*Digitaria californica* was taken at 460 m at Tule Well, 115 km west-northwest.) Unlike *Aristida ternipes* and *Digitaria californica*, the range of *Muhlenbergia porteri* does not reach the lower elevations in the southern portions of the Sonoran Desert. The Puerto Blanco individuals growing at 550 m are below the 610-m lower limit presented in Gould (1951) and Kearney and Peebles (1964). A collection from Meneger's Dam on the Papago Indian Reservation 32 km east at 460 m is the lowest record for Arizona. Additional low-elevation (490 m) collections were from the Sierra Estrella near Phoenix and below Rampart Cave in the Grand Canyon.

Grasses found only along the washes include *Bromus arizonicus* (= ephemeral *Bromus carinatus*), *B. rubens*, *Echinochloa colona* (jungle rice), *Eragrostis cilianensis* (stink grass), *Leptochloa filiformis*, *Panicum hirticaule*, *Phalaris minor* (littleseed canary grass), and *Sorghum halapense* (Johnson

grass). A single *Bouteloua repens* at 535 m in the wash below the Ajo Loop sites is 75 m lower than the previous elevational record for the species in Arizona from Diablo Canyon just east of the midden site. Most of these species increase with livestock disturbance, and *B. rubens*, *E. colona*, *E. cilianensis*, *P. minor*, and *S. halapense* are introduced exotics. Organ Pipe Cactus National Monument was grazed heavily from the late 1930s until 1979.

Fossil record. Plant macrofossils from twenty packrat midden assemblages from 550 to 605 m elevation in the Puerto Blanco Mountains provide a rich record of the local flora since the last glacial period (Van Devender, 1987; Table 14.1). Fossils in the 14.1-ka late Wisconsin sample include *Juniperus californica* (California juniper) and *Yucca brevifolia* (Joshua tree), with *Yucca whipplei* (Whipple yucca) and *Salvia mohavensis* (Mojave sage) and no Sonoran Desert plants. Similar xeric woodlands are found today at the upper edge of the Mojave Desert in southern California and west-central Arizona.

Three early Holocene assemblages dated between 10.5 and 9.1 ka record relatively xeric communities dominated by *Encelia farinosa* and *Acacia greggii* (catclaw acacia) with *Juniperus californica* and *Carnegiea gigantea* in association. By 8.8 ka juniper apparently had retreated to higher elevations and many Sonoran Desert plants had arrived. However, certain trees and shrubs such as *A. greggii*, *Celtis pallida* (desert hackberry), *Cercidium floridum* (blue palo verde), and *Prosopis velutina* (velvet mesquite) that are now restricted to washes were thriving on open, south-facing slopes. Reasonable modern analogs to this desertscrub occur today at the lower edge of desert grassland, xeric woodland, or interior chaparral delimiting the northern margin of the Sonoran Desert in areas with greater summer rainfall, substantial winter rainfall, and more winter freezes.

At 5.2 ka the vegetation was more similar to modern associations with *Carnegiea gigantea* and *Cercidium microphyllum* as dominants, although *Acacia greggii*, *C. floridum*, and *Prosopis velutina* were still growing on slopes. The modern desertscrub communities, and by inference modern climate, were established by about 4 ka at the beginning of the late Holocene. Subsequent minor changes in community composition may reflect climatic fluctuations at 1 ka. Today desertscrub communities near all of the midden shelters have fewer species than were seen in the midden samples. The vegetation appears to be recovering from a severe drought some time in this century, perhaps as recently as 1974–75. Evidently the modern vegetation reflects one of the driest climates of the present interglacial.

Fossil grasses. Fifteen species of grasses were identified from the packrat middens (Fig. 14.3), with 53% still occurring in the area but comprising only 50% of the native modern grasses. For eight to ten thousand years or more, *Erioneuron pulchellum*, a short-lived perennial, and ephemerals, including *Aristida adscensionis*, *Bouteloua aristidoides*, *B. barbata*, and *Muhlenbergia microsperma*, have grown on the rocky slopes. In the middle Holocene medium-sized perennial grasses, including *Setaria macrostachya*, *Digitaria californica*, *Stipa speciosa*, and *Bothriochloa barbinodis* (cane beardgrass), were somewhat more common than today, suggesting wetter conditions. *Hordeum pusillum* (little barley), found in the 8-ka sample, is a C3 spring ephemeral of disturbed soils that is known today no closer than Fresnal Canyon, Baboquivari Mountains, Pima County, 105 km east of the midden site.

The 1-ka sample indicates a short, wetter period with seven species of grasses, including three that have not been found in the monument: *Eriochloa acuminata* (Presl.) Kunth (= *E. gracilis* and *E. lemmoni*, southwestern cupgrass; Shaw and Webster, 1987), *Bouteloua trifida*, and *Panicum* cf. *hallii* (Hall panic grass). *Eriochloa acuminata* is a C4 summer ephemeral grass restricted to moist soils at lower elevations; it is a weed in disturbed agricultural areas. The nearest modern collection, 32 km southeast of the site, is from Meneger's Dam just east of the Ajo Mountains. It is common in lowland habitats in the Sierra Pinacate 50 km to the southwest (R. S. Felger, personal communication, 1987).

Bouteloua trifida is a small C4 perennial of shallow soils, often on limestone, that can flower throughout the year. Chihuahuan Desert populations extend from Coahuila to Texas and New Mexico. In Arizona its range

is fragmented and extends from the southeastern corner to the Waterman Mountains and northwest to the Grand Canyon. In Yuma County it has been collected in the Buckskin Mountains (365 m) and in the Castle Dome Mountains (470–560 m). Isolated collections in Sonora are from the limestone Sierra del Viejo (630 m) and from south of Tule Tank in the Pinacate Mountains (200 m). A single Baja California collection was on granite in San Matias Pass near San Felipe (Gould and Moran, 1981). Apparently its range was more extensive in the past.

Panicum hallii, a small C4 perennial often found growing in crevices in limestone, is a common Chihuahuan Desert plant that ranges across Arizona in a northwest-southeast belt below the Mogollon Rim, north to the Grand Canyon (915–1710 m). The nearest modern Arizona population is in the Waterman Mountains (760–975 m) 135 km to the east-northeast. A recent collection from the Sierra del Viejo extends its range into Sonora (670 m; see discussion in Yatskievych and Fischer, 1983).

Tinajas Altas Mountains

Modern vegetation. The Tinajas Altas Mountains are a granitic range on Luke Air Force Range west of Cabeza Prieta National Wildlife Refuge, Yuma County, Arizona (155 km west-northwest of the Puerto Blanco Mountains). Estimated precipitation for 365–580 m elevation is 125–155 mm/yr with 40% in summer.

The modern Sonoran desertscrub on the slopes above Tinajas Altas is dominated by *Cercidium microphyllum* with a rich mixture of shrubs, including *Ambrosia dumosa*, *A. ilicifolia* (hollyleaf bursage), *Bursera microphylla* (elephant tree), *Encelia farinosa*, *Eriogonum fasciculatum* and *E. wrightii* (wild buckwheats), *Fouquieria splendens* (ocotillo), *Hyptis emoryi*, *Jatropha cuneata*, and *Rhus kearneyi* (Kearney sumac). Succulents include *Agave deserti* (desert agave), *Ferocactus acanthodes* (California barrel cactus), *Nolina bigelovii* (Bigelow beargrass), *Opuntia acanthocarpa* (staghorn cholla), and *O. bigelovii*. *Carnegiea gigantea* is uncommon on slopes and along washes. None of the mountain ranges in the vicinity is high enough to support desert grassland, chaparral,

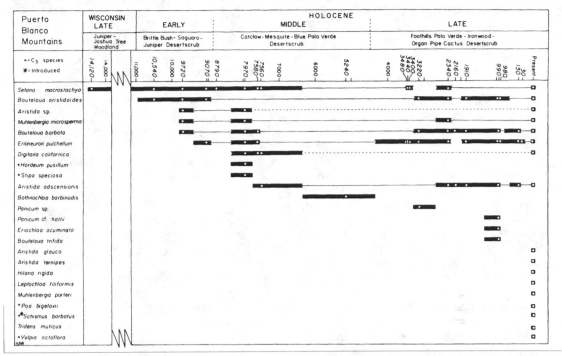

Fig. 14.3. Grasses from a chronological sequence of packrat middens and the modern flora at midden sites in the Puerto Blanco Mountains, Pima County, Arizona (after Van Devender, 1987).

or woodland.

The modern grass flora of the rocky slopes in the Tinajas Altas Mountains is very similar to that described for the Puerto Blanco Mountains. Short-lived perennials dominate, including *Aristida glauca, Erioneuron pulchellum,* and *Tridens muticus. Tridens muticus* reaches its lowest elevation (440 m) for Arizona in the study area. At Tinaja Chivos in the Sierra Pinacate of Sonora (about 100 km southeast) it grows at 245 m elevation.

Ephemeral grasses collected in the Tinajas Altas Mountains include *Aristida adscensionis, Muhlenbergia microsperma, Schismus barbatus,* and *Vulpia octoflora. Bouteloua aristidoides* and *B. barbata* have been collected from Yuma to the Fortuna Mine in the Gila Mountains. They probably occur in the study area when conditions are favorable.

Digitaria californica, Muhlenbergia porteri, and *Poa bigelovii* were not found in the Tinajas Altas Mountains, although *Heteropogon contortus, Hilaria rigida,* and *Stipa speciosa* were. The nearest collection site of *Digitaria californica* is from 460 m elevation at Tule Tank in the Cabeza Prieta Mountains (30 km

to the east). *Heteropogon contortus* is a medium-to-large perennial grass that reaches its lowest desert elevations on these granite slopes and in the Sierra Pinacate (R. S. Felger, personal communication, 1987). *Hilaria rigida* is common with *Larrea divaricata* and *Ambrosia dumosa* in sandy lowlands and on rocky slopes. A few relictual *Stipa speciosa,* the common Mojave Desert C3 perennial, were found in a shady, protected canyon. Unlike the Puerto Blanco Mountains, additional grass species were not encountered in the washes. By March 1986 the aggressive introduced perennial *Pennisetum ciliare* (buffel grass) had recently colonized a dirt road in the Tinajas Altas Mountains.

Fossil record. Plant macrofossils from twenty-one packrat midden assemblages from 330 to 580 m elevation provide a detailed record of the species turnovers for the last eighteen thousand years (Van Devender, *this volume* chap. 8). Radiocarbon dates for the thirteen samples that contained grass remains are presented in Table 14.1. The late Wisconsin (15.7 and 11 ka) woodlands with *Juniperus californica* and *Pinus monophylla* in the Tinajas Altas Mountains (365–580 m) were more mesic than those of the Puerto Blanco Mountains. Tinajas Altas samples from 330 m dated at 16.2 and 15.1 ka record a California juniper–Joshua-tree woodland very similar to the Puerto Blanco sample at 565 m. Modern analogs to this paleocommunity can be found in Hidden Valley (1280 m; TRV, personal observation, 1977) and Covington Flats (Rowlands, 1978) in Joshua Tree National Monument, California.

In the early Holocene *J. californica* was associated with various Sonoran and Mojave desertscrub and riparian trees and shrubs until 9 ka. The middle Holocene was apparently warmer and wetter than today in the Tinajas Altas Mountains until 4 ka. Trees now restricted to desert riparian habitats lived on rocky slopes. At all sites species richness was much greater than today. The modern desertscrub communities formed in the late Holocene.

Fossil grasses. Twelve species of grasses were identified from thirteen midden samples dating between 15.7 and 4 ka from granite rock shelters at 365–580 m elevation (Fig. 14.4). A total of 83.3% of these grasses occur in the area today, comprising 81.8% of the native modern grass flora.

Muhlenbergia microsperma was found in the late Wisconsin sample. All of the ephemeral grasses found today in the Puerto Blanco Mountains study area and discussed above were well established in the Tinajas Altas Mountains by the early Holocene. *Vulpia microstachys* (Nutt.) Gmel. (small fescue), identified in the 11-ka samples, is a C3 spring ephemeral that no longer occurs in the area. It now is a western North American grass that enters Arizona south of the Mogollon Rim at 800–1615 m elevation, extending as far east as the Rincon Mountains near Tucson (Bowers and McLaughlin, 1987). The nearest localities to the midden site are the Ajo Mountains, Pima County, 135 km to the east-southeast, and at 1035 m in the Harquahala Mountains, Yuma County, 180 km to the north-northeast.

Among the perennial grasses, C4 *Setaria macrostachya* occurs in the 11-ka sample but not subsequently. The C3 *Stipa speciosa* in the same sample is found throughout the record and is rare in the area today. The earli-

est record for *Hilaria rigida* is 10.1 ka. *Heteropogon contortus* was present by 7.9 ka. *Tridens muticus* is the only native grass found in the study area today that was not identified from the middens.

Butler Mountains

Modern vegetation. The granitic Butler Mountains are a small, isolated outlier of the Tinajas Altas Mountains on the Yuma Mesa surrounded by the sands of the Gran Desierto. Estimated precipitation is 100–105 mm/year with 35% in summer.

The vegetation of the Butler Mountains is low, sparse, and limited to plants highly adapted to aridity. It is very similar to the vegetation of the granitic Sierra del Rosario of Sonora (70 km south; Felger, 1980). *Ambrosia dumosa* and *Larrea divaricata* dominate the granitic slopes as well as the sandy flats. Other common plants on the slopes include *Asclepias albicans* (ocotillo milkweed), *Eriogonum wrightii*, *Peucephyllum schottii* (pygmy cedar), *Hofmeisteria pluriseta* (arrowleaf), and *Trixis california* (a woody composite). *Encelia farinosa*, a common rocky-slope dominant of the Sonoran and eastern Mojave deserts, is rare. *Cercidium microphyllum*, the widespread Arizona Upland dominant, is absent as are *Bursera microphylla* and *Jatropha cuneata*. *Ambrosia ilicifolia*, *Hyptis emoryi*, *Encelia farinosa*, and *Fouquieria splendens* are mostly confined to washes at these low elevations. The only succulents on the slopes are a few *Agave deserti*, *Carnegiea gigantea*, *Ferocactus acanthodes*, and *Mammillaria tetrancistra* (fishhook cactus).

The grass flora of the Butler Mountains is less varied than that of the Tinajas Altas Mountains. *Aristida glauca*, *Erioneuron pulchellum*, and *Hilaria rigida* are the only perennials. The area apparently is too arid for *Heteropogon contortus*, *Stipa speciosa*, and *Tridens muticus*. *Aristida californica* is locally common in sandy habitats on the Yuma Mesa between the Butler Mountains and Yuma. Ephemerals collected include *Aristida adscensionis*, *Muhlenbergia microsperma*, and *Schismus barbatus*. *Bouteloua aristidoides* and *B. barbata* probably grow in the study area when conditions are warm and wet. *Vulpia octoflora* has been taken at Tina-

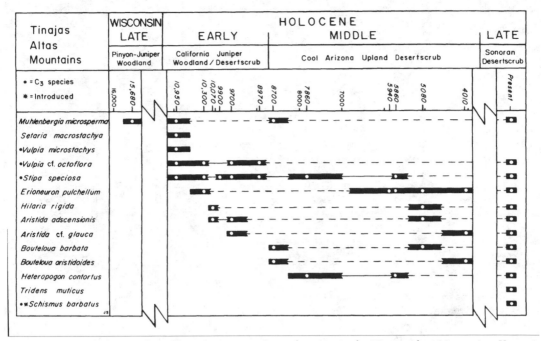

Tinajas Altas Mountains	WISCONSIN LATE	HOLOCENE		LATE
	Pinyon-Juniper Woodland	EARLY	MIDDLE	Sonoran Desertscrub
		California Juniper Woodland / Desertscrub	Cool Arizona Upland Desertscrub	

Fig. 14.4. Grasses from a chronological sequence of packrat middens and the modern flora at the midden sites in the Tinajas Altas Mountains, Yuma County, Arizona.

jas Altas but not on the Yuma Mesa to the west. In Sonora it has been collected many times in the Sierra Pinacates but not in the Gran Desierto (Felger, 1980). The Butler Mountains apparently are near or just west of its local range limit.

Fossil record. Nine packrat midden samples dating between 11.3 and 0.2 ka from 240–255 m elevation (Table 14.1) disclose changes in the woody flora. Here *Juniperus californica* reached its known lower elevational limit (240 m) at 11.3 ka. It was in association with such lower Colorado River valley or Mojavean species as *Larrea divaricata*, *Encelia farinosa*, and *Ambrosia dumosa*. The assemblages do not suggest a typical juniper woodland. *Juniperus californica* lingered in the Butler Mountains until at least 10.4 ka, and the remainder of the community changed little until after 8.2 ka. A relatively modern desertscrub including *Olneya tesota* was present at 3.8 ka. In all cases, including samples dated at 0.7 and 0.6 ka, the modern flora at the sites comprises fewer species than were found in the midden assemblages.

Fossil grasses. Eleven species of grasses identified from the midden samples in the Butler Mountains record the same basic desert-adapted grass flora found at the other study areas for the entire Holocene and probably the late Wisconsin as well (Fig. 14.5). A floret of the introduced *Schismus barbatus* in the 11.3-ka sample is presumably a contaminant. Eighty percent of the native species identified in the middens are still present and represent the entire modern grass flora. *Stipa speciosa*, found in the 10.4- and 3.8-ka samples, no longer occurs locally. *Hilaria rigida* first appears in the 10.4-ka sample.

The specimen of *Bouteloua repens* is of special interest because it records a westward expansion of a subtropical grass that today is limited in the United States to the southern parts of Texas, Arizona, and New Mexico. In Pima County, Arizona, it is found in Sonoran Desert and desert-grassland habitats between 610 and 1525 m elevation. The population nearest to the Butler Mountains is at 535 m in the Puerto Blanco Mountains 160 km to the east-southeast. Differences between the

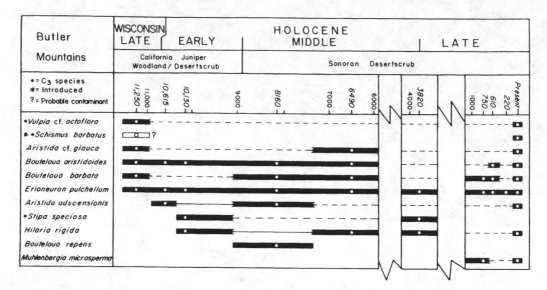

Fig. 14.5. Grasses from a chronological sequence of packrat middens and the modern flora at the midden sites in the Butler Mountains, Yuma County, Arizona.

modern rainfall estimates for these sites indicate that by 8.2 ka in the early Holocene warm, subtropical conditions with about 78 mm summer rainfall (an increase of 220% over today) were established (Van Devender, *this volume*, chap. 8).

Whipple Mountains

Natural vegetation. The Whipple Mountains are located in San Bernardino County, California, across the Colorado River from Parker, Arizona, at the northwestern edge of the Sonoran Desert (Shreve, 1964). The study area is 195 km north of Yuma. Estimated precipitation between 320 and 520 m elevation is 139–182 mm with about 27% in summer.

The vegetation on rhyolite near the midden rock shelters is an open desertscrub dominated by *Larrea divaricata* and *Peucephyllum schottii*, typical of low elevations in both the lower Colorado River valley and the Mojave Desert. *Ambrosia dumosa*, *Encelia farinosa*, *Bebbia juncea* (chuckwalla's delight), *Hyptis emoryi*, *Hofmeisteria pluriseta*, *Trixis californica*, and *Viguiera deltoidea* (Parish goldeneye) are present at many sites. Succulents found in the area include *Agave deserti*, *Ferocactus*

acanthodes, *Opuntia acanthocarpa*, *O. basilaris* (beavertail cactus), and *O. bigelovii*. Subtropical Sonoran Desert plants, including *Carnegiea gigantea*, *Cercidium floridum*, and *Olneya tesota*, are found in the Whipple Mountains below about 320 m elevation (Munz, 1974).

From our limited observations, grasses are uncommon in the Whipple Mountains, and those that occur there appear to be typical of the lower Colorado River valley. *Aristida adscensionis*, *A. glauca*, *Erioneuron pulchellum*, and *Muhlenbergia porteri* were collected or seen in the midden study area. The 320-m site is very low for *M. porteri* (595 m lower than the general elevational range in Munz, 1974). The only other individual seen in the study area was growing at 520 m. *Bouteloua trifida*, *Poa bigelovii*, *Tridens muticus*, and *Vulpia octoflora* have been collected in Whipple Wash, a relatively mesic canyon nearby (A. Sanders, personal communication, 1985). This is an exceptionally low-elevation and disjunct population of *B. trifida*.

Fossil record. Plant macrofossils from twenty-nine packrat middens from 320 to 520 m elevation provide a record of biotic change for the last 13.8 ka (Van Devender, *this vol-*

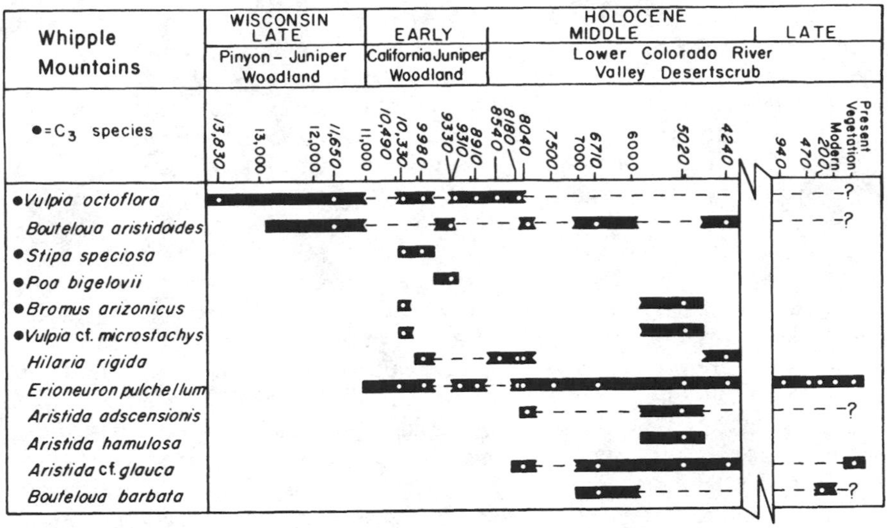

Whipple Mountains

	WISCONSIN LATE	EARLY	HOLOCENE MIDDLE	LATE
	Pinyon–Juniper Woodland	California Juniper Woodland	Lower Colorado River Valley Desertscrub	

●=C₃ species

Date columns (ka, left to right): 13,830 · 13,000 · 11,650 · 12,000 · 11,000 · 10,490 · 10,330 · 9980 · 9930 · 9330 · 8910 · 8540 · 8180 · 8060 · 8040 · 7500 · 7000 · 6710 · 6000 · 5020 · 4240 · 940 · 470 · 200 · Present Vegetation Modern

Taxa:

- ● *Vulpia octoflora*
- *Bouteloua aristidoides*
- ● *Stipa speciosa*
- ● *Poa bigelovii*
- ● *Bromus arizonicus*
- ● *Vulpia* cf. *microstachys*
- *Hilaria rigida*
- *Erioneuron pulchellum*
- *Aristida adscensionis*
- *Aristida hamulosa*
- *Aristida* cf. *glauca*
- *Bouteloua barbata*

Fig. 14.6. Grasses from a chronological sequence of packrat middens and the modern flora at the midden sites in the Whipple Mountains, San Bernardino County, California.

ume, chap. 8). Radiocarbon dates on samples yielding grass remains are presented in Table 14.1. The late Wisconsin vegetation was an open California juniper woodland with *Yucca brevifolia*, *Salvia mohavensis*, and *Nolina bigelovii*. *Pinus monophylla* grew as low as 510–520 m at 13 and 11.4 ka (Van Devender and Spaulding, 1979; Van Devender et al., 1985). The late Wisconsin assemblages also contain a few remains of *Artemisia tridentata*-type (big sagebrush) and *Atriplex confertifolia* (shadscale), important dominants in Great Basis desertscrub.

Juniperus californica persisted in the Whipple Mountains until 8.9 ka (Van Devender, 1977), when a relatively modern Mojave desertscrub dominated by *Larrea divaricata* and *Peucephyllum schottii* associated with *Encelia farinosa* and *Ambrosia dumosa* was established.

Fossil grasses. Twelve species of grasses were identified from nineteen midden samples dating between 13.8 ka and modern (Fig. 14.6). Only two species (16.7%) were observed near the midden sites. Four other species potentially occur there as well for a combined total of 50% of the modern taxa. Additional species were not found, although a few others

(*Bouteloua trifida*, *Muhlenbergia porteri*, *Tridens muticus*) are known from the Whipple Mountains.

Two late Wisconsin samples dated at 13.8 and 11.7 ka yielded only ephemerals (*Vulpia octoflora* and *Bouteloua aristidoides*). Six early Holocene samples from between 10.5 and 8.9 ka contain eight species of grasses, including a C3 perennial (*Stipa speciosa*), C4 perennials (*Erioneuron pulchellum* and *Hilaria rigida*), C3 ephemerals (*Bromus arizonicus*, *Poa bigelovii*, *Vulpia* cf. *microstachys*, and *V. octoflora*), and a C4 ephemeral (*Bouteloua aristidoides*). Only *Bouteloua aristidoides* and *Erioneuron pulchellum* occur at the midden sites today. This is the only Sonoran Desert midden record for *Bromus arizonicus*, a C3 annual found from Baja California to central California and east to Arizona. In Arizona it is found mostly below 1525 m elevation (Gould, 1951) down to about 305 m in agricultural areas in the Phoenix-Tempe area. The nearest Arizona collection is from 700 m elevation in the Harcuvar Mountains in La Paz County. Another isolated western Arizona collection is from around 610 m at the Horse Tanks in the Castle Dome Mountains, Yuma County.

Ten grasses were found in seven middle Holocene samples dating between 8.5 and 4.2 ka. Most early Holocene grasses except *Stipa speciosa* and *Poa bigelovii* persisted, with the addition of the C4 perennials *Aristida* cf. *glauca* and *A. hamulosa*, and the C4 ephemerals *A. adscensionis* and *Bouteloua barbata*. *Aristida hamulosa* is a perennial three awn with a major gap in its distribution in the lower Colorado River valley. It reaches the limit of the eastern portion of its range in western Arizona above 675 m elevation. In southern California it occurs from the coast as far east as Riverside (Munz, 1974; A. Sanders, personal communication, 1986). *Hilaria rigida* no longer grows on rocky slopes near the midden shelters, and it was not observed in lowland desert areas around the base of the Whipple Mountains but may occur there. The presence of these grasses suggests that summer rainfall was greater in the middle Holocene than today.

In contrast, the late Holocene samples contained only *Erioneuron pulchellum* and *Bouteloua barbata*. These fossil records and the present rarity of grasses suggest that drought conditions were more extensive.

DISCUSSION

Paleoecology of Desert Grasses

Nine of the twenty-three species of grasses from Sonoran Desert midden samples were eleven thousand years old or older. Most of these ice age grasses are members of the drought-adapted flora of the lower Colorado River valley; e.g., *Aristida* cf. *glauca*, *Bouteloua aristidoides*, *B. barbata*, *Erioneuron pulchellum*, *Muhlenbergia microsperma*, and *Vulpia octoflora*. *Oryzopsis hymenoides* and *Stipa speciosa* from Picacho Peak area middens (Cole, 1986) are the only species that no longer grow near the midden sites. *Setaria macrostachya* from the Puerto Blanco Mountains is an interesting late Wisconsin record of a C4 perennial found from the Great Plains west into the eastern Sonoran Desert in Arizona. Although it does occasionally flower in response to spring rainfall it is predominantly a summer grass.

Thirteen species of grasses identified from early Holocene middens include all of the late Wisconsin species except *Oryzopsis hymenoides*. Additional grasses include *Vulpia mi-*

crostachya from the Tinajas Altas Mountains and *Bromus arizonicus*, *Hilaria rigida*, and *Poa bigelovii* from the Whipple Mountains; none of these species occurs near the midden sites today. *Stipa speciosa* no longer occurs near the Picacho Peak or Whipple Mountain sites. *Setaria macrostachya* now occurs east of the Tinajas Altas Mountains.

At first impression the late Wisconsin and early Holocene grasses of these Sonoran Desert habitats appear to have been less affected by ice-age climates than the trees, shrubs, and succulents in the same area. Indeed, the grasses usually occur in desertscrub habitats well below modern analog woodlands, and few high-elevation grasses have been found displaced into desert lowlands. However, most of the species found in late Wisconsin middens range up into pinyon-juniper woodland in the higher mountains of the eastern Mojave Desert (Thorne et al., 1981). *Setaria macrostachya*, which does not reach California, possibly associates with *Juniperus californica* around the base of the Hualapai Mountains in west-central Arizona, where *J. californica* receives more summer rainfall than in California.

Most of the grasses in the early Holocene *Juniperus californica* middens now range into juniper or pinyon-juniper woodlands in the eastern Mojave Desert. The exceptions may occur in woodlands in other areas. At Covington Flats in Joshua Tree National Monument, *Hilaria rigida* is a common associate of *Pinus monophylla*, *J. californica*, and *Yucca brevifolia* (Rowlands, 1978). While *Poa bigelovii* occurs mostly at desert elevations in the eastern Mojave Desert ranges, it has been collected in pine-oak woodlands as high as 1890 m in the Santa Rita Mountains of Arizona and 2200 m in the Sacramento Mountains of New Mexico. While *Bromus arizonicus* of Arizona, Baja California, and coastal California occurs mostly below woodlands, in the Rincon Mountains of Arizona it ranges up to 1370 m in pine-oak woodland (Bowers and McLaughlin, 1987). The only area where *B. arizonicus* might contact *J. californica* would be in the Hualapai Mountains area. Moreover, *B. arizonicus* may be only an ephemeral form of *B. carinatus* (California brome), a common perennial of high-elevation woodland and forest.

The late Wisconsin and early Holocene paleoclimates indicated by midden grasses present an interesting discrepancy with other plants in the samples. While the C3 grasses are obligate winter-spring species (33.3% and 38.5% respectively), the C4 grasses are all summer species with a secondary spring season. In contrast, the associated ephemeral plants in the middens are predominately winter-spring obligates with few summer species (Van Devender, *this volume*, chap. 8). Rather than indicating summer rainfall, the presence of C4 grasses is probably due to their ability to respond at low elevations to available moisture at relatively cool temperatures.

Five grass species in middle Holocene middens no longer occur in the study areas: *Bothriochloa barbinodis, Digitaria californica,* and *Hordeum pusillum* in the Puerto Blanco Mountains; *Bouteloua repens* in the Butler Mountains; and *Aristida hamulosa* in the Whipple Mountains. *Heteropogon contortus* was found in two Tinajas Altas Mountains samples (5.7 and 7.9 ka) and still occurs in the area.

Bouteloua trifida, Eriochloa acuminata, and *Panicum* cf. *hallii,* all found in a 1-ka sample from the Puerto Blanco Mountains, no longer occur in the area. Grasses that grow near the midden sites today but were not identified from the middens include *Aristida ternipes, Leptochloa filiformis, Muhlenbergia porteri,* and *Tridens muticus.*

Most grasses of the lower elevations in the Arizona Upland and in the lower Colorado River valley communities are either ephemerals or small-to-medium-sized perennials adapted to moisture fluctuations in harsh desert climates. This basic drought-selected grass flora includes the ephemerals *Aristida adscensionis, Bouteloua aristidoides, B. barbata, Muhlenbergia microsperma,* and *Vulpia octoflora,* and the perennials *Erioneuron pulchellum* and *Aristida glauca.* These taxa have been important in these areas for at least ten thousand years. With the exception of *Vulpia octoflora,* all are C4 type and flower rapidly in response to available moisture in summer, fall, or winter-spring. In the early and middle Holocene, and for a short period in the late Holocene, the presence of riparian trees and shrubs on open, south-facing slopes suggests that rainfall was substantially greater

than today. A few medium-sized perennials such as *Stipa speciosa, Setaria macrostachya, Digitaria californica,* and *Heteropogon contortus* expanded their ranges into desert lowlands. The 7.9- and 5.9-ka records of *Heteropogon contortus* from the Tinajas Altas Mountains confirm it as a native species, not an introduced exotic (Correll and Johnston, 1970). It is surprising that larger perennial grasses such as *Aristida ternipes* and *Muhlenbergia porteri* were not found in the middens. The 8.2-ka record of *Bouteloua repens* from the Butler Mountains is the only example of a range expansion from subtropical areas. The 5.2-ka record of *Bothriochloa barbinodis* in the Puerto Blanco Mountains is the only example of a C4 perennial expanding into the area from eastern desert grasslands. The 5-ka record of *Aristida hamulosa* from the Whipple Mountains is a possible example of a C4 perennial extending its range westward into California, but its range may have expanded eastward in the early Holocene. *Aristida ternipes,* a common grass in Arizona Upland communities and subtropical areas in Mexico, probably entered the Puerto Blanco Mountains in the late Holocene along with *Cercidium microphyllum, Olneya tesota,* and *Stenocereus thurberi.* Perhaps many of the larger perennial grasses in the Arizona Upland should be viewed as desert-grassland and/or thornscrub plants that enter the Sonoran Desert locally from areas with more summer rainfall.

Sonoran Savanna Grassland

The historical ecology of grasses in these Sonoran Desert communities is of interest in considering the Sonoran savanna grassland proposed by Brown (1982). At elevations above desert from southwestern Arizona to trans-Pecos Texas, shrub invasion and degradation of grasslands have been linked to overgrazing by cattle (Hastings and Turner, 1965; York and Dick-Peddie, 1969; Martin and Turner, 1977). Brown (1982) extended the overgrazing hypothesis into the lowlands of the Sonoran Desert. He proposed that certain areas in Sonora that support Sinaloan thornscrub or Sonoran desertscrub in the Plains of Sonora and in the Altar and Santa Cruz valleys of Arizona were subtropical, fire-climax grasslands. Old photographs and notes from early biologists depicted dense grass cover in areas later characterized by shrubs and bare soil. In the Altar Valley at 860–1160 m elevation, where desert-grassland communities are dominated by perennial grasses (*Aristida hamulosa, Bouteloua curtipendula, B. gracilis, B. hirsuta, B. repens, B. rothrockii, Eragrostis intermedia, Hilaria belangeri,* and *Sporobolus cryptandrus*) shrub invasion probably has occurred (Griffiths, 1912; F. R. Reichenbacher, personal communication, 1985; D. Junkin, personal communication, 1987). The shrub invasion of desert grasslands near Arivaca just east of the Altar Valley (Cable and Martin, 1973) was similar to that observed in higher areas from Texas to Arizona. Many of the dominant grasses in the Altar Valley also are important in the outstanding grasslands on the Sonoita Plain 75 km to the east-northeast (Bock and Bock, 1986). *Aristida ternipes, A. glauca, Erioneuron pulchellum,* and *Muhlenbergia porteri,* the common perennials in the Arizona Upland, are not common in the Altar Valley grasslands. In contrast, the grasses reported from the Sonoran areas do not form the dense communities with relatively continuous perennial cover found in grasslands at elevations above desert. At least one of the principal grasses cited by Brown (1982), *B. rothrockii,* undergoes major changes in abundance tracking short-term fluctuations in rainfall. Two others, *B. aristidoides* and *B. barbata,* are ephemerals that have been important in the Sonoran Desert for the last ten thousand years, and they have not increased as perennials declined due to recent grazing, as suggested by Brown (1982). Thus, the Sonoran areas probably did not support a climax grassland in historic times.

We suggest that Brown's Sonoran savanna grassland be reexamined considering differences in demographics and response times to climatic fluctuations among plant life-forms that make up the vegetation. Most of the grasses in the area increase their population sizes dramatically with a few years of good summer rainfall, and decrease in numbers and population sizes even more rapidly during drought. Even the populations of the largest, longest-lived perennial grasses in the area will respond to five-to-ten-year climatic fluctua-

tions. Common small shrubs in the area such as *Encelia farinosa, Isocoma tenuisecta* (burroweed), and *Ambrosia cordifolia* (Sonoran bursage) also can increase rapidly given optimum conditions. In contrast, some other trees, shrubs, and columnar cacti may reproduce only a few times in fifty to one hundred years, yet still dominate the landscape. Grasses are present in the desert lowlands and can be important at times, but they do not dominate the more stable, long-term structure of the communities. Although the term savanna has been used broadly in the Old World to mean any subtropical vegetation that contains grasses, classification in the Southwest has more often focused on the trees and shrubs (Brown et al., 1979). The Sonoran portion of the "Sonoran savanna grassland" is best considered desertscrub or thornscrub, whereas the Altar Valley is a desert grassland. However, the introduction and periodic burning of *Pennisetum ciliare* for livestock forage is rapidly converting many areas into a savanna very similar to those of the Old World.

ACKNOWLEDGMENTS

George L. Bradley, Stephen F. Hale, W. Eugene Hall, Jim I. Mead, Barbara G. Phillips, W. Geoffrey Spaulding, Frank W. Reichenbacher, and Rebecca K. Van Devender helped with the fieldwork. John R. and Charlotte G. Reeder helped with fossil and modern grass determinations and current nomenclature. Andrew C. Sanders, University of California at Riverside, and Kathryn L. Tomlinson, Rancho Santa Ana Botanical Garden, provided information on the California distribution of *Aristida hamulosa.* Sanders also shared his grass records from the Whipple Mountains. Richard S. Felger made many helpful comments on the manuscript and the grasses of northwestern Sonora. The careful editing of Paul S. Martin and Julio L. Betancourt improved the manuscript. The herbarium specimens in the University of Arizona Herbarium under the care of Charles T. Mason, Jr., and Rebecca K. Van Devender are an invaluable source of information. Tandem accelerator mass spectrometer radiocarbon dates were run at the National Science Accelerator Facility at the University of Arizona. Funds for the

Puerto Blanco Mountains study were provided by the Southwest Parks and Monuments Association through Organ Pipe Cactus National Monument. Additional funds were provided from National Science Foundation grants DEB 79-23840 to Paul S. Martin, University of Arizona, and BSR 82-14939 to P. S. Martin and T. R. Van Devender. Helen A. Wilson and Jacqui Soule drafted the diagrams. Jean Morgan typed the manuscript.

REFERENCES

Beatley, J. C. (1967). "Survival of winter annuals in the northern Mojave Desert." *Ecology* 48, 745–50.

———. (1970). "Perennation in *Astragalus lentiginosus* and *Tridens pulchellus* in relation to rainfall." *Madroño* 20, 326–32.

Bock, J. H., and Bock, C. E. (1986). "Habitat relationships of some native perennial grasses in southeastern Arizona." *Desert Plants* 8, 3–14.

Bowers, J. E. (1980). "Flora of Organ Pipe Cactus National Monument." *Journal of the Arizona-Nevada Academy of Sciences* 15, 1–11, 33–47.

Bowers, J. E., and McLaughlin, S. P. (1987). "Flora and vegetation of the Rincon Mountains, Pima County, Arizona." *Desert Plants* 8, 51–94.

Brown, D. E. (1982). "Sonoran savanna grassland." *Desert Plants* 4, 137–41.

Brown, D. E.; Lowe, C. H., Jr.; and Pase, C. P. (1979). "A digitized classification system for the biotic communities of North America, with community (series) and association examples for the Southwest." *Journal of the Arizona-Nevada Academy of Sciences* 14, 1–16.

Brown, W. V. (1977). "The Kranz syndrome and its subtypes in grass systematics." *Memoirs of the Torrey Botanical Club* 23, 1–97.

Cable, D. R., and Martin, S. C. (1973). "Invasion of semidesert grassland by velvet mesquite and associated vegetation changes." *Journal of the Arizona Academy of Sciences* 8, 127–34.

Cole, K. L. (1986). "The lower Colorado Valley: A Pleistocene desert." *Quaternary Research* 25, 392–400.

Correll, D. S., and Johnston, M. C. (1970). *Manual of the vascular plants of Texas*. Renner: Texas Research Foundation.

Ehleringer, J. (1978). "Implications of quantum yield differences on the distribution of C_3 and C_4 grasses." *Oecologia* (Berlin) 31, 255–67.

Ehleringer, J., and Bjorkman, O. (1977). "Quantum yields for CO_2 uptake in C_3 and C_4 plants: Dependence on temperature, CO_2, and O_2 concentration." *Plant Physiology* 59, 86–90.

Ezcurra, E., and Rodrigues, V. (1986). "Rainfall patterns in the Gran Desierto, Sonora, Mexico." *Journal of Arid Environments* 10, 13–28.

Felger, R. S. (1980). "Vegetation and flora of the Gran Desierto, Sonora, Mexico." *Desert Plants* 2, 87–114.

Gould, F. W. (1951). *Grasses of southwestern United States*. Tucson: University of Arizona Press.

Gould, F. W., and Moran, R. (1981). *The grasses of Baja California, Mexico*. San Diego Society of Natural History Memoirs no. 12.

Griffiths, D. (1912). "The grama grasses: *Bouteloua* and related genera." *Contributions to the United States National Herbarium* 14, 343–428.

Gutierrez, M.; Gracen, V. E.; and Edwards, G. E. (1974). "Biochemical and cytological relationships in C_4 plants." *Planta* (Berlin) 119, 279–300.

Harris, G. A. (1967). "Some competitive relationships between *Agropyron spicatum* and *Bromus tectorum*." *Ecological Monographs* 37, 89–111.

Hastings, J. R., and Turner, R. M. (1965). *The changing mile*. Tucson: University of Arizona Press.

Hatch, M. D., and Osmund, C. B. (1976). "Compartmentation and transport in C_4 photosynthesis." In *Transport in plants*. Volume 3, *Intracellular interactions and transport processes*. C. R. Stocking and V. Heber, eds., pp. 144–84. New York: Springer-Verlag.

Hattersley, P. W., and Watson, L. (1976). "C_4 grasses: An anatomical criterion for distinguishing between NADP-Malic enzyme species and PCK or NAD-Malic enzyme species." *Australian Journal of Botany* 24, 297–308.

Huning, J. R. (1978). *A characterization of the climate of the California Desert*. Riverside: Desert Planning Staff, Bureau of Land Management.

Kearney, T. H., and Peebles, R. H. (1960). *Arizona flora*. Berkeley: University of California Press.

Kemp, P. R. (1983). "Phenological patterns of Chihuahuan Desert plants in relation to the timing of water availability." *Journal of Ecology* 71, 427–36.

Laetsch, W. M. (1974). "The C_4 syndrome: A structural analysis." *Annual Review of Plant Physiology* 25, 27–52.

Lehr, J. H. (1978). *A catalogue of the flora of Arizona*. Phoenix: Desert Botanical Garden.

Lorimer, G. H. (1981). "The carboxylation and oxygenation of ribulose 1,5-bisphosphate: The primary events in photosynthesis and photorespiration." *Annual Review of Plant Physiology* 32, 349–83.

Martin, S. C., and Turner, R. M. (1977). "Vegetation change in the Sonoran Desert region, Arizona and Sonora." *Journal of the Arizona Academy of Sciences* 12, 59–69.

Mott, J. J. (1974). "Factors affecting seed germination in three annual species from an arid region of western Australia." *Journal of Ecology* 62, 699–709.

Mulroy, T. M., and Rundel, P. W. (1977). "Annual plants: Adaptations to desert environments." *BioScience* 27, 109–14.

Munz, P. A. (1974). *A flora of southern California.* Berkeley: University of California Press.

Neilson, R. P. (1986). "High-resolution climatic analysis and Southwest biogeography." *Science* 232, 27–34.

Nobel, P. S. (1980). "Water vapor conductance and CO_2 uptake for leaves of a C_4 desert grass, *Hilaria rigida.*" *Ecology* 61, 252–58.

———. (1981). "Spacing and transpiration of various sized clumps of a desert grass, *Hilaria rigida.*" *Journal of Ecology* 69, 735–42.

Osmond, C. B.; Winter, K.; and Ziegler, H. (1982). "Functional significance of different pathways of CO_2 fixation in photosynthesis." In *Physiological plant ecology.* Volume 2 of *Encyclopedia of Plant Physiology*, New ser. O. L. Lange, P. S. Nobel, C. B. Osmond, and H. Ziegler, eds., 12B: 479–547. New York: Springer-Verlag.

Rowlands, P. G. (1978). "The vegetation dynamics of Joshua tree (*Yucca brevifolia* Engelm.) in the Southwestern United States of America." Ph.D. diss., University of California, Riverside.

Sala, O. E., and Laurenroth, W. K. (1982). "Small rainfall events: An ecological role in semiarid regions." *Oecologia* (Berlin) 53, 301–4.

Shaw, R. B., and Webster, R. D. (1987). "The genus *Eriochloa* (Poaceae: Paniceae) in North and Central America." *Sida* 12, 165–207.

Shmida, A., and Burgess, T. L. (1988). "Plant growth-form strategies and vegetation types in arid environments." In *Plant form and vegetation structure.* N. J. A. Werger, P. J. M. van der Aart, H. J. During, and J. T. A. Verhoeren, eds., pp. 211–41. The Hague: SPB Academic Publishers.

Shreve, F. (1911). "The influence of low temperatures on the distribution of giant cactus." *Plant World* 14, 136–46.

———. (1964). "Vegetation of the Sonoran Desert." In *Vegetation and flora of the Sonoran Desert.* F. Shreve and I. L. Wiggins, eds., pp. 6–186. Stanford: Stanford University Press.

Smith, B. N. (1976). "Evolution of C_4 photosynthesis in response to changes in carbon and oxygen concentrations in the atmosphere through time." *Biosystems* 8, 24–32.

Soreng, R. J., and Van Devender, T. R. (1988). "Late Quaternary fossils of *Poa fendleriana* (muttongrass): Holocene expansion of apomicts." *Southwestern Naturalist* 34, in press.

Svertsen, J. P.; Nickell, G. L.; Spellenberg, R. W.; and Cunningham, G. L. (1976). "Carbon reduction pathways and standing crop in three Chihuahuan Desert plant communities." *Southwestern Naturalist* 21, 311–20.

Teeri, J. A., and Stowe, L. G. (1976). "Climatic patterns and the distribution of C_4 grasses in North America." *Oecologia* 23, 1–12.

Tevis, L., Jr. (1958). "Germination and growth of ephemerals induced by sprinkling a sandy desert." *Ecology* 39, 681–88.

Thomasson, J. R.; Nelson, M. E.; and Zakrzewski, R. J. (1986). "A fossil grass (Gramineae: Chloridoideae) from the Miocene with Kranz anatomy." *Science* 233, 876–78.

Thorne, R. F.; Prigge, B. A.; and Henrickson, J. (1981). "A flora of the higher ranges and the Kelso Dunes of the eastern Mojave Desert." *Aliso* 10, 71–186.

Turnage, W. V., and Hinckley, A. L. (1938). "Freezing weather in relation to plant distribution in the Sonoran Desert." *Ecological Monographs* 8, 529–50.

Turner, R. M., and Brown, D. E. (1982). "Sonoran desertscrub." *Desert Plants* 4, 181–221.

Van Devender, T. R. (1977). "Holocene woodlands in the southwestern deserts." *Science* 198, 189–92.

———. (1987). "Holocene vegetation and climate in the Puerto Blanco Mountains, southwestern Arizona." *Quaternary Research* 27, 51–72.

Van Devender, T. R.; Martin, P. S.; Thompson, R. S.; Cole, K. L.; Jull, A. J. T.; Long, A.; Toolin, L. J.; and Donahue, D. J. (1985). "Fossil packrat middens and the tandem accelerator mass spectrometer." *Nature* 317, 610–13.

Van Devender, T. R., and Spaulding, W. G. (1979). "The development of vegetation and climate in the southwestern United States." *Science* 204, 701–10.

Van Devender, T. R.; Thompson, R. S.; and Betancourt, J. L. (1987). "Vegetation history of the deserts of southwestern North America: The nature and timing of the late Wisconsin–Holocene transition." In *North America and adjacent oceans during the last deglaciation.* W. F. Ruddiman and H. E. Wright, Jr., eds., pp. 323–52. Boulder: Geological Society of America.

Waller, S. S., and Lewis, J. K. (1979). "Occurrence of C_3 and C_4 photosynthetic pathways in North America grasses." *Journal of Range Management* 32, 12–28.

Went, F. W. (1949). "Ecology of desert plants II. The effect of rain and temperature on germination and growth." *Ecology* 30, 1–13.

Wright, R. G., and Van Dyne, G. M. (1976). "Environmental factors influencing semidesert grassland perennial grass demography." *Southwestern Naturalist* 21, 259–74.

Yatskievych, G., and Fischer, P. C. (1983). "New plant records from the Sonoran Desert." *Desert Plants* 5, 180–85.

York, J. C., and Dick-Peddie, W. A. (1969). "Vegetation changes in southern New Mexico during the past hundred years." In *Arid lands in perspective*. W. G. McGinnies and B. J. Goldman, eds., pp. 155–66. Tucson: University of Arizona Press.

Chapter 15

Late Quaternary Mammals from the Chihuahuan Desert: Paleoecology and Latitudinal Gradients

Thomas R. Van Devender   George L. Bradley

Plant macrofossils preserved in ancient packrat (*Neotoma* sp.) middens have allowed detailed reconstructions of rocky-slope vegetation of the last thirty thousand years or more for many places in the southwestern United States (Van Devender and Spaulding, 1979; Van Devender, Thompson, and Betancourt, 1987). Packrat middens have recorded the expansion of *Pinus edulis* (Colorado pinyon), *P. remota* (papershell pinyon), *Juniperus* spp. (junipers), and *Quercus* spp. (shrub oaks) woodlands throughout the modern Chihuahuan Desert during the Wisconsin glacial period (Van Devender, 1986; *this volume*, chap. 7). In addition to the abundant plant macrofossils in middens, there are also animal remains, especially those of ground-dwelling arthropods such as scorpions, millipedes, and beetles (Hall et al., *this volume*, chap. 16), and packrat fecal pellets. Bones of small vertebrates, including amphibians, reptiles, birds, and mammals, are less common but of special interest because of the associated floras and radiocarbon dates. Midden vertebrates have been reported from the Grand Canyon of Arizona (Van Devender et al., 1977; Cole and Mead, 1981; Mead, 1981; Mead and Phillips, 1981) and the Sonoran Desert of southwestern Arizona and adjacent California (Van Devender and Mead, 1978; Mead et al., 1983). Mammal remains from an extensive midden

series spanning the last forty-two thousand years from the Hueco Mountains just east of El Paso, Texas, in the Chihuahuan Desert were reported by Van Devender, Bradley, and Harris (1987). Here we present the mammals identified from packrat middens south of the Hueco Mountains in Trans-Pecos Texas and the Bolson de Mapimi in Mexico (Fig. 15.1) and discuss their paleoecological and biogeographical implications.

METHODS AND RESULTS

Fossil packrat middens are hard, dark, organic deposits preserved in dry rock shelters. They contain abundant plant macrofossils collected within the packrats' 30–50-m foraging range. Radiocarbon dating of midden materials allows the fossil assemblages to be placed within the overall chronological framework.

In general, bones are relatively uncommon in middens, although they can be more common in rock shelters where raptors roost. Most midden bones are isolated, small elements from different individual small vertebrates. Breakage, dissolution, pitting, and small size suggest that the bones are transported to the rock shelter by raptors or small predators rather than collected by the packrat. Although packrat midden faunas contain few specimens and taxa compared to cave deposits, each specimen provides a robust fossil

Fig. 15.1. Area discussed in text. Chihuahuan Desert in stipple after Schmidt (1986). Solid triangles are packrat midden sites; solid circles are cave or sedimentary sites.

record when combined with the associated flora and radiocarbon date. Differences in the sampling areas of raptors and packrats indicate that some midden vertebrates may not be directly associated with the local flora.

A total of thirty-five packrat and porcupine middens from seven study areas in trans-Pecos Texas and three areas in the Bolson de Mapimi in Coahuila and Durango (Fig. 15.1) yielded eighty-nine identifiable mammalian specimens (Table 15.1). Twenty-two taxa in nine families were associated with radiocarbon ages from > 44,900 to 600 B.P. (radio-

carbon years before 1950). Specimens were identified using reference skeletons in Van Devender's personal collection and in the University of Arizona mammal collection. Fossils were deposited in the University of Arizona Laboratory of Paleontology vertebrate paleontological collection (UALP). Sierra de la Misericordia #1 and #14 were porcupine (*Erethizon dorsatum*) middens rather than packrat middens. Betancourt et al. (1986) compare packrat and porcupine middens.

The Hudspeth County sites are in the Quitman Mountains (9 km south of Sierra Blanca

Table 15.1. *Radiocarbon dates and locality numbers for packrat middens from trans-Pecos Texas and the Bolson de Mapimí, Mexico, yielding identifiable mammalian remains.*

Midden Name	Elevation (m)	Radiocarbon Date (yr B.P.)	Laboratory No.	Material Dated	Locality No.	Paleovegetation
Hudspeth County						
Quitman Mts. #1	1430	10,910 ± 170	A-1612	*Juniperus* sp. twigs	8515	Pinyon-juniper-oak woodland
Steeruwitz Hills #1(3)	1430	14,290 ± 290	A-1844	*Juniperus* sp. twigs	8516	Pinyon-juniper-oak woodland
Steeruwitz Hills #1(4)	1430	18,060 ± 1320	A-1623	*Juniperus* sp. twigs	8517	Pinyon-juniper-oak woodland
Presidio County						
Bennett Ranch #1	1035	18,190 ± 380	A-1831	*Neotoma* sp. fecal pellets	8519	Pinyon-juniper-oak woodland
Bennett Ranch #4	1035	12,130 ± 170	A-1826	*Juniperus* sp. twigs	8520	Pinyon-juniper-oak woodland
Bennett Ranch #5	1035	12,600 ± 500	A-1836	*Juniperus* sp. twigs	8521	Pinyon-juniper-oak woodland
Shafter #1A	1310	15,950 ± 900	A-1845	*Juniperus* sp. twigs	8518	Pinyon-juniper-oak woodland
Shafter #1B	1310	15,695 ± 230	A-1581	*Juniperus* sp. twigs	[a] UTEP 102	Pinyon-juniper-oak woodland
Brewster County						
Baby Vulture Den #1	635	600 ± 140	A-2962	Midden debris	8531	Chihuahuan desertscrub
Baby Vulture Den #4	605	5500 ± 190	A-3169	*Agave lechuguilla* leaf	8532	Chihuahuan desertscrub
Baby Vulture Den #5B(1)	615	26,430 ± 4600	A-3138	*Juniperus* sp. twigs	8533	Pinyon-juniper-oak woodland
Baby Vulture Den #5B(2)	615	19,600 ± 800	A-3133	*Juniperus* sp. twigs	8534	Pinyon-juniper-oak woodland
Baby Vulture Den #10A	585	5560 ± 120	A-4250	Midden debris	8564	Chihuahuan desertscrub
Ernst Tinaja #2A	835	18,760 ± 500	A-2978	*Neotoma* sp. fecal pellets	8523	Pinyon-juniper-oak woodland
Ernst Tinaja #2B(2)	835	16,220 ± 400	A-2982	*Juniperus* sp. twigs	8524	Pinyon-juniper-oak woodland
[b] Terlingua #1	910	15,000 ± 440	WK-175	*Juniperus* sp. twigs	7817	Pinyon-juniper-oak woodland
Tunnel View #1	670	24,050 ± 770	Av. A-2958 & A-3126	*Juniperus* sp. twigs, *Opuntia phaeacantha* seeds	8525	Juniper desert grassland
Tunnel View #3A	680	15,060 ± 450	A-2938	*Juniperus* sp. twigs	8526	Pinyon-juniper-oak woodland
Tunnel View #5A	680	8980 ± 130	Av. A-2936 & A-3128	*Lycium puberulum*, *Opuntia phaeacantha* seeds	8527	Desert grassland
Tunnel View #6	670	30,000 + 3200 − 2300	A-4247	Midden debris	8560	Juniper desert grassland
Tunnel View #8A(2)	670	43,000 + 2400 − 1800	AA-1671	*Juniperus* sp. twigs	8528	Juniper desert grassland
Tunnel View #8C	670	>44,900	AA-1672	*Juniperus* sp. twigs	8529	Juniper desert grassland
Tunnel View #10	680	19,960 ± 700	A-3286	*Juniperus* sp. twigs	8530	Pinyon-juniper-oak woodland

Locality	Paleovegetation	Material	Sample #	Date (B.P.)	Elevation	A-number
Tunnel View #11	Pinyon-juniper-oak woodland	Neotoma sp. fecal pellets	8561	11,800 ± 250	730	A-4232
Tunnel View #12	Juniper desert grassland	Neotoma sp. fecal pellets	8562	9870 ± 150	685	A-4237
Tunnel View #17	Juniper desert grassland	Midden debris	8563	10,500 ± 190	670	A-4248
Coahuila						
Cañon de la Fragua #2A	Pinyon-juniper woodland	Juniperus sp. twigs	8571	12,380 ± 260	930	A-4137
Cañon de la Fragua #3D	Chihuahuan desertscrub	Neotoma sp. fecal pellets	8574	2180 ± 90	900	A-4136
Cañon de la Fragua #3E	Chihuahuan desertscrub	Neotoma sp. fecal pellets	8575	1600 ± 170	900	A-4135
Puerto de Ventanillas #12A	Chihuahuan desertscrub	Neotoma sp. fecal pellets	8578	5000 ± 90	1370	A-4162
Puerto de Ventanillas #12B	Chihuahuan desertscrub	Neotoma sp. fecal pellets	8579	3500 ± 100	1370	A-4149
Durango						
Sierra de la Misericordia #1	Chihuahuan desertscrub	Erethizon dorsatum fecal pellets	8582	820 ± 70	1310	A-4145
Sierra de la Misericordia #12B		Midden debris	8584	7700 ± 180	1310	A-4126
Sierra de la Misericordia #13	Chihuahuan desertscrub	Midden debris	8585	7770 ± 210	1310	A-4142
Sierra de la Misericordia #14	Chihuahuan desertscrub	Erethizon dorsatum pellets	8586	780 ± 90	1310	A-4122

[a] UTEP = University of Texas at El Paso Laboratory for Environmental Biology.
[b] Except for this sample, all Brewster County samples are from Big Bend National Park.

on Texas 1111) and in the Streeruwitz Hills (28 km west-northwest of Van Horn on I-10). Presidio County sites are on the Bennett Ranch above the Rio Grande just southeast of the Jeff Davis County line and in the Livingston Hills, 8 km south-southwest of Shafter on U.S. 67 (Van Devender et al., 1978). The Terlingua #1 site in Brewster County is in Carrizo Canyon, just west of Terlingua. The remainder of the Brewster County samples are from within a few kilometers of Rio Grande Village in Big Bend National Park. Cañon de la Fragua is on the southeastern edge of the Cuatro Cienegas Basin, Coahuila (Van Devender, Thompson, and Betancourt, 1987). Puerto de Ventanillas (43 km northeast of San Pedro de las Colonias, Coahuila) and Sierra de la Misericordia, near Bermejillo, Durango, are discussed in Van Devender and Burgess (1985).

SYSTEMATIC ACCOUNTS

The following accounts include the midden name, elements identified in chronological order from oldest to youngest, and the UALP and University of Texas at El Paso Laboratory for Environmental Biology (UTEP) catalogue numbers. Associated radiocarbon dates, elevations, paleovegetations, and locality numbers are in Table 15.1. Summaries of the fossil records for the taxa are presented in Harris (1985) and Van Devender, Bradley, and Harris (1987). Time abbreviations are: MW = middle Wisconsin (22,000 − > 44,900 B.P.), LW = late Wisconsin (11,000 − 22,000 B.P.), EH = early Holocene (8900 − 11,000 B.P.), MH = middle Holocene (4000 − 8900 B.P.), and LH = late Holocene (4000 B.P. to present); ka = thousands of years.

Order Insectivora: Family Soricidae

Sorex sp. (shrew). Material: SM#12B, RI[1], LI[1] (16878). Age: MH, 7.7 ka. The incisors resemble *Sorex* and differ from *Notiosorex crawfordi* (desert shrew) in: (1) a narrower, less-curved root; (2) much darker pigment on distal half of enamel grading from reddish-brown to nearly black; (3) dorsal edge of enamel on labial surface convex rather than concave; and (4) lingual surface of larger lobe broader and more concave. In *Cryptotis parva* (least shrew) the two lobes of the incisors are separated by a much less acute angle than in *Notiosorex* or *Sorex*.

The ranges of two *Sorex* species approach the Sierra de la Misericordia site in northeastern Durango (Hall, 1981). *Sorex milleri* (Sierra del Carmen shrew) is in the Sierra Madre Oriental in eastern Coahuila, while *S. saussurei* (Saussure's shrew) is found on the Mexican Plateau as far north as southern Coahuila. The nearest Durango record is from 2 km west of San Luis, 80 km south of the midden site (Hall, 1981). Findley (1953) identified *S. saussurei* and *S. cinereus* (masked shrew) from late Wisconsin deposits in San Josecito Cave, Nuevo Leon, 480 km east-southeast of Sierra de la Misericordia. *Sorex cinereus*, a Rocky Mountain shrew characteristically found in mesic hydrosere habitats (Brown, 1967), was a common associate of *Microtus pennsylvanicus* (meadow vole) in both mesic-montane and lowland-hydrosere habitats in the Wisconsin of the southwestern United States (A. H. Harris, personal communication, 1987). Sierra de la Misericordia is probably too xeric for any *Sorex* and, in association with *Microtus* cf. *pennsylvanicus*, indicates wetter conditions than today as late as 7700 B.P. in the middle Holocene.

Order Chiroptera

Family Vespertilionidae. *Antrozous pallidus* (pallid bat). Material: BR#5, LC_1 (16596). Age: LW, 12.6 ka.

Order Lagomorpha

Family Leporidae. *Lepus* sp./cf. *Lepus* (jackrabbit). Material: TV#8A(2), R. astragalus (16608), RP_3 (16609); TV#10, LM^1 (16613). Age: MW, 43 ka; LW, 20 ka.

Sylvilagus cf. *auduboni* (desert cottontail). Material: TV#6, RP^4 (16819); BVD#5B(1), LP_3 (16620); ET#2B(2), LP_3 (16603); QM#1, distal and R. radius (16582); BVD#10A, LP^4 (16829); BVD#1, LM^2 (16614). Age: MW, 26.4–30.0 ka; LW, 10.9–16.2 ka; MH, 5.6 ka; LH, 0.6 ka.

Sylvilagus cf. *floridanus* (eastern cottontail). Material: TV#8A(2), LM^2 (16610); BR#1, LP_3 (16592); TE#1, L. calcaneus (16597). Age: MW, 43 ka; LW, 15.0–18.2 ka. These specimens are from a large cottontail similar in size to *S. f. robustus* (Davis Mountains cottontail) of the pinyon-juniper-oak woodlands in the Guadalupe, Davis, Chinati, and Chisos mountains, and larger than *S. f. chapmani* of the Edwards Plateau

and the local *S. auduboni* (Schmidly, 1977).

Sylvilagus sp./cf. *Sylvilagus* (cottontail). Material: ET#2A, patella (16601); BR#1, L. astragalus (16594); TE#1, distal end R. ulna (16598), metatarsals (16599–16600); QM#1, proximal end L. femur (16581); BVD#4, LM_3 (16618). Age: LW, 10.9–18.8 ka; MH, 5.5 ka.

Order Rodentia

Family Sciuridae. *Spermophilus* cf. *spilosoma* (spotted ground squirrel). Material: TV#1, partial dentary with P_4–M_1 (16604); SH#1(4), RM_1 (16587); SH#1(3), RM_1 (16585). Age: MW, 24.1 ka; LW, 14.3–18.1 ka.

Cynomys cf. *gunnisoni* (white-tailed prairie dog). Material: SH#1(4), maxillary fragment with RM^{1-2} (16586).

Cynomys sp. Material: SH#1(4), RM_2 (16588), LM_1 (16589). Age: LW, 18.1 ka. The maxilla was similar in size to *C. gunnisoni* and smaller than *C. ludovicianus* (black-tailed prairie dog), which occurred in the area historically (Schmidly, 1977). The specimens were from a late Wisconsin (18,060 B.P.) woodland assemblage from the Streeruwitz Hills on the south end of the Sierra Diablo west of Van Horn. *Cynomys gunnisoni* is the prairie dog of the Colorado Plateau in Arizona, Utah, Colorado, and New Mexico. The nearest population to the midden site is about 340 km to the northwest in western New Mexico (Findley et al., 1975).

The Pleistocene record of *Cynomys* in the northern Chihuahuan Desert is confusing. *Cynomys gunnisoni* was reported from late Wisconsin sediments in Williams Cave (Ayer, 1936) and Lower Sloth Cave (Logan, 1983) in the Guadalupe Mountains, Culberson County, Texas. Harris (1977) reported *C.* cf. *gunnisoni* from 8 km south of Alpine in Brewster County but later referred it to *C.* cf. *ludovicianus*. Middle Wisconsin *C. ludovicianus* were reported from Dry Cave, Eddy County, New Mexico (Harris, 1977). Late Wisconsin *C. ludovicianus* have been reported from Blackwater Draw, Roosevelt County (Slaughter, 1975), Burnet Cave in the Guadalupe Mountains (Schultz and Howard, 1935), and Eddy County and Conklings Cavern on Bishop's Cap, Doña Ana County, New Mexico (Smartt, 1977). A specimen was identified as *C.* cf. *ludovicianus* from an early Holocene (10,750 B.P.) packrat midden from the Hueco

Mountains, El Paso County, Texas (Van Devender, Bradley, and Harris, 1987). The late Wisconsin record of *C. ludovicianus* from Fowlkes Cave in the Apache Mountains (Dalquest and Stangl, 1984) is only 70 km east of the Streeruwitz Hills midden site. Additional study is needed to clarify the Wisconsin species of *Cynomys*.

Family Geomyidae. *Thomomys* sp. (pocket gopher). Material: TV#5A, LP_4 (16606); CF#3D, RM_1 or M_2 (16887). Age: EH, 9 ka; LH, 1.6 ka.

Family Heteromyidae. *Dipodomys merriami/ordi* (Merriam's or Ord's kangaroo rat). Material: SH#1(3), RM^1 (16584); SM#12B, RM^1 (16879); BVD#10A, LM_2 (16831), LM_3 (16830); BVD#4, maxillary fragment with M^{1-2} (16615), LM_2 (16616), RP_4 (16617); SM#14, RP_4 (16886). Age: LW, 14.3 ka; MH, 5.5–7.7 ka; LH, 0.8 ka. Due to the fragmentary condition of the specimens we were unable to distinguish between *D. merriami* and *D. ordi*.

Perognathus flavescens/flavus (plains or silky pocket mouse). Material: TV#8A(2), R. dentary (16607); SM#13, dentary with RM_{2-3} (16881), dentary with RP_4 and RM_3 (16882), maxilla with RP^4 (16883); PV#12B, L. edentulous dentary (16887). Age: MW, 43 ka; MH, 7.8 ka; LH, 3.5 ka. The specimens are from very small pocket mice.

Perognathus penicillatus-type (desert pocket mouse). Material: SF, dentary (UTEP 102-9); TV#12, LP^4 (16882). Age: LW, 16 ka; EH, 9.9 ka. The specimens are from medium-sized *Perognathus* similar to *P. penicillatus*, *P. intermedius* (rock pocket mouse), and *P. nelsoni* (Nelson's pocket mouse).

Family Muridae. *Microtus* cf. *pennsylvanicus* (meadow vole). Material: SM#12B, dentary with LM_{1-2} (16876); SM#13, LM_1 (16880). Age: MH, 7.7–7.8 ka. The teeth are probably from a large southern race of *M. pennsylvanicus* (A. H. Harris, personal communication, 1985; Messing, 1986; Smartt, 1977), although *M. mexicanus* (Mexican vole) cannot be excluded without larger samples (C. A. Repenning, personal communication, 1985).

Microtus pennsylvanicus has been found in late Wisconsin–early Holocene deposits in the northern Chihuahuan Desert at Dry (Harris, 1977, 1985), Muskox (Logan, 1981), An-

thony (Smartt, 1977), and Howell's Ridge (Smartt, 1977) caves, New Mexico. The only previous Mexican fossil record for *M. pennsylvanicus* is from a late Wisconsin/ early Holocene deposit in Jimenez Cave, Chihuahua (Messing, 1986).

Neotoma cf. *albigula* (white-throated packrat). Material: TV#8C, RM^{1-3} (16612); TV#6, RM^1 (16818); BVD#5B(2), LM_3 (16622); ET#2B(2), RM_3 (16602); SF#1A, LM_3 (16591); TV#3A, RM_3 (16605); CF#2A, RM_1, LM_1 (16871–72); BR#4, RM^2 (16595); TV#17, RM_3 (16824); TV#12, LM^1 (16823), LM_3 (16821); BVD#10A, dentary fragment with LM_1 (16828); BVD#1, LM_2 (17104); PV#12A, RM_1 (16868); CF#3E, LM_2 (16873). Age: MW, 30.0–> 44.9 ka; LW, 12.1–19.6 ka; EH, 9.9–10.5 ka; MH, 5.0–5.6 ka; LH, 1.6 ka. The fossils resemble *N. albigula* in morphology and size. The similar *N. micropus* (southern plains packrat) is relatively larger (Dalquest and Stangl, 1984) and lives primarily in deep-soiled areas away from rocky habitats (Schmidly, 1977).

Neotoma sp. (packrat). Material: All samples contained fecal pellets.

Onychomys sp. (grasshopper mouse). Material: BR#1, RM^1 (16593); SM#12B, LM_1 (16877). Age: LW, 18.2 ka; MH, 7.7 ka.

Peromyscus sp. (deer mouse). Material: BVD#5B(1), L. edentulous dentary (16621); TV#11, R. edentulous dentary (16820); SM#13, LM^1 (16884); BVD#10A, LM_2 (16827). Age: MW, 26.4 ka; LW, 11.8 ka; MH, 5.6–5.8 ka.

Reithrodontomys sp. (harvest mouse). Material: SM#13, LM^1 (16885). Age: MH, 7.8 ka. This tooth is from a *Reithrodontomys* similar in size to *R. megalotis* and *R. montanus* (western and mountain harvest mice), smaller than typical *R. fulvescens* (fulvous harvest mouse), and larger than *Baiomys taylori* (pygmy mouse).

Sigmodon sp. (cotton rat). Material: BVD#10A, partial dentary with LM_{1-3} (16825), LM_1 (16826); BVD#4, LM_1 (16619); PV#12B, RM_1 (16869). Age: MH, 5.6–5.5 ka; LH, 3.5 ka. Isolated teeth of *Sigmodon* are difficult to identify to species. The Baby Vulture Den specimens from near Rio Grande Village in Big Bend National Park resemble *S. hispidus* (hispid cotton rat). The tooth from Puerto de Ventanillas, Coahuila, was less robust and best resembled *S. ochrognathus* (yellow-nosed cotton rat).

These middle and late Holocene specimens of *Sigmodon* sp. are the first midden records for the Chihuahuan Desert, although the genus was recorded previously from two Sonoran Desert middens from Arizona (Mead et al., 1983). *Sigmodon hispidus* has an extensive Wisconsin fossil record in Florida, Georgia, and central Texas (references in Lundelius et al., 1983). In the Chihuahuan Desert it was reported from middle Wisconsin (29,290 B.P.) and the latest Wisconsin–early Holocene (< 12,000 B.P.) deposits in Dry Cave, New Mexico (Harris, 1977), and late Wisconsin deposits in Fowlkes Cave, Texas (Dalquest and Stangl, 1984). The apparent collecting bias against *Sigmodon* in the midden record is puzzling considering its wide range, locally high densities, and abundance in predator-accumulated cave faunas (Dalquest and Stangl, 1984).

Family Erethizontidae. *Erethizon dorsatum* (Porcupine) Material: SM#1, 16 quills (16875), fecal pellets; SM#14, fecal pellets. Age: LH, 0.8 ka.

Order Artiodactyla

Family Cervidae cf. *Odocoileus* (Deer) Material: SH#1(4), patella (16590). Age: LW, 18.1 ka.

PALEOENVIRONMENTS

The packrat middens in the study areas record local floras over forty-five thousand years of the last glacial-interglacial cycle. In the late Wisconsin glacial a pinyon-juniper-oak woodland occurred over the entire elevational gradient now occupied by the northern Chihuahuan Desert (Lanner and Van Devender, 1981; Van Devender, *this volume*, chap. 7). A similar woodland extended back at least forty-two thousand years ago in the middle Wisconsin in the Hueco Mountains (Van Devender, Bradley, and Harris, 1987; Van Devender, Thompson, and Betancourt, 1987). Harris (1987) used the extensive vertebrate faunas of Dry and U-Bar caves to reconstruct middle Wisconsin environments in southern New Mexico. However, middle Wisconsin samples from Rio Grande Village in Big Bend National Park recorded juniper in a desert-grassland association relatively drier than late

Wisconsin pinyon-juniper-oak woodlands. Middle and late Wisconsin middens from Maravillas Canyon Cave in Black Gap Wildlife Management Area and from higher elevations in the park showed few differences among the woodlands (Wells, 1966; Van Devender, *this volume*, chap. 7).

After the close of the Wisconsin at about 11,000 B.P., the vegetation and climate began a series of steps culminating in the modern environment. A transitional oak-juniper woodland was present from 11,000 B.P. until 8000–9000 B.P. in the early Holocene. The middle Holocene was a warm, wet period lasting until about 4500 B.P., not the hot, dry altithermal of Antevs (1948). In the northern Chihuahuan Desert a desert grassland preceded the formation of relatively modern Chihuahuan desertscrub in the late Holocene. Middle Holocene desertscrub communities in the Big Bend were merely more mesic than modern ones.

MAMMALIAN PALEOECOLOGY

The midden mammal fossils document a basically conservative fauna, with most of the Wisconsin taxa still occurring near the midden sites. *Sylvilagus* cf. *floridanus* and *Cynomys* cf. *gunnisoni* in late Wisconsin samples and *Microtus* cf. *pennsylvanicus* and *Sorex* sp. in middle Holocene samples are the only animals that indicate range changes.

Sylvilagus f. robustus, the large mountain cottontail, was apparently widespread in the ice-age woodlands, with middle Wisconsin records for the Hueco Mountains (Van Devender, Bradley, and Harris, 1987) and late Wisconsin records from Fowlkes Cave (Dalquest and Stangl, 1984) and the Rio Grande lowlands (Bennett Ranch, Terlingua, and Rio Grande Village middens). Dalquest and Stangl (1984) interpret the retraction of the range of *S. f. robustus* into the mountains of trans-Pecos Texas as a response to the increasing aridity of early Recent (early Holocene) climates and not a temperature limitation. Paleoclimatic reconstructions based on packrat midden plant remains from the Chihuahuan Desert reflect not a greater aridity at the end of the Wisconsin but a gradual warming of summers and a shift from predominantly winter rainfall to strongly developed summer monsoons after 9 ka (Van Devender et al.,

1984; Van Devender, *this volume,* chap. 7). Annual precipitation was probably greater than today from at least 45,000 B.P. in the middle Wisconsin until after 4000–5000 B.P., suggesting that the range retraction of *S. f. robustus* at the beginning of the Holocene was probably related to warmer summers rather than a decrease in moisture. Isolation of the *S. f. robustus* montane populations probably occurred in several steps. The disappearance of *Pinus remota* (papershell pinyon) from the desert lowlands at the end of the Wisconsin probably marks a range contraction for *s. f. robustus* as well. Harris (1985) suggests that *S. floridanus* expanded its range into New Mexico in the latest Wisconsin and early Holocene as *S. nuttalli* (Rocky Mountain cottontail) retreated northward from its southernmost late Wisconsin range. However, with oak-juniper woodlands at lower elevations in the early Holocene, well-developed montane corridors probably connected the Guadalupe, Delaware, Sierra Diablo, and Apache mountains of Culberson County. The Davis Mountains populations were probably connected with the eastern Big Bend through the Glass and Dead Horse mountains. The Glass Mountains also provided a corridor to the east for *S. f. robustus* to contact *S. f. chapmani* of the Edwards Plateau race. The final isolation of *S. f. robustus* into isolated montane populations in New Mexico and Texas probably occurred about nine thousand years ago. On the southern Stockton and Edwards plateaus east of the climatic Chihuahuan Desert (Schmidt, 1986) but within the Chihuahuan biotic province (Blair, 1950), *S. f. chapmani* remained in semiarid desert grasslands through the middle Holocene. Hulbert (1984) recorded continuous *S. floridanus* populations in Val Verde County caves (Bonfire Shelter, Centipede, Damp, and Eagle caves) until 5000 B.P. and again after 3000 B.P. The range of *S. f. chapmani* retracted farther east only during the driest part of the late Holocene. With repetition of this environmental sequence culminating in interglacial isolation many times in the 1.8 million years of the Pleistocene (Imbrie and Imbrie, 1979), intense selection has almost differentiated *s. f. robustus* to the species level.

While there are questions about the identity of the Wisconsin *Cynomys* of the north-ern Chihuahuan Desert, open, grassy habitats appear to have been present in deep-soiled areas away from mountains throughout the middle and late Wisconsin. Reexamination of the fossils may show that *C. gunnisoni* expanded its range southeastward during the late Wisconsin, and *C. ludovicianus* replaced it in the latest Wisconsin or early Holocene (A. H. Harris, personal communication, 1987). The isolation of populations of *C. ludovicianus* in northeastern Mexico leading to the evolution of *C. mexicanus* (Mexican prairie dog) may have occurred in the Holocene (McCullough et al., 1987).

Microtus pennsylvanicus is widespread in mesic sedge-grass habitats from the northeastern United States through much of Canada to Alaska. A spur of its range extends down the Rocky Mountains into north-central New Mexico, with isolated populations in the western portion of the state (Findley et al., 1975; Hall, 1981). Specimens from Ojo de Array, near Galeana, Chihuahua, described as *M. p. chihuahuensis* (Bradley and Cockrum, 1968), represent the only known Mexican population.

Martin (1968) suggests that *M. pennsylvanicus* is limited by warm summer temperatures. The average July temperature is 19.4°C for its entire range and 23.9°C on the range's southern edge. However, the average July temperature for Galeana, Chihuahua, 5 km north-northeast of Ojo de Array, is 26°C (Garcia, 1981). July averages for fossil sites discussed below at Jimenez, Chihuahua, and Sierra de la Misericordia, Durango, are 30.4°C and about 28.0°C, respectively. In the Wisconsin much of its northern range was covered by a continental glacier, and boreal forests and *M. pennsylvanicus* were displaced southward into northern Florida, northern Texas, and southern New Mexico (Harris, 1985; Martin, 1968; Smartt, 1977). The fossil record from Jimenez Cave in southern Chihuahua (Messing, 1986) extended its Wisconsin range farther south. Warmer and drier climates in the Holocene fragmented the southern portion of its ice-age range. It apparently persisted in the Rio Grande Valley of New Mexico into Holocene times (Harris, 1985; Smartt, 1977). *M. pennsylvanicus* teeth, identified from nine stratigraphic levels in Howell's Ridge Cave in southwestern New Mexico, suggest extirpation of some populations after 4000 B.P. in the late Holocene (Van Devender and Worthington, 1977; R. A. Smartt, personal communication, 1975). The 7700 B.P. and 7770 B.P. midden specimens from Sierra de la Misericordia, 590 km southeast of the relict Chihuahuan population, also indicate its survival into the middle Holocene at the southernmost locality known for the species. Martin (1968) interprets the late Pleistocene range expansions of M. pennsylvanicus as indicating cooler summers in the humid southeastern United States and cooler summers and increased moisture in the Southwest, where moisture is limiting. The Holocene fossil records and the surviving Chihuahuan population suggest that it can live in relatively warm summer climates if sedge-grass marsh habitats are exceptionally well developed. Ultimately, most Mexican populations were extirpated by habitat loss as the springs dried.

Neotoma albigula has maintained its dominance of the rocky-slope habitats for the last forty-two thousand years in the Hueco Mountains in the northern Chihuahuan Desert (Van Devender, Bradley, and Harris, 1987) and for the last forty-five thousand years in the Big Bend. Neotoma albigula probably constructed most of the fossil packrat middens in the Chihuahuan Desert. This rodent has readily adapted its diet as the local vegetation shifted from woodland to desert grassland to succulent desertscrub and must be viewed as a dietary generalist (Vaughan, this volume, chap. 2). The expansion of N. cinerea (bushy-tailed packrat) into areas now occupied by other species of Neotoma, recorded for more northern areas (Harris, 1984), was not found over most of the Chihuahuan Desert.

LATITUDINAL TRENDS

General latitudinal gradients in temperature and rainfall are complicated by regional elevational gradients and circulation patterns. Base levels fall from above 1525–1675 m in central New Mexico to 1220–1675 m in the northern Chihuahuan Desert. East of the Guadalupe and Davis mountains rainfall increases at lower elevations, counter to the typical elevational gradient in the Southwest (Schmidt, 1986). Base-level elevations along the Rio Grande range from 1220 m near Las Cruces to

560 m at Rio Grande Village on the eastern edge of Big Bend National Park. Downstream the gradient reverses, and rainfall increases by 170% as elevation drops 255 m at Del Rio. The southern Chihuahuan Desert in the Bolson de Mapimi is a series of internal drainage basins mostly above 1200 m in Coahuila, Durango, and Chihuahua (Morafka, 1977). Elevations on the Mexican Plateau continue to rise to the south, where the rain shadows of the Sierra Madre Oriental and Occidental cause relatively dry climates.

Paleofloras in Chihuahuan Desert packrat middens reflect a latitudinal trend with differences between Wisconsin and modern floras decreasing to the south (Van Devender, 1986; this volume, chap. 7). The pinyon-juniper-oak woodlands of the Hueco Mountains (32° N) were almost devoid of Chihuahuan desertscrub plants, while Agave lechuguilla (lechuguilla) and Koeberlinia spinosa (allthorn) were common in similar woodlands in the Rio Grande Village area (29°16′ N). Late Wisconsin woodlands in the Bolson de Mapimi, Coahuila and Durango (26° N), in the southern Chihuahuan Desert were simply juniper or papershell pinyon and juniper growing in a succulent desertscrub community (Van Devender and Burgess, 1985; Van Devender, Thompson, and Betancourt, 1987). This Wisconsin latitudinal pattern was due primarily to the usual gradient of increasing temperatures at lower latitudes, relatively lower increases in winter precipitation at greater distances from the Pacific Ocean, and blockage of air flow from the Gulf of Mexico by the Sierra Madre Oriental.

The small mammals from Chihuahuan Desert packrat middens show a similar pattern of decreasing differences between Wisconsin and modern faunas at lower latitudes for the Chihuahuan Desert along the Rio Grande trough. The Hueco Mountains (1270–1495 m) just east of El Paso are a southern extension of the massive Sacramento Mountains. The only extralimital mammals in twenty-three taxa identified from the middens were a few northern/montane species (Geomys bursarius, Microtus sp., and Spermophilus cf. tridecemlineatus; Van Devender, Bradley, and Harris, 1987). Great Basin and mesic northern/montane mammals were absent, although a few leaves of Great Basin plants (e.g., *Artemisia triden-tata*-type, *Ceratoides lanata*, and *Oryzopsis hymenoides*) were found. The middens from farther south in the trans-Pecos (585–1430 m) yielded even fewer extralimital mammals. *Cynomys* cf. *gunnisoni* from 1430 m elevation in the Streeruwitz Hills was the only potential northern species. *Sylvilagus* cf. *floridanus* from 670–1035 m in the Rio Grande lowlands was the only local montane species (see above). The absence of *Marmota flaviventris* (yellow-bellied marmot), *Microtus* spp., *Neotoma mexicana* (Mexican packrat), *Sorex* spp., and *Tamias* spp. (chipmunks) suggests that the Wisconsin woodlands in the modern Chihuahuan Desert in the Big Bend of Texas were probably too dry or warm for many northern small mammals.

The mammals identified from Coahuila and Durango middens and Jimenez Cave, Chihuahua (1450 m; Messing, 1986), suggest that the Wisconsin latitudinal trend extended into the Mexican portion of the Chihuahuan Desert. The only northern mammal was *Microtus pennsylvanicus*, which probably extended its range southward from western New Mexico in sedge-grass marsh habitats associated with pluvial lakes, rivers, and springs along the northeastern base of the Sierra Madre Occidental, rather than through the Rio Grande–Rio Conchos corridor in the Chihuahuan Desert.

The midden mammals fit into a general latitudinal pattern reflected in Chihuahuan Desert cave faunas. Limestone caves in the Guadalupe Mountains (a southeastern extension of the Sacramento Mountains) on the northeastern edge of the Chihuahuan Desert have yielded exceptionally rich vertebrate faunas. Several discrete middle and late Wisconsin deposits in Dry Cave (1280 m), New Mexico, contained a number of extralimital mammals, including Great Basin (*Lagurus curtatus*), northern (*Lepus townsendi*, *Sylvilagus nuttalli*, *Spermophilus tridecemlineatus*, and *Pitymys ochrogaster*), and northern/montane (*Marmota* cf. *flaviventris*, *Microtus longicaudus*, *Neotoma cinerea*, *Sorex nanus*, and *S. vagrans*; Harris, 1970, 1977, 1985) species. A fauna from Muskox Cave (1600 m), Eddy County, New Mexico, dated between 25,500 and 18,140 B.P., contained some of the same extralimital species (*Marmota flavi-*

ventris, *Neotoma cinerea*, *Pitymys ochrogaster*, *Sorex vagrans*, and *Sylvilagus* cf. *nuttalli*), with additional northern/montane elements (*Sorex cinerus*, *S. palustris*, and *Tamiasciurus hudsonicus*) and species now found at higher elevations in the Guadalupe Mountains (*Neotoma mexicana* and *Sylvilagus* cf. *floridanus*; Logan, 1981). A fauna from Lower Sloth Cave (2000 m) dated at 11,590 B.P. in the Texas portion of the Guadalupe Mountains contained many of the same mammals (*Marmota flaviventris*, *Microtus mexicanus*, *Mustela frenata*, *Neotoma cinerea*, *N. mexicana*, *Sorex cinereus*, *S. vagrans*, and *Tamiasciurus hudsonicus*) as well as *Cynomys* cf. *gunnisoni* (Logan, 1983). The extensive late Wisconsin small-mammal fauna from Fowlkes Cave in the Apache Mountains, which is connected to the Guadalupe Mountains by the Delaware Mountains–Sierra Diablo corridor, contained many of the local desert-grassland mammals and relatively fewer extralimital northern/montane (*Marmota flaviventris*, *Sorex palustris*, and *S. vagrans*) and montane (*Microtus mexicanus*, *Neotoma mexicana*, *Sylvilagus floridanus*, and *Tamias cinereicollis*) species (Dalquest and Stangl, 1984).

One interesting record from the Big Bend of Texas is a small fauna recovered from a depth of 98 m in a cinnabar mine in Terlingua. Skulls and jaws of *Desmodus stocki* (an extinct vampire bat) and *Macrotus californicus* (California leafnosed bat), a subtropical phyllostomatid no longer found in Texas, were associated with fossils of *Bison* sp. (bison), *Equus* sp. (horse), and *Ovis* sp. (bighorn sheep; Ray and Wilson, 1979).

Ultimately, the topography of the Mexican Plateau, with the elevation of the xeric lowlands rising to the south and the influence of the bounding Sierra Madre, modified the latitudinal trend in Wisconsin small-mammal faunas. The vertebrate fauna from San Josecito Cave (2250 m) above Aramberri, Nuevo Leon, provides a rich record from the Sierra Madre Oriental southeast of the Chihuahuan Desert (references in Harris, 1977, 1985). Harris (1977) tentatively assigns the deposit an age of 25,000–12,500 B.P. based on faunal grounds. *Marmota* sp., *Sorex cinereus*, and *Synaptomys cooperi* (southern bog lemming) are northern elements in the fauna. *Marmota flaviventris* and *Sorex cinereus* may have dispersed from the Rocky Mountains via the mountains of Trans-Pecos Texas and northern Coahuila or possibly through lowland hydrosere or riparian habitats. *Marmota monax* (ground hog) of the eastern and central United States has not been recorded in the rich vertebrate deposits in limestone caves on the Edwards Plateau or sedimentary sites from north-central Texas or the Texas Panhandle (Lundelius et al., 1983) and probably was not the San Josecito marmot. The record for *S. cinereus* represents a remarkable southern range extension of 1280 km (Findley, 1953). In contrast, *Synaptomys cooperi* apparently moved southward from the Midwest through central Texas into northeastern Mexico. However, San Josecito Cave yielded a number of Mexican tropical montane animals (*Coendou* sp., *Cryptotis mexicana*, *Liomys irroratus*, *Microtus mexicanus*, and *Sorex saussurei*) as well as nominally extinct species (*Desmodus stocki*, *Mustela reliquus*, *Orthogeomys onerosus*, and *Plecotus tetralophodon*).

DISCUSSION

Bones in packrat middens and cave deposits provide a record of the changes in the small-mammal fauna of the Chihuahuan Desert for forty-five thousand years of the last glacial-interglacial cycle. In general, mammals were affected less than plants by glacial climates. The differences between glacial and modern faunas increased to the north, with more species from the Great Basin Desert and Rocky Mountain forest and mesic hydrosere habitats. Desert mammals were increasingly important in ice-age woodlands to the south. Sonoran Desert middens have yielded records of vertebrates (mostly reptiles) living in plant communities not inhabited today (Van Devender, *this volume*, chap. 8). Similar examples from the northern Chihuahuan Desert include *Microtus* sp. (vole) and *Xerobates agassizi* (desert tortoise) in pinyon-juniper-oak woodland on Bishop's Cap and in the Hueco Mountains (Van Devender, Bradley, and Harris, 1987; Van Devender, *this volume*, Chap. 7). Farther south unusual vertebrate habitat associations are not apparent for several reasons, including resolution at the generic level for many taxa. More important, the differences between desertscrub and

woodland are blurred on the eastern edges of the Chihuahuan Desert in the lowlands along the Rio Grande toward Del Rio in Texas (Van Devender and Bradley, in press) and in the lower elevations of the Sierra Madre Oriental in Coahuila and Nuevo Leon. In these areas increasing rainfall at lower elevations and equable temperature regimes combine to allow mixtures of desert and woodland plants and animals that mostly are separated into elevationally zoned habitats in the drier areas to the west. The late Wisconsin and early Holocene woodlands from the Big Bend southward were also mixed biotically, suggesting equable paleoclimates.

The range adjustments following climatic changes in the postglacial were serial in the north, beginning with the demise of Wisconsin woodlands, and other adjustments occurred later in the Holocene. In the southern Chihuahuan Desert, where desertscrub developed in the early Holocene without a transitional woodland, most range adjustments were probably completed by ten thousand years ago. In all of these areas mesic/montane animals disappeared first in open-slope habitats and last in relict hydrosere sedge-marsh cienega habitats near springs, rivers, or canyon streams. Concomitantly, more arid-adapted mammals increased in abundance and expanded their ranges to higher elevations and latitudes as hotter, drier habitats developed. Similar changes in species distribution and the composition of the small-mammal fauna probably occurred in each of the fifteen to twenty glacial-interglacial cycles, although the species in earlier cycles were different due to relatively high evolutionary rates.

ACKNOWLEDGMENTS

Jan Bowers, Tony Burgess, Ben Everitt, Daryl Frost, Gene Hall, Ron Neilson, Larry Toolin, and Becky Van Devender helped with the field work. Steve Hale collected the Terlingua #1 midden. Yar Petryszyn and E. Lendell Cockrum provided access to the University of Arizona mammal collection. Charles A. Repenning examined the *Microtus* specimens. Richard A. Smartt identified the *Microtus* teeth from Howell's Ridge Cave. Review comments by Arthur H. Harris, Walter W. Dalquest, and Julio L. Betancourt helped the manuscript. Collecting permits were issued through Big Bend National Park and Consejo Nacional de Ciencia y Tecnologia. Funds were provided by the Big Bend Natural History Association through Rick L. LoBello, and through National Science Foundation grants DEB 80-22773 and BSR 84-00797 to T. R. Van Devender. Helen Wilson drafted the map. Jean Morgan typed the manuscript.

REFERENCES

Antevs, E. (1948). "The Great Basin, with emphasis on glacial and post-glacial times: Climatic changes and pre-white man." *Bulletin of the University of Utah* 38, 168–91.

Ayer, M. Y. (1936). "The archaeological and faunal material from Williams Cave, Guadalupe Mountains, Texas." *Proceedings of the Academy of Natural Sciences of Philadelphia* 88, 599–618.

Betancourt, J. L.; Van Devender, T. R.; and Rose, M. (1986). "Comparison of plant macrofossils in woodrat (*Neotoma* spp.) and porcupine (*Erethizon dorsatum*) middens from the western United States." *Journal of Mammalogy* 67, 166–73.

Blair, W. F. (1950). "The biotic provinces of Texas." *Texas Journal of Science* 2, 93–117.

Bradley, W. G., and Cockrum, E. L. (1968). "A new subspecies of the meadow vole (*Microtus pennsylvanicus*) from northwestern Chihuahua, Mexico." *American Museum Novitates* 2324, 1–7.

Brown, L. M. (1967). "Ecological distribution of six species of shrews and comparison of sampling methods in the central Rocky Mountains." *Journal of Mammalogy* 48, 617–23.

Cole, K., and Mead, J. I. (1981). "Late Quaternary animal remains from packrat middens in the eastern Grand Canyon, Arizona." *Journal of the Arizona-Nevada Academy of Sciences* 16, 24–25.

Dalquest, W. W., and Stangl, F. B., Jr. (1984). "Late Pleistocene and early Recent mammals from Fowlkes Cave, Culberson County, Texas." In *Contributions in Quaternary vertebrate paleontology: A volume in memorial to John Guilday.* H. H. Genoways and M. R. Dawson, eds., pp. 432–55. Special Publication of the Carnegie Museum of Natural History no. 8.

Findley, J. S. (1953). "Pleistocene Soricidae from San Josecito Cave, Nuevo Leon, Mexico." *University of Kansas Publications of the Museum of Natural History* 5, 633–39.

Findley, J. S.; Harris, A. H.; Wilson, D. E.; and Jones, C. (1975). *Mammals of New Mexico.* Albuquerque: University of New Mexico Press.

Garcia, E. (1981). *Modificaciones al sistema de clasificacion climatica de Köppen.* Talleres de Offset Larios, Mexico.

Hall, E. R. (1981). *The mammals of North America.* New York: Wiley and Sons.

Harris, A. H. (1970). "The Dry Cave mammalian fauna and late pluvial conditions in southeastern New Mexico." *Texas Journal of Science* 22, 3–27.

———. (1977). "Wisconsin age environments in the northern Chihuahuan Desert: Evidence from the higher vertebrates." In *Transactions of the Symposium on the Biological Resources of the Chihuahuan Desert Region, United States and Mexico.* R. H. Wauer and D. H. Riskind, eds., pp. 23–52. Washington, D.C.: National Park Service Transactions and Proceedings Series no. 3.

———. (1984). "*Neotoma* in the late Pleistocene of New Mexico and Chihuahua." In *Contributions in Quaternary vertebrate paleontology: A volume in memorial to John Guilday.* H. H. Genoways and M. R. Dawson, eds., pp. 164–78. Special Publications of the Carnegie Museum of Natural History no. 8.

———. (1985). *Late Pleistocene vertebrate paleoecology of the West.* Austin: University of Texas Press.

———. (1987). "Reconstruction of mid-Wisconsin environments in southern New Mexico." *National Geographic Research* 3, 142–51.

Hulbert, R. C., Jr. (1984). "Latest Pleistocene and Holocene leporid faunas from Texas: Their composition, distribution, and climatic implications." *Southwestern Naturalist* 29, 197–210.

Imbrie, J., and Imbrie, K. P. (1979). *Ice ages: Solving the mystery.* Hillside, N.J.: Enslow.

Lanner, R. M., and Van Devender, T. R. (1981). "Late Pleistocene piñon pines in the Chihuahuan Desert." *Quaternary Research* 15, 278–90.

Logan, L. E. (1981). "The mammalian fossils of Muskox Cave, Eddy County, New Mexico." *Proceedings of the Eighth International Congress of Speleology* 1, 159–60.

———. (1983). "Paleoecological implications of the mammalian fauna of Lower Sloth Cave, Guadalupe Mountains, Texas." *National Speleological Society Bulletin* 45, 3–11.

Lundelius, E. L., Jr.; Graham, R. W.; Anderson, E.; Guilday, J.; Holman, J. A.; Steadman, D. W.; and Webb, S. D. (1983). "Terrestrial vertebrate faunas." In *Late-Quaternary environments of the United States.* Volume 1, *The late Pleistocene.* S. C. Porter, ed., pp. 311–53. Minneapolis: University of Minnesota Press.

McCullough, D. A.; Chesser, R. K.; and Owen, R. D. (1987). "Immunological systematics of prairie dogs." *Southwestern Naturalist* 68, 561–68.

Martin, R. A. (1968). "Late Pleistocene distribution of *Microtus pennsylvanicus.*" *Journal of Mammalogy* 49, 265–71.

Mead, J. I. (1981). "The last 30,000 years of faunal history within the Grand Canyon, Arizona." *Quaternary Research* 15, 311–26.

Mead, J. I., and Phillips, A. M. III. (1981). "The late

Pleistocene and Holocene fauna and flora of Vulture Cave, Grand Canyon, Arizona." *Southwestern Naturalist* 26, 257–88.

Mead, J. I.; Van Devender, T. R.; and Cole, K. L. (1983). "Late Quaternary small mammals from Sonoran Desert packrat middens, Arizona and California." *Journal of Mammalogy* 64, 173–80.

Messing, H. J. (1986). "A late Pleistocene–Holocene fauna from Chihuahua, Mexico." *Southwestern Naturalist* 31, 277–88.

Morafka, D. J. (1977). *A biogeographical analysis of the Chihuahuan Desert through its herpetofauna.* The Hague: W. Junk.

Ray, C. E., and Wilson, D. E. (1979). *Evidence for Macrotus californicus from Terlingua, Texas.* Occasional Papers of the Museum, Texas Tech University no. 57.

Schmidly, D. J. (1977). *The mammals of Trans-Pecos Texas.* College Station: Texas A&M University Press.

Schmidt, R. H., Jr. (1986). "Chihuahuan climate." In *Second Symposium on the Resources of the Chihuahuan Desert Region, United States and Mexico.* J. C. Barlow, A. M. Powell, and B. N. Timmermann, eds., pp. 40–63. Alpine, Texas: Chihuahuan Desert Research Institute.

Schultz, C. B., and Howard, E. B. (1935). "The fauna of Burnet Cave, Guadalupe Mountains, New Mexico." *Proceedings of the Academy of Natural Sciences of Philadelphia* 87, 273–98.

Slaughter, B. H. (1975). "Ecological interpretation of the Brown Sand Wedge local fauna." In *Late Pleistocene environments of the southern high plains.* F. Wendorf and J. J. Hester, eds., pp. 179–92. Publications of the Fort Burgwin Research Center no. 9.

Smartt, R. A. (1977). "The ecology of late Pleistocene and Recent *Microtus* from south-central and southwestern New Mexico." *Southwestern Naturalist* 22, 1–19.

Van Devender, T. R. (1986). "Pleistocene climates and endemism in the Chihuahuan Desert flora." In *Second Symposium on the Resources of the Chihuahuan Desert Region, United States and Mexico.* J. C. Barlow, A. M. Powell, and B. N. Timmermann, eds., pp. 1–19. Alpine, Texas: Chihuahuan Desert Research Institute.

Van Devender, T. R.; Betancourt, J. L.; and Wimberly, M. (1984). "Biogeographic implications of a packrat midden sequence from the Sacramento Mountains, south-central New Mexico." *Quaternary Research* 22, 344–60.

Van Devender, T. R.; Bradley, G. L.; and Harris, A. H. (1987). "Late Quaternary mammals from the Hueco Mountains, El Paso and Hudspeth Counties, Texas." *Southwestern Naturalist* 32, 179–95.

Van Devender, T. R., and Bradley, G. L. (In press). "The paleoecology and historical biogeography of the herpetofauna of Maravillas Canyon, Texas." *Copeia.*

Van Devender, T. R., and Burgess, T. L. (1985). "Late Pleistocene woodlands in the Bolson de Mapimi: A refugium for the Chihuahuan Desert?" *Quaternary Research* 24, 346–53.

Van Devender, T. R.; Freeman, C. E.; and Worthington, R. D. (1978). "Full-glacial and Recent vegetation of the Livingston Hills, Presidio County, Texas." *Southwestern Naturalist* 23, 289–302.

Van Devender, T. R., and Mead, J. I. (1987). "Early Holocene and late Pleistocene amphibians and reptiles in Sonoran Desert packrat middens." *Copeia* 1987, 464–75.

Van Devender, T. R.; Phillips, A. M. III; and Mead, J. I. (1977). "Late Pleistocene reptiles and small mammals from the lower Grand Canyon of Arizona." *Southwestern Naturalist* 22, 49–66.

Van Devender, T. R., and Spaulding, W. G. (1979). "Development of vegetation and climate in the southwestern United States." *Science* 204, 701–10.

Van Devender, T. R.; Thompson, R. S.; and Betancourt, J. L. (1987). "Vegetation history of the deserts of southwestern North America: The nature and timing of the late Wisconsin–Holocene transition." In *North America and adjacent oceans during the last deglaciation.* W. F. Ruddiman and H. E. Wright, Jr., eds., pp. 323–52. Boulder: Geological Society of America.

Van Devender, T. R., and Worthington, R. D. (1977). "The herpetofauna of Howell's Ridge Cave and the paleoecology of the northwestern Chihuahuan Desert." In *Transactions of the Symposium on the Biological Resources of the Chihuahuan Desert Region, United States and Mexico.* R. H. Wauer and D. H. Riskind, eds., pp. 85–106. Washington, D.C.: National Park Service Transactions and Proceedings Series, no. 3.

Wells, P. V. (1966). "Late Pleistocene vegetation and degree of pluvial climatic change in the Chihuahuan Desert." *Science* 153, 970–75.

Chapter 16

Arthropod History of the Puerto Blanco Mountains, Organ Pipe National Monument, Southwestern Arizona

W. Eugene Hall   Thomas R. Van Devender   Carl A. Olson

Well-preserved plant remains in hundreds of ancient packrat (*Neotoma* spp.) middens have provided evidence that glacial-age woodlands of juniper, pinyon, and/or oak occupied what are now the warm deserts of North America prior to eleven thousand years ago (Van Devender and Spaulding, 1979; Van Devender et al., 1987). In the Sonoran Desert drier California juniper (*Juniperus californica*) woodlands persisted for several thousand years into the Holocene interglacial in the Whipple Mountains of southeastern California, and the Kofa, Butler, and Tinajas Altas mountains of southwestern Arizona (Van Devender, 1977; Van Devender et al., 1985, 1987). Packrat middens from the Picacho Peak area in California just north of Yuma reveal an invasion of Mojave Desert plants and the continued presence of desertscrub communities in the Colorado River lowlands in the late Wisconsin (Cole, 1986).

Animal remains in packrat middens are less common than plant macrofossils, but they represent an exciting new source of fossils that are associated with radiocarbon-dated floras. Bones and teeth of small vertebrate animals from middens have been reported from a number of areas, including the Sonoran Desert (Mead et al., 1983; Van Devender and Mead, 1978; Van Devender, *this volume*, chap. 15). Although fossil arthropods occur in virtu-

ally every midden sample, they have hardly been studied. Ashworth (1973) reported a few beetle remains from late Wisconsin middens collected in the Big Bend of Texas. Elias (1987) identified the insect remains from a series of packrat middens from the San Andres and Fra Cristobal mountains in south-central New Mexico. Ashworth (1976) presented a brief summary of his analyses of insects from Sonoran Desert middens. Spilman (1976) described a new species of spider beetle (*Ptinus priminidi*) from Wisconsin middens from the Whipple Mountains and the Artillery and Kofa mountains of Arizona. Morgan et al. (1983) discussed packrat middens as a source of fossil insects in the Southwest and mentioned a 14,400 B.P. specimen of a darkling beetle (Tenebrionidae: *Stibia tuckeri*) from the Kofa Mountains. Packrat midden arthropods were reported from Sonoran Desert packrat middens from the Tinajas Altas and Butler mountains of southwestern Arizona and several areas in northwestern Sonora (Hall et al., 1988). Here we present the fossil arthropods identified from packrat middens collected in the Puerto Blanco Mountains of southwestern Arizona.

METHODS AND RESULTS

Packrats collect plants for food and building within about 30 m of their dens. In dry rock shelters their middens sometimes persist as hard, dark, shiny organic deposits cemented together by urine and preserved for thousands of years. The plant remains in the middens are ideal for radiocarbon dating. In the Puerto Blanco Mountains, Pima County, Arizona (31°58–59' N, 112°47–48' W; Fig. 16.1), twenty-one midden samples were collected in south-facing rhyolitic rock shelters at 535–605 m elevation from three areas (Ajo Loop, Twin Peaks, and Cholla Pass) within a few kilometers of the Organ Pipe Cactus National Monument Visitor Center (Fig. 16.2A). Radiocarbon dates on the samples range from 14,120 B.P. to 30 B.P. (radiocarbon years before 1950; Table 16.1). Analyses and interpretations of plant macrofossils are presented in Van Devender (1987) and chapter 7 in this volume.

Hall identified a total of 43 arthropod taxa representing 12 orders, 26 families, and 27 genera from the midden samples using modern comparative material in the entomology collection at the University of Arizona (Table 16.2, Figs. 16.3 and 16.4). Eleven taxa were

identified to species. Of a total of 715 specimens identified, 5 to 123 specimens were found in each midden ($\bar{x} = 34$). One to 16 taxa occurred in each sample, with an average of 6.3. The numbers of specimens per taxon may not reflect local abundances because of differences in the ease of identification. For example, millipede segments are especially easy to recognize, and a single individual could leave many fragments. In many specimens the structures necessary for identification to the species level were missing.

The taphonomy of arthropods in packrat middens has been largely ignored. Unlike the plant remains, most of the arthropod material was not collected by the packrat for food, curiosity, or protection. We have seen occasional pieces of beetle elytra and other chitinous fragments inside fecal pellets, indicating that arthropods form some portion of the diet of the predominantly herbivorous packrats. Packrat parasites such as kissing bugs (Reduviidae: *Triatoma*), ticks, and botflies (Cuterebridae: *Cuterebra*) may be incorporated into the deposit. Packrats unintentionally collect

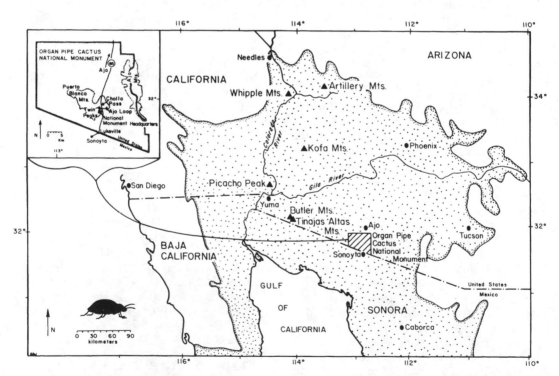

Fig. 16.1. Map of study area and places mentioned in text. Sonoran Desert in stipple after Shreve (1964). Solid triangles are packrat midden sites. Pinacate beetle is *Eleodes armata*.

16.2A

16.2B

Fig. 16.2. (A) South-facing slope of Twin Peaks near Visitor Center, Organ Pipe Cactus National Monument. Packrat middens collected from rock shelters in upper center of photo. (B) Modern packrat den composed of teddy bear cholla (*Opuntia bigelovii*) joints on open slope in the Cholla Pass area about two kilometers north of Visitor Center.

adult insects on the vegetation carried into the rock shelter (i.e., beetles of the families Cleridae, Curculionidae, Elateridae, Salpingidae, etc.), or immature stages that emerge from wood or fruit (i.e., bruchid and anobiid beetles; White, 1965). Other arthropod remains may arrive in the rock shelter in the pellets of raptorial birds or the feces of small mammalian predators such as the ringtail (*Bassariscus astutus*). Many of the midden

Table 16.1. *Radiocarbon dates on packrat middens containing fossil arthropods from the Puerto Blanco Mountains, Organ Pipe Cactus National Monument, Pima County, Arizona.* Asterisk indicates the sample was dated on the tandem accelerator mass spectrometer.

Sample Site	Elevation (m)	Radiocarbon Date (B.P.)	Lab. No.	Material Dated
Twin Peaks #1	565	14,120 ± 260	A-3986	*Juniperus californica* twigs
Cholla Pass #1A	605	10,540 ± 250	A-4276	*Neotoma* fecal pellets
Twin Peaks #7	565	9860 ± 180	A-4212	*Neotoma* fecal pellets
Twin Peaks #5	560	9720 ± 460	*AA-625	*Encelia farinosa* twigs
Cholla Pass #1B	605	9070 ± 170	A-4278	*Neotoma* fecal pellets
Cholla Pass #1C	605	8790 ± 210	A-4277	*Neotoma* fecal pellets
Ajo Loop #1D	550	7970 ± 130	A-3981	*Neotoma* fecal pellets
Twin Peaks #6	580	7580 ± 100	A-3980	*Neotoma* fecal pellets
Ajo Loop #1G	550	7560 ± 170	A-3985	*Neotoma* fecal pellets
Ajo Loop #1A	550	5240 ± 90	A-4213	*Neotoma* fecal pellets
Ajo Loop #5B	535	3480 ± 80	A-4233	*Neotoma* fecal pellets
Cholla Pass #2	590	3440 ± 90	A-4279	*Neotoma* fecal pellets
Ajo Loop #6C	550	3400 ± 100	A-4235	*Neotoma* fecal pellets
Ajo Loop #6B	550	3220 ± 100	A-4234	*Neotoma* fecal pellets
Twin Peaks #2	580	2340 ± 70	A-3973	*Neotoma* fecal pellets
Twin Peaks #3	550	2160 ± 60	A-3982	Midden debris
Twin Peaks #4	555	1910 ± 50	A-3998	*Neotoma* fecal pellets
Ajo Loop #3A	550	990 ± 50	A-3972	*Neotoma* fecal pellets
Ajo Loop #4	550	980 ± 70	A-4214	*Neotoma* fecal pellets
Ajo Loop #2C	550	130 ± 50	A-3971	*Neotoma* fecal pellets
Ajo Loop #2B	550	30 ± 50	A-3970	*Neotoma* fecal pellets

arthropod remains are from small, ground-dwelling forms that live in the rock shelters. Even rock shelters with bare mineral floors support a surprisingly diverse fauna (scorpions, darkling beetles, etc.). In shelters where packrats have accumulated extensive floor deposits commensals and parasites may be abundant. The arthropod remains are cemented into the hardened midden by packrat urine, which can rehydrate during humid periods and function as a trap for small arthropods such as spider beetles (Ptinidae) and ants or their wind-blown remains. Few, if any, of these rock-shelter inhabitants can be considered true cavernicoles. The hardness of the middens, their saturation with urine and plant chemicals, lack of visible signs of burrowing, and the fragmentary nature of fossils suggest that most arthropod specimens are not intrusive.

Twenty pitfall traps were set in the midden study area from March to August of 1985 to sample the modern arthropod community

(Table 16.3). The traps were plastic cups (ca. 100 mm diameter at top and in depth) with a smaller cup of ethylene glycol preservative in the bottom and a conical funnel at the top. The trap was buried with the top at ground level and covered with a slightly elevated board. Traps were set in rock shelters, at cliff bases, on rocky slopes, and on the flats at the bases of hills. The sampling period spanned both the spring and summer flowering seasons, with about six hundred trap nights per month for five months. The relatively long sampling periods were especially productive of secretive, uncommon, ground-dwelling species. Traps of this size occasionally captured small vertebrates; i.e., single specimens of the banded gecko (*Coleonyx variegatus*) and side-blotched lizard (*Uta stansburiana*), but no rodents. Specimens also were collected monthly at night from screen doors in a residence at the base of Twin Peaks. The modern specimens have been deposited in the collections at Organ Pipe Cactus National Monument and the University of Arizona.

ENVIRONMENTAL SETTING

Organ Pipe Cactus National Monument is a unique area in southwestern Arizona where a suite of subtropical Sonoran Desert plants, including organ pipe cactus (*Stenocereus thurberi*), Mexican jumping bean (*Sapium biloculare*), and limber bushes (*Jatropha cinerea, J. cuneata*), enter the United States (Bowers, 1980). The massive Ajo Mountains 9 km northeast of the midden sites rise to 1465 m elevation and support desert grassland and sparse chaparral. The vegetation near the sites is a xeric Sonoran desertscrub dominated by brittlebush (*Encelia farinosa*), foothills palo verde (*Cercidium microphyllum*), Mexican jumping bean, and organ pipe cactus (Shreve, 1964; Turner and Brown, 1982). The area has a characteristic Sonoran Desert climate with biseasonal rainfall, hot summers, and infrequent winter freezes. Total annual rainfall ranges from 233 mm/yr at the Visitor Center at 510 m elevation (Sellers and Hill, 1974) to an estimated 457–510 mm/yr on the top of Mount Ajo.

ARTHROPOD COMMUNITY

The 105 arthropod taxa identified from the pitfall traps provide a representative but not exhaustive sample of the ground-dwelling or wandering arthropod community in the rocky habitats near the packrat midden sites (Fig. 16.2A). Although arthropods are fairly abundant and diverse in all habitats, gentler areas at the bases of the hills support a greater arthropod fauna than the steeper rocky slopes or cliffs. Arthropod diversity is probably even greater along the larger riparian washes. With sixty species from many families, the beetles (Coleoptera) are the most common arthropods. In terms of numbers of species (seventeen) and individuals, especially of the pinacate (*Eleodes subnitens* and *E. wickhami*) and ironclad (*Centrioptera muricata* and *C. variolosa*) beetles, and *Argoporus alutacea*, the darkling beetles are the most common group. Darkling beetles are most common on the flatter areas but were found in all habitats. Spider beetles (Ptinidae) live in rock shelters inhabited by mammals or birds but were rare or very elusive. Velvet ants and wasps (Mutillidae) were also common; at

least five taxa were identified. Ants (Formicidae), including nine species in eight genera, were common in all habitats.

Scorpions (Scorpionida) were abundant in the traps in most areas from barren flats to rocky cliffs. The stripe-tailed scorpion (*Vejovis* sp.) was the dominant form, while bark scorpions (*Centruroides exilicauda*) were less common. Small ridgeback or crested millipedes (Diplopoda: Lysiopetalidae) were taken in a single rock shelter on Twin Peaks. The common desert millipede (*Orthoporus ornatus*) was not collected, although old, bleached segments were seen in packrat dens in rock shelters and under shrubs on desert flats. These large, conspicuous detritivores are common during the summer rainy season in Alamo Canyon in the Ajo Mountains a few hundred meters higher. Their rarity in the Puerto Blanco Mountains study area suggests a recent decline at the lower elevational limit.

Numerous spiders (Araneae), harvestmen (Phalangida), and red velvet mites (Trombidiidae: *Dinothrombium*) were found in the traps, especially during the middle of the summer. Bristletails (Thysanura) were abundant from April to June. A small centipede (*Scolopendra*) was found in rock shelters with dirt floors and less commonly in other habitats. Larval antlions (Neuroptera: Myrmeleontidae) commonly build conical pit traps in the dirt floors in the rock shelters.

The organic materials accumulated by packrats in their houses or dens provide special habitats for many arthropods who feed on the dung or rotting vegetation (White, 1983). A number of species have been reported to live in packrat houses, including dung beetles (*Onthophagus velutinus*; Howden and Cartwright, 1963), spider beetles (*Ptinus feminalis*), and scarab beetles (*Trox carinatus* and *T. punctatus*), and kissing bugs (*Triatoma rubida*). The results of this study and that by Hall et al. (1988) indicate that the narrow-waisted bark beetle (Salpingidae: *Cononotus bryanti*), endemic to Arizona, must have close ties to packrat dens. The pitfall traps provided excellent samples of the ground-dwelling detritivores, especially tenebrionid beetles. Specimens of two tenebrionid beetles (*Cryptoglossa angularis* and *Stibia tuckeri*) represent significant Arizona range extensions. *Stibia tuckeri*, a rare beetle species of a

Table 16.2. *Numbers of identified specimens of fossil arthropods from Puerto Blanco Mountains packrat middens.*
Sample ages in thousands of years (after Table 16.1). Distribution code: N = near site today; F = on flats below sites; E = not observed in study area but occurs elsewhere in the Monument; U = distribution in monument unknown.

	Distribution	14.1	10.5	9.9	9.7	9.1	8.8	8.0	7.6
ARACHNIDA									
Scorpionida genus indet.	N						1		
Buthidae									
Centruroides sp. (bark scorpion)	N	6							
Vejovidae									
Vejovis sp. (stripe-tailed scorpion)	N								
Chelonethida (pseudoscorpions) genus indet.	F					1			
CRUSTACEA									
Isopoda (sow bugs) genus indet.	N								
Diplopoda (millipedes)									
cf. Lysiopetalidae (crested millipede)	N		1			92			
genus indet.	U	50	1		24	11		7	11
INSECTA									
Neuroptera (antlions and relatives)									
Myrmeleonidae (antlion) genus indet.	N					1			
Orthoptera (grasshoppers and relatives)									
genus indet.	N								
Homoptera (hoppers and relatives)									
Cicadidae (cicadas) genus indet.	U	6							
Hemiptera (true bugs)									
Cydnidae (burrower bugs) cf. *Melanaethus* sp.	N					1			
Reduviidae (assassin bugs) *Triatoma* sp.	N								
Lepidoptera (moths and butterflies)									
Saturniidae (silkmoths) cf. *Hemileuca tricolor* (Packard)	E								
Coleoptera (beetles)									
Carabidae (ground beetles) genus indet.	N				1		1		
Scarabaeidae (dung and June beetles)									
Cremastocheilus sp.	U								
Onthophagus sp.	N					2			
Melolonthinae indet.	N								
genus indet.	U			3	3		4	3	1
Elateridae (click beetles) genus indet.	N			1	1				
Dermestidae (carpet beetles) genus indet.	U								
Anobiidae (deathwatch beetles) genus indet.	E								

Table 16.2. *(Continued)*

7.6	5.2	3.5	3.4	3.4	3.2	2.3	2.2	1.9	1.0	1.0	0.1	0.0
		1		3					1		1	
						1						
1					1							
						1						
6	105	25	8	15	10	43		27	6	28	1	2
												1
						4						
			1									
									3	4		3
						1						
						6						
			1									
					1	10						
			1									
												1
			1									

Ptinidae (spider beetles)									
Niptus cf. ventriculus Le Conte	U								
Ptinus feminalis Fall	N								
Ptinus cf. priminidi Spilman	U								
Bostrichidae (branch and twig borers)									
Bostrichinae indet.	E				1				
Cleridae (checkered beetles)									
Cymatodera sp.	E								
Tenebrionidae (darkling beetles)									
Argoporis sp.	N								
genus indet.	U	17	5	1	12	12	13	2	1
Salpingidae (narrow-waisted bark beetles) Cononotus bryanti Van Dyke	F								
Nitidulidae (sap beetles) Carpophilus sp.	E								
Cerambycidae (longhorn beetles) genus indet.	F								
Bruchidae (bean weevils) genus indet.	N								1
Curculionidae (weevils)									
Ophryastes sp.	U								1
genus indet.	N				1	3			
Scolytidae (bark beetles) Cactopinus hubbardi Schwarz	U								
Diptera (flies)									
Stratiomyidae (soldier flies) cf.									
Cyphomyia sp.	U								
Hymenoptera (ants, bees, wasps)									
Formicidae (ants)									
cf. Acromyrmex	U				1				
Camponotus sp.	N						1		
Crematogaster sp.	N								
Neivamyrmex nigrescens (Cresson)	U								
Novomessor sp.	N								
Pheidole sp.	N				2				
Pogonomyrmex sp.	U								
Solenopsis cf. aurea Wheeler	N								
Mutillidae (velvet ants) genus indet.	N								
Number of taxa: 43		4	3	3	9	8	5	3	5
Number of identified specimens: 715		79	7	5	46	123	18	12	15

genus found mostly in Baja California, was previously known only from the type specimens from near Tucson (Casey, 1924) and a 14,400 B.P. packrat midden from the Kofa Mountains of western Arizona (Morgan et al., 1983). The ant *Pheidole vistana* was collected for the first time in Arizona.

Examination of the distribution patterns of the arthropods in the Puerto Blanco Mountains collection (Table 16.3) is enlightening. The largest groups are species now widespread in the southwestern United States (57.7%) and also found in Mexico (53.8%). Twenty-five of the Mexican species (33.3%) occur on the Baja California peninsula. Van Dyke (1942) previously pointed out close ties between the arthropod faunas of Sonora and Baja California, especially Baja California del

								2			1	
		3	8	3				1	2	7	1	3
								1				
			1	2								
	1											
1	5	4			4	3			4	5		
						1						
				1	1						1	
						1				1		
1	2	2	7	3	3	2		1	11	6	1	
						1			1			
		3										
			1				11		1			
	1											
1			1			7			2	1		
						1				1		
												2
			1			2						
						9		2		1		
								1				
				3								
										2		
5	5	6	10	7	5	16	1	6	10	10	6	6
10	114	37	31	30	20	91	11	34	32	56	11	12

Sur, in his study of metallic wood-boring beetles (Buprestidae). Although Sonoran distributions were not as clearly recognized, the border is only 10 km to the south; undoubtedly most of the Puerto Blanco Mountains arthropods also occur there.

Six or seven species collected in the Puerto Blanco Mountains are dominant among the tenebrionid fauna. The remainder of the seven- teen species probably represent less-common species with low populations or transients. Thomas (1983) reached similar conclusions in his extensive study of tenebrionid beetles in the eastern Mojave Desert in southern Nevada and northwestern Arizona; that is, there is a nucleus of common species and a number of transients. The steep, rocky slopes in our study area do not tend to accumulate litter in

Table 16.3. *Collections of arthropods from March to August 1985 in the Puerto Blanco Mountains, Pima County, Arizona.*
Abbreviations: states of the United States follow U.S. Postal Service two-letter code; AG, Argentina; BC, Baja California; BCS, Baja California del Sur; CAM, Central America; CH, Chihuahua; CM, Colombia; EC, Ecuador; HO, Honduras; MR, Mississippi River; MX, Mexico; NI, Nicaragua; RM, Rocky Mountains; SI, Sinaloa; SO, Sonora; US, United States; VC, Vera Cruz; WS, widespread; e, eastern; n, northern; s, southern; se, southeastern; sw, southwestern; w, western.

Taxa		Geographical Distribution
Class Arachnida		
Order Scorpionida	*Centruroides exilicauda* (Wood)	NM-seCA, MX(SO, BC)
	Vejovis sp.	
Class Insecta		
Order Coleoptera		
Family Anobiidae	*Tricorynus gibbulus* (Fall)	TX-AZ
Family Bostrichidae	*Amphicerus teres* (Horn)	TX-CA
	Apatides fortis (LeConte)	UT, TX-CA, MX (BCS), NI, CM
	Dendrobiella aspersa (LeConte)	AZ, CA, MX (BCS)
Family Bruchidae	*Algarobius prosopis* (LeConte)	TX-AZ, MX (BC)
	Mimosestes amicus (Horn)	TX-AZ, MX (BC)
Family Buprestidae	*Acmaeodera quadrivittata* Horn	CO, UT, TX-CA, MX (BCS)
Family Carabidae	*Anisotarsus* sp.	
	Apristus laticollis LeConte	AZ-sCA
	Bembidion transversale Dejean	
	Lebia sp.	
	Schizogenius sp.	
	Selenophorus sp.	
	Tetragonoderus fasciatus Haldeman	WS
Family Cerambycidae	*Achryson surinamum* Linnaeus	AZ, MX (BCS), CAM-AG
	Anelaphus brevidens (Schaeffer)	AZ, MX (BCS)
	Peranoplium hoferi (Knull)	AZ, MX
	Peranoplium simile (Schaeffer)	AZ, MX
	Peranoplium subdepressum (Schaeffer)	AZ, MX (SO, BC)
Family Chrysomelidae	*Pachybrachys snowi* Bowditch	AZ
Family Cleridae	*Cymatodera oblita* Horn	AZ
	Enoclerus quadrisignatus (Say)	TX-CA, MX (SO, BCS)
	Phyllobaenus cribripennis (Fall)	
Family Curculionidae	*Eucyllus saesariatus* Sleeper	AZ
	Ophryastes varius LeConte	
	Smicronyx sp.	
Family Dytiscidae	*Copelatus chevrolati renovatus* Guignot	swUS
Family Histeridae	*Euspilotus* sp.	
Family Meloidae	*Lytta magister* Horn	swUT-seCA, MX (SO)
Family Melyridae	*Rhadalus testaeus* LeConte	AZ
Family Monommidae	*Hyporhagus* sp.	
Family Nitidulidae	*Conotelus mexicanus* Murray	TX-sCA, MX (BC), CAM
	Thalycra sp.	
Family Ostomidae	*Temnochila aerea* (LeConte)	NM-CA, MX (BC), HO
Family Ptinidae	*Ptinus feminalis* Fall	AZ
Family Salpingidae	*Cononotus bryanti* Van Dyke	AZ
Family Scarabaeidae	*Ataenius desertus* Horn	UT, AZ, CA, MX (BCS)

	Diplotaxis morens Schaeffer	swUS, MX (BC)
	Diplotaxis subangulata LeConte	swUS, MX (BC)
	Onthophagus velutinus Horn	CO, TX, AZ, MX (BCS)
	Phyllophaga lenis (Horn)	AZ, s.CA
	Trox carinatus Loomis	TX-eAZ, MX(CH)
	Trox punctatus Germar	MR-CA, MX (BC)
Family Tenebrionidae	*Argoporus alutacea* Casey	AZ, CA, MX (SO)
	Centrioptera muricata LeConte	AZ, CA
	Centrioptera variolosa Horn	AZ, MX (BC)
	Conibius reflexus Horn	AZ, CA, MX (BC)
	Cryptoglossa angularis Horn	AZ, CA, MX (BC)
	Eleodes subnitens LeConte	AZ
	Eleodes wickhami Horn	NM, AZ
	Eusophulus castaneus Horn	AZ
	Helops arizonensis Horn	AZ
	Hymenorus sp.	
	Merotemnus filiformis (Castelnau)	AZ, CA
	Metoponium sp.	
	Nocibiotes granulatus (LeConte)	AZ, CA, MX (BC)
	Stibia tuckeri Casey	AZ
	Telabis aliena Casey	AZ
	Telabis sp.	
	Triorophus histrio Casey	AZ
Order Dictyoptera		
Family Mantidae	*Yersiniops solitarium* (Scudder)	AZ-RM
Family Polyphagidae	*Arenivaga* sp.	
Order Hemiptera		
Family Cydnidae	*Melanaethus subglaber* (Walker)	UT, NV, TX-CA, MX (SO, BC), EC
	Tominotus conformis (Uhler)	UT, AZ, CA, MX (BCS)
Family Lygaeidae	*Exptochiomera oblonga* (Stal)	swUS
	Ligyrocoris delitus Distant	
	Ligyrocoris rubricatus Barber	
	Lygaeus lateralis Dallas	swUS
Family Miridae	*Eustictus* sp.	
	Phytocoris brevicornis Knight	AZ
	Phytocoris ramosus Uhler	NV, AZ, CA
Family Pentatomidae	*Mecidea minor* Ruckes	KA-UT, TX-CA
	Thyanta pallidovirens Ruckes	wUS
Family Reduviidae	*Apiomerus longispinnis* Champion	swUS
	Triatoma rubida (Uhler)	TX-CA, MX (VC, SI, SO, BC)
Order Homoptera		
Family Cixiidae	*Oecleus* sp.	
Family Cicadellidae	*Paraphlepsius lascivius* (Ball)	
	Texanus neomexicanus (Baker)	
Order Hymenoptera		
Family Andrenidae	*Nomadopsis australior* (Cockerell)	sOK, NB, CO, UT, TX, AZ, MX (CH)
Family Anthophoridae	*Ericrocis lata* (Cresson)	FL-sCA
Family Apidae	*Apis mellifera* Linnaeus	WS
Family Formicidae	*Aphaenogaster cockerelli* (Andre)	TX-sCA, MX
	Camponotus ochreatus Emery	NV, NM-sCA, MX
	Crematogaster sp.	
	Cyphomyrmex rimosus (Spinola)	FL-CA, MX-AG
	Ephebomyrmex imberbiculus (Wheeler)	OK, CO, NV, TX-CA, MX
	Ephebomyrmex pima (Wheeler)	AZ, MX

Table 16.3. *(Continued)*

Taxa		Geographical Distribution
	Pheidole vistana Forel	sAZ-sCA, MX (BC)
	Pogonomyrmex rugosus Emery	OK, CO, TX-CA, MX
	Solenopsis aurea Wheeler	wTX-sCA, MX
Family Halictidae	*Agapostemon* sp.	
Family Mutillidae	*Dasymutilla magnifica* Mikel	NM-CA, MX (BC)
Family Pompilidae	*Aporinellus* sp.	
	Pompilius sp.	
Order Neuroptera		
Family Mantispidae	*Plega* sp.	
Family Myrmeleontidae	*Eremolgon nigribasis* Banks	UT, NM, AZ, MX (BCS)
	Paranthaclisis congener (Hagen)	
Order Orthoptera		
Family Acrididae	*Melanoplus arizonae* Scudder	KS, OK, TX-AZ, MX
Family Gryllidae	*Gryllus assimilis* (Fabricius)	WS
Family Tettigoniidae	*Atelopus schwarzi* Caudell	sAZ
	Insara covilleae Rhen & Hebard	AZ, MX (SO)
	Insara elegans Scudder	TX, CO-CA
Order Thysanura		
Family Machilidae	*Machilis* sp.	

comparison with gentler alluvial bajadas or relatively flat areas in valleys. Low litter availability limits the number of desert detritivores, especially flightless tenebrionids, in permanent residency. The vagility of these insects helps account for the many apparent transients in the fauna.

Eight species in the Puerto Blanco Mountains fauna (10.3%) enter the United States only in Arizona, with six more also reaching southern California and one as far as New Mexico, for a total of 19.2%. Fourteen species (17.9%) are known only from Arizona. An additional four species known only from Arizona and California brings the total of regional endemics to 23.1%!

VEGETATION HISTORY

At 14,120 B.P. in the late Wisconsin, plant fossils of California juniper and Joshua tree (*Yucca brevifolia*) prevailed in the absence of Sonoran Desert plants (Van Devender, 1987). Between 10,540 B.P. and 8900 B.P. in the early Holocene, brittlebush and catclaw acacia (*Acacia greggii*) were abundant in association with California juniper, Mojave sage (*Salvia mohavensis*), Mormon tea (*Ephedra nevadensis*), velvet mesquite (*Prosopis velutina*), and saguaro (*Carnegiea gigantea*). California juniper lingered in the area until about 8900 B.P.

The middle Holocene midden yielded abundant brittlebush and saguaro mixed with riparian trees (catclaw acacia and velvet mesquite) and blue palo verde (*Cercidium floridum*). Foothills palo verde appeared in the 5240 B.P. sample to dominate all younger samples. The modern vegetation regime was established by about 4000 B.P. as catclaw acacia, velvet mesquite, and blue palo verde were confined to the washes, and ironwood (*Olneya tesota*), Mexican jumping bean, and organ pipe cactus increased in importance. Although it presently is restricted to flatter areas, ironwood grew on south-facing slopes from 5240 B.P. to 1910 B.P. This and other changes at the sites suggest that the modern vegetation is more xeric than at any time in the last 10,600 years.

ARTHROPOD HISTORY

The packrat middens record the history of the arthropods in the Puerto Blanco Mountains since the last glacial period (Figs. 16.3 and 16.4). Ground-dwelling arthropods make up the majority of arthropods found in the midden samples. Beetles, millipedes, and ants dominate the samples.

Late Wisconsin. The 14,120 B.P. sample yielded only remains of millipedes, scorpions, darkling beetles, and a cicada (Cicadidae). Per-

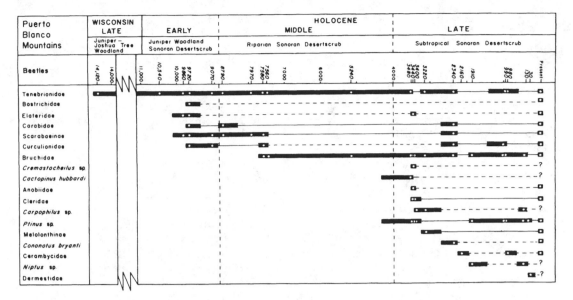

Fig. 16.3. Chronological summary of fossil beetles recovered from Puerto Blanco Mountains packrat middens. Question marks indicate possible occurrence of taxon near midden sites, although it was not collected near midden sites.

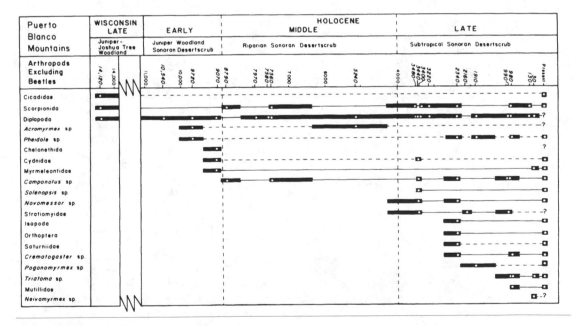

Fig. 16.4. Chronological summary of fossil arthropods exclusive of beetles from Puerto Blanco Mountains packrat middens. Question marks indicate possible occurrence of taxon near midden sites although it was not collected near midden sites.

haps the glacial paleoclimate with only winter rainfall and very cool summers did not support a diverse arthropod fauna.

Early Holocene. Four samples dating from 10,540 B.P. to 9070 B.P. show a marked increase in abundance and species richness in the arthropod fauna with a total of twelve taxa. Darkling beetles and millipedes still dominate the samples. The 10,540 B.P. sample contains segments from a crested millipede. The 9720 B.P. sample is the first assemblage large enough to reflect the arthropod community composition. In addition to darkling beetles and millipedes, heads and gasters of two genera of ants (*Pheidole* and cf. *Acromyrmex*), a branch-and-twig-borer (Bostrichidae) pronotum, dung beetle sternites, and weevil (Curculionidae) elytra were found. The 9070 B.P. sample contains the "horned" head of a male dung beetle (*Onthophagus* sp.), which is closely associated with mammal burrows. The darkling beetle remains represent two tribes in the Tenebrionidae (Pedini and Saurini). In addition, the sample yielded a pseudoscorpion body, a burrowing Hemiptera (Cydnidae) head, crested millipede segments, darkling beetle parts, and an antlion larva mandible. An early Holocene climate of slightly increased annual and summer rainfall and warmer summers would have been more favorable to arthropods.

Middle Holocene. The final disappearance of juniper in the Puerto Blanco Mountains at about 8900 B.P. marked the end of the early Holocene. Middle Holocene climate was essentially modern, with biseasonal rainfall and hot summers, although annual rainfall appears to have been about 30% greater than today (Van Devender, 1987). There were fewer droughts, and the freeze frequency was great enough to prevent more subtropical plants from living in the area. Five samples from 8790 B.P. and 5240 B.P. yielded ten arthropod taxa dominated by darkling beetles and millipedes. New animals include a carpenter ant (*Camponotus* sp.), a stripe-tailed scorpion "stinger" or telson, a ground beetle head, and various remains of bean weevils (Bruchidae). The appearance of the latter matches the increase in blue palo verde, a common host. A weevil exoskeleton in the 7580 B.P. sample was identified as *Ophryastes* sp. (Fig. 16.5A). A darkling beetle body in the 5240 B.P. sample

was identified as *Argoporis* sp., a beetle common in the area today.

Late Holocene. About four thousand years ago the climate became essentially modern with fewer hard freezes, biseasonal rainfall, and a well-developed early-summer drought. The arrival of more subtropical species and the development of more diverse desertscrub communities apparently was reflected in a larger and more diverse arthropod fauna. Eleven middens yielded twenty new arthropods for a total of thirty-two taxa. Millipedes, bean weevils, and spider beetles (*Ptinus* and *Niptus*) dominated the late Holocene fauna, while darkling beetles were less common. Soldier fly larvae (Stratiomyidae: *Cyphomyia* sp.) in three samples (Fig. 16.5B) were probably living in decaying plant material (James, 1981).

The saguaro bark beetle (Scolytidae: *Cactopinus hubbardi*) in the 3480 B.P. sample (Fig. 16.5C) is known only from saguaro (Wood, 1982), the stately columnar cactus that dominates much of southern Arizona. The sample also contains organ pipe cactus, another potential host. A beetle elytron fragment in the 3440 B.P. sample is from *Cremastocheilus* sp., an unusual scarab that lives in ant burrows. The 3400 B.P. sample yielded two checkered beetle (Cleridae: *Cymatodera* sp.; Fig. 16.5D) bodies and an ant (*Solenopsis* cf. *aurea*) head. A sap beetle (Nitidulidae: *Carpophilus* sp.) body was recovered from the 3220 B.P. sample (Fig. 16.5E).

The 2340 B.P. sample, with at least sixteen arthropod taxa, illustrates the great variety of arthropods in the late Holocene, including a number of interesting arthropods. A silkmoth larva (Saturniidae) appears to represent the genus *Hemileuca* (Fig. 16.5F). Today the larvae of *H. tricolor* feed on palo verde and catclaw acacia in the study area. The sample also contains a bark scorpion (Buthidae: *Centruroides* sp.; Fig. 16.5G) telson, a narrow-waisted bark beetle (Salpingidae: *Cononotus bryanti*) body, and four genera of ants (*Camponotus*, *Crematogaster*, *Novomessor*, and *Pheidole*).

The spider beetles from the 1910 B.P. sample represent two species: *Niptus* cf. *ventriculus* and *Ptinus feminalis*. An exceptionally large, hairy *Ptinus* in the 990 B.P. sample (Fig. 16.5H) appears to be *P. priminidi*. This recently described species was previously known from

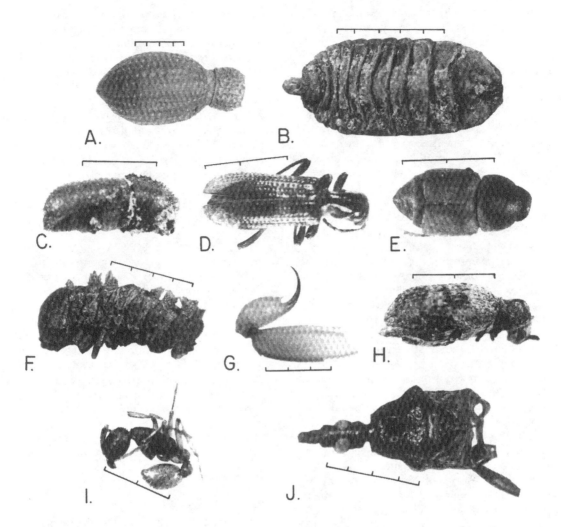

Fig. 16.5. Arthropod remains from Puerto Blanco Mountains packrat middens. Divisions on scale bars are in millimeters. (A) Weevil (Curculionidae: *Ophryastes* sp.) from Twin Peaks #6 dated at 7580 B.P. (B) Larva of a soldier fly (Stratiomyidae) from Ajo Loop #5B dated at 3480 B.P. (C) Bark beetle (Scolytidae: *Cactopinus hubbardi*) from Ajo Loop #5B dated at 3480 B.P. (D) Checkered beetle (Cleridae: *Cymatodera* sp.) from Ajo Loop #6C dated at 3400 B.P. (E) Sap beetle (Nitidulidae: *Carpophilus*) from Ajo Loop #6B dated at 3220 B.P. (F) Larva of a silkmoth (Saturniidae: *Hemileuca* cf. *tricolor*) from Twin Peaks #2 dated at 2340 B.P. (G) Telson of a bark scorpion (Buthidae: *Centruroides* sp.) from Twin Peaks #2 dated at 2340 B.P. (H) Spider beetle (Ptinidae: *Ptinus* cf. *priminidi*) from Ajo Loop #3A dated at 990 B.P. (I) Ant (Formicidae: *Neivamyrmex nigrescens*) fom Ajo Loop #2B dated at 30 B.P. (J) Head/pronotum of kissing bug (Reduviidae: *Triatoma* sp.) also from Ajo Loop #2B.

three Wisconsin packrat middens dated at 12,330 B.P., 13,400 B.P., and > 30,000 B.P. from southeastern California and western Arizona (Spilman, 1976). This "fossil" species probably remains to be discovered in the modern Sonoran Desert.

The 990–980 B.P. samples show a moderate increase in the abundance of arthropods during a short wet period. Both the species richness and the number of specimens decline in the younger (130 B.P. and 30 B.P.) samples. The 30 B.P. sample yielded an ant (*Neivamyrmex nigrescens*; Fig. 16.5I) exoskeleton, a kissing bug (*Triatoma* sp.; Fig. 16.5J) head/pronotum, and a larval carpet beetle (Dermestidae) exoskeleton. The diversity and abundance of arthropods in the midden rock shelters is probably much lower today than two thousand to four thousand years ago.

DISCUSSION

Organ Pipe Cactus National Monument, established in 1937, protects a unique subtropical Sonoran desertscrub vegetation as well as a diversified arthropod community. Fossils preserved in ancient packrat middens provide a detailed record of the development of the modern biota since the last glacial period (Van Devender, 1987). Unlike the dynamic plant community, the midden arthropod fauna is basically conservative in that all taxa recovered from the middens may be expected to occur within a few kilometers of the midden sites today. With the exception of a few species, the beetles identified in the middens today are mostly found, or are more common, on the flats below the rocky slopes. Identification of arthropod remains to a relatively high taxonomic level, inadequate sampling of the modern fauna, and small Wisconsin and early Holocene samples open the possibility of more significant range changes than we detected. Few arthropod species's distributions are as well documented as those of virtually all species of Sonoran Desert plants.

The plant remains record the development of a Sonoran desertscrub dominated by brittlebush and riparian trees by 8900 B.P. Although the precipitation was probably 30% greater than today in the middle Holocene, the arthropod fauna was only moderately diverse. The arthropod fauna increased dramatically in the late Holocene as more subtropical plants and a warmer climate became established about four thousand years ago (Figs. 16.3 and 16.4). The increase in arthropods matches a general increase in total plant taxa, especially herbaceous plants. The species richness in the arthropod fauna apparently is strongly correlated with the frequency of winter freezes, and secondarily with the amount of summer precipitation. Many of the more temperature-sensitive arthropods probably dispersed into the study area from Sonora during the last four thousand years.

The arthropod record in a series of middens from northwestern Sonora resembles that of the Puerto Blanco Mountains in the increase in species richness in the late Holocene (Hall et al., 1988). In contrast, a relatively modern arthropod fauna was established in the early Holocene soon after 11,000 B.P. in the Tinajas Altas Mountains, even though California juniper woodland was much better developed than in the Puerto Blanco Mountains. Additional fossils identified to species may help to resolve whether Sonoran Desert arthropod faunas were conservative or changed in concert with the flora and were disguised by generic-level determinations.

ACKNOWLEDGMENTS

We thank Harold J. Smith, superintendent, and Bill Mikus, resource manager, Organ Pipe Cactus National Monument, for their help and encouragement. George L. Bradley, University of Arizona, and Steve Prchal, Arizona-Sonora Desert Museum, helped with the fieldwork. Floyd G. Werner of the University of Arizona Department of Entomology helped with arthropod identification. Careful reviews by Paul S. Martin and Floyd G. Werner, University of Arizona, and Scott A. Elias, University of Colorado, helped the manuscript. Funds were provided by Southwest Parks and Monuments Association. Jean Morgan and Justine Collins typed the manuscript. Helen A. Wilson drafted the figures.

REFERENCES

Ashworth, A. C. (1973). "Fossil beetles from a fossil wood rat midden in western Texas." *Coleopterists' Bulletin* 27, 139–40.

———. (1976). "Fossil insects from the Sonoran Desert, Arizona." *American Quaternary Association Abstracts* 1976, 122.

Bowers, J. E. (1980). "Flora of Organ Pipe Cactus National Monument." *Journal of the Arizona-Nevada Academy of Sciences* 15, 1–11, 33–47.

Casey, T. L. (1924). "Additions to the known Coleoptera of North America." In *Memoirs on the Coleoptera*, vol. 11, pp. 296–97. Lancaster, Pa.: Lancaster Press.

Cole, K. L. (1986). "The lower Colorado Valley: A Pleistocene desert." *Quaternary Research* 25, 392–400.

Elias, S. A. (1987). "Paleoenvironmental significance of late Quaternary insect fossils from wood rat middens in south-central New Mexico." *Southwestern Naturalist* 32, 383–90.

Hall, W. E.; Van Devender, T. R.; and Olson, C. A. (1988). "Late Quaternary arthropod remains from Sonoran Desert packrat middens, southwestern Arizona and northwestern Sonora." *Quaternary Research* 29, 1–18.

Howden, H. F., and Cartwright, O. L. (1963). "Scarab beetles of the Genus *Onthophagus* Latreille north of Mexico (Coleoptera: Scarabaeidae)." *Proceedings of the United States National Museum* 114, 1–135.

James, M. T. (1981). "Stratiomyidae." In *Manual of Nearctic Diptera*, vol. 1. S. F. McAlpine, B. V. Peterson, G. E. Shewell, H. S. Taskay, S. R. Vockeroth, and D. M. Wood, eds., pp. 497–511. Ottawa: Research Branch Agriculture Canada Monograph no. 27.

Mead, J. I.; Van Devender, T. R.; and Cole, K. L. (1983). "Late Quaternary small mammal remains from Sonoran Desert packrat middens, Arizona and California." *Journal of Mammalogy* 64, 173–80.

Morgan, A. V.; Morgan, A.; Ashworth, A. C.; and Matthews, J. V., Jr. (1983). "Late Wisconsin fossil beetles in North America." In *Late-Quaternary environments of the United States*. Volume 1, *The late Pleistocene*. S. C. Porter, ed., pp. 354–63. Minneapolis: University of Minnesota Press.

Sellers, W. D., and Hill, R. H. (1974). *Arizona climate, 1931–1972*. Tucson: University of Arizona Press.

Shreve, F. (1964). "Vegetation of the Sonoran Desert." In *Vegetation and flora of the Sonoran Desert*. F. Shreve and I. L. Wiggins, eds., pp. 259–93. Stanford: Stanford University Press.

Spilman, T. J. (1976). "A new species of fossil *Ptinus* from fossil wood rat nests in California and Arizona (Coleoptera, Ptinidae), with a postscript on the definition of a fossil." *Coleopterists' Bulletin* 30, 239–44.

Thomas, D. B., Jr. (1983). "Tenebrionid beetle diversity and habitat complexity in the eastern Mojave Desert." *Coleopterists' Bulletin* 37, 135–47.

Turner, R. M., and Brown, D. E. (1982). "Sonoran desertscrub." *Desert Plants* 4, 181–221.

Van Devender, T. R. (1977). "Holocene woodlands in the southwestern deserts." *Science* 198, 189–92.

———. (1987). "Holocene vegetation and climate in the Puerto Blanco Mountains, southwestern Arizona." *Quaternary Research* 27, 51–72.

Van Devender, T. R.; Martin, P. S.; Thompson, R. S.; Cole, K. L.; Jull, A. J. T.; Long, A.; Toolin, L. J.; and Donahue, D. J. (1985). "Fossil packrat middens and the tandem accelerator mass spectrometer." *Nature* 317, 610–13.

Van Devender, T. R., and Mead, J. I. (1978). "Early Holocene and late Pleistocene amphibians and reptiles in Sonoran Desert packrat middens." *Copeia* 1978, 464–75.

Van Devender, T. R., and Spaulding, W. G. (1979). "Development of vegetation and climate in the southwestern United States." *Science* 204, 701–10.

Van Devender, T. R.; Thompson, R. S.; and Betancourt, J. L. (1987). "Vegetation history in the Southwest: The nature and timing of the late Wisconsin–Holocene transition." In *North American and adjacent oceans during the last deglaciation*. W. F. Ruddiman and H. E. Wright, Jr., eds., pp. 323–52. Boulder: Geological Society of America.

Van Dyke, E. C. (1942). "Contributions toward a knowledge of the insect fauna of lower California. No. 3 Coleoptera: Buprestidae." *Proceedings of the California Academy of Sciences* 24, 97–132.

White, R. E. (1965). "A revision of the genus *Tricorynus* of North America (Coleoptera: Anobiidae)." *Miscellaneous Publications of the Entomological Society of America* 4, 1–113.

White, R. E. (1983). *Field Guide to the Beetles of North America*. Boston: Houghton Mifflin.

Wood, S. L. (1982). *Bark and ambrosia beetles of North and Central America (Coleoptera: Scolytidae), a taxonomic monograph*. Great Basin Naturalist Memoirs no. 6.

Chapter 17

Deuterium Variations in Plant Cellulose from Fossil Packrat Middens

Austin Long   Lisa A. Warneke   Julio L. Betancourt   Robert S. Thompson

INTRODUCTION

Several studies have demonstrated that oxygen, hydrogen, and carbon isotope variations in modern plant cellulose are related to spatial and temporal patterns in temperature, humidity, seasonality, and amount of rainfall (Pearman et al., 1976; Epstein and Yapp, 1976; Grey and Thompson, 1976, 1977; Epstein et al., 1977; Ferhi and Letolle, 1979; Burk and Stuiver, 1981; Leavitt and Long, 1983; Lawrence and White, 1984; Edwards and Fritz, 1986; Ramesh et al., 1985, 1986; Friedman and Gleason, 1980). A few investigators have extended the modern plant studies to derive paleoclimatic information (Wilson and Grinsted, 1975; Epstein and Yapp, 1976; van der Straaten, 1981; Brenninkmeijer et al., 1982). These studies show the potential of hydrogen and oxygen stable isotopes from fossil plant material as valuable paleoclimatic indicators. They are limited, however, by time span (in the case of tree rings), spatial distribution (bogs), and sample density (fossil trees), thus restricting their value in verifying results from global circulation models. The paleoclimatic potential of the plant stable-isotope system is greatest if the fossil plant sites are broadly distributed and have maximum time depth.

Plant cellulose preserved for more than forty thousand years abounds in packrat middens, the primary source of paleovegetation data in the arid interior of western North America. More than one thousand dated middens are currently (1989) available. The distribution maps and frequency diagrams displayed by Webb and Betancourt (*this volume*, chap. 6) indicate the potential for stable-isotope studies across broad temporal and spatial ranges. Thus far, only two stable-isotope studies have included samples from packrat middens. Arnold (1979) analyzed ^{13}C in juniper cellulose from Sonoran Desert middens, and Siegel (1983) examined variations in deuterium values (δD) and $\delta^{13}C$ from middens dating from modern to 34 ka (thousands of years ago) in the Snake Range in east-central Nevada. These studies indicate that δD is a more accurate indicator of past climate than $\delta^{13}C$.

In this report we relate δD in modern juniper twigs and leaves to local climate at twenty-six sites in Texas, New Mexico, Arizona, Utah, and Nevada. These data are used to interpret δD values of Pleistocene and Holocene plant remains from middens in the Snake Range, the Canyon Lands of southeastern Utah, and Chaco Canyon in northwestern New Mexico (see Thompson and Betancourt, *this volume*, chaps. 10 and 12, for vegetation summaries for these regions). An important priority of paleohydrology–stable-isotope studies is to measure the effects of long-term climatic variability on isotopic composition within the hydrologic cycle. The use of radio-isotopes as solute tracers in ground-water studies requires knowledge about their past variability. Here we compare the plant cellulose chronology with deuterium variations in ground water from New Mexico and Nevada. We also compare our δD chronology with paleovegetation summaries from the regional midden record, pluvial lake histories, global and regional deglaciation sequences, and grid predictions from global circulation models.

The abundance of midden cellulose from the Pleistocene-Holocene transition provides a unique opportunity to determine how quickly vegetation responds after a major climatic change. During the transition different rates of local extinction and immigration could result from varied mechanisms of dispersal, competition, and other biotic factors (Davis, 1986) triggered by a single climatic change. However, synchronous departures and arrivals for different species also may stem from the individualistic responses of species to a continually changing climate (Webb, 1986). In the eastern Grand Canyon, for example, claims for vegetational inertia and migrational lag (Cole, 1985, *this volume*, chap. 11) might be better supported if principal changes in deuterium abundance antedated periods of high species turnover by one to several millennia. Contemporaneity between major changes in stable-isotope values and floral composition would support the counterpart view—that plant distributions rapidly achieve equilibrium in response to climatic changes (Markgraf, 1986).

Because of its sensitivity to growing-season temperatures, δD from plant cellulose also can be used to determine the regional expression of a postulated worldwide middle Holocene (8–4 ka) warm interval. Although once widely accepted, this interval is under close scrutiny because it appears out of phase with the insolation maximum at about 10 ka, and because it would serve as an analog for imminent carbon dioxide warming (Webb and Wigley, 1985).

DEUTERIUM/HYDROGEN IN PRECIPITATION

The most abundant isotope molecules of natural water are $H_2{}^{16}O$, $H_2{}^{18}O$, and $HD{}^{16}O$. Their distribution in any water sample is primarily a function of the temperature difference between the oceanic source of the air mass and the site of condensation. Processes of evaporation and condensation occurring along the air-mass trajectory to the site and at the site also affect isotopic distributions. These are all features of climate. Isotopic depletion occurs in the water vapor of an air mass, or in the liquid water droplets of a cloud, as moist air moves inland, cools adiabatically or by mixing with cooler air, and loses water by precipitation. Liquid- and solid-phase water are isotopically heavier than the water vapor from which they condense, and, as another phase leaves the system, the remaining water vapor becomes isotopically lighter. Only the temperature of the site at ground surface would determine the δD value of precipitation if the following conditions were true: (1) the source of the atmospheric moisture (primarily the tropical oceans) must be temporally constant in temperature and isotopic composition; (2) progressive isotopic depletion should follow exactly the Rayleigh condensation model, condensing under isotopic equilibrium conditions; and (3) condensed water droplets did not undergo significant evaporation. If a real system closely meets these conditions, the δD of the precipitation should correlate well with the mean annual temperature of the site. This is what Dansgaard (1964) found for Atlantic coastal sites: 5.6‰/°C.

Similar stable-isotopic/paleotemperature studies have found significant correlations between these two variables, but not always the same coefficient. For example, Yurtsever (1975) found a slightly but not significantly larger coefficient between mean monthly temperatures and weighted average monthly δD values at warmer continental sites. Cooler continental sites showed coefficients significantly lower than Dansgaard's 5.6‰/°C. Long et al. (personal communication, 1988) found the average δD of precipitation at Tucson to be 9‰ lighter in winter than in summer. The average temperature difference in Tucson between October–March and April–September is 12°C. The relation between mean annual temperature and δD of precipitation is insignificant for many sites due to a variety of factors, including seasonality of precipitation, different air-mass trajectories at different seasons, and evaporation (Friedman et al., 1964).

Extrapolation of the modern spatial temperature–δD relation to the interpretation of δD values in paleoclimatic terms involves the following critical assumptions: (1) the temperature and isotopic composition of the oceanic source of the moisture have remained constant or are known during the time range of the study (about 40 ka to present); and (2) the air-mass trajectories have remained unchanged during the same time. For the first assumption, nearly all (about 80%; Sellers, 1965) atmospheric moisture derives from the tropical oceans. CLIMAP (1976) concluded that the tropical ocean temperatures at 18 ka (fullglacial time) were within 2°–6°C of the tropical ocean temperatures today. The stable-isotopic composition of surface tropical ocean water varies about 6‰ today from place to place due to extreme evaporation or influx of river water, but tropical moisture derives from a vast spatial area and thus well represents the average ocean at any given time. The average ocean today is about 8‰ lighter isotopically than at 18 ka owing to the large amount of isotopically light ice stored on the continents when glaciers were at their most recent maximum. The first assumption thus seems manageable.

The second assumption requires more consideration. Different air-mass trajectories can bring isotopically distinctive precipitation to the same locality (Gat and Issar, 1974). Friedman and Smith (1972) found the δD of successive winters' snow in the Sierra Nevada to be characteristic of contrasting prevailing winter cyclonic patterns. On a considerably longer time scale, Winograd et al. (1985) suggested that the rising orographic barrier of the Sierras Nevada (about 600 m during the past 2 million years) has progressively lowered the δD values of ground water in the Death Valley region, which is recharged primarily by winter cyclonic storms originating in the Pacific. δD in spring and summer precipitation in the Great Basin may be further complicated by latitudinal shifts in the polar or subtropical

jet stream around (north or south) the Sierra Nevada, which brings Pacific moisture, and by cross-equatorial circulation that contributes summer moisture from the Atlantic and Pacific oceans. Figure 17.1 shows that the direction of movement of the precipitable water vapor is locally perpendicular to the lines of equal isotope in the study area, from higher to lower δD values. Note that the continental divide separates the relative influences of the Gulf of Mexico and the Pacific Ocean.

DEUTERIUM IN PLANTS

δD values in plant cellulose are systematically related to the source water, which is withdrawn from the soil by the roots, passes up through the stem, and transpires through the leaves. Cellulose is a carbohydrate composed of polymeric chains of glucose rings. Of the two classes of chemically bound hydrogen, O–H and C–H, only the hydroxyl hydrogen can detectably modify its isotopic composition by exchange with water. The C–H hydrogen is more strongly bound and represents the originally synthesized hydrogen.

The isotopic composition of the water in the mesophyll leaf cells, where photosynthesis occurs, is affected not only by the isotopic composition of water drawn up by the leaves but also by evaporative processes and by exchange with atmospheric water vapor. Figure 17.2 illustrates that changes in local precipitation (root water) move the δD and $^{18}O/^{16}O$ along the meteoric line slope of about 8 ($\delta D = 8\delta^{18}O + 10$; Craig, 1961). Because of leaf water transpiration (Yapp and Epstein, 1982), changes in humidity can move the composition along a slope of about 3. The effect of humidity on isotopic composition imposes certain ambiguities in inferring paleotemperatures from δD values alone. Field measurements of the relation between temperature and δD values range from +2 to +13‰/°C (see above; Yurtsever, 1975), compared to Dansgaard's (1964) value of 5.6‰/°C. Differences are owing to coastal versus inland sites and seasonal versus annual data collection schemes. Theoretical and experimental values of Allison et al. (1985) give a precise humidity effect of −7.4‰ per 10% increase in relative humidity. Thus both warmer and drier conditions would increase the δD of plant cellulose. This would compound the ambiguity

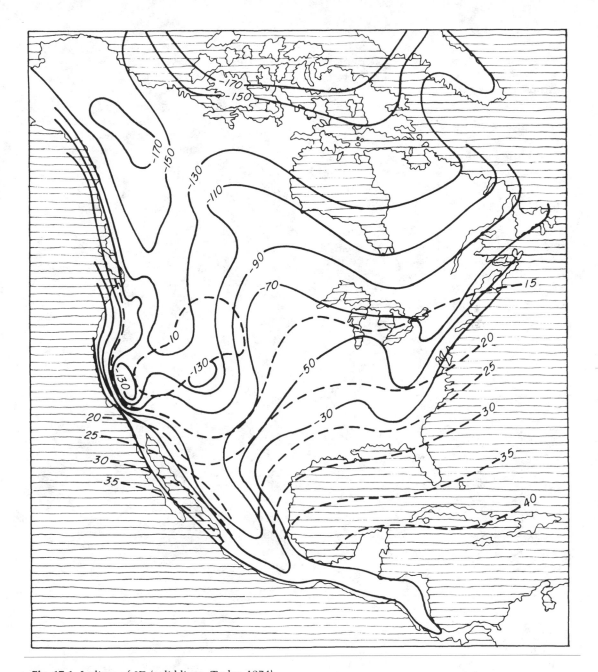

Fig. 17.1. Isolines of δD (solid lines; Taylor, 1974)
and precipitable water vapor (dashed lines; Reitan,
1960) for North America.

already inherent in air-mass sources and trajectories. A necessary part of this research was the collection and analysis of modern plants growing in areas that might approximate conditions under which the fossil plants grew.

METHODS

The control samples consisted of living trees (all *Juniperus* sp.) growing over a wide range of climatic conditions and plants from modern packrat middens at the fossil packrat midden sites. Figure 17.3 is a map of the living-tree sample localities and the fossil midden sites. Sites for the living tree collections were selected for open canopies and no nearby high-altitude-derived water or air drainage that might bring in isotopically light water or unusually cold air. Leaves were collected equally from the north, east, south, and west sides of the tree at about 2 m above the ground. Field and laboratory methods for mid-

den collection and plant species separation are described elsewhere (Spaulding et al., *this volume*, chap. 5). Water wells were generally not available for collection of local ground water.

We chose two geographic areas for fossil-plant isotopic analysis, based on their abundance of middens ranging in age from the present to the middle Wisconsin. Figure 17.3 shows these areas, as well as the areas of Siegel's (1983) study in east-central Nevada, whose data are included here.

Living and fossil plants received identical chemical treatment. Our procedures generally followed the methods described in Epstein et al. (1976). Laboratory processing consisted of cellulose extraction, nitration to nitrocellulose (which eliminates exchangeable O–H hydrogen), combustion, reduction of the water-combustion product to hydrogen, and

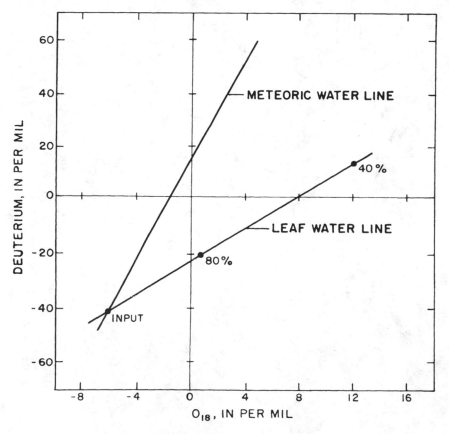

Fig. 17.2. Behavior of stable isotopes of water in leaves (after Allison et al., 1985). The diagonal lines are trends of meteoric water (upper line) and leaf water (lower line) at different relative humidities.

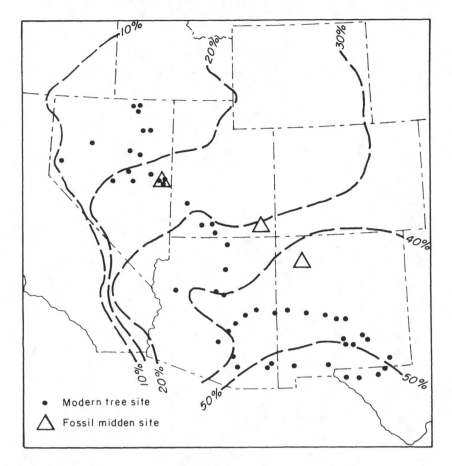

Fig. 17.3. Living tree and fossil plant localities, with percentage of summer precipitation contours.

isotopic analysis of the hydrogen on a VG 602 isotope ratio mass spectrometer. As is conventional, the stable isotopic results were reported as parts-per-thousand difference from the SMOW reference standard (R_s):

$$\delta D \ (‰) = (R_x/R_s - 1)1000$$

where R_x is the δH of the sample and R_s the δD of the reference, SMOW.

RESULTS

Figures 17.4A and 17.4B show plots of the living-tree δD data as a functional mean annual and average growing-season (March–June) temperatures. Temperature records from the nearest meteorological station to the modern tree site were averaged over the fifteen years previous to collection. These stations usually were 1–3 km from the plant collection sites. The plant leaves sampled,

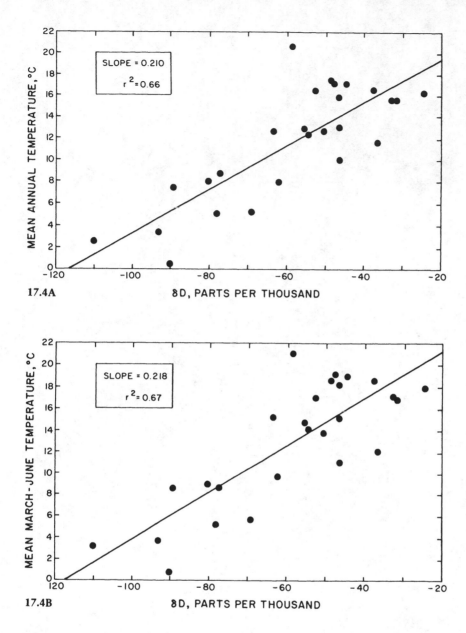

Fig. 17.4. Living tree data: δD as a function of (A) mean annual temperature and (B) mean growing-season temperature in the Southwest and Great Basin.

mostly juniper, represent several years' growth, and we believe fifteen years to be a reasonable estimate of that time span. In cases where the meteorological station was not at the same altitude as the sampled plants, the temperature was adjusted according to the local lapse rate.

We attempted linear correlations with rainfall amounts and with proportion of winter (frontal) to summer (monsoonal) rainfall as well as with temperature. The temperature correlations were the most significant: $r^2 = 0.67$ for growing season, and $r^2 = 0.66$ for average annual temperature. Less significant was the correlation between percentage of summer precipitation and δD ($r^2 = 0.50$). Multiple linear regression analysis found that either mean annual temperature or mean growing-season temperature explained two-thirds of the δD variance, and percentage of summer precipitation explained the remaining one-third. The quasi-parallel nature of δD in precipitation contours (Fig. 17.1) and lines marking zones of the same percentage of summer precipitation (Fig. 17.3) suggest that a temperature component is concealed in the 0.50 correlation coefficient. Perhaps coincidentally, the temperature as compared with δD slope in this study is essentially the same as Dansgaard (1964) found for mean annual precipitation and temperature at coastal stations. The greater noise in the δD (plant) data is at least in part due to the fact that these are continental measurements. Atmospheric moisture over continents during the summer growing season, especially in semiarid zones, derives significantly from continental sources (evaporation and transpiration).

Table 17.1 lists the δD data on fossil plants from packrat middens. Each radiocarbon date and its accompanying ± figure is the conventional date (5568 yr half-life) and standard deviation. It does not account for calendric time calibration, ^{14}C date reproducibility within each midden, or the possibility that some dates do not represent true averages of the time span of midden deposition (see Spaulding et al., *this volume*, chap. 5). Nine of the δD values first reported here are averages of replicate analyses from the same plant sample. Our average difference was 4.6‰, with a range of 0.9‰ to 9.0‰. The data from the Snake Range (Siegel, 1983) were all replicated.

Siegel's experimental reproducibility was ± 3.7‰. The data in Table 17.1 include a variety of species, but the table does not yet incorporate different species from the same midden, which usually were not available in sufficient quantities for isotopic analysis.

In order to combine on the same plot data from sites of different environments (and consequently exposed simultaneously to different δD values), we have adjusted the Nevada and New Mexico site values by adding 36‰ to the δD data from the east-central Nevada site and subtracting 14‰ from those at the New Mexico site (far-right column, Table 17.1). These adjustments offset numbers (36‰ and 14‰) derived from comparisons of δD values in juniper from modern packrat middens from each of the three sites.

DISCUSSION

Figure 17.5 shows the adjusted δD data from all three localities plotted as a function of radiocarbon age. The curve fitted through the points is based on the regression technique of Cleveland (1979). This smoothing technique locally weights the data points. The curve shown in Figure 17.5 uses a relatively stiff smoothing function ($F = 0.60$). Although some real signal may be lost with such an extreme degree of stiffness, this removes most biological noise and local and short-term variation. The curve thus represents the major regional δD changes in these tree species over the past thirty-five thousand years. On the basis of field and laboratory experiments in the present study, and those cited above, we assume here that these δD trends are responses to paleoclimate. It is noteworthy that the same major trend is apparent regardless of differences in tree species and locality. Relatively light δD values, about 20‰ lighter than modern, characterize the late Wisconsin before about 15–13 ka. From 13 to 8 ka δD values are essentially modern, increasing sharply by 10‰ to 20‰ between 8 and 4 ka, then dropping back to modern levels.

The simplest and most straightforward climatic interpretation of the δD trends is in terms of growing-season temperature. Applying the δD versus growing-season-temperature relationship obtained from the living trees (see Fig. 17.3), a 20‰ change corresponds to 3.5°C. As discussed above, however, humidity

Table 17.1. Summary of stable hydrogen values in fossil plant material from fossil packrat middens (SMOW = standard mean ocean water).

Sample Code	Midden ID	State	Latitude	Longitude	Elevation (meters)	Species Analyzed	Radiocarbon Dates	D/H SMOW (‰)	D/H Adjust (‰)
FM10B	Fishmouth Cave	Utah	37°25′45″N	109°39′W	1585	*Juniperus osteosperma*	modern	−60.2	
FM10A	Fishmouth Cave	Utah	37°25′45″N	109°39′W	1585	*Juniperus osteosperma*	modern	−61.8	−60.0
LC-M	Ladder Cave, Snake Range	Nevada	39°21′N	114°05′W	2060	*Juniperus osteosperma*	modern	−96.0	−60.3
SCMA	Sheep Camp Can., Chaco Canyon	N. Mex.	36°05′N	107°59′W	1980	*Juniperus monosperma*	modern	−46.3	−59.4
MC4C	Mockingbird Can., Chaco Canyon	N. Mex.	36°05′N	107°59′W	1930	*Juniperus monosperma*	1230 ± 60	−45.4	−46.8
MC1A	Mockingbird Can., Chaco Canyon	N. Mex.	36°05′N	107°59′W	1930	*Juniperus monosperma*	1,990 ± 90	−32.8	
AC10B	Atlatl Cave, Chaco Canyon	N. Mex.	36°05′N	107°59′W	1910	*Juniperus monosperma*	2,270 ± 110	−51.0	−65.0
MC9A	Mockingbird Can., Chaco Canyon	N. Mex.	36°05′N	107°59′W	1940	*Juniperus monosperma*	2,550 ± 70	−44.5	−58.5
SC2AB	Sheep Camp Can., Chaco Canyon	N. Mex.	36°05′N	107°59′W	1980	*Juniperus monosperma*	3,030 ± 130	−47.2	−61.2
FM6	Fishmouth Cave	Utah	37°25′45″N	109°39′W	1585	*Juniperus osteosperma*	3,550 ± 60	−40.0	
FM7	Fishmouth Cave	Utah	37°25′45″N	109°39′W	1585	*Juniperus osteosperma*	3,740 ± 70	−25.2	
FA9	Falling Arches Cave	Utah	37°24′20″N	109°38′W	1460	*Juniperus scopulorum*	3,770 ± 60	−56.0	
CC1BC	Casa Chiquita, Chaco Canyon	N. Mex.	36°05′N	107°59′W	1865	*Juniperus monosperma*	3,940 ± 110	−29.5	−43.5
CH-1B	Council Hall Cave, Snake Range	Nevada	39°20′N	114°06′W	2040	*Juniperus osteosperma*	4,220 ± 60	−85.0	−49.0
GW2B	Gallo Wash, Chaco Canyon	N. Mex.	36°05′N	107°59′W	1910	*Juniperus monosperma*	4,480 ± 90	−48.0	−62.0
CC2C	Casa Chiquita, Chaco Canyon	N. Mex.	36°05′N	107°59′W	1865	*Juniperus monosperma*	4,780 ± 90	−25.0	−39.0
FM4C	Fishmouth Cave	Utah	37°25′45″N	109°39′W	1585	*Juniperus osteosperma*	6,100 ± 100	−37.0	−39.0
CH-1A	Council Hall Cave, Snake Range	Nevada	39°20′N	114°06′W	2040	*Juniperus osteosperma*	6,120 ± 80	−75.0	
SV-3	Streamview Rock-shelter, Snake Range	Nevada	39°20′N	114°06′W	1860	*Juniperus osteosperma*	6,490 ± 190	−76.0	−40.0
SC1A	Sheep Camp Can., Chaco Canyon	N. Mex.	36°05′N	107°59′W	1980	*Juniperus monosperma*	6,670 ± 200	−31.3	−45.3

AC-8	Arch Cave, Snake Range	Nevada	39°17'N	114°05'W	2010	*Juniperus osteosperma*	7,350 ± 250	−57.0	−21.0
WR9BB	Weritos Rincon, Chaco Canyon	N. Mex.	36°05'N	107°59'W	2025	*Juniperus monosperma*	7,940 ± 100	−57.7	−71.7
AC11	Atlatl Cave, Chaco Canyon	N. Mex.	36°05'N	107°59'W	1910	*Juniperus monosperma*	8,300 ± 240	−31.6	−46.1
FA2C	Falling Arches Cave	Utah	37°24'20"N	109°38'W	1460	*Juniperus osteosperma*	9,200 ± 180	−54.6	
BA-1	Bloody Arm Cave, Snake Range	Nevada	39°08'N	114°25'W	2060	*Symphoricarpos* sp.	9,680 ± 700	−98.0	−62.0
FM5	Fishmouth Cave	Utah	37°25'45"N	109°39'W	1585	*Juniperus osteosperma*	9,700 ± 110	−46.9	
AC4BC	Atlatl Cave, Chaco Canyon	N. Mex.	36°05'N	107°59'W	1910	*Juniperus monosperma*	10,030 ± 150	−35.0	−49.0
SC-2	Smith Creek Cave, Snake Range	Nevada	39°21'N	114°05'W	2060	*Symphoricarpos* sp.	10,450 ± 290	−96.0	−60.0
FM2	Fishmouth Cave	Utah	37°25'45"N	109°39'W	1585	*Juniperus scopulorum*	10,540 ± 180	−55.5	
LC-6	Ladder Cave, Snake Range	Nevada	39°21'N	114°05'W	2060	*Symphoricarpos* sp.	11,080 ± 115	−92.0	−56.0
LC-4	Ladder Cave, Snake Range	Nevada	39°21'N	114°05'W	2060	*Pinus longaeva*	12,100 ± 150	−95.0	−59.0
SC-4	Smith Creek Cave, Snake Range	Nevada	39°21'N	114°05'W	2060	*Pinus longaeva*	12,235 ± 150	−96.0	−60.0
CW1	Cottonwood Cave	Utah	37°19'15"N	109°34'30"W	1387	*Pinus flexilis*	12,690 ± 230	−81.5	
FM1B	Fishmouth Cave	Utah	37°25'45"N	109°39'W	1585	*Juniperus scopulorum*	12,770 ± 140	−61.0	
FA8B	Falling Arches Cave	Utah	37°24'20"N	109°38'W	1460	*Juniperus scopulorum*	13,060 ± 210	−47.0	
FA8C	Falling Arches Cave	Utah	37°24'20"N	109°38'W	1460	*Juniperus scopulorum*	13,060 ± 210	−46.1	
LC-2C	Ladder Cave, Snake Range	Nevada	39°21'N	114°05'W	2060	*Pinus longaeva*	13,100 ± 210	−88.0	−52.0
SC-5	Smith Creek Cave, Snake Range	Nevada	39°21'N	114°05'W	2060	*Pinus longaeva*	13,340 ± 430	−86.0	−50.0
FM12	Fishmouth Cave	Utah	37°25'45"N	109°39'W	1585	*Pinus flexilis*	13,800 ± 320	−87.5	
CW3	Cottonwood Cave	Utah	37°19'15"N	109°34'30"W	1387	*Picea pungens*	15,660 ± 320	−71.0	
SV-2	Streamview Rock-shelter, Snake Range	Nevada	39°20'N	114°06'W	1860	*Pinus longaeva*	17,350 ± 435	−118.0	−82.0
LC-2B	Ladder Cave, Snake Range	Nevada	39°21'N	114°05'W	2060	*Pinus longaeva*	17,960 ± 1100	−119.0	−83.0
FA6C	Falling Arches Cave	Utah	37°24'20"N	109°38'W	1460	*Pinus flexilis*	19,590 ± 460	−90.8	
LC-2A	Ladder Cave, Snake Range	Nevada	39°21'N	114°05'W	2060	*Juniperus osteosperma*	27,280 ± 970	−110.0	−74.0
AC-6	Arch Cave, Snake Range	Nevada	39°17'N	114°05'W	2010	*Juniperus osteosperma*	29,920 ± 2000	−114.0	−78.0
AC-2A	Arch Cave, Snake Range	Nevada	39°17'N	114°05'W	2010	*Symphoricarpos* sp.	34,040 ± 4200	−104.0	−68.0

during photosynthesis also can affect the δD of leaf water and likely affects the δD in cellulose. Moreover, a shift in relative importance of summer precipitation (enriched in deuterium) with respect to winter precipitation (depleted in deuterium) may cause changes in δD in leaf cellulose. Leavitt and Long (1982) showed that one-seed juniper (*J. monosperma*) irrigated regularly on the University of Arizona campus continued growing into October. Colorado pinyon (*Pinus edulis*), though it responds predominantly to winter precipitation, will exhibit leaf growth throughout the summer (H. Fritts, oral communication; Douglas and Erdman, 1967).

Regarding the humidity effect, most plants adapted to semiarid climates shut down photosynthetic activity when the humidity is very low in order to conserve water (Fritts, 1976). Under dry conditions these trees may produce glucose only in early mornings while the environmental relative humidity is high compared to that at midday. Therefore it would not be surprising if semiarid trees are poor recorders of relative humidity.

Comparison with Other Records

The transition to modern values occurred some two or three thousand years sooner than the time of greatest floral turnover (between 11 and 10 ka). Betancourt (*this volume*, chap. 12) does identify early turnovers in midden floras from southeastern Utah. At Cottonwood Cave, for example, dominance shifts from blue spruce (*Picea pungens*) in a midden dated at 15.7 ka to Douglas fir (*Pseudotsuga menziesii*), limber pine (*Pinus flexilis*), and Rocky Mountain juniper (*Juniperus scopulorum*) in the next-youngest sample at 12.7 ka. However, the spruce from the 15.7-ka midden yielded a δD value of −71.0‰, while limber pine from the younger midden was actually lighter (−81.5‰). Only 10 km away at Falling Arches and Fishmouth Cave, the relation is reversed. At Falling Arches a midden dated at 19.6 ka yielded a δD value of −90.8‰, compared with −46.6‰ in a sample dated at 13.1 ka. At Fishmouth Cave a midden dated at 13.8 ka yielded a δD value of −87.5‰, compared with −61‰ at 12.7 ka.

The two-to-three-thousand-year delay between the transition to near-modern δD values (13 ka) and major floral turnovers (11–12

ka), both measured from middens, may be pertinent to questions about the rates of vegetation response after a major climatic change. Davis (1986) argued that northern range limits in the eastern United States were not in equilibrium with climate for several thousand years. Cole (1985) made a similar case for prolonged changes in community composition in the eastern Grand Canyon. Both authors suggested that the rapid expansion of species at the end of the Pleistocene may have been limited by biotic factors such as seed dispersal, establishment on immature soils, or the persistence of soils favorable to preexisting vegetation and competition from already-established species. Webb (1986) presents the alternate view that the transitional climate itself was unstable and affected species differently, resulting in a complicated record of vegetation change. The asynchroneity implied in our δD chronology might be taken as evidence of substantial vegetational change inertia. However, because of the complications with moisture source and relative humidity, different climates can produce a similar range in δD values. The spatial coherence of vegetation changes at 11–10 ka, not only in the southwestern United States but also nationwide, argues strongly that the climate changed significantly at this time. For whatever reasons, this change is simply not reflected in our δD chronology. On the other hand, the δD values do change significantly at 13 ka, even though midden floras from some areas (Van Devender, *this volume*, chaps. 7 and 8; Thompson, *this volume*, chap. 10) show stability or complacency at this time.

The early transition to heavier δD values could be associated with the early phases of deglaciation for both continental and alpine glaciers. The initiation of rapid melt-water flux down the Mississippi River has now been precisely dated to 14 ka (Broecker et al., 1987). Just north of Chaco Canyon and east of the southeastern Utah sites, the San Juan Mountain ice sheet in southwestern Colorado melted about 15 ka (Carrara et al., 1984). Similarly, the Yellowstone ice cap melted by 14–13 ka (Porter et al., 1983). Dates on the till-bog interface at two sites in northern Utah suggest that deglaciation of alpine glaciers began prior to 12.5 ka (Mehringer et al., 1971; Madsen and Currey, 1979). In the

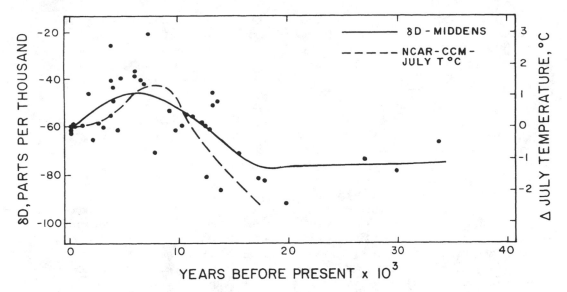

Fig. 17.5. Fossil plant δD + NCAR-CCM mean July temperature simulations as a function of ^{14}C age.

Ruby Mountains of northeastern Nevada a similar till-bog interface has been dated at 13 ka (Wayne, 1984). In addition, pluvial lakes in the Great Basin did not achieve their maximum lake stands until about 18 ka, and began their recessions by 14–13 ka (Benson and Thompson, 1987; Thompson, *this volume*, chap. 10).

Figure 17.5 shows general agreement between our δD chronology and trends evident in July surface temperatures in the southwestern United States (area averages for four grid cells centered at 37.8° N, 112.5° W; 37.5° N, 105.0° W; 33.3° N, 112.5° W; and 33.3° N, 105.0° W) for 18, 15, 12, 9, 6, and 3 ka, simulated by the National Center for Atmospheric Research (NCAR) Community Climate Model (CCM) (see Kutzbach, 1987). Taking into account the slight offset in time, which could be accounted for by differences between astronomical time and uncalibrated ^{14}C time, the coincidence between the curves is remarkable. However, the largest simulated difference between modern and past July temperatures, about −2.5°C for 18 ka, seems too low. This is lower than even the most conservative estimates for growing-season and mean annual temperatures from other paleoclimatic evidence (see Spaulding et al., 1983; Spaulding, 1985). Also, the midden curve peaks between 8 and 4 ka rather than at 10 ka, during the time of maximum solar insolation in July. If the plants were using winter moisture, the midden data may reflect warmer winters in the middle Holocene. Early Holocene winters may have been cooler compared to summers (see Thompson, *this volume*, chap. 10).

Other deuterium measurements over similar time periods in the Southwest are in conflict with our δD chronology from fossil middens. Epstein and Yapp (1976) obtained δD values from fossil wood (not leaves, as in the present study) from the Guadalupe Mountains in Eddy County, southeast New Mexico. The wood was dated at 11,850 ± 350, and yielded a δD value of −52‰. Other wood samples from elsewhere in the Yapp and Epstein (1976) study in the age range of 10 to 13 ka gave values from −114‰ to −37‰. They made no adjustments for altitude differences. These data were part of a large-scale survey of geographic and time variability in δD in fossil wood. The growth environments of these samples, mostly from buried deposits, are essentially unknown and uncontrolled. In the present study we have attempted to minimize the number of localities and normalize for altitudinal differences.

In the central San Juan Basin just north of Chaco Canyon, Phillips et al. (1986) measured δD values in ground waters radiocarbon-dated from 35 ka to modern. Ground water older than 7.3 ka was about 25‰ lighter than modern, except for samples dated between 19 and 17 ka (full glacial), which were anomalously heavy. Phillips et al. (1986) interpreted these to mean a brief period during the full glacial in which mean annual temperatures were about 3°C warmer than the rest of the late Wisconsin.

In Figure 17.6 we have plotted the San Juan Basin ground-water values and applied the same smoothing technique as used for the midden data for direct comparison. Salient differences include a full-glacial interval of relatively heavy δD values in the San Juan Basin ground-water chronology, the persistence of light δD values in ground water until 7.3 ka rather than a shift toward modern values by 13 ka in the midden data, and the lack of a discernible middle Holocene warm period in the ground-water data. The absence of ground water in the San Juan Basin with heavier-than-modern values may simply reflect the lack of net recharge under a summer-rainfall regime.

Assuming that the full-glacial warm event recorded in ground water is real, a reconciliation is needed between it and the midden data. Ground water in the San Juan Basin represents the heavy precipitation/runoff events, whereas the δD in plant-leaf cellulose samples shows a restricted part of annual moisture. Only two data points, both at 19 ka, support this event, which is subdued by rigorous smoothing of the data (Fig. 17.6). Their average δD is −97.5‰, only 13‰ heavier than the average of the other thirteen data points on the "full-glacial" trend line. An unusual few decades with high summer precipitation during the full glacial could affect the δD of recharged water yet be invisible to the plant cellulose. The difference in timing of the observed δD changes in the two studies may be owing to differences in confidence normally placed on ground-water ^{14}C dates compared to terrestrial plant material. Phillips et al. (1986) discussed the problems and uncertainties in the dating of ground water.

The midden data also can be compared to deuterium values from ground water in the Amargosa Desert, Nevada (Claassen, 1985, 1986). Note that the Amargosa chronology runs from 31.8 to 9.1 ka, with a range of only 10.5‰, compared to a range of 32‰ in ground

water from the San Juan Basin and 70‰ for the midden data. Though the sample points are few, the smoothed curve for the Amargosa ground water parallels that of the midden data (Fig. 17.6).

CONCLUSIONS

1. δD in the C–H hydrogen of fossil plants from packrat middens in the Great Basin of the southwestern United States yields paleoclimatic information. The data suggest that growing-season temperatures were 3°–4°C cooler than present during the full glacial and 3°–4° warmer during the mid-Holocene than now. Possibly other environmental/climatic changes were involved as well, which would modify these conclusions.

2. The δD paleoclimatic chronology from plant cellulose is out of phase with floristic turnovers defined by midden sequences throughout the Southwest. The δD chronology suggests major warming at about 13 ka, coincident with deglaciation of regional ice sheets and alpine glaciers and with desiccation of lakes in the Great Basin. The major floristic changes at 11–10 ka are not reflected in the δD chronology, while only minor floristic turnovers occur at 13 ka, when δD values become significantly heavier. This asynchroneity between vegetation and stable-isotopic changes could be interpreted to mean vegetational inertia, or it could mean that isotopic variations are insensitive to climatic changes that induce floral turnovers.

3. Paleoclimatic trends from this study have the same general trends as July average temperatures simulated by the NCAR-CCM for the same geographic area. However, the NCAR-CCM amplitudes are less than those in this and other studies. Heavier-than-modern deuterium values characterize the middle Holocene rather than the early Holocene, the time of maximum solar insolation in July. One explanation is that the plants were utilizing moisture that fell during winters that were warmer after 8 ka than at 9–10 ka.

4. The present data disagree somewhat with other isotopic studies in the Southwest, significantly with those from ground water. Additional studies should resolve these differences.

ACKNOWLEDGMENTS

This work was performed under National Sci-

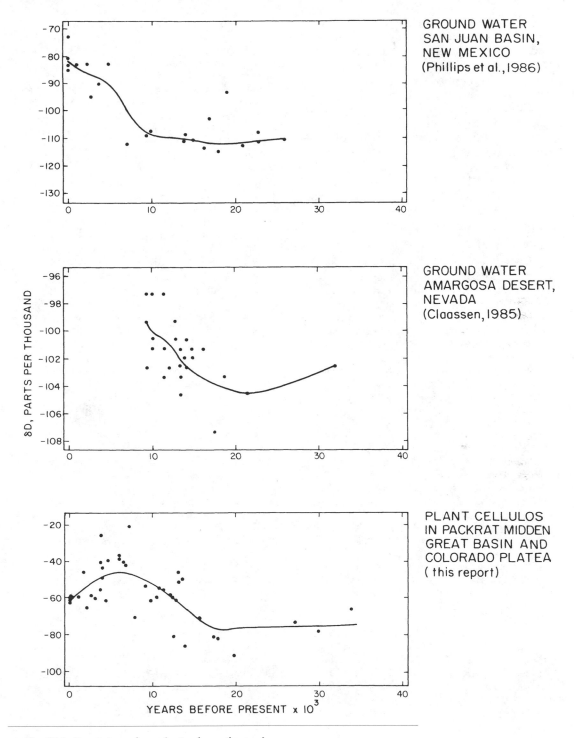

GROUND WATER
SAN JUAN BASIN,
NEW MEXICO
(Phillips et al., 1986)

GROUND WATER
AMARGOSA DESERT,
NEVADA
(Claassen, 1985)

PLANT CELLULOS
IN PACKRAT MIDDEN
GREAT BASIN AND
COLORADO PLATEA
(this report)

Fig. 17.6. Deuterium chronologies from plant cellulose in packrat middens (this study) compared with ground water (modified from Phillips et al., 1986; Claassen, 1985).

ence Foundation grant ATM 8313099. Songlin Cheng helped in data presentation, Robert H. Webb suggested the smoothing technique, and Robin L. Sweeney helped in sample preparation and retrieved the meteorological data. Pat Behling and Rich Selin of the Center for Climatic Research, University of Wisconsin, provided the new COHMAP runs using the NCAR-CCM. Chuck Sternberg drafted the illustrations. I. Winograd and I. Friedman reviewed the manuscript and made useful suggestions. The authors are most grateful to David C. Steinke for keeping the mass spectrometer and other arcane laboratory equipment functioning properly.

REFERENCES

Allison, G. B.; Gat, J. R.; and Leaney, F. W. J. (1985). "The relationship between deuterium and oxygen-18 delta values in leaf water." Chemical Geology (Isotope Geoscience Section) 58, 145–56.

Arnold, L. D. (1979). "The climatic response in the partitioning of the stable isotopes of carbon in juniper trees from Arizona." Ph.D. diss., University of Arizona, Tucson.

Benson, L. V., and Thompson, R. S. (1987). "The physical record of lakes in the Great Basin." In North America and adjacent oceans during the last deglaciation. W. F. Ruddiman and H. E. Wright, Jr., eds., pp. 241–60. The Geology of North America, vol. K-3. Boulder: Geological Society of America.

Brenninkmeijer, C. A. M.; van Geel, B.; and Mook, W. G. (1982). "Variations in the D/H and O^{18}/O^{16} ratios in cellulose extracted from a peat bog core." Earth and Planetary Science Letters 61, 283–90.

Broecker, W. S.; Andree, J.; Wolfi, W.; Oeschger, H.; Bonani, G.; Peteet, D.; and Kennett, J. (1987). "The chronology of the last deglaciation: Implications to the cause of the Younger Dryas event." Paleoceanography 3, 1–19.

Burk, R. L., and Stuiver, M. (1981). "Oxygen isotope ratios in trees reflect mean annual temperature and humidity." Science 211, 1417–19.

Carrara, P. E.; Mode, W. N.; Rubin, M.; and Robinson, S. W. (1984). "Deglaciation and postglacial timberline in the San Juan Mountains, Colorado." Quaternary Research 21, 42–55.

Claassen, H. C. (1985). Sources and mechanisms of recharge for ground water in the west-central Amargosa Desert, Nevada. U.S. Geological Survey Professional Paper 712-F.

———. (1986). "Late Wisconsin paleohydrology of the west-central Amargosa Desert, Nevada, U.S.A." Chemical Geology 58, 311–23.

Cleveland, W. S. (1979). "Robust locally weighted

regression and smoothing scatterplots." Journal of American Statistical Association 74, 829–36.

CLIMAP Project Members (1976). "The surface of the ice-age earth." Science 191, 1131–37.

Cole, K. L. (1985). "Past rates of change, species richness, and a model of vegetation inertia in the Grand Canyon, Arizona." American Naturalist 125, 289–302.

———. (1986). "In defense of inertia." American Naturalist 127, 727–28.

Craig, H. (1961). "Isotopic variations in meteoric waters." Science 133, 1702–3.

Dansgaard, W. (1964). "Stable isotopes in precipitation." Tellus 16, 436–38.

Davis, M. B. (1986). "Climatic instability, time lags, and community disequilibrium." In Community ecology. J. Diamond and T. J. Case, eds., pp. 269–84. New York: Harper and Row.

Douglas, C. L., and Erdman, J. A. (1967). "Development of terminal buds in pinyon pine and Douglas-fir trees." Pearce-Sellards Series 8, 5–19.

Edwards, T. W. D., and Fritz, P. (1986). "Assessing meteoric water composition and relative humidity from O^{18} and H^2 in wood cellulose, paleoclimatic implications for southern Ontario, Canada." Applied Geochemistry 1, 715–23.

Epstein, S.; Thompson, P.; and Yapp, C. J. (1977). "Oxygen and hydrogen isotopic ratios in plant cellulose." Science 198, 1209–15.

Epstein, S., and Yapp, C. J. (1976). "Climatic implications of the D/H ratio of hydrogen in C–H groups in tree cellulose." Earth and Planetary Science Letters 30, 252–61.

Epstein, S.; Yapp, C. J.; and Hall, J. H. (1976). "The determination of the D/H ratio of nonexchangeable hydrogen in cellulose extracted from aquatic and land plants." Earth and Planetary Science Letters 30, 241–51.

Ferhi, A., and Letolle, R. (1979). "Relation entre le milieu climatique et les teneurs en oxygene-18 de la cellulose des plantes terrestres." Physiologie Vegetale 17(1), 107–17.

Friedman, I., and Gleason, J. (1980). "Deuterium content of lignin and of the methyl and OH hydrogen of cellulose." Proceedings of the International Meeting on Stable Isotope Tree-Ring Research, New Paltz, New York, May 22–24, 1979, pp. 50–55. Washington, D.C.: U.S. Department of Energy.

Friedman, I.; Redfield, A. C.; Schoen, B.; and Harris, J. (1964). "The variation of the deuterium content of natural waters in the hydrologic cycle." Review Geophysics 2, 117–224.

Friedman, I., and Smith, G. I. (1972). "Deuterium content of snow as an index to winter climate in the Sierra Nevada area." Science 176, 790–93.

Fritts, H. C. (1976). Tree rings and climate. New York: Academic Press.

Gat, J. R., and Issar, A. (1974). "Desert isotope hydrology: Water sources in the Sinai Desert." *Geochimica et Cosmochimica Acta* 38, 1117–31.

Grey, J., and Thompson, P. (1976). "Climatic information from O^{18}/O^{16} ratios of cellulose in tree rings." *Nature* 262, 481–82.

Grey, J., and Thompson, P. (1977). "Climatic information from O^{18}/O^{16} analysis of cellulose, lignin and whole wood from tree rings." *Nature* 270, 708–9.

Kutzbach, J. E. (1987). "Model simulations of the climatic patterns during the deglaciation of North America." In *North America and adjacent oceans during the last deglaciation*. W. F. Ruddiman and H. E. Wright, Jr., Eds., pp. 425–46. *The Geology of North America*, vol. K-3. Boulder: Geological Society of America.

Lawrence, J. R., and White, J. W. C. (1984). "Growing season precipitation from D/H ratios of eastern white pine." *Nature* 311, 558–60.

Leavitt, S. W., and Long, A. (1982). "Evidence for C^{13}/C^{12} fractionation between tree leaves and wood." *Nature* 298, 742–44.

Leavitt, S. W., and Long, A. (1983). "An atmospheric C^{13}/C^{12} reconstruction generated through removal of climate effects from tree-ring C^{13}/C^{12} measurements." *Tellus* 35B, 91–102.

Long, A.; Cheng, S.; and Kalin, R. (1988). "Isotope hydrogeochemistry of the Tucson Basin, I." MS.

Markgraf, V. (1986). "Plant inertia reassessed." *American Naturalist* 127, 726.

Madsen, D. B., and Currey, D. R. (1979). "Late Quaternary glacial and vegetation changes, Little Colorado Canyon area, Wasatch Mountains, Utah." *Quaternary Research* 12, 254–70.

Mehringer, P. J., Jr.; Nash, W. P.; and Fuller, R. H. (1971). "A Holocene volcanic ash from northwestern Utah." *Utah Academy of Sciences, Arts, and Letters* 48, 46–51.

Pearman, G. I.; Francey, R. J.; and Fraser, P. J. B. (1976). "Climatic implications of stable carbon isotopes in tree rings." *Nature* 260, 771–73.

Phillips, F. M.; Peeters, L. A.; and Tansey, M. (1986). "Paleoclimatic inferences from an isotopic investigation of ground water in the central San Juan Basin, New Mexico." *Quaternary Research* 26, 179–93.

Porter, S. C.; Pierce, K. L.; and Hamilton, T. D. (1983). "Late Wisconsin mountain glaciation in the western United States." In *Late-Quaternary Environments of the United States*. Volume 1, *The late Pleistocene*. S. C. Porter, ed., pp. 71–114. Minneapolis: University of Minnesota Press.

Ramesh, R.; Bhattacharya, S. K.; and Gopalan, K. (1985). "Dendroclimatological implications of isotope coherence in trees from Kashmir Valley, India." *Nature* 317, 802–4.

Ramesh, R.; Bhattacharya, S. K.; and Gopalan, K. (1986). "Climatic correlations in the stable isotope records of silver fir (*Abies pindrow*) trees from Kashmir, India." *Earth and Planetary Science Letters* 79, 66–74.

Reitan, C. H. (1960). "Distribution of precipitable water vapor over the continental United States." *Bulletin of the American Mineralogical Society* 41, 79–87.

Sellers, W. D. (1965). *Physical climatology*. Chicago: University of Chicago Press.

Siegel, R. D. (1983). "Paleoclimatic significance of D/H and C^{13}/C^{12} ratios in Pleistocene and Holocene wood." Master's thesis, University of Arizona, Tucson.

Sonntag, C.; Munnich, K. O.; Jacob, H.; and Rosanski, K. (1983). "Variations of deuterium and oxygen-18 in continental precipitation and ground water and their causes." In *Variations in the global water budget*. A. Street-Perrot, M. Berau, and R. Ratcliffe, eds., pp. 107–25. Boston: D. Reidel.

Spaulding, W. G. (1985). *Vegetation and climates of the last 45,000 years in the vicinity of the Nevada Test Site, south-central Nevada*. U.S. Geological Survey Professional Paper 1329.

Spaulding, W. G.; Leopold, E. B.; and Van Devender, T. R. (1983). "Late Wisconsin paleoecology of the American Southwest." In *The late Pleistocene of the United States*. S. C. Porter, ed., pp. 259–93. Minneapolis: University of Minnesota Press.

Taylor, H. P., Jr. (1974). "The application of oxygen and hydrogen isotope studies to problems of hydrothermal alteration and ore deposition." *Economic Geology* 69, 843–83.

van der Straaten, C. M. (1981). "Deuterium in organic matter." Ph.D. diss., Rijksuniversiteit to Groningen.

Wayne, W. J. (1984). "Glacial chronology of the Ruby Mountains–East Humboldt Range, Nevada." *Quaternary Research* 21, 286–303.

Webb, T. III (1986). "Is vegetation in equilibrum with climate? How to interpret late Quaternary pollen data." *Vegetatio* 67, 75–91.

Webb, T. III, and Wigley, T. M. L. (1985). "What past climates can indicate about a warmer world." In *The potential climatic effects of increasing carbon dioxide*. M. C. MacCracken and F. M. Luther, eds., pp. 239–57. Washington, D.C.: U.S. Department of Energy Report DOE/ER-0237.

Wilson, A. T., and Grinsted, M. J. (1975). "Palaeotemperatures from tree rings and the D/H ratio of cellulose as a biochemical thermometer." *Nature* 257, 387–88.

Winograd, I. J.; Szabo, B. J.; Coplen, T. B.; Riggs, A. C.; and Kolesar, P. T. (1985). "Two-million-year record of deuterium depletion in Great Basin ground waters." *Science* 227, 519–22.

Yapp, C. J., and Epstein, S. (1982). "A reexamination

of cellulose carbon-bound hydrogen D/H measurements and some factors affecting plant-water D/H relationships." *Geochemica et Cosmochimica Acta* 46, 955–65.

Yurtsever, Y. (1975). "Worldwide survey of stable isotopes in precipitation." In *Stable isotope hydrology, deuterium and oxygen-18 in the water cycle*. J. R. Gat and R. Gonfiantini, eds., p. 117. Vienna: International Atomic Energy Agency.

PART IV
Middens Abroad

Hyrax (Procaviidae) and Dassie Rat (Petromuridae) Middens in Paleoenvironmental Studies in Africa

Louis Scott

INTRODUCTION

Plant remains are rarely preserved in flood-plain or lacustrine deposits of arid and semi-arid regions. Packrat middens preserved in caves or rock shelters have provided the best source of datable fossil plant material in arid North America and have assumed prime importance in paleoenvironmental studies in the region. Since packrats (*Neotoma*) are restricted to North and Central America, the discovery of other species with similar habits is crucial if the method is to be adopted in other arid regions. In the case of Africa some obvious candidates can be identified, such as Procaviidae (hyraxes or dassies, Fig. 18.1) and Petromuridae (dassie rats, Fig. 18.2), both of which build middens. The potential value of such middens in paleoenvironmental studies justifies further investigation.

PROCAVIIDAE

While its broader significance seems to have gone unappreciated, one of the first uses of radiocarbon-dated fossil middens in a paleoenvironmental reconstruction was accomplished not in arid America but in North Africa. In 1958 Pons and Quezel reported "guano" of fossil hyrax (*Procavia rufipes*) dated at 4680 ± 300 B.P. from the Hoggar Mountains in the Sahara. A rich pollen assemblage with a relatively strong representa-tion of Mediterranean forms suggested that the climate in the area had become more arid since the formation of the midden deposit.

The Procaviidae have been subdivided into three genera, *Dendrohyrax*, *Heterohyrax*, and *Procavia*, each represented by one species, according to Haltenorth and Diller (1977). However, Bothma (1971) recognizes three species in each of the former two genera and eleven in *Procavia*. The Procaviidae are rabbit-sized subungulates with short, rounded ears; adults weigh around 3 kg (Fig. 18.1). They are widely distributed over Africa, excluding the northern and northwestern parts, and also are found in Arabia, Israel, and Lebanon (Fig. 18.3). *Dendrohyrax* is an arboreal genus confined to humid, forested parts of Africa. In dry tropical and subtropical parts, as well as in cooler southern and high-elevation areas, *Heterohyrax* and *Procavia* generally occur in rocky habitats. Since midden preservation is more likely in rock shelters of relatively dry zones, the latter two genera promise more for paleoenvironmental studies. Despite poor thermoregulatory qualities of Procaviidae (Louw et al., 1972), *Procavia* enjoy a wide altitudinal range. In South Africa they are abundant at elevations from sea level up to roughly 2500 m. On Mount Kenya they are found at elevations as high as 4000–5500 m (Haltenorth

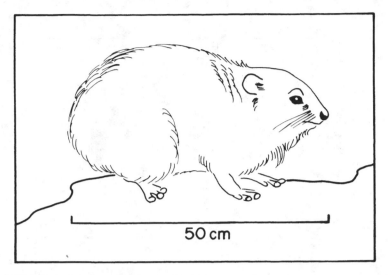

Fig. 18.1. Cape hyrax or dassie, *Procavia capensis*, redrawn after Smithers (1983).

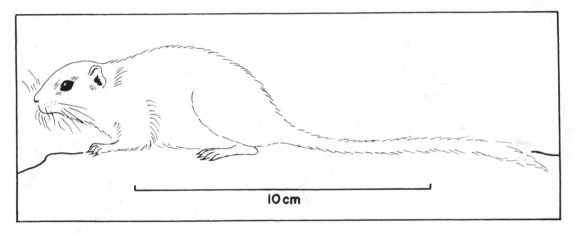

Fig. 18.2. Dassie rat, *Petromus typicus*, redrawn after de Graaff (1981).

and Diller, 1977). In view of the abundance of dassies and their adaption to a wide range of ecological conditions, it is unlikely that past climatic oscillations would have had a significant effect on their distribution during the Quaternary. Therefore, the chances are good that fossil hyrax middens may be easily obtainable over large parts of Africa and neighboring territories.

Dassies are herbivores consuming leaves, grass, berries, fruit, and bark of a wide diversity of species. While not forming an important part of their diet, dassies are known to eat flowers (Fourie, 1983). Dassies leave their shelters to feed only for very short periods each day. They are not known to carry food or other plant material to their shelters. They graze and climb trees to browse within about 50–100 m from their shelters but hurry back to seek the protection of these rocky areas, where they remain fairly inactive for most of the day. Up to 75% of each daylight period is spent basking (Fourie, 1983). The fecal pellets of hyraxes, which are about 1 cm long and

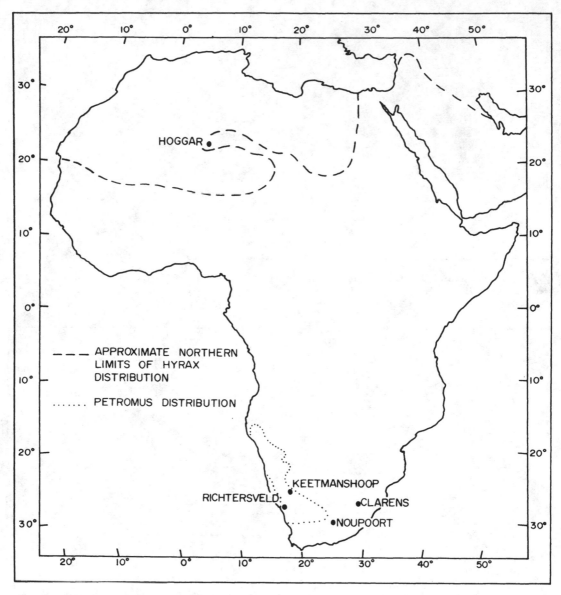

Fig. 18.3. Distribution map for hyraxes and *Petromus typicus*.

nearly spherical, accumulate in large piles near the entrances of their refuges. In dry sites on inaccessible rock ledges these piles may form solid middens. In moist sites the pellets remain loose and will eventually decompose into soil. Hyraxes are well adapted to drought and have very efficient kidney action (Louw et al., 1972) so that they produce highly concentrated urine, which on evapora- tion hardens to form an amber-colored sub- stance at least analogous to the "amberat" of Orr (1957). Often this so-called hyracium ce- ments fecal pellets together and allows exten- sive middens to form at sites that are used continuously for many years. Smithers (1983) mentions the existence of a deposit approxi- mately 3 m wide and 1 m deep in the Ceder- berg, Cape Province. In some parts of Africa hyracium is prized by natives as an important ingredient in folk medicine, and it is likely that the destruction and disappearance of middens through the years is not strictly a natural process.

Natural enemies of dassies such as lynx, leopard, jackal, and eagle have been exterminated in some regions and their numbers reduced in others. This is believed to have led to widespread local overpopulation of dassies despite continued hunting by natives. It is possible that middens have accumulated more rapidly in recent years than in the past.

Unlike the middens of packrats and dassie rats, those attributed to hyraxes may not contain plant material. When present in small numbers the plant fragments may not have been brought in by the hyrax. Parts of the dungheaps seem to have an orderly stratification, with the oldest material at the bottom. In some cases the middens extend horizontally away from the rock ledges on which they rest to form overhanging platforms. This lateral growth process occurs when urine flows over the edge of the midden and solidifies. Where large midden deposits accumulate, extensive lobes of hyracium usually flow down the steep rock slopes below the middens, behaving in the same way as stalactites and tufas. Small "stalactites" of hyracium hanging from such a platform have been found at Clarens. Like packrat middens, dassie middens can be dated by radiocarbon, and a measurement of 1130 ± 80 b.p. (Beta-14658) was obtained on a deposit from Noupoort (Fig. 18.3), indicating that relatively old middens are potentially available in the area.

Apart from that found in fecal pellets and hairs, only small numbers of seeds and other macrobotanical remains have been found in dassie middens. The middens are ideal traps for the natural pollen rain, pollen from the adjacent vegetation, and pollen derived from flowers swallowed and excreted by the dassies. It is difficult to determine the contributions of these different sources and this may complicate palynological interpretations.

An advantage of midden pollen spectra over swamp spectra is that they provide direct insights into plant communities on rocky slopes. These spectra are not masked by high proportions of undiagnostic hygric elements. Preliminary results suggest that the study of

pollen in dassie middens can be rewarding, and fresh fecal pellets have been collected over a wide range of habitats in South Africa. These include some from the winter-rainfall macchia region, the western desert area and semidesert karoo shrubland, and the subhumid grassland area of the highlands. Four large midden accumulations were found in sandstone shelters in the latter two habitats. Preliminary investigations show that the pollen spectra in dassie feces of the different areas can clearly be related to their respective vegetation types.

One of two middens at Clarens (Fig. 18.3) was studied in some detail. It occurs in an east-facing rock shelter in Cave Sandstone on a wooded slope in the subhumid Highland Sourveld grassland (Acocks, 1953) at about 2000 m elevation. The shelter is still in use and the formation of the midden is apparently still in progress. The midden is more than 1 m wide and has a vertical thickness of about 15 cm. A horizontal overhanging midden platform has formed, and large sheets of hyracium flow down the vertical cliffs below it. A section was obtained by sawing a 5-cm slice from the overhanging platform, transecting the midden from the modern surface down to bedrock. Thirteen subsamples crossing a diagonal section of around 25 cm, with clear stratigraphical lines, were obtained.

A pollen diagram of the midden sequence is shown in Figure 18.4. The average pollen concentration in the midden samples is about 100,000 grains per gram (Fig. 18.5). Sample 7 produced about 640,000 grains per gram. Fossil *Neotoma* middens typically yield a million grains per cubic centimeter (O. K. Davis, personal communication). In view of the high percentage of *Buddleia* pollen in sample 7 (Fig. 18.4), it is likely that the large concentration results from the chance inclusion of flowers of this plant in the sample. The bottom sample (no. 1), resting on the rock ledge, was dated to either 1980 or 1963 A.D. (Pta-4232). The presence of such exotic elements as *Pinus* and Myrtaceae (*Eucalyptus*) in most parts of the upper section above sample 1 supports an age no greater than the time of introduction of these plants into the region about one hundred years ago. No further increase in the thickness of the midden is noticeable in the five years since its discovery (1981), and an

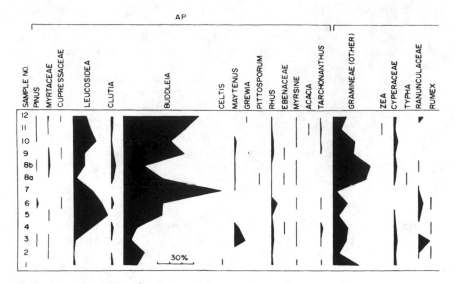

Fig. 18.4. Pollen diagram of the dassie midden sequence from Clarens. AP, arboreal pollen; NAP, non-arboreal pollen.

age of only six or twenty-three years is surprisingly young. The sequence seems to reflect changes over a longer period. The relatively young radiocarbon date could be the result of contamination. The relatively humid climate of the area causes slow evaporation and downward movement of unsolidified material, resulting in the lobes of hyracium below the midden. Because the studied section contains enough insoluble macroremains, however, contamination in such a way seems unlikely.

The pollen spectra imply that relatively open vegetation existed on the slope during the early history of the midden (samples 1–4). The spectra suggest that shrubby vegetation, including a high proportion of Compositae of the *Pentzia*-type and later of Ericaceae and *Anthospermum*, prevailed. During the next phase (samples 5–7) woody plants, especially *Buddleia* and *Leucosidea*, became more prominent. A decline in these woody elements occurs in samples 8a and 8b, while the pollen of the woody shrub *Clutia*, grasses, Compositae (undifferentiated) and Cheno-

podiaceae/Amarantheae (Cheno-Am) increase. In the youngest samples (9–12) the latter three types remain prominent, but the woody elements *Leucosidea* and *Buddleia* return in fair numbers.

Historical vegetation changes depend on long-term climatic fluctuations and differing land uses by indigenous tribes, early settlers, and modern farmers. At present, mountain slopes in the Orange Free State grasslands are densely wooded. Photographs of the veld at the end of the nineteenth century and the early part of this century show that in many cases kopjes (dolerite inselbergs) and other mountain slopes in this region were almost denuded of woody vegetation. More open-slope vegetation probably existed earlier, as suggested by the paintings of Thaba Nchu and Bloemfontein made by the explorer Thomas Baines around 1850 and 1851. The open vegetation suggested by the pollen spectra in the oldest part of the midden sequence was expected to correspond with the end of this phase, but the relatively young radiocarbon date poses new questions concerning the re-

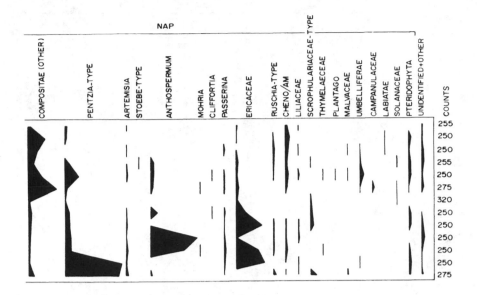

cent development of vegetation on the slope.

PETROMURIDAE

In the Namib desert region nests that appear
to be very similar to packrat middens have
been observed in granite and dolerite cavities
in the Richtersveld (just south of the Orange
River) and Keetmanshoop (SWA-Namibia)
(Fig. 18.3). These middens consist mainly of
dry plant material, while some elongated,
ratlike fecal pellets (about 7 mm long) occur
in and around them. In the Richtersveld a
relatively old nest, apparently still in use, has
been found in a rock crevice about 50 cm
wide (Figs. 18.6 and 18.7). Rat-sized tunnels
entering the nest also were noticed. The plant
material in the nest seems to be cemented
together by urine with a musky scent. Several smaller nests can be seen amongst low
dolerite blocks in the "Kokerboomwoud"
(Aloe dichotoma forest) near Keetmanshoop,
but these apparently are not saturated with
urine. Although the inhabitants of the nests
in both areas were not seen in or near them,
it is fairly certain that they are dassie rats

(Petromus typicus), which belong to the
monospecific family Petromuridae (de Graaff,
1981; Smithers, 1983).

Dassie rats are smaller than dassies, with a
mean mass of about 220 g (Fig. 18.2); Neotoma
adults range from 100 to 400 g (Vaughan, this
volume, chap. 2). Dassie rats are confined to
the extremely arid zone of southwestern Africa (Namibia) between the southernmost
part of Angola and the northwestern part of
the Cape Province (Fig. 18.3). They are found
in rocky habitats similar to those of dassies,
and in some areas they share the same territory. Like dassies, dassie rats will climb
trees. However, unlike dassies, they will take
leaves to their shelters to be eaten in safety
(Smithers, 1983). Their diet also consists of
stems and flowering heads of grasses and,
when available, flowers of Compositae (de
Graaff, 1981). In the Augrabies Falls National
Park the fecal pellets consist predominantly
of grass remains, while those of dassies consist of 80% Schotia afra rests and only 20%
grass remains (Smithers, 1983).

Pollen analysis of the Petromus midden

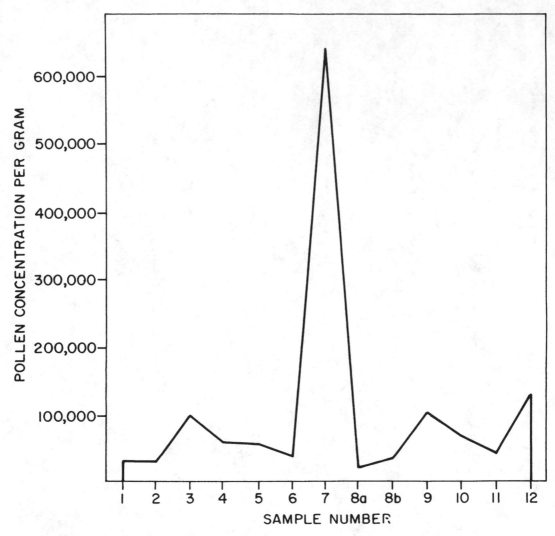

Fig. 18.5. Pollen concentration of the dassie midden sequence from Clarens.

from the Richtersveld shows high proportions of succulent Aizoaceae-type pollen (including Mesembryanthemaceae), which is consistent with the vegetation of the area (Namaqualand Broken Veld and Western Mountain Karoo; Acocks, 1953). These succulents apparently are scarce as macrorests. The dassie rat midden from the Richtersveld consists mainly of grass. Other middens that consist partly of *Aloe* twigs have been observed elsewhere within the area in which dassie rats are found. Dassie rat middens seem to be more suitable for macrobotanical analysis than dassie mid-

dens. The strong selection for grasses and the rats' collection and ingestion of flowers may complicate paleoenvironmental interpretations. In extremely arid southwestern Africa both dassie rat and dassie middens provide a potential new source of information in an area where deposits suitable for late Quaternary paleoenvironmental study seemed to be limited.

OTHER FAUNA

Because porcupines (*Erethizon*) produce mid-

Fig. 18.6. Locality of the dassie rat midden in granites of the Richtersveld.

dens in the United States (Betancourt et al., 1986), it is possible that in relatively dry regions African porcupines (*Hystrix* species) may accumulate resistant middens. In addition, different cave-dwelling hyenas in Africa and the Middle East are known to collect carrion and bones of their prey. The bones accumulate in deposits together with hyena feces. Brown hyena (*Hyaena brunnea*) coprolites in deposits from Taung, representing the late Pleistocene and Holocene, have provided useful pollen spectra (Scott, unpublished data). These unconsolidated hyena deposits

differ from the urine-based dassie and dassie rat middens in that their pollen content is low and they contain virtually no other plant remains.

Old, solidified bird guano attached to a dolerite outcrop near Tarkastad (Karoo) has been investigated and found to contain an unusual pollen spectrum. Bird deposits, therefore, seem to have potential use in paleoenvironmental studies, but dating of avian guano is problematic in view of its relatively low organic carbon content. Defining the spatial limits of pollen spectra derived from

Fig. 18.7. A sample of the dassie rat midden from the Richtersveld consisting mainly of grass and other remains and some fecal pellets.

highly mobile birds presents a further problem.

DISCUSSION

Preliminary indications are that hyrax and *Petromus* middens are important sources of plant remains for paleoenvironmental studies in Africa. Although hyrax middens are not as rich a source of macrobotanical fossils as dassie rat middens, their pollen contents appear to be sensitive to environmental changes. Pollen analysis of stratigraphic layers over close intervals in a dassie midden at Clarens indicates environmental changes

in the recent past. Leaching of urine and detritus in dassie middens in relatively humid areas may result in minimum values for radiocarbon age determinations. The "stalactites" and lobes of hyracium appearing below midden deposits are likely to be products of leaching. In both dassie and dassie rat middens, macrobotanical and pollen studies should be combined in order to overcome constraints involving selective representation of plant macrofossils and accidental over-representations of pollen grains. Further control may be possible with complementary studies of middens of both dassies and dassie rats from the same area and time range, as in the case of packrat and porcupine middens from the southwestern United States (Betancourt et al., 1986).

ACKNOWLEDGMENTS

Paul Martin (University of Arizona) and Julio Betancourt (U.S. Geological Survey) made possible the radiocarbon dating of a dassie midden from Noupoort, while John Vogel (Pretoria) dated the midden from Clarens. Martin Wessels (Clarens) and Britt Bousman (Southern Methodist University, Dallas) helped with locating and sampling of dassie middens. Palynological research in the Clarens area is supported by the Foundation for Research Development of the Council for Scientific and Industrial Research (National Programme for Weather Climate and Atmosphere Research).

REFERENCES

Acocks, J. P. H. (1953). "Veld types of South Africa." *Memoirs of the Botanical Survey, S.A.* 28, 1–192.

Betancourt, J. L.; Van Devender, T. R.; and Rose, M. (1986). "Comparison of fossil packrat (*Neotoma* spp.) and porcupine (*Erethizon dorsatum*) middens from the western United States." *Journal of Mammalogy* 67, 266–73.

Bothma, J. du P. (1971). "Order Hyracoidea." In *The mammals of Africa, an identification manual*. J. Meester and H. W. Setzer, eds., pp. 1–8. Washington, D.C.: Smithsonian Institution Press.

de Graaff, G. (1981). *The rodents of southern Africa*. Pretoria: Butterworth.

Fourie, L. J. (1983). "The population dynamics of the rock hyrax, *Procavia capensis* (Pallas, 1766), in the Mountain Zebra National Park." Ph.D. diss., Rhodes University, Grahamstown.

Haltenorth, T., and Diller, H. (1977). *A field guide to the mammals of Africa including Madagascar*. London: Collins.

Louw, E.; Louw, G. N.; and Retief, C. P. (1972). "Thermoability, heat tolerance and renal function in the dassie or hyrax, *Procavia capensis*." *Zoologica Africana* 7, 451–69.

Orr, P. C. (1957). "On the occurrence and nature of 'amberat.' *Observations Western Speleological Institute* 2, 1–3.

Pons, A., and Quezel, P. (1958). "Premières remarques sur l'étude palynologique d'un guano fossile du Hoggar." *Comptes rendus des séances de l'Académie des Sciences* 244, 2290–92.

Smithers, R. H. N. (1983). *The mammals of the southern African subregion*. Pretoria: University of Pretoria Press.

Chapter 19

Fossil Hyrax Middens from the Middle East: A Record of Paleovegetation and Human Disturbance

Patricia L. Fall   Cynthia A. Lindquist   Steven E. Falconer

INTRODUCTION

Paleoecological studies of the world's deserts are constrained by a paucity of suitable environments for plant fossil preservation. Palynological analysis of lacustrine deposits, a window to the past in temperate regions, is limited in desert environments. In the North American deserts fossil packrat midden analysis has supplanted palynology for reconstructing paleovegetation. Over twenty years ago in the deserts of Nevada, fossil packrat nests revealed twigs and seeds of plants that grow today at much higher elevations (Wells and Jorgensen, 1964). Radiocarbon analysis has shown these nests to be more than ten thousand years old, demonstrating the paleoecological potential of this approach. In contrast to the insensitivity of pollen stratigraphies from playa lakes in the American Southwest, packrat midden analysis illuminates the varied history of Holocene climates (Spaulding et al., 1983), plant migrations (Thompson, 1984), and human disturbance (Betancourt and Van Devender, 1981).

Packrats, cricetid rodents in the genus *Neotoma*, are prolific collectors of plant material, faunal remains, and other debris used for the construction of their dens. This material, when deposited in dry caves and crevices, is preserved by aridity and allelochemical substances in packrat urine known as amberat.

Over time, plant macrofossils and pollen become cemented into a rock-hard matrix that is preserved for thousands of years as an "indurated" midden.

Now it appears that animals in other deserts create deposits that may be of comparable value in reconstructing past environments (Fig. 19.1). Middens of the stick-nest rat, *Leporillus*, have been collected and dated from Australia (Green et al., 1983; Nelson et al., this volume, Chap. 20). Deposits from an unidentified animal have been dated to the middle Holocene in Argentina. Fossil nests from South Africa are attributed to dassie rats (*Petromus*) and hyraxes (Procaviidae) (Scott, *this volume*, chap. 18). "Guano" from hyrax deposits in the Sahara was found to contain a rich pollen assemblage of Mediterranean flora, demonstrating that a different flora flourished in North Africa under wetter climatic conditions during the mid-Holocene (Pons and Quezel, 1958).

Plant and animal husbandry and human population growth during the Holocene have devastated many natural landscapes. Nowhere is this more evident than in southwestern Asia, where a deeply humanized landscape has evolved through millennia of farming, grazing, burning, and fuel harvesting. Isolated trees, often preserved intention-

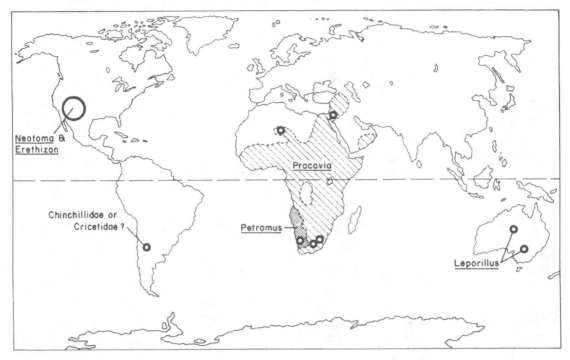

Fig. 19.1. Worldwide distribution of fossil middens and animals to which they are attributed. Dense hatching is the distribution of *Petromus*; light hatching is distribution of *Procavia*.

ally or in places inaccessible to man, testify to formerly widespread forests and woodlands. While modern remnant vegetation offers a few tantalizing clues, it cannot reveal the distribution of plants and animals on the presettlement landscape.

Based on the success of packrat midden research in the American Southwest, we suspected that hyrax middens would provide analogous paleobotanical insight into the paleoecology of arid southwestern Asia. The Syrian rock hyrax nests in the mountains and rocky terrain of Sinai, Palestine, Transjordan, and Syria. The Valley of Petra in southern Jordan was chosen for this initial study because of its long and varied history of land use. Situated within the modern distribution of hyraxes, Petra has numerous sandstone caves and crevices that lend themselves to preservation of midden deposits.

Here we report on pollen and plant macrofossils in hyrax middens from southern Jordan and discuss their potential to provide detailed information on past vegetation, climate, and human ecology in the deserts of the Middle East. Data from four middens col-

lected near the ancient city of Petra reflect vegetation changes spanning 1800 years that are attributed to human deforestation and excessive animal browsing.

ENVIRONMENTAL SETTING

The Wadi Araba and the Jordan Valley, from Aqaba north to the Sea of Galilee, form the northern extension of the African rift system, dropping to 396 m below sea level at the Dead Sea, the lowest point on Earth (Raikes, 1984). The ancient city of Petra lies in the mountains of Edom (the southernmost region of Transjordan), which bound the eastern side of the rift. Immediately to the west, the Wadi Araba averages 100–200 m elevation. The Valley of Petra lies at about 1000 m elevation (Fig. 19.2), while the hills to the north reach 1736 m between Petra and Shobak. The Syrian Desert to the east of Petra averages 700 m elevation.

Modern Climate

The climate of southern Jordan is a Saharan variant of a Mediterranean climate characterized by mild, rainy winters and hot, dry sum-

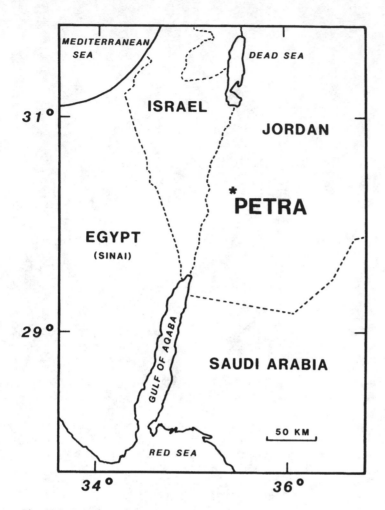

Fig. 19.2. Location of the ancient city of Petra in
southern Jordan.

mers. Fragmentary meteorological data for
Petra come largely from observations by researchers in the region (e.g., Zohary, 1940a;
Raikes, 1966; Kirkbride, 1984) and from nearby
weather stations (e.g., Shobak at 1400 m).
Winter temperatures usually fall below 0°C,
with snow recorded almost every year, while
summer temperatures reach 35°–40°C. Precipitation is concentrated between November
and April. Based on four years of data, Zohary
(1940a) estimates a mean annual rainfall of
167 mm, while Raikes (1966) reports 170 mm
of annual precipitation over a twenty-year
period at the nearby town of Wadi Musa.
Mean annual rainfall in the hills of Shobak
is 298 mm (Al-Eisawi, 1984). Accordingly,

Fig. 19.3. Degraded oak-pistachio forest-steppe in the hills northeast of Petra.

the hill slopes above Petra to the east are estimated to receive 200–300 mm of rainfall per year. Zohary (1960) places Petra in the sub-Mediterranean vegetation belt, which requires at least 200 mm rainfall per year.

Modern Vegetation

The modern vegetation of Edom includes floristic elements from the four major vegetation regions in southwestern Asia. Petra lies near the southern limit of the Mediterranean and Irano-Turanian vegetation territories and at the northern edge of the Saharo-Arabian and Sudanian vegetation regions (Zohary, 1973).

This paper adopts the plant taxonomy of *Flora Palaestina* (Zohary, 1966, 1972; Feinbrun-Dothan, 1978, 1986), and Eig's (1946) terminology for Mediterranean vegetation and stages of degradation. "Maquis" refers to the first step in degradation of the climax Mediterranean *Quercus calliprinos–Pistacia palestina* forest. "Garigue" implies

further alteration of the oak-pistachio forest with invasion of shrubs, for example, *Cistus* spp. "Batha" describes the final stage of degradation before the vegetation is reduced to steppe, and is characterized by the presence of *Sarcopoterium*. These stages of degradation from forest to steppe result from direct human disturbance or destruction by livestock.

The highlands of Edom above Petra (1400–1700 m) are covered with a remnant Mediterranean forest. This impoverished maquis or garigue of *Quercus calliprinos* and *Pistacia atlantica* represents the southern terminus of Mediterranean vegetation in southwestern Asia (Fig. 19.3). *Daphne linearifolia* and *Crategeus aronia* are common Mediterranean shrubs in this forest (Zohary, 1973). *Juniperus phoenicea* and *Pistacia atlantica* also grow in the hills of Edom on sandstone substrates. Irano-Turanian elements form the understory. *Artemisia herba-alba* is the dominant dwarf shrub, and *Centaurea damascena*, *Retama raetam*, and *Noaea mucronata* are common shrub species (Zohary, 1973). *Artemisia herba-alba* "is the first [shrub] to penetrate into Mediterranean arboreal communities of the borderland affected by man and to make up the undergrowth in these cleared woods" (Zohary, 1940a, p. 75). Isolated trees of *Cupressus sempervirens, Ceratonia siliqua*, and *Olea europaea* survive in the maquis of Edom.

Juniperus phoenicea forms an open steppe forest on Nubian Sandstone in the Petra region and surrounding hills (700–1400 m). Occasional trees of *Pistacia atlantica* and *Cratageus aronia* also can be found on these slopes. These trees attest to a past extension of Mediterranean forests in this region. Dwarf shrubs typical of Mediterranean bathas (e.g., *Sarcopoterium spinosum*) indicate a seral community of the Mediterranean arboreal climax vegetation (Zohary, 1982). *Thymelaea hirsuta, Ephedra campylopoda*, and *Ononis natrix* are common elements of the semi-steppe batha. Steppe elements include *Artemisia herba-alba, Hammada salicornia*, and *Anabasis articulata* (Fig. 19.4). *Noaea mucronata* and *Peganum harmala* invade ruined *Artemisia* steppes (Zohary, 1973).

Desert vegetation includes *Zygophyllum dumosum, Ephedra alte, Salsola tetrandra, Suaeda asphaltica, S. aegyptiaca*, and *Helianthemum* spp., with *Retama raetam* growing along wadi beds.

The vegetation of Petra includes a diverse array of plants. *Juniperus phoenicea*, the most common tree in the valley, occurs with the Mediterranean maquis species *Pistacia palestina*. In addition, the showy shrubs *Nerium oleander* and *Capparis cartilaginea* are common locally along drainages.

Ecology of the Syrian Rock Hyrax

The rock hyrax, *Procavia capensis* (order Hyracoidea), ranges throughout Africa, Arabia, Sinai, and the Syrian Desert from sea level to 4500 m elevation (Fig. 19.5; Walker, 1975). The subspecies *Procavia capensis syriaca* (Schreber) lives in Sinai, Jordan, Palestine, Lebanon, and Syria (Olds and Shoshani, 1982). The rock hyrax is a small, subungulate mammal, similar in size to a pika or a small marmot. Its closest evolutionary relative, however remote, is the elephant. It inhabits rocky terrain and favors steep mountains with numerous crevices and caves. The

hyrax's unusual feet have naked elastic pads on the ends of each digit that contract into suction cups and permit climbing on precipitous exposures (Harrison, 1968).

The gregarious rock hyrax lives in colonies that normally average about ten individuals but may include hundreds (Lynch, 1983). However, they are quite shy of humans and generally remain hidden from local inhabitants. The rock hyrax is primarily a diurnal browser or grazer that "will feed on almost any plant, preferring to feed near to the crevices" where it lives; "it can even ingest some [plants] that are poisonous to most other animals such as Solanaceae and Euphorbiaceae and leaves of the fig tree" (Harrison, 1968, p. 320). In Egypt and Sinai staple foods include *Acacia* leaves and pods and *Rumex vesicarius* (Osborn and Helmy, 1980). Hyraxes need very little water to survive and can travel more than 600 m to drink (Dorst, 1970). Their middens are made up of numerous plant parts, faunal remains, and fecal pellets. These deposits, which can be quite large, are hardened and preserved by hyracium, a substance in the animals' urine. Local Bedouin brew a tea from fragments of indurated hyrax middens that is purported to have medicinal value. Similar medicinal use of hyrax middens is reported in South Africa (Brain, 1981).

History of Human Settlement in the Petra Area

The cultural history of the Petra area extends back to the Paleolithic (Schyle and Uerpmann, 1988) and is marked by dramatic fluctuations in human population and subsistence practices. Natufian settlement at the ancient village of Beidha, 8 km north of Petra, began in the eighth millennium B.C. (Kirkbride, 1966; Byrd, 1988). While experimentation with potential domesticates may have started at this time, the earliest evidence for local plant and animal domestication comes from subsequent Pre-Pottery Neolithic deposits (about 7000–6500 B.C.) (Helbaek, 1966; Perkins, 1966). Recent excavations at the site of Basta (Gebel et al., 1988), southeast of Petra, augment this evidence from Beidha. Together these communities attest to a substantial local Neolithic population. Hyrax bones recovered from Neolithic Beidha suggest this animal's presence in the Petra area during the last nine

Fig. 19.4. Steppe batha overlooking the hills of Beidha and, in the distant background, the Wadi Araba.

thousand years of human settlement (Perkins, 1966; Kirkbride, 1984).

Deforestation could have begun very early in the Holocene. The large (12.5 ha maximum) Pre-Pottery Neolithic community of Ain Ghazal, near modern Amman, included numerous houses with large timbers and massive branches for roof supports, and very substantial plastered floors and walls (Rollefson, 1985; Banning and Byrd, 1984). An average house incorporated 3.3 metric tons of quicklime, and the firing of each ton of this material may have required up to 4 tons of fuelwood (G. O. Rollefson, unpublished data). The scale of this plastering points to a substantial human impact on the vegetation surrounding Ain Ghazal (Rollefson and Köhler-Rollefson, in press). Similar plaster construction at Beidha (Kirkbride, 1966) and Basta, as well as reliance on sheep and goat husbandry, could have initiated a long-term process of deforestation and overgrazing in the greater Petra environment.

Evidence of settlement in the Petra area between the Neolithic and Nabataean periods is sporadic and poorly understood. Late Iron Age settlements (around 800–600 B.C.) at Tawilan and atop the hill of Umm al-Biyara relate to the Edomites of the Old Testament (Bennett, 1964, 1966, 1971). Increasing local populations in this and subsequent periods most likely renewed the clearing of forests and grazing of livestock. The Bible provides historical reference to the hyrax at this time: "the high mountains are for the wild goat; the rocks are a refuge for the conies [hyraxes]" (Psalms 104:18).

The earliest historical account of the Nabataeans was authored by Diodorus of Sicily around 312 B.C. (Negev, 1986; Hammond, 1973). By this time Petra (ancient Reqem), the Nabataean capital, already had emerged as a commercial entrepôt between the Mediterranean basin, Arabia, and the Orient (Fig. 19.6). Accordingly, the local economy became tethered to the larger exchange system of the Ro-

Fig. 19.5. Procavia sp. (Photo by Roy Pinney;
Walker, 1983, p. 1143, photo B; reproduced with
permission of Johns Hopkins University Press.)

man Empire. The Nabataean Kingdom was
annexed formally by Rome to the province of
Arabia in A.D. 106, and it remained under Roman administration until the establishment
of the Byzantine Empire in A.D. 395. Under
Byzantine imperial rule the lands of the eastern Mediterranean experienced population
growth on a scale surpassed only in the twentieth century (Avi-Yonah, 1958, 1966; Broshi,
1979).

As the peripheral Nabataean Kingdom
flourished, Petra's population grew to several
thousand, despite the harsh desert environment of southern Jordan. The city and its gardens and orchards were supported by elaborate water-conservation methods (see Evenari
et al., 1971, on Nabataean hydraulic engineering). Rock-cut channels and terra cotta pipes
carried springwater and rainfall runoff to
cisterns chiseled out of the surrounding sandstone hills. Although Petra was a distant colo-
nial city in both the Roman and Byzantine

empires, the impact of its inhabitants on
the natural environment would have been
enormous.

Byzantine authority collapsed rapidly following the Battle of the Yarmouk in A.D. 636,
and subsequent Muslim administration of the
Levant was guided by very different commercial and political interests (Sharon, 1969).
Southern Jordan was no longer a nexus for
trade with the Mediterranean, nor was it a
border area worthy of government subsidy.
Now denied larger imperial investment,
Petra's inhabitants became increasingly re-
liant on sheep and goat pastoralism and local
resources. From the seventh century to the
present the Petra area was characterized by a
diminished, less-sedentary population, and
intensive grazing continued its impact on the
landscape. Modern technology induced a very
pronounced episode of deforestation during
the First World War, when the Ottoman government decimated the remnant woodlands

Fig. 19.6. The Valley of Petra, including the paved main street of the Roman city, the remains of the Nabataean temple Qasr al-Bint Faroun, and the mountain Umm al-Biyara in the background.

of Edom for fuelwood to be used by the Hejaz Railway (Mousa, 1982).

METHODS

In 1985 we found the hardened nests of hyraxes on the more remote buttes around Petra (Fig. 19.7). A modern midden (Petra 1) was readily accessible along the Siq, a narrow rock gorge that forms the eastern entrance to the valley. We recovered eight indurated middens from crevices in the mountain of Amm Seseban (Fig. 19.8) with the help of our Bedouin hosts. Three of these fossil middens (Petra 2, Petra 6, and Petra 9) and the un-indurated modern midden (Petra 1) and one surface soil sample from Amm Seseban have been analyzed.

The deposits from Petra are attributed to hyraxes for three reasons. First, Bedouin informants verified that the deposits were made by hyraxes. In addition, fecal pellets from the Jordanian middens were comparable in size and shape to modern fecal pellets from South African hyraxes (provided by Louis Scott). Finally, radioimmunoassay analysis showed that proteins in crystallized urine from Petra 2 reacted with anti-hyrax serum (W. E. Rainey and J. M. Lowenstein, personal communication, 1987).

Laboratory Methods

The outer rinds of the indurated hyrax middens were scraped and brushed to remove possible contamination. Samples of each midden, weighing 200–250 g, were soaked in a solution of water and dilute HC1 to disaggregate the midden and free the plant fragments. After soaking, the plant fragments and sediments were washed through a 1-mm mesh screen. Material remaining on the screen was dried and sorted for plant fragments, seeds, faunal remains, and insects. Samples of fecal pellets, weighing approximately 8.5 g each, were used for radiocarbon analyses.

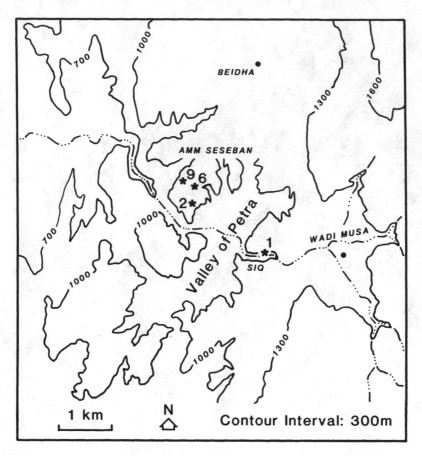

Fig. 19.7. Midden localities in the Petra study area.

Multiple pollen samples from each indu- rated midden (10–30 g) were soaked in water to dissolve the cementing matrix, sieved to remove larger plant fragments, and extracted using standard palynological techniques (Faegri and Iversen, 1975). The entire loose modern midden was soaked in water. The re- sulting liquid and sediment were concen- trated and processed for pollen analysis. Unknown and unidentifiable pollen grains are included in the total pollen sums. Tablets of *Lycopodium* spores (Lund Batch no. 414831) added to each sample allowed calculation of pollen concentrations.

Assumptions

Several assumptions have been made in this initial research on fossil hyrax middens from the Middle East: (1) hyraxes randomly sample the vegetation near their dens; (2) plant frag- ments and pollen from hyrax deposits can be used to characterize the vegetation near the den at the time the midden was created; (3) relative pollen frequencies reflect pri- marily "extralocal" or "regional" vegetation and do not result simply from variability within the deposits; and (4) arboreal pollen deposition in soil samples is generally compa-

Fig. 19.8. Rock-cut Nabataean tomb, ed-Deir, on the south face of Amm Seseban. Tomb facade stands 40 m high. Arrow indicates rock crevice from which midden #2 was recovered.

rable to arboreal pollen deposition in hyrax middens.

In spite of twenty-five years of packrat midden analyses in North America, neither midden macrofossil assemblages nor pollen spectra have yet been calibrated to modern vegetation. Thompson (1985) suggests that pollen in packrat middens represents the surrounding vegetation in a broad sense but may be skewed by selective gathering. Although Thompson has shown midden pollen to be highly variable, he concludes that it may accurately reflect changes "of the magnitude of a shift from woodland to desertscrub" (Thompson, 1985, p. 103). In this initial study we limit our use of hyrax midden pollen to this context. For discussion of vegetation surrounding hyrax middens we adopt Jacobson and Bradshaw's (1981) definitions of "local" (within 20 m), "extralocal" (20 m to hundreds of meters), and "regional" (greater than several hundred meters) vegetation.

RESULTS AND DISCUSSION

Hyrax middens from Jordan appear to be analogous to packrat middens in western North America in composition and paleobotanical potential. Indurated deposits containing abundant pollen (Table 19.1) and plant macrofossils (30–100 taxa per midden) have been preserved in the dry caves of Petra for at least 1800 years. Midden pollen is well preserved in concentrations of 20,000–100,000 grains/g dry weight in the fossil middens, and 4500 grains/g in the modern, unindurated midden. Pollen concentrations are comparable to those from packrat middens (Thompson, 1985) and dassie rat middens (Scott, *this volume*, chap. 18).

Radiocarbon Dating

Radiocarbon ages ranging from 1750 B.P. to 1120 B.P. place the indurated middens in periods of distinctly different local economy and land use (see Table 19.2). Calibration to ca-

Table 19.1. *Pollen taxa in hyrax middens from southern Jordan.*

Trees	Shrubs	Herbs
Pinus	*Nerium*	*Alkanna*-type
Juniperus	Cistaceae	*Anchusa*-type
Quercus	*Artemisia*	Caryophyllaceae
Pistacia	*Ephedra*	Chenopodiineae
Ziziphus	*Poterium*-type	*Ambrosia*-type
Olea/Phillyrea	Thymelaeaceae	Anthemidae
Celtis	*Capparis*	Liguliflorae
Acacia	*Salix*	Tubuliflorae
Tamarix	Rhamnaceae	*Centaurea*-type
*Eucalyptus	*Prunus*-type	*Carduus*-type
*Rutaceae	*Rhus*	*Echinops*
		Cruciferae
		Cyperaceae
		Gramineae
		Liliaceae
		Dipsaceae
		Papilionaceae
		Plantago
		Polygala
		Rubiaceae
		Umbelliferae
		Urtica
		Typha-Sparganium
		Solanaceae
		Labiatae
		Rumex
		Thalictrum

*Introduced taxa

lendric years (based on Pearson, 1983) suggests that Petra 6 and Petra 9 might date to the Roman and Byzantine periods, respectively. However, overlapping 95% confidence intervals show that the ages of these middens are not significantly different. Therefore, middens 6 and 9 are both interpreted with reference to Petra's Roman/Byzantine imperial affluence, while midden 2 reflects the city's postimperial decline, and midden 1 serves as a modern analog for comparison of past and present plant material. Petra 1 is considered modern because it contains pollen from *Eucalyptus*, a tree introduced in the twentieth century.

Variability of Pollen Spectra within Middens

Rather than assuming that pollen deposition is homogeneous within hyrax middens, we compared pollen spectra from subsamples of each fossil midden (Petra 2A, 2B, 2C; Petra 6A, 6B, 6C; Petra 9A, 9B). Ninety-five percent confidence intervals are calculated for each taxon in each subsample, and for each midden as a whole, using Mosimann's (1965) equation for pollen data and nomograms published by Maher (1972). Significant pollen frequency differences within middens (i.e., between subsamples in a midden and between subsamples and pooled data for the midden as a whole) are identified in pairwise comparisons using Student's t. Mutually exclusive (i.e., nonoverlapping) confidence intervals indicate variability in pollen deposition within a midden that is significant at $p < .05$ (Table 19.3).

Homogeneous frequencies of some taxa and heterogeneous frequencies of others may indicate different means of pollen deposition and reflect vegetation at varying distances from the hyrax nests. Pronounced heterogeneity in Papilionaceae, Umbelliferae, Urticaceae, Boraginaceae, and *Ephedra* fre-

Table 19.2. *Radiocarbon ages of hyrax middens from southern Jordan.*

Midden #	Radiocarbon Years B.P.	Calendar Years A.D. (1 Sigma)
Petra 2	1120 ± 110 (A-4529)	790–1020
Petra 9	1580 ± 65 (GX-13060)	420–550
Petra 6	1750 ± 60 (GX-13059)	160–390

Table 19.3. *Number of significant differences (at p < .05) in pairwise comparisons between subsamples and pooled data for each midden. Maximum number of significant differences is 3.*

Taxa	Midden 6	Midden 9	Midden 2
Pistacia	—	—	—
Juniperus	1	—	—
Olea	—	—	1
Ephedra	1	—	—
Tubuliflorae	—	—	—
Gramineae	—	—	—
Liliaceae	—	—	—
Chenopodiineae	—	—	—
Thymelaeaceae	—	—	1
Artemisia	—	—	2
Papilionaceae	2	1	—
Umbelliferae	—	—	3
Urticaceae	—	—	2
Boraginaceae	1	—	1

quencies (Fig. 19.9) may indicate localized concentrations of flowers within these middens and, therefore, hyrax preferences for food or nesting material in the local vegetation. In contrast, relatively consistent percentages of the anemophilous taxa *Juniperus, Pistacia, Olea,* Tubuliflorae, Gramineae, and Chenopodiineae may constitute a "background" pollen rain that may reflect extralocal and regional vegetation and is somewhat less biased by local hyrax collecting behavior.

Comparisons of Pollen Spectra between Middens

Comparison of pollen frequencies between middens reveals past vegetation changes in the Petra area. Figures 19.10–19.13 present 95% confidence intervals for seventeen pollen taxa in four middens. Means and confidence intervals were calculated using pooled data from all subsamples within each midden. Total arboreal pollen (i.e., *Juniperus, Pistacia, Olea, Quercus, Pinus,* and *Ziziphus*) occurs at very similar frequencies in the Roman/Byzantine middens (Petra 6 and Petra 9), and in the Islamic and modern middens (Petra 2 and Petra 1, respectively). However, a drastic decline in arboreal pollen marks the aftermath of imperial city life in Petra (Fig. 19.10). When compared to Petra 6 or Petra 9, middens 2 and 1 reflect significant decreases in specific taxa: *Pistacia, Juniperus,* and *Olea.* This pervasive decrease in arboreal pollen may be the legacy of several centuries of Roman/Byzantine wood harvesting to supply a substantial urban population with building material and fuel. The drop in pollen from *Olea,* an important cultivated tree, also may implicate reduced maintenance and eventual abandonment of orchards and complex irrigation after the seventh century A.D.

Below we compare these data with modern pollen data, and hypothesize that the uplands of Edom were covered with a Mediterranean forest-steppe—a garigue—from approximately A.D. 160 to A.D. 550. Subsequently this vegetation was reduced to steppe or batha, with only remnants of the fomer Mediterranean forests in the mountains.

Pollen from shrubs and herbs also diminishes through the sequence of middens (Fig. 19.11). This vegetation, preferred by grazing and browsing animals, includes grasses, the woody shrubs *Artemisia* and *Ephedra,* and species in the subfamilies Papilionaceae and Tubuliflorae. The data for *Artemisia,* Tubuliflorae, and Gramineae are assumed to reflect background vegetation, and Papilionaceae and *Ephedra* may provide at least a partial background signal. Thus a second component of Petra's natural vegetation was decimated over a span of centuries, probably by widespread foraging of domestic livestock.

Figures 19.12 and 19.13 include pollen taxa that show significant increases after the Roman and Byzantine periods. Taxa that may have benefited from disturbance are Liliaceae, Umbelliferae, Urticaceae, Chenopodiineae, and Thymelaeaceae (Fig. 19.12). The extremely high frequencies of Liguliflorae,

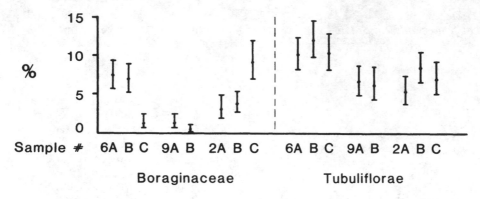

Fig. 19.9. Selected pollen taxa that illustrate variable (e.g., Boraginaceae) and relatively consistent (e.g., Tubuliflorae) frequencies within middens 2, 6, and 9.

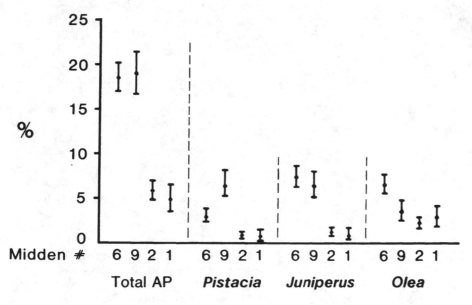

Fig. 19.10. Arboreal pollen (AP) percentages for each of the four middens studied showing means and 95% confidence intervals. Total arboreal pollen includes *Juniperus, Pistacia, Olea, Quercus, Pinus,* and *Ziziphus.*

Boraginaceae, and *Capparis* in the modern midden (Fig. 19.13) are assumed to reflect anomalous hyrax collections of flowers from these taxa.

Comparison of Pollen from Hyrax Middens with Modern Surface Samples

Because palynological analysis of the Holo-cene has been limited in the southern Levant, we take the liberty here of comparing pollen spectra from hyrax middens in southern Jordan with modern pollen data from surface samples collected in Syria and Lebanon (Table 19.1; Bottema and Barkoudah, 1979). Our intent is to characterize the vegetation palynologically in the broadest sense. There-

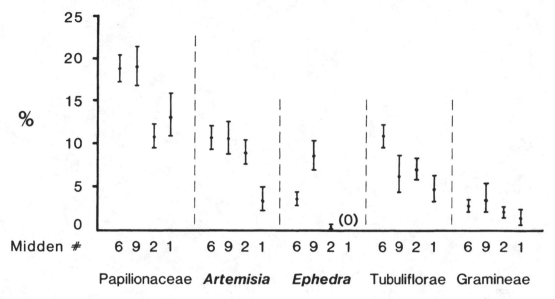

Fig. 19.11. Mean shrub and herb pollen percentages and 95% confidence intervals. Papilionaceare includes three taxa.

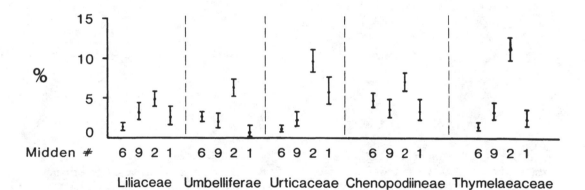

Fig. 19.12. Disturbance pollen taxa that show significant increases in middens 2 and 1. Means and 95% confidence intervals are shown.

fore we offer only general comparisons between pollen spectra from major plant zones (e.g., Mediterranean forest, steppe-forest, and steppe). Many of the same species and virtually all of the most abundant genera are common to both regions (southern Jordan and Syria-Lebanon).

Bottema and Barkoudah (1979) present pollen data for eighty-five modern surface samples according to vegetation zones, which include several variations of Mediterranean forests, degraded Mediterranean forest, steppe-forest, forest-steppe, and steppe and desert-steppe. The most abundant arboreal pollen types in each of these broad zones include *Quercus* (two types), *Pistacia*, *Olea*, *Pinus*, and *Juniperus*, the same genera used to calculate the total arboreal pollen sums for the Petra hyrax middens. The averaged arboreal pollen frequencies in modern surface samples from Syria and Lebanon are compared with arboreal pollen percentages from

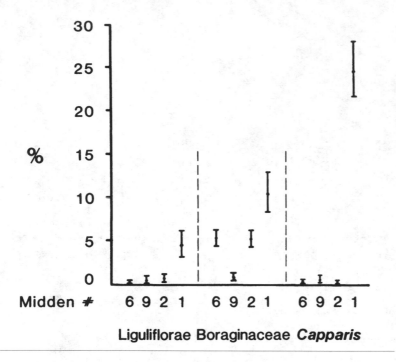

Fig. 19.13. Pollen taxa that show significantly different percentages in the modern midden. Means and 95% confidence intervals are shown.

middens in southern Jordan. We assume that data for these pollen types in the Petra middens represent extralocal vegetation and are not skewed significantly by the activities of hyraxes.

Bottema and Barkoudah (1979) found that total arboreal pollen percentages ranged from 40% to 85% in the Mediterranean forest zones. (Unusually high *Pinus* pollen percentages in one of their zones account for the value of 85% arboreal pollen.) Average arboreal pollen is about 25% in "degraded" Mediterranean forests, and 22% and 18% in the steppe-forest and forest-steppe zones, respectively; arboreal pollen averages 10% or less in the steppe and desert-steppe zones (Bottema and Barkoudah, 1979).

Based on these modern analogs, total arboreal pollen percentages of 17% and 19% in

hyrax middens 6 and 9 suggest that the vegetation during the Roman and Byzantine occupation of Petra was a degraded Mediterranean forest or forest-steppe. Total arboreal pollen drops to 6% in midden 2, dated approximately between A.D. 790 and A.D. 1020, and to 5% in the modern midden. Similarly, a modern surface pollen sample collected from Amm Seseban (1100 m) contains less than 9% arboreal pollen (*Juniperus* accounts for almost 7%). These middens, which postdate A.D. 550, are most analogous to surface samples from steppe or desert-steppe vegetation in Syria and Lebanon. Thus we suggest that intense utilization of the environment by the people of Petra, and increased pastoralism (promoting intensive browsing by goats) following the collapse of Byzantine rule further reduced the vegetation around Petra to the denuded steppe or batha of today.

Modern vegetation on the hill slopes above Petra clearly represents a ruined Mediterranean oak-pistachio forest. Steppe elements have invaded and now form the understory of this batha. Closely trimmed lower tree branches suggest the continued heavy impact of browsing goats. Understory taxa are heavily grazed, permitting few seedlings to survive in this denuded environment. In Petra trees cling to cliffs and crevices, away from agile ungulates, while toxic species such as *Nerium oleander* and *Capparis cartilaginea* flourish along dry stream beds and in seeps in the sandstone walls.

Comparison of Pollen and Plant Macrofossil Abundances

We also compare relative abundances of pollen and plant macrofossils in each midden. Counts of plant fragments (e.g., seeds, twigs, leaves, etc.) are ranked according to a six-part scale modified from Van Devender (1973): 0 = absent; 1 = rare (1); 2 = uncommon (2–20); 3 = common (20–30); 4 = abundant (30–100); and 5 = very abundant (> 100). To facilitate comparison of pollen and plant macrofossil data we also rank pollen frequencies on a 0-to-5 abundance scale (Thompson, 1985). This scale is based on natural breaks in the distribution of pollen percentages for the entire data set, as follows: 0 = 0%; 1 = 0.1%–0.9%; 2 = 1.0%–2.9%; 3 = 3.0%–5.9%; 4 = 6.0%–9.9%; and 5 = 10% or greater.

Three taxa are most abundant in both pollen and plant macrofossils: *Juniperus*, *Ephedra*, and Thymelaeaceae (plant macrofossils are from *Thymelaea hirsuta*) (Fig. 19.14). We infer that hyraxes favor these taxa for food or nest building. The relative abundances of *Juniperus* and *Ephedra* pollen and plant macrofossils decrease from the oldest midden (Petra 6) to the modern midden (Petra 1), perhaps indicating reduced availability of these plants. In contrast, Thymelaeaceae pollen and macrofossils increase in middens 9 and 2, then drop off in the modern midden. Both pollen and plant macrofossil data for these taxa reflect parallel trends over time and may be tracking the same vegetation signal.

Because of limited reference material from

Jordan, the analysis of plant macrofossils is not yet complete. However, a few points relevant to this discussion can be made. *Ficus* seeds are found in all middens and are particularly abundant (100 or greater) in Petra 2 and Petra 1. No pollen of this cleistogamous taxon was found. *Ficus carica*, the domesticated fig, is cultivated widely in the region and common in the canyons of Petra. An old *F. carica* tree grows about 100 m from the Petra 1 locality, which probably accounts for abundance of *Ficus* seeds in this midden. A single *Vitis vinifera* seed was found in Petra 2, but no *Vitis* pollen was found in any of the samples. Grape also has been an important cultigen in the Middle East since the Early Bronze Age (D. Zohary and Spiegel-Roy, 1975).

No plant fragments of *Pinus*, *Quercus*, or *Ziziphus* were found, and only one seed of *Olea europaea* and a few Anacardiaceae (cf. *Pistacia*) seeds were found in the analyzed hyrax middens. As mentioned above, *Juniperus* macrofossils are common in the hyrax middens, and pollen deposition from this tree may be, in part, influenced by the hyrax. However, we suggest that pollen from the remaining arboreal taxa does not directly reflect hyrax collecting behavior and most likely results from wind transport to the middens. While further identification of plant fragments in the middens may eventually detail the composition of local plant communities in Petra, the conclusions we reach in this chapter rest primarily upon changes in arboreal pollen percentages.

Comparisons between Neotoma and Procavia Middens

Given a suitable environment for preservation, there is every reason to anticipate that hyrax middens can provide a fossil record of vegetation in the Middle East much as packrat middens have in the deserts of North America. Although the rock hyrax (order Hyracoidea) and packrat (order Rodentia) are taxonomically quite distinct, similarities between the floras of western North America and the Middle East are reflected by the taxa found in their middens. For example, both hyrax and packrat middens contain abundant *Juniperus* macrofossils. *Juniperus* seeds and twigs are preferred by at least one species

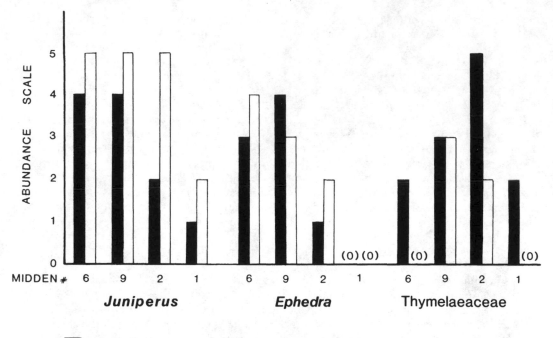

POLLEN ABUNDANCE
MACROFOSSIL ABUNDANCE

Fig. 19.14. Comparison between pollen and plant macrofossil abundances for each of the four middens. The three taxa shown are the most abundant types.

of *Neotoma* for food and nesting material (Vaughan, 1980). *Artemisia* pollen, common in packrat middens (Cole, 1981), also is abundant (5% – 15%) in the Petra hyrax middens.

While the number of taxa represented in hyrax middens is similar to that of packrat deposits, abundances of common pollen taxa differ widely. *Neotoma* middens usually contain relatively high percentages (20% to > 50%) of major pollen taxa (e.g., *Pinus, Juniperus*, Chenopodiineae, and Tubuliflorae [Thompson, 1985]). In contrast, *Procavia* middens have very few pollen taxa in frequencies greater than 10% (e.g., *Artemisia* and *Juniperus*), and none that exceed 15% of the total pollen sum (except for *Capparis* in the modern midden). On a related note, our comparison of pollen data from *Procavia* middens in Jordan and from a midden attributed to Procaviidae in South Africa (Scott, *this volume*, chap. 18) shows very few shared plant taxa (with the exception of nonspecific types such as Gramineae and Compositae), presumably due to the very dissimilar floras of the two regions.

CONCLUSIONS

This study presents the first paleobotanical study of both plant macrofossils and pollen from fossil hyrax middens. The discussions above compare hyrax midden pollen with pollen from modern surface samples, examine variability within and between middens, and compare relative abundances of pollen and plant macrofossils between middens. These data suggest a plausible explanation for the demise of arboreal vegetation in the Petra area. We hypothesize that human disturbance degraded the Mediterranean forests of Edom to a maquis or garigue prior to the second century A.D. Intensified browsing by livestock and continued human impact further reduced this vegetation to a Mediterranean batha or steppe following the collapse of Byzantine town life at Petra.

We propose that hyrax middens are sources of paleobotanical data generally analogous to the packrat middens of the North American deserts. Although taphonomic questions regarding pollen and plant macrofossils in both hyrax and packrat middens must be addressed in greater detail, we argue that plant fragments in hyrax middens most likely reflect local vegetation; and although hyrax midden pollen is quite variable, at least some pollen taxa are derived primarily from extralocal and regional vegetation. Specifically, we have used pollen data to develop hypotheses concerning the legacy of changing plant communities and human exploitation of the natural environment in southern Jordan. More generally, this study introduces hyrax midden analysis as a means to help unravel the history of vegetation in southwestern Asia and the human impact it has endured over millennia.

ACKNOWLEDGMENTS

Many individuals have been instrumental in this study. We thank Dakhilallah Qublan and Harun Dakhilallah for guiding us to the hyrax deposits in Petra, and Hal Lindquist for financial support and field assistance. Austin Long, University of Arizona Radiocarbon Laboratory, provided one radiocarbon date; Timothy Linick, National Science Foundation Regional Accelerator Facility for Isotope Dating, University of Arizona, calibrated the dates; and William E. Rainey and J. M. Lowenstein, University of California, Berkeley, conducted the immunological analysis of the hyracium. The Department of Antiquities of Jordan issued us a permit to collect middens in Petra, and the American Center of Oriental Research provided a jeep and research facilities in Amman. We thank all these individuals and organizations for their kind assistance.

Pollen identifications could not have been made without the generous assistance of Elise Van Campo and Madeleine Van Campo at the Laboratoire de Palynologie, Montpellier, and Raymonde Bonnefille and Guy Riollet at the Laboratoire de Geologie du Quaternaire, Marseille. We thank Charles Mason and Rebecca Van Devender at the University of Arizona Herbarium for their help and use of facilities. Many people aided us with identifications of plant fragments. These include Thomas Van Devender, Tony Burgess, Karen Adams, Richard Felger, and Charles Miksicek. Julio Betancourt and Mary Kay O'Rourke provided editorial comments.

REFERENCES

Al-Eisawi, D. M. (1984). "Vegetation in Jordan." In *Studies in the history and archaeology of Jordan*, vol. 2. A. Hadidi, ed., pp. 45–57. Amman: Department of Antiquities, Jordan.

Avi-Yonah, M. (1958). "The economics of Byzantine Palestine." *Israel Exploration Journal* 8(1), 39–51.

———. (1966). *The Holy Land from the Persian to the Arab conquests (536 B.C. to A.D. 640). A historical geography.* Grand Rapids: Baker Book House.

Banning, E. B., and Byrd, B. F. (1984). "The architecture of PPNB Ain Ghazal, Jordan." *Bulletin of the American Schools of Oriental Research* 255, 15–20.

Bennett, C.-M. (1964). "Chronique archeologique: Umm el-Biyara." *Revue biblique* 71, 250–53.

———. (1966). "Fouilles d'Umm el-Biyara: Rapport preliminaire." *Revue biblique* 73, 372–403.

———. (1971). "A brief note on excavations at Tawilan, Jordan, 1968–1970." *Levant* 3, v–vii.

Betancourt, J. L., and Van Devender, T. R. (1981). "Holocene vegetation in Chaco Canyon, New Mexico." *Science* 214, 656–58.

Bottema, S., and Barkoudah, Y. (1979). "Modern pollen precipitation in Syria and Lebanon and its relation to vegetation." *Pollen et Spores* 21(4), 427–80.

Brain, C. K. (1981). *The hunters or the hunted? An introduction to African cave taphonomy.* Chicago: University of Chicago Press.

Broshi, M. (1979). "The population of western Palestine in the Roman-Byzantine period." *Bulletin of the American Schools of Oriental Research* 236, 1–10.

Byrd, B. F. (1988). "The Natufian of Beidha. Report on renewed field research." In *The prehistory of Jordan. The state of research in 1986.* A. N. Garrard and H. G. Gebel, eds., pp. 175–97. Oxford: British Archaeological Reports, International Series, vol. 396.

Cole, K. L. (1981). "Late Quaternary environments in the eastern Grand Canyon: Vegetational gradients over the last 25,000 years." Ph.D. diss., University of Arizona, Tucson.

Dorst, J. (1970). *A field guide to larger mammals of Africa.* Boston: Houghton Mifflin.

Eig, A. (1946). "Synopsis of the phytosociological units of Palestine." *Palestine Journal of Botany* 3(4), 183–247.

Evenari, M.; Shanan, L.; Tadmor, N.; and Itzhaki, Y. (1971). *The Negev: The challenge of a desert.* Cambridge: Harvard University Press.

Faegri, K., and Iversen, J. (1975). *Textbook of pollen analysis.* 3d ed. New York: Hafner.

Feinbrun-Dothan, N. (1978). *Flora Palaestina.* Part 3, *Ericaceae to Compositae.* Jerusalem: Israel Academy of Sciences and Humanities, Jerusalem

Academic Press.

———. (1986). *Flora Palaestina*. Part 4, *Alismataceae to Orchidaceae*. Benei Beraq: Israel Academy of Sciences and Humanities, Monoline Press.

Gebel, H. G.; Muheisen, M. S.; Nissen, H. J.; Qadi, N.; and Starck, J. M. (1988). "Preliminary report on the first season of excavations at the Late Aceramic Neolithic site of Basta." In *The prehistory of Jordan: The state of research in 1986*. A. N. Garrard and G. G. Gebel, eds., pp. 101–34. Oxford: British Archaeological Reports, International Series, vol. 396.

Green, N.; Caldwell, J.; Hope, J.; and Luly, J. (1983). "Pollen from an 1,800 year old stick-nest rat (*Leporillus* sp.) midden from Gualta, Western New South Wales." *Quaternary Australasia* 1, 31–41.

Hammond, P. C. (1973). *The Nabataeans—Their history, culture, and archaeology*. Volume 37 of *Studies in Mediterranean Archaeology*. Gothenburg, Sweden: Paul Aströms.

Harrison, D. L. (1968). *The Mammals of Arabia*, vol. 2. London: Ernest Benn.

Helbaek, H. (1966). "Pre-Pottery Neolithic farming at Beidha." *Palestine Exploration Quarterly*, 61–66.

Jacobson, G. L., and Bradshaw, R. H. W. (1981). "The selection of sites for paleovegetational studies." *Quaternary Research* 16, 80–96.

Kirkbride, D. (1966). "Five seasons at the Pre-Pottery Neolithic village of Beidha in Jordan." *Palestine Exploration Quarterly*, 8–61.

———. (1984). "The environment of the Petra region during the Pre-Pottery Neolithic." In *Studies in the history and archaeology of Jordan*, vol. 2. A. Hadidi, ed., pp. 117–24. Amman: Department of Antiquities, Jordan.

Lynch, C. D. (1983). *The mammals of the Orange Free State*. Memoirs Van Die Nasionale Museum Bloemfontein, no. 18.

Maher, L. J., Jr. (1972). "Nomograms for computing 0.95 confidence limits of pollen data." *Review of Palaeobotany and Palynology* 13, 85–93.

Mosimann, J. E. (1965). "Statistical methods for the pollen analyst: Multinomial and negative multinomial techniques." In *Handbook of paleontological techniques*. B. Kummel and D. Raup, eds., pp. 636–73. San Francisco: Freeman.

Mousa, S. (1982). "Jordan: Towards the end of the Ottoman Empire, 1841–1918." In *Studies in the history and archaeology of Jordan*, vol. 1. A. Hadidi, ed., pp. 385–91. Amman: Department of Antiquities, Jordan.

Negev, A. (1986). *Nabatean archaeology today*. New York: New York University Press.

Olds, N., and Shoshani, J. (1982). "*Procavia capensis*." *Mammalian Species* 171, 1–7.

Osborn, D. J., and Helmy, I. (1980). *The contempo-rary land mammals of Egypt (including Sinai)*. Fieldiana Zoology, new series, no. 5. Chicago: Field Museum of Natural History.

Pearson, G. W. (1983). "The development of high precision ¹⁴C measurement and its application to archaeological time scale problems." Ph.D. diss., Queen's University, Belfast.

Perkins, D., Jr. (1966). "The fauna from Madamagh and Beidha." *Palestine Exploration Quarterly*, 66–67.

Pons, A., and Quezel, P. (1958). "Premières remarques sur l'étude palynologique d'un guano fossile du Hoggar." *Comptes rendus des séances de l'Académie des Sciences* 244, 2290–92.

Raikes, R. L. (1966). "Beidha, prehistoric climate and water supply." *Palestine Exploration Quarterly*, 68–72.

———. (1984). "The character of the Wadi Araba." In *Studies in the history and archaeology of Jordan*, vol. 2. A. Hadidi, ed., pp. 95–101. Amman: Department of Antiquities, Jordan.

Rollefson, G. O. (1985). "The 1983 season at the early Neolithic site of Ain Ghazal." *National Geographic Research* 1(1), 44–62.

Rollefson, G. O., and Köhler-Rollefson, I. (In press). "The collapse of PPNB settlements in the Southern Levant." In *Proceedings of the Second Symposium on Upper Paleolithic, Epipaleolithic, and Neolithic Populations in Europe and the Mediterranean Basin*.

Schyle, D., and Uerpmann, H.-P. (1988). "Palaeolithic sites in the Petra area." In *The prehistory of Jordan. The state of research in 1986*. A. N. Garrard and H. G. Gebel, eds., pp. 39–65. Oxford: British Archaeological Reports, International Series, vol. 396.

Sharon, M. (1969). "The history of Palestine from the Arab conquest until the Crusades (A.D. 633–1099)." In *A history of the Holy Land*. M. Avi-Yonah, ed., pp. 185–222. Toronto: MacMillan.

Spaulding, W. G.; Leopold, E. B.; and Van Devender, T. R. (1983). "Late Wisconsin paleoecology of the American Southwest." In *The late Pleistocene of the United States*. S. C. Porter, ed., pp. 259–92. Minneapolis: University of Minnesota Press.

Thompson, R. S. (1984). "Late Pleistocene and Holocene environments in the Great Basin." Ph.D. diss., University of Arizona, Tucson.

———. (1985). "Palynology and *Neotoma* middens." In *Late Quaternary vegetation and climates of the American Southwest*. B. F. Jacobs, P. L. Fall, and O. K. Davis, eds., pp. 89–112. American Association of Stratigraphic Palynologists, Contribution Series, 16.

Van Devender, T. R. (1973). "Late Pleistocene plants and animals of the Sonoran Desert: A survey of ancient packrat middens in southwestern Arizona." Ph.D. diss., University of Arizona, Tucson.

Vaughan, T. A. (1980). "Woodrats and picturesque juniper." In *The aspects of vertebrate history*. L. L. Jacobs, ed., pp. 387–402. Flagstaff: Museum of Northern Arizona Press.

Walker, E. P. (1964). *Mammals of the world*, vol. 2, 3rd edition. Baltimore: Johns Hopkins University Press.

Wells, P. V., and Jorgensen, C. D. (1964). "Pleistocene wood rat middens and climatic change in Mohave Desert: A record of juniper woodlands." *Science* 143, 1171–74.

Zohary, D., and Spiegel-Roy, P. (1975). "Beginning of fruit growing in the Old World." *Science* 187, 319–27.

Zohary, M. (1940a). "Geobotanical analysis of the Syrian Desert." *Palestine Journal of Botany* (Jerusalem series) 2, 46–96.

———. (1940b). "Forests and forest remnants of *Pistacia atlantica* Desf. in Palestine and Syria." *Palestine Journal of Botany* (R. series) 3(1–2), 156–61.

———. (1960). "The maquis of *Quercus calliprinos* in Israel and Jordan." *Bulletin of the Research Council of Israel* 90, 51–72.

———. (1966). *Flora Palestina*. Part 1, *Equisetaceae to Moringaceae*. Jerusalem: Israel Academy of Sciences and Humanities, Jerusalem Academic Press.

———. (1972). *Flora Palestina*. Part 2, *Platanaceae to Umbelliferae*. Jerusalem: Israel Academy of Sciences and Humanities, Goldberg's Press.

———. (1973). *Geobotanical foundations of the Middle East*. Stuttgart: Gustav Fischer Verlag.

———. (1982). *Vegetation of Israel and adjacent areas*. Beihefte zum Tubinger Atlas des Vorderen Orients. Wiesbaden: Dr. Ludwig Reichert Verlag.

Chapter 20

Analysis of Stick-Nest Rat (*Leporillus*: Muridae) Middens from Central Australia

Desmond J. Nelson   Robert H. Webb   Austin Long

INTRODUCTION

The analysis of fossil middens deposited by animals other than packrats (Cricetidae: *Neotoma*) provides a means for evaluating biotic changes on other continents. Analyses of hyrax (Procaviidae) and dassie rat (Petromuridae) middens in Africa (Scott, *this volume*, chap. 18) and hyrax middens in the Middle East (Fall et al., *this volume*, chap. 19) demonstrate the enduring preservation of organic deposits accumulated by arid-region mammals. Here we report preliminary analyses of stick-nest rat (Muridae: *Leporillus*) middens in Finke Gorge National Park, Northern Territory, central Australia (Fig. 20.1).

Stick-nest rats are about the size of packrats but have a rabbitlike appearance with blunt noses and large ears (Fig. 20.2). Both species of stick-nest rats (*Leporillus conditor* and *L. apicalis*) were extirpated from continental Australia in the 1930s, although the greater stick-nest rat (*L. conditor*) is extant on Franklin Island off the south coast (Troughton, 1967; Strahan, 1985). The lesser stick-nest rat (*L. apicalis*; Fig. 20.2), also called the white-tailed stick-nest rat, is believed to be extinct (Troughton, 1967). The cause of the extinction is unknown but is believed to have resulted from a combination of domestic livestock grazing, rabbit plagues, and the introduction of European foxes and domestic cats

into Australia (Newsome and Corbett, 1975; Strahan, 1983; Troughton, 1967). The two species were sympatric over most of their ranges, but only *L. apicalis* lived in the vicinity of Finke Gorge National Park.

Stick-nest rats build characteristic shelters in caves, rock shelters, or crevices as well as in the open (Walker, 1975; Watts and Eves, 1976). One shelter found in Finke Gorge National Park shows the characteristic external shell of sticks (Fig. 20.3) that covers an internal core midden of grasses, small sticks, and herbaceous material cemented with urine and feces. Unlike packrats, pairs or colonies of stick-nest rats may inhabit a single shelter (Watts and Eves, 1976).

Chemical analyses of probable fossil stick-nest rat middens were reported by Maitland in 1905 (Bridge, 1975). The paleoecological significance of stick-nest rat middens was first reported by Green et al. (1983), who extracted fossil pollen and chemically analyzed a midden radiocarbon dated at 1840 ± 100 B.P. from New South Wales. They found that the midden contained 30% soluble organic material, 65% inorganic minerals, and only 5% insoluble organic materials that might contain paleoecological information. Of nineteen identifiable pollen types preserved in the midden, only one type was from a genus not

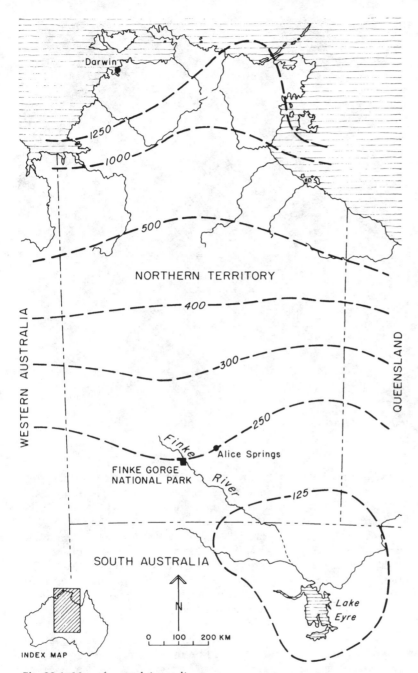

Fig. 20.1. Map of central Australia.

present within several kilometers of the site. Green et al. (1983) did not analyze macrofossils from the midden they collected.

METHODS

A large (greater than 20 kg) stick-nest rat midden and several other middens built by unidentified species were collected from an outcrop of Hermannsburg Sandstone at the Amphitheatre in Finke Gorge National Park. Because the smaller middens were composed almost entirely of soluble, concentrated urine,

Fig. 20.2. Photograph of *Leporillus conditor* (courtesy of Steve Doyle).

only the large midden was examined further. This midden lacked a woody exterior shell, although nearby deposits (Fig. 20.3) were covered with a blanket of sticks. Protein analysis of the urine indicated a good reaction with Muridae and Cricetidae antisera (W. E. Rainey and J. M. Lowenstein, University of California at Berkeley, personal communication, 1986), which eliminates a marsupial species as the builder.

The large midden (60 × 20 × 20 cm) was so indurated that a gasoline-powered saw normally used to cut metal was required to bisect it. One-half of the midden is preserved at the Desert Laboratory of the University of Arizona at Tucson as a type specimen. The other half of the midden was separated into three stratigraphic samples labeled SNM-1, SNM-2, and SNM-3. Sample SNM-1 was the lowest and potentially the oldest stratigraphic unit, whereas SNM-2 was the youngest. Each sample was soaked in a weak HC1 solution to

remove inorganic carbonates and soluble organic material. The remaining residue, after drying in an oven, was only 10%–20% of the prewashed weight of the samples. The residue consisted mainly of sandstone fragments with small fragments of plant, animal, and insect parts and fecal pellets. Samples SNM-2 and SNM-3 were subdivided to obtain samples for radiocarbon analysis, whereas the residue from sample SNM-1 was too small to subdivide.

The residue of the samples was washed gently in a fine-mesh sieve to remove any remaining soluble, nonvegetative, or animal matter and was then dried. The material was spread thinly in a shallow tray and examined under a binocular microscope. Fecal pellets, 5–10 mm in length, were much less numerous than would be expected in packrat middens (see Spaulding et al., *this volume*, chap. 5). Fossil fragments including stem and leaf parts, floral and fruit particles, hairs, seeds,

Fig. 20.3. *Leporillus* den in northern south Australia, illustrating why *Leporillus* is known as the stick-nest rat. Its propensity to collect large sticks is similar to that of the bushy-tailed woodrat (*Neotoma cinerea*) in western North America (courtesy of Peter Copley).

vertebrate remains such as bone, feather particles, and hairs, and insect parts were separated.

The vegetative matter removed from the middens was of a much smaller size than the macrofossils recovered from packrat middens (Spaulding et al., *this volume*, chap. 5). Consequently, we used methods reported by Chippendale (1962, 1964, 1968a, 1968b) for the analysis of rumen contents of cattle and kangaroo stomachs. A modified point-frame analysis (Chamrad and Box, 1964) was used to estimate the abundances of the macrofossils. Use of a high-powered light microscope was required in the identification phase.

A reference collection of the modern flora, including herbarium sheets and reference specimens of leaves, stems, floral and fruiting parts, and seeds, was obtained from the Amphitheatre for comparison with fossil fragments. Other fragments were identified by comparison with specimens in the herbarium of the Northern Territory Arid Zone Research Institute in Alice Springs. Matches in hair shapes and patterns or epidermal characteristics such as venation patterns, stomatal size and patterns, and color on vegetation parts enabled identification of some species. For example, the striation pattern on phyllodes of *Acacia kempeana* is a unique and positive identifying characteristic.

Identification of insect and animal remains was of a general nature. Fecal pellets were either dark and contained mainly invertebrate material or were light and contained mainly plant material. Most fecal pellets were dark. In addition to the remains found in pellets, vertebrate and invertebrate remains also occur within the matrix of the midden.

RESULTS AND DISCUSSION

Analysis of samples of the stick-nest rat midden revealed little change in vegetation over the last 2800 years in the Amphitheatre.

Radiocarbon analyses of washed but unsorted midden samples SNM-2 and SNM-3 yielded ages of 1660 ± 100 B.P. (A-3989) and 2780 ± 100 B.P. (A-3990), respectively, in accord with the relative ages deduced from stratigraphic position. Although about one hundred plant species were collected from within several hundred meters of the midden site, only sixteen species were identified from the midden samples (Table 20.1). Of these species only a few were abundant enough to be counted using the modified point-frame analysis. This result may suggest either a restricted foraging strategy for *Leporillus* or possibly a change in the flora after the midden was deposited. One of the species identified in all midden samples, *Acacia kempeana* (Table 20.1), is a common dominant species on alluvial slopes

Table 20.1. *Plant species and percentages from stick-nest rat middens from the Amphitheatre, Finke Gorge National Park, Northern Territory.*
Asterisk indicates plant species not found near the stick-nest rat midden.
Percentages determined by modified point-frame method (Chamrad and Box, 1964).

	SNM-1 (undated)		SNM-2 (1660 ± 100 B.P.)		SNM-3 (2780 ± 100 B.P.)	
MIDDEN NUMBER	Presence	%	Presence	%	Presence	%
WOODY PLANTS						
Abutilon leucopetalum	X	0	—	0	—	0
Acacia kempeana	X	10	X	23	X	9
Cassia artemisioides	X	15	—	0	X	8
*Cassia nemophila	X	5	—	0	X	6
Not identified		24		17		19
GRASSES						
Total		16		17		17
Aristida sp. nitidula?	—	—	X	—	—	—
Eragrostis falcata	—	—	—	—	X	—
Paspalidium sp. (?)	X	—	X	—	X	—
FORBS						
Boerhavia sp. coccinea?	—	0	X	0	—	0
Cleome viscosa	—	0	X	0	—	0
*Daucus glochidiatus	X	0	X	0	—	0
*Glossogyne tenuifolia	—	0	X	0	—	0
Helipterum saxatile	X	3	X	3	X	1
Mukia maderaspatana	—	0	X	0	—	0
*Rutidosis helichrysoides	X	0	X	0	—	0
Portulaca oleracea	—	0	—	0	X	0
Solanum sp.	—	0	—	0	X	0
Not identified	—	10	—	19	—	21
FECAL PELLETS						
Dark color		7		8		5
Light color		4		4		2
INSECT AND ANIMAL REMAINS						
Beetle parts	X	4	X	3	X	9
Slaters	X	1	X	2	X	1
Feathers	X	1	X	2	X	1
Hair	—	0	X	2	X	1
Parts of ants	X	0	X	0	—	0
Body of false scorpion	X	0	—	0	X	0
Bones	X	0	X	0	X	0

in the Amphitheatre (Latz, 1981) but does not occur within 30 m of the midden. Fragments of only three of the sixteen species recorded within a 30-m radius of the midden site were identified in the midden samples.

Five species identified in the middens were not found near the midden site (Table 20.1). One of these, *Cassia nemophila*, is a common perennial shrub in the Amphitheatre and central Australia (Dunlop and Bowman, 1986). We think the absence of *Cassia nemophila* in the local flora is the result of recent drought or fire. Other species found in the midden but not in the local flora are usually widespread during wet years, and the absence of these species from nearby vegetation assemblages does not have paleoecological significance. For example, *Glossogyne tenuifolia*, a forb, has not been collected from Finke Gorge National Park (Latz, 1981) but is found during wet seasons in central Australia.

Examination of fecal pellets revealed some limited dietary information on *Leporillus*. Most of the pellets were dark-colored and contained invertebrate remains, mainly beetles of unknown species. The presence of insect remains in the midden contrasts with the findings of Green et al. (1983), who reported no insect remains in the midden they analyzed. The discovery of pellets consisting mainly of plant fragments suggests that *Leporillus* was an omnivorous forager.

The results of this study suggest the usefulness of stick-nest rat middens for understanding historic ecology and paleoecology in central Australia. These middens reportedly are common in central Australia, particularly in the Gibson Desert (Strahan, 1983). Analysis of middens and fecal pellets may yield clues as to the former distribution, ecological habits, and the reason for the extirpation of *Leporillus*. As suggested in our scant data, *Leporillus* relied heavily on insects in its diet. Destruction of palatable vegetation by livestock may have eliminated both an important direct forage for stick-nest rats and the food source for insects that also were important to the rats' diets. Stick-nest rat middens contain both pollen and macrofossils, and analysis of the two types of paleoecological information may be useful in reconstructing central Australian climate during the late Quaternary. The extremely stable landforms in central Australia (e.g., Pickup et al., 1988) should allow preservation of ancient stick-nest rat middens. Finally, the impact of historic grazing on vegetation assemblages can be evaluated by analyzing these valuable paleoecological repositories.

ACKNOWLEDGMENTS

The authors thank P. S. Martin for his encouragement to pursue stick-nest rat middens and G. Griffin for his assistance in locating the middens in Finke Gorge National Park.

REFERENCES

Bridge, P. J. (1975). "Urea from Wilgie Mia Cave, W. A., and a note on the type locality of urea." *Western Australian Naturalist* 13, 85–86.

Chamrad, A. D., and Box, T. W. (1964). "A point frame for sampling rumen contents." *Journal of Wildlife Management* 28, 473–77.

Chippendale, G. M. (1962). "Botanical examination of kangaroo stomach contents and cattle rumen contents." *Australian Journal of Science* 25, 21–22.

———. (1964). "Determination of plant species grazed by open range cattle in central Australia." *Proceedings of the Australian Society for Animal Production* 5, 256–57.

———. (1968a). *A study of the diet of cattle in central Australia as determined by rumen samples.* Technical Bulletin no. 1, Primary Industries Branch. Darwin, N.T.: Northern Territory Administration.

———. (1968b). "The plants grazed by red kangaroos, *Magaleia rufa* (Desmarest), in central Australia." *Proceedings of the Linnean Society of New South Wales* 93, 98–110.

Dunlop, C. R., and Bowman, D. M. J. S. (1986). *Atlas of the vascular plant genera of the Northern Territory.* Australian Flora and Fauna Series no. 6. Canberra: Australian Government Publishing Service.

Green, N.; Caldwell, J.; Hope, J.; and Luly, J. (1983). "Pollen from an 1800-year old stick-nest rat (*Leporillus* sp.) midden from Gnalta, western New South Wales." *Quaternary Australasia* 1, 31–41.

Latz, P. K. (1981). *Checklist of vascular plants of Palm Valley, Finke Gorge National Park.* Northern Territory Botanical Bulletin no. 4. Darwin: Northern Territory Government, Division of Agriculture and Stock, Department of Primary Production.

Newsome, A. E., and Corbett, L. K. (1975). "Rodents in desert environments." In *Monographiae biologicae*, vol. 28. J. Prakash and B. Ghosh, eds., pp. 117–54. Dr. W. Junk Publishers.

Pickup, G.; Allan, G.; and Baker, V. R. (1988). "History, palaeochannels, and palaeofloods of the Finke River, central Australia." In *Fluvial Geomorphology of Australia*. R. F. Warner, ed., pp. 177–200. Sydney: Academic Press Australia.

Strahan, R., ed. (1985). *The Australian Museum complete book of Australian mammals*. Sydney: Angus and Robertson.

Troughton, E. (1967). *Furred animals of Australia*.

Sydney: Angus and Robertson.

Walker, E. P. (1975). *Mammals of the world*. 3d ed., vol. 2. Baltimore: Johns Hopkins University Press.

Watts, C. H. S., and Eves, B. M. (1976). "Notes on the nests and diet of the white-tailed stick-nest rat, *Leporillus apicalis*, in northern South Australia." *South Australian Naturalist* 51, 9–12.

Chapter 21

Synthesis and Prospectus

Julio L. Betancourt   Thomas R. Van Devender   Paul S. Martin

In some circles the paleoecologist is considered an unfortunate ecologist, one who has the vantage of time but lacks too many pieces of the puzzle for a coherent view. By their own admission, however, ecologists study processes that often operate on temporal scales exceeding the researcher's own life span, not to mention the duration of his funding. One could say that ecologists study detailed snapshots, still images that the fossil record blurs but nevertheless animates. Because both perspectives have shortcomings, each stands to gain from the other.

Some ecological theory holds that natural communities converge to stable equilibria, which are maintained by competition and predation. In this view of community dynamics, history, chance, and long-term environmental variability would play only minor roles. The fossil record of the last ice age offers a perspective of community dynamics against a backdrop of considerable climatic changes. The overwhelming evidence for climatic instability on time scales of decades to millennia would alone argue against classic equilibrium theory, especially when it involves communities dominated by long-lived perennials.

Equilibrium theory may be further compromised if the ability to track an environmental perturbation varies drastically among species,

a corollary to the familiar individualistic concept of the plant association (Shreve, 1914; Gleason, 1926). A demographic response slower than the rate of environmental change might lead to community inertia (resistance to change) and migrational lag, distorting community patterns in both time and space. Paleoecological evidence supporting such distortions was first presented in 1962 at the First International Palynological Congress in Tucson (Smith, 1965). The cause has been furthered by the elegant studies of Davis (1986) and others, though not without dissent in the paleoecological community (Webb, 1986; Markgraf, 1986). If species distributions and abundances slowly or rarely equilibrate with environmental changes, does this necessarily compromise paleoclimatic interpretations from vegetation histories?

In the Southwest vast differences in life-history strategies and immense environmental heterogeneity pose a formidable challenge to stable-equilibria models. Some have even argued that classic succession does not occur in deserts (Shreve, 1942; Muller, 1940; see reviews by Lathrop and Rowlands, 1983; and Webb et al., 1988). The packrat midden record offers ample evidence for individualistic responses as glacial-interglacial climates displaced different species to different latitudes and elevations. This uneven displacement produced at least some anomalous associations, a distinctive altitudinal zonation, and heterochroneity that begs the questions of inertia and migrational lag (Cole, 1985; Spaulding et al., 1983; Van Devender et al., 1987).

To help organize a synthesis, we offer a map of late Wisconsin vegetation (Fig. 21.1). Discussions of major vegetation types highlight midden contributions to contemporary issues in historical biogeography and paleoclimatology. The map, which includes the Great Basin, the southwestern United States, and northernmost Mexico, should be compared to an earlier effort by Martin and Mehringer (1965; *this volume*, chap. 1, Fig. 1.1). Some caveats are in order. Understandably, our map is based on the midden record, and fossil middens are largely restricted to rocky terrain including cliffs. Details of topography and vegetational gradients are inevitably lost at the scale employed. Because the middens themselves show little change during the late Wisconsin, our map serves eleven thousand years. It does not reflect the melting of mountain glaciers or desiccation of pluvial lakes by around 13 ka. Finally, we can adopt the structural terms "desertscrub," "woodland," and "montane conifer forest" on floristic grounds only. It is worth emphasizing that structure as such cannot be determined independently from midden macrofossils. The middens can disclose peculiar and anomalous mixtures of species in an ancient community, but they cannot disclose vegetation structure directly. For example, they would not discriminate a krumholtz form of spruce from a closed-canopy forest or, less plausible perhaps, a closed-canopy pygmy forest of dense pinyon-juniper from a grassland sparsely punctuated by a few trees. Unlike the assumption of community composition, the view that modern vegetation structure is reliably projected into the world of the late Wisconsin is rarely challenged (Martin, 1959).

To construct the map we extrapolated from the elevational ranges of key plants indicated by midden records. Though there is evidence for higher desertscrub assemblages, especially in the southwestern Great Basin, the interval from sea level to 300 m was chosen to indicate widespread desertscrub on rocky terrain. We are confident that desertscrub also covered alluvial slopes and basin floors within

this elevational range. The elevation interval between 300 and 1500 m incorporates both chaparral and pygmy conifer woodlands, including juniper, pinyon-juniper, and pinyon-juniper-oak woodlands. The importance of grasslands or steppe (*Artemisia*) on fine-grained soils is unspecified. Except on limestone, we truncated woodlands at 1400 m in areas north of the Arizona-Utah border based on the prevalence of montane conifer assemblages near that elevation on the Colorado Plateau.

In the Great Basin we reserved the term "subalpine woodlands" for late Wisconsin communities dominated by bristlecone pine (*Pinus longaeva*) or limber pine (*P. flexilis*) and lacking other trees. The map reflects the prevalence of these species in fossil deposits over most of the Great Basin, from the upper limits of juniper or pinyon-juniper woodlands to upper treeline (1600–2900 m). On the Colorado Plateau and in the central and southern Rockies, we split this elevational range to discriminate between montane conifer (1500–2200 m) and boreal (2200–2900 m) communities, which we will henceforth term "forests." Montane conifer forests include assemblages dominated by two or more of the following conifers: limber pine, southwestern white pine (*Pinus strobiformis*), Douglas fir (*Pseudotsuga menziesii*), white fir (*Abies concolor*), blue spruce (*Picea pungens*), common juniper (*Juniperus communis*), and Rocky Mountain juniper (*J. scopulorum*). Ponderosa pine (*Pinus ponderosa*) was probably an important component of montane communities in the highlands south of 35° W, and probably was most prevalent in the Sierra Madre Occidental. The primary conifers in the boreal forests were Engelmann spruce (*Picea engelmannii*), subalpine fir (*Abies lasiocarpa*), lodgepole pine (*Pinus contorta*), limber pine, and Rocky Mountain bristlecone pine (*P. aristata*). In the northern portion of the area on the map these trees may have ranged no farther west than the Ruby Mountains. We have placed upper treeline uniformly at about 2900 m. However, we realize that upper treeline in the northernmost areas may have been up to 300 m lower than to the south, as it is today.

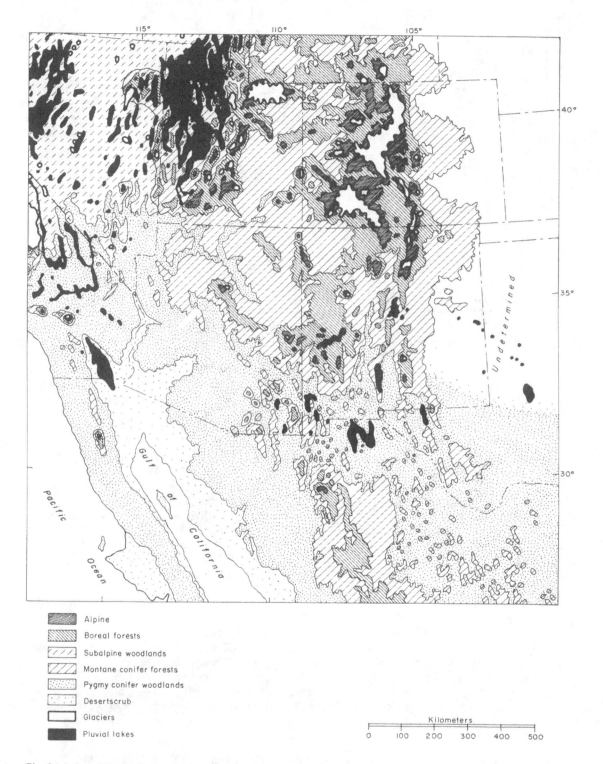

Fig. 21.1. Late Wisconsin vegetation of rocky terrain in the arid interior of western North America (see explanation in text).

DESERTS

The age and origin of the southwestern deserts is a recurrent theme in North American biogeography that can be reexamined through the midden record. The high degree of endemism and the unique morphologies of desert plants suggest great antiquity. The arid Southwest has more endemic genera and families than all of North America east of the Rockies (Axelrod, 1979). About 5500 species occur in the Southwest; of these, about 1200 are considered endemic (McLaughlin, 1986). Surely the bizarre anatomies—the succulence of cacti, the chlorophyllous bark of trees, and the drought-deciduous stems of shrubs—imply age-old adaptations to aridity. This need not mean that the deserts have always been where they are now. Perhaps their distributions shifted radically with Plio-Pleistocene uplift of the Sierra Nevada and with the waxing and waning of each ice age. To biogeographers contemplating such matters, the midden record has become a source of indispensable facts (e.g., Johnson, 1968; Axelrod, 1979; Thorne, 1986). The emerging view is that although many desert plants may have evolved locally, "our present desert plant communities are not only very recent, they are highly dynamic, indeed kaleidoscopic in content and location" (Thorne, 1986, p. 647).

Few would dispute that late Wisconsin woodlands and chaparral were replaced by modern desertscrub over vast areas of the North American deserts. Opinions differ, however, over the extent of treeless communities during the last glacial. Desertscrub assemblages of full-glacial age are known from about 425 m in Death Valley (Wells and Woodcock, 1985; Woodcock, 1986) and from about 900 m elevation in southern Nevada (Spaulding, *this volume*, chap. 9); they bear little resemblance to modern Mojave communities. Those from Death Valley contain remains of chaparral species, and those from southern Nevada are dominated by Great Basin shrubs. Mojavean plants such as Joshua tree (*Yucca brevifolia*) and Mojave sage (*Salvia mohavensis*) were widespread in woodlands of what is now the Sonoran Desert. Mojave shrubs without trees grew below 300 m in the Colorado River delta and possibly on the coastal plains of Sonora and Baja California.

Creosote bush (*Larrea divaricata*), thought by Wells and Hunziker (1976) to be a Holocene immigrant from South America, was present on the Arizona-Sonora border during the full glacial. However, the fossil record currently sheds little light on the timing for its immigration to the Chihuahuan Desert, or for its differentiation into three chromosomal races with diploids in the Chihuahuan Desert, tetraploids in the Sonoran Desert, and hexaploids in the Mojave Desert. In the present interglacial Chihuahuan and Sonoran desert populations probably did not come into contact until perhaps 4500 years ago, when a corridor became established across the Continental Divide.

During the last glacial age, many plants typical of the Sonoran Desert (*Carnegiea, Cercidium, Olneya*, etc.) were presumably confined to southern refugia, possibly in central Sonora. Tropical areas to the south may have been drier during the late Wisconsin, making room for Sonoran Desert plants. Because of wetter conditions during the past ten thousand years and conditions favorable to termite activity and other reducers in caves, Pleistocene middens from tropical refugia may be difficult to find. The search for the glacial refugia of Chihuahuan Desert plants presents other problems. Many Chihuahuan Desert plants, particularly succulents, occurred locally in association with pinyon pines, junipers, and oaks. If some Chihuahuan Desert endemics were displaced to the south, which has not been demonstrated, Wells (1977) suggests refugia in the Valle del Mesquital (Hidalgo) or in eastern Queretaro. These localities now represent the southernmost outposts of many Chihuahuan Desert plants. Persistence of Pleistocene middens in these tropical, high-elevation areas also is questionable; the most likely sites would be in intense rain-shadow valleys, provided that air-mass trajectories were like today. Overall, the midden record of desert plants, limited as it is to the last forty thousand years, records much more movement than had been imagined by most biogeographers operating without macrofossils.

SOUTHWESTERN GRASSLANDS

Today plains and semidesert grasslands, flowering primarily in late summer, cover extensive areas west of the one hundredth meridian. Little is known about the late Quaternary history of these grasslands, particularly when compared with the prairie-forest ecotone in the Midwest (Webb et al., 1983). Based on the relative abundance of spruce and pine pollen in late Wisconsin sediments, various authors (Martin and Mehringer, 1965; Wendorf and Hester, 1975; Wells, 1983a) thought that trees invaded the Great Plains and southern High Plains during the late Wisconsin. However, buried soils on the plains lack evidence of podzolization characteristic of forest soils, and the pollen evidence has unrecognized limitations (Holliday, 1987).

Even though middens can yield a rich macrofossil record of grasses, they lack resolution in reconstructing the spatial scale and structure of grasslands. Middens represent records of local vegetation in rocky settings, not the open expanses normally occupied by grasslands. As Webb and Betancourt note in chapter 6, middens from the occasional scarp on the Great Plains would miss the sea of grass for the few trees. So what, if anything, can middens contribute to knowledge about the history of southwestern grasslands?

One of the primary issues in the biogeography of western grasslands involves the relative importance of C3 and C4 grasses in subregional vegetation. For example, C4 grasses (chloridoids in short-grass prairie to the west and panicoids in tall-grass prairie to the east) dominate grasslands in the southern portions of the Great Plains; C3 grasses, primarily festucoids, are more common to the north. A similar north-south gradient is evident in grasslands west of the Continental Divide, where the proportion of C3 to C4 grasses also increases with elevation.

The composition of western grasslands probably has been unstable throughout the late Quaternary, but paleoecological evidence is scant or lacks taxonomic resolution. Though grasslands may have a characteristic pollen spectrum, pollen grains of grasses are too similar to allow identification below the family level. Classes of opal phytoliths may correspond to subfamilies, but even this is questionable (D. A. Brown, 1984). Fossil grass cuticles abound in some lake sediments and other deposits, but they are seldom deter-

minable to species. That leaves middens as the only regional source of grass macrofossils identifiable to species (Van Devender et al., this volume, chap. 14). Given the close correspondence of C4 grasses to minimum July temperature (Teeri and Stowe, 1976), we imagine use of their former latitudinal and elevational distributions as a paleothermometer, much like Livingstone and Clayton (1980) envisioned for tropical Africa.

PYGMY CONIFER WOODLANDS

The Quaternary macrofossil record of pygmy conifer woodlands in the American West is unrivaled worldwide for woodland associations. During the late Wisconsin, species of pinyon pine, junipers, and oaks dominated midden floras at what are now desert elevations; the average upper limit of pinyon pines was at 1500 m. A similar upper limit now confines many thermophilous species of the southern deserts, like creosote bush. The topographic break that occurs between the warm southern deserts and the cold northern steppe of the Great Basin and Colorado Plateau occurs at about 1500 m. Similarly, the late Wisconsin ranges of woodland trees were truncated by high terrain in the northern parts of the Southwest. One exception is Utah juniper (*Juniperus osteosperma*), which extended well into the Great Basin, with upper ranges of around 2000 m in subalpine woodlands of the northern Mojave Desert. Its northern and upper limits also exceed those of the pinyons today.

Pygmy conifer woodlands were remarkably stable during the middle and late Wisconsin and did not experience major turnovers until 11 ka. The lengthy records from localities as far separated as west Texas and the Great Basin offer unique opportunities for stable isotope/paleotemperature studies. It will be interesting to see whether or not the stability evident in the midden floras is reflected by deuterium abundance from plant cellulose. In other words, will the deuterium record from west Texas reflect the same warming by 13 ka recorded in deuterium chronologies from the Great Basin and Colorado Plateau (Long et al., this volume, chap. 17)?

Postglacial expansion of pinyons northward and upward involved shifts in the dominant taxa. *Pinus edulis* and *P. monophylla* exchanged roles with *P. remota* and *P. californiarum* var. *fallax* as the dominant species across the Pleistocene-Holocene boundary. This phenomenon has spawned considerable discussion about the taxonomic and evolutionary status of the northern pinyons, particularly single-needle pinyon in Arizona, which has been variously regarded as *P. edulis* var. *fallax*, *P. monophylla* var. *tenuis*, and, most recently, as *P. californiarum* subsp. *fallax* (Lanner and Van Devender, 1974; Wells, 1987; Bailey, 1987). If Bailey (1987) is right, the dominance of *P. californiarum* subsp. *fallax* in the Sonoran Desert may have been part of a large-scale eastward expansion of California chaparral into Arizona. If so, modern populations of chaparral species along the Mogollon Rim, where this pinyon is now common, are probably Pleistocene relicts. The paleoclimatic significance of the role reversal in the pinyons will remain uncertain until their physiological differences are studied. Perhaps lower carbon dioxide concentrations during the last glacial selected for those species best suited to lower photosynthetic rates, irrespective of shifts in rainfall seasonality.

Surprisingly, much more is known about the Wisconsin range of woodland trees than about their migrational dynamics during the Holocene. Pinyons vacated desert lowlands at about 11 ka. The rate at which they achieved their modern elevational ranges is unknown. Few middens have been collected from localities above 1500 m in the Sonoran and Chihuahuan deserts. In the Guadalupe Mountains of southern New Mexico, Colorado pinyon occurred at 2000 m by 10.9 ka, suggesting rapid colonization of its modern elevational range.

More is known about the northern migration of pinyon pines. *Pinus monophylla* replaced limber pine between 13.2 and 11.7 ka at the Eleana Range on the southern fringes of the Great Basin, but it was not until about 6 ka that it reached the central Great Basin. Colorado pinyon was present in the western Grand Canyon during the full glacial and along the Little Colorado River (Wupatki National Monument) north of Flagstaff by at least 13.7 ka (Cinnamon and Hevly, 1988). However, it is not recorded in middens from the eastern Grand Canyon until 10.5 ka; at

Chaco Canyon, in the eastern part of the Colorado Plateau, it arrived some time between 10 and 8.3 ka. Future investigations will better define the northern and upper ranges of the pinyons during the late Wisconsin and trace their northward and upward migrations during the Holocene. Such knowledge will be critical to evaluating claims of vegetational inertia and migrational lag during the transition from glacial to interglacial conditions.

Contrasting paleoclimatic interpretations have been offered for the persistence of junipers and oaks at desert elevations well into the early Holocene. In chapters 7 and 8 Van Devender argues for continued winter rainfall dominance with only modest summer precipitation; he suggests that monsoonal circulation intensified during the middle Holocene, supporting Martin's (1963) earlier claim for a wet altithermal. In chapter 9 Spaulding counters that, in the early Holocene, maximum solar insolation in July and increased summer rainfall (due to the increased thermal and pressure contrasts between land and sea) adequately explain an increase in succulents in the Mojave Desert and maintenance of "relict" woodlands in the lowlands to the south and southeast. Refilling of pluvial lakes, including Searles Lake, Lake Lahontan (Russell Shoreline), Lake Bonneville (Gilbert Shoreline), and Lake San Agustin, between 11 and 10 ka suggests that at least the beginning of the early Holocene was much wetter than any other time during the Holocene (a Younger Dryas in the Southwest?).

General circulation models offer a way of translating hypotheses into a form testable by paleoenvironmental data (COHMAP, 1988). This function of the models will likely improve as they gain sophistication and spatial resolution. Kutzbach and Guetter's (1986) latest runs for 9 ka, using the NCAR CCM (National Center for Atmospheric Research, Climate Community Model) simulates warmer summers and more intense monsoonal circulation than at any other time during the Holocene (see also COHMAP, 1988). A newer model by Mitchell et al. (1988), which incorporates interactive soil moisture and sea-surface temperatures, was used to simulate conditions at 9 ka with and without a remnant Laurentide ice sheet. The long-term retreat of the ice sheet probably begat a complicated response of mid-latitude atmospheric circulation to orbital forcing. During maximum glaciation the ice acted as an orographic barrier, splitting the jet stream. Its role as a reflective surface gained importance as the ice thinned faster than it shrank in area. The simulation without the ice sheet, which could be considered a near-analog for the middle Holocene, produced warmer summers and winters, weaker westerlies (reduced winter precipitation), and more intense subtropical flow than the run with the ice sheet present. Thus, the COHMAP model is supported by Spaulding's interpretation; the simulations by Mitchell et al. (1988) are supported by Van Devender's interpretation.

Both single-needle pinyon and Colorado pinyon may still be on the move, though some of the advances have been attributed to fire suppression and overgrazing. West (1984) surmises that one-half of the area now occupied by pinyon and juniper in the Great Basin is of rather recent, historical derivation. The northernmost outpost of *P. edulis* along the Colorado Front Range became established some four hundred years ago (Betancourt, 1987). Today Utah juniper reaches as far northeast as the Big Horn Basin in Wyoming. Apparently, little now prevents it from migrating south along the Front Range. Its southward expansion into eastern Colorado probably will invoke claims of fire suppression and overgrazing. But as Mehringer and Wigand have shown for western juniper (*Juniperus occidentalis*) in chapter 13, such expansions could occur prior to historic land-use activity. Hence, directional change in climate and subsequent migrational dynamics may be invoked to explain both prehistoric and modern invasions.

PONDEROSA PINE FORESTS

No aspect of southwestern vegetation history has been more drastically reinterpreted than the role of ponderosa pine. In retrospect there was nothing radical about Martin and Mehringer's (1965) mapping of full-glacial, yellow pine parkland across the southern Great Basin, the Southwest, and the southern High Plains (*this volume*, chap. 1, Fig. 1.1). After all, it was based on an analog approach still much in vogue. Elevated pine pollen percentages in full- and late-glacial lacustrine sediments were duplicated only in modern ponderosa pine stands attaining forest densities. Martin and Mehringer (1965) assumed that the fossil pollen was that of ponderosa or some other yellow pine.

The assumption proved incorrect. At Dead Man Lake in the Chuska Mountains, morphological separation of fossil pine pollen showed the preponderance of limber pine during the full glacial (Hansen and Cushing, 1973). Wright et al. (1973) surmised that limber pine may have been a major rather than a minor component of the ponderosa pine forest, or that it may have replaced ponderosa in part during the late Wisconsin, much like it does today in Wyoming. The latter view is supported by the virtual absence of ponderosa pine and the omnipresence of limber pine in middens from the northern Mojave Desert, the southern Great Basin, and the Colorado Plateau. The extensive ponderosa stands that now are the trademark of southwestern uplands may be a Holocene phenomenon.

The puzzling absence of ponderosa pine from most of the interior West during the last glaciation is central to several biogeographic and bioclimatological issues. Minimal gene exchange now occurs between eastern (*P. ponderosa* var. *scopulorum*) and western (var. *ponderosa*) populations in central Montana. Even this limited contact was unlikely during the late Wisconsin. Protracted separation punctuated by east-west migrations may explain differentiation of the five discrete races of ponderosa pine and closely related species (*P. jeffreyi* and *P. washoensis*) during the late Cenozoic.

Paleoclimatic explanations for the marked contrast between glacial and postglacial distributions of ponderosa pine have thus far been elusive. *Pinus ponderosa* now extends from northern Montana to central Mexico over a latitudinal range of about twenty-three degrees. Hence it is able to tolerate a broad suite of climates. It seems incomprehensible that a handful of other conifers, such as Douglas fir, thrived in northern parts of the Southwest during the full glacial while ponderosa pine did not. Several explanations have been offered, including pronounced summer drought, reduced fire activity, and anomalously cool temperatures forestalling seed germination in the fall. At Yellowstone Park lake sediments assigned to oxygen isotope stage 5e (the Sangamon interglacial) contained ponderosa pine macrofossils about 400 m higher than its maximum elevations today. Baker (1986) interpreted this to mean warmer temperatures than at any time during the present interglacial. A warmer-than-present Sangamon also was inferred from remains of subtropical reptiles and amphibians near Rancho La Brisca in northern Sonora (Van Devender et al., 1985).

Large distributional shifts during glacial-interglacial cycles recommend ponderosa pine as a primary target for models linking regional vegetation response to projected future climates. In states like Arizona ponderosa pine is the dominant forest type in the headwaters of major rivers and provides up to three quarters of the commercial timber. Among southwestern conifers it will likely wield the greatest economic impact in responding to future climatic change.

Few mammals are as closely associated with a plant as Abert's squirrel (*Sciurus aberti*) is with ponderosa pine. Disjunct populations of this squirrel from southern Wyoming to the Sierra Madre Occidental are thought by some to have developed vicariantly through fragmentation of ponderosa pine forests during interglacials. The fossil record for ponderosa pine suggests otherwise. Abert's squirrels either tracked the northward, postglacial migration of ponderosa pine or thrived in other habitats during the last glaciation (Lomolino et al., 1989). This example also illustrates a complication for the paleontologist who routinely infers vegetation from animal remains (Van Devender et al., *this volume*, chap. 15). Would Abert's squirrel in a Pleistocene cave fauna necessarily mean that ponderosa pine was growing nearby?

SUBALPINE WOODLANDS, MONTANE FORESTS, AND BOREAL FORESTS

During the late Wisconsin, subalpine woodlands and montane forests dominated vast areas now occupied by pygmy conifer woodlands. Midden records of these communities are far more abundant in the Great Basin and Colorado Plateau than in the southern deserts, where the focus has been on desert elevations. The poor record from modern pygmy

conifer woodlands in the southern deserts also may have a physical basis. Today these woodlands tend to be associated with higher precipitation levels than those to the north, constraining midden preservation. In the Great Basin and northern Mojave Desert, woodlands dominated by bristlecone and limber pines dominated full- and late-glacial landscapes between 2400 and 1600 m elevation. Limber pine was more prevalent in the southern Great Basin–northern Mojave Desert; it gained greater importance to the north during the late glacial. Expansion of these white pines may have been aided by the absence of other montane and boreal conifers. Bristlecone pine, and possibly whitebark pine (*Pinus albicaulis*), may have been the only trees near upper treeline in ranges west of the Ruby Mountains. Though Engelmann spruce occurred locally in the Snake Range, the nearest records for most other conifers (Douglas fir, white fir, Rocky Mountain juniper) are in the northern Mojave Desert (Clark Mountain, the Sheep Range) and southeastern Great Basin (Meadow Valley Wash).

Montane communities were far more diverse eastward on the Colorado Plateau, where a single glacial-age midden might contain as many as five different conifers. In the absence of ponderosa pine and pinyon pine, these montane communities graded into Utah juniper woodland at about 1400 m in the eastern Grand Canyon. Because of the high minimum elevations of the plateau, this ecotone may have been restricted to the gorge of the Colorado River; on limestone Utah juniper may have extended to the mouth of the Green River. Middens from 2200 m in the eastern Grand Canyon and the Abajo Mountains of southeastern Utah record the position of the ecotone between spruce-fir forests and montane forests.

As the case for Abert's squirrel suggests, the midden record bears directly on vicariance versus long-distance dispersal as explanations for insular distributions of subalpine and montane communities. Whether or not mountaintop habitats behave as ecological islands hinges on the degree of isolation in the recent geologic past. Thus the midden record can temper applications of island biogeographic theory to mountaintops in the Great Basin (J. H. Brown, 1978; Wells, 1983b;

Grayson, 1987; Thompson, *this volume*, chap. 10) and the southern Rockies (Patterson, 1980; Lomolino et al., 1989; Betancourt, *this volume*, chap 12).

Long-distance dispersal and vicariance have played different roles in the two regions, as might be expected from differences in physiography and vegetation history. Contrasting views have been presented for the Great Basin. Wells (1983b) thinks that full-glacial displacement of bristlecone pine was sufficient to establish continuous populations across intermontane basins. The lack of significant genetic divergence in modern populations would tend to support him (Hiebert and Hamrick, 1983). On the other hand, Thompson, in chapter 10, points to shrub dominance in pollen spectra and the character of alluvial soils in arguing for substantial discontinuities. Seed dispersal by birds and pollen dispersal by wind may explain gene flow between populations up to 100 km apart. Wells (1983b), Thompson (*this volume*, chap. 10), and Spaulding (*this volume*, chap. 9) agree that montane conifers (e.g., ponderosa pine, Douglas fir, white fir, and Rocky Mountain juniper) probably achieved their Holocene distributions by long-distance dispersal or island hopping. This may account for their haphazard occurrence on isolated ranges in the Great Basin and in the Mojave Desert.

Most montane and boreal conifers on Sonoran Desert mountains were also established by long-distance dispersal. Dispersal across more than 100 km is the only reasonable explanation for late Wisconsin records of Rocky Mountain juniper at Organ Pipe Cactus National Monument (Van Devender, *this volume*, chap. 8). In Arizona unusual disjuncts such as the small population of subalpine fir in the Santa Catalina Mountains and Engelmann spruce in the Pinalenos and Chiricahua mountains probably resulted from similar Wisconsin dispersals. The precarious nature of such disjuncts suggests that many conifers suffered local extinctions on isolated mountaintops. Though evidence is lacking, expansion of limber pine into Sonoran and Chihuahuan desert highlands would be consistent with its dominance in lowlands to the north. Though most of these populations would have been extirpated during the Holocene, their genetic signature may still be imprinted on isolated hybrid swarms currently identified as southwestern white pine (*Pinus strobiformis*).

A case for continuous Wisconsin distribution and Holocene fragmentation (i.e., vicariance) can be made for the southern Rockies, where a lower limit of 1400–1500 m would have integrated montane communities across the lowlands. Also, augmented perennial flow in the highly dissected drainages of the Rio Grande and Colorado River would have stretched the lower limits of most conifers well beyond mountain piedmonts. In the southern Rockies, including submogollon areas, dispersal along riparian corridors may have linked populations that otherwise were isolated. Engelmann spruce, subalpine fir, and Rocky Mountain bristlecone pine were probably the only conifers with significantly isolated populations during the late Wisconsin.

The midden record has contributed the following insights about mountaintop biotas and island biogeographic theory. According to Wells (1983b), species-area patterns for Great Basin conifers are largely determined by Holocene immigration to most mountains plus the balance between extinction and immigration on the summits of smaller ranges. For poor dispersers such as mammals, pulses of colonizations may not have kept pace with extinctions to maintain dynamic equilibrium during the Holocene. However, in their analysis of the southern Rockies, Lomolino et al. (1989) found evidence for long-range dispersals of montane forest mammals across woodland habitats today and during the last glacial. They argued for a more dynamic system than Brown (1978) found for Great Basin mammals. The same pattern may hold for conifers in the Great Basin and southern Rockies.

The present decrease in conifer diversity from the Rocky Mountains to the Great Basin was mirrored during the late Wisconsin. The gradient includes differences in summer precipitation and bedrock geology, which may explain conifer diversity better than simple distance from propagule sources. In part the Colorado Rockies are now more diverse than isolated ranges in the region because they have suffered fewer extinctions at high elevations while receiving a host of southern immigrants at lower elevations. Compared to now, the Colorado Rockies were biologically impoverished during the glacial, except for greatly expanded treeless communities above boreal forests. The latter probably extended to the base of the mountains (Betancourt, *this volume*, chap. 12).

MIDDEN RESEARCH IN THE FUTURE

When the idea for this book was first conceived, we thought it might serve the dual purpose of reviewing what we know as well as what we have failed to learn from the midden record. Also, we hoped it would guide new discoveries and applications in the future. Four mammalian ecologists (Vaughan, Finley, Dial, and Czaplewski) were invited to consider how packrat behavior affects midden contents and thus the quality of the fossil record. Their preliminary studies confirm some of our own misgivings about the link between species abundance in middens and in nearby plant communities.

Vaughan points out in chapter 2 that packrats in forests or mesic woodlands tend to be dietary generalists, whereas those in deserts and dry woodlands—chaparral are specialists. We are not sure how this affects midden contents, which incorporate both food and construction materials. Contrary to Vaughan's expectations, Holocene midden floras from the Chihuahuan and Sonoran deserts are typically more diverse than those of woodlands and forests to the north. Perhaps a graver problem is the partitioning that accompanies sympatry. In southern California both *Neotoma fuscipes* and *N. lepida* prefer *Quercus turbinella* when allopatric, but when they occur together *N. lepida* switches to *Juniperus californica* (Cameron, 1971). A similar conclusion resulted from the removal experiment at Woodhouse Mesa, where reoccupation of a den by a different species radically altered midden contents (Dial and Czaplewski, *this volume*, chap. 4).

Unfortunately, ways of determining the rat species that built a given midden have not been established. In chapter 3 Finley reviews potential biochemical and molecular techniques that may eventually be helpful. Radioimmunoassay of serum albumin preserved in the crystallized urine seems especially promising for studies trying to verify the agents of midden construction abroad. Development of antisera to more rapidly evolving proteins

may be needed for *Neotoma* identifications, because the albumin of different packrat species may not be distinguishable electrophoretically (William Rainey, personal communication). One by-product would be a detailed biogeography of packrats, with each occurrence linked with a known local flora. Mapping of former packrat distributions in association with known floras would thus test the common assumption that habitat affinities remain stable through time.

Other aspects of midden methods need refinement. Considerable disagreement exists as to what portions of the den become indurated middens, exactly how much time is involved in midden deposition and formation, what measures of species abundance should be used and should they be standardized, and what fraction should be sampled for radiocarbon dating (Spaulding et al., *this volume*, chap. 5). We recognize the need for experimental studies of midden taphonomy and exhaustive comparisons between the contents of modern middens and the local plant community.

In years to come it will be interesting to witness the evolution in temporal and spatial distributions of radiocarbon dates from middens. Are middle Holocene middens less common than latest Wisconsin—early Holocene ones, or is the bimodal peak noted by Webb and Betancourt in chapter 6 purely an artifact of past researcher bias? What are the physical limits for midden preservation, and are they of paleoclimatic value? Already middens are being reported from as far afield as British Columbia, where the bushy-tailed woodrat (*Neotoma cinerea*) must have followed plant invasions onto formerly glaciated terrain (Hebda and Warner, 1988). These middens could help determine whether or not stands of Rocky Mountain juniper in British Columbia represent northern advances during the altithermal, as suggested by Adams (1983).

In years to come we anticipate increasing use of the midden record by population and evolutionary biologists. At one time or another most paleoecologists are confronted with questions about the genetic integrity of their fossil identifications. Have there been any Quaternary extinctions or originations of plant species? Are paleoclimatic interpretations compromised by genotypic variations presently unresolvable in the fossil record? Is preservation of macrofossils good enough to permit evolutionary studies at the molecular level?

Relatively little has been learned about plant extinctions and speciations during the past 2 million years. The prevailing view is that of an unchanging flora (Tidwell et al., 1972). However, systematists of herbaceous and other diverse taxa openly speculate about accelerated rates of speciation during postglacial times (e.g., Stutz, 1978; Reveal, 1979; McLaughlin, 1986). To date, the midden record has yielded only one example of what may be an extinct plant species, *Chrysothamnus pulchelloides* (Anderson, 1981). The heavy losses of ground sloths, saber-toothed tigers, and other large mammals that struck the Southwest around eleven thousand years ago (Martin, 1987) occurred at a time when extralocal plants still held sway and the readjustment to modern glacial conditions had just begun.

Plenty of opportunities for adaptive radiation must have developed with intermingling of Eurasian and North American steppe floras as the Bering Strait opened and closed repeatedly over the last two million years. Late Quaternary radiations, for example, may explain North American species of *Corispermum*, an annual herb of Eurasian origin with a relatively rich fossil record in the American Southwest (Betancourt et al., 1984). Because of overreliance on negative evidence, however, one would have to conclude that the fossil record is blind when it comes to Quaternary speciations. It also is insensitive to genotypic variations not evident in macrofossil morphologies. One promising aspect of research is extraction of DNA from fossil materials. In the first such study involving macrofossils from packrat middens, Rogers and Bendich (1985) found that the DNA from forty-five-thousand-year-old juniper tissue was degraded but still contained high-molecular-weight DNA. They suggested the possibility of digesting these DNAs with restriction enzymes and cloning the fragments for evolutionary studies.

Finally, the discovery of middens abroad described in chapters 18, 19, and 20 holds great promise, even if these deposits are not as ubiquitous and ancient as packrat middens in the American Southwest. To cite but one example, hyrax middens could yield a spectacular macrofossil record for the early Holocene expansion of savanna and thornscrub into what is now the hyperarid interior of North Africa.

REFERENCES

Adams, R. P. (1983). "Infraspecific terpenoid variation in *Juniperus scopulorum*: Evidence for Pleistocene refugia and recolonization in western North America." *Taxon* 32, 30–46.

Anderson, L. C. (1981). "A new species of fossil *Chrysothamnus* (Asteraceae) from New Mexico." *Great Basin Naturalist* 40, 351–52.

Axelrod, D. I. (1979). *Age and origin of Sonoran Desert vegetation.* California Academy of Sciences Occasional Paper 132.

Bailey, D. K. (1987). "A study of *Pinus* Subsection Cembroides I: The single-needle pinyons of the Californias and the Great Basin." *Notes from the Royal Botanical Garden Edinburgh* 44, 275–310.

Baker, R. G. (1986). "Sangamonian (?) and Wisconsinan paleoenvironments in Yellowstone National Park." *Geological Society of America Bulletin* 97, 717–36.

Betancourt, J. L. (1987). "Paleoecology of pinyon-juniper woodlands: Summary." *Proceedings of the Pinyon-Juniper Conference,* pp. 129–39. U.S. Department of Agriculture, Forest Service, General Technical Report INT 215.

Betancourt, J. L.; Long, A.; Donahue, D. J.; Jull, A. J. T.; and Zabel, T. (1984). "Pre-Columbian age for North American *Corispermum* L. (Chenopodiaceae) confirmed by accelerator radiocarbon dating." *Nature* 311, 653–56.

Brown, D. A. (1984). "Prospects and limits of a phytolith key for grasses in the central United States." *Journal of Archeological Science* 11, 345–68.

Brown, J. H. (1978). "The theory of island biogeography and the distribution of boreal birds and mammals." *Great Basin Naturalist Memoirs* 2, 209–27.

Cameron, G. N. (1971). "Niche overlap and competition in woodrats." *Journal of Mammalogy* 52, 288–96.

Cinnamon, S. K., and Hevly, R. H. (1988). "Late Wisconsin macroscopic remains of pinyon pine on the southern Colorado Plateau." *Current Research in the Pleistocene* 5, 47–48.

COHMAP Members (1988). "Climatic changes of the last 18,000 years: Observations and model simulations." *Science* 241, 1043–52.

Cole, K. L. (1985). "Past rates of change, species richness, and a model of vegetational inertia in the Grand Canyon, Arizona." *American Naturalist* 125, 289–303.

———. (1986). "The lower Colorado River valley: A Pleistocene desert." *Quaternary Research* 25, 392–400.

Davis, M. B. (1986). "Climatic instability, time lags, and community disequilibrium." In *Community Ecology*, J. Diamond and T. J. Case, eds., pp. 269–84. New York: Harper and Row.

Gleason, H. S. (1926). "The individualistic concept of the plant association." *Torrey Botanical Club Bulletin* 53, 7–26.

Grayson, D. K. (1987). "The biogeographic history of small mammals in the Great Basin: Observations from the last 20,000 years." *Journal of Mammalogy* 68, 359–75.

Hansen, B. S., and Cushing, E. J. (1973). "Identification of pine pollen of late Quaternary age from the Chuska Mountains, New Mexico." *Geological Society of America Bulletin* 84, 1181–99.

Hebda, R. J., and Warner, B. G. (1988). "Pollen and macroscopic plant remains from fossil woodrat (*Neotoma cinerea*) middens in Canada." American Quaternary Association, Program and Abstracts of the Tenth Biennial Meeting, p. 74. Amherst: University of Massachusetts.

Hiebert, R. D., and Hamrick, J. L. (1983). "Patterns and levels of genetic variation in Great Basin bristlecone pine, *Pinus longaeva*." *Evolution* 37, 302–10.

Holliday, V. T. (1987). "A reexamination of late-Pleistocene boreal forest reconstructions for the southern High Plains." *Quaternary Research* 28, 238–44.

Johnson, A. W. (1968). "The evolution of desert vegetation in western North America." In *Desert biology*. G. W. Brown, ed., pp. 101–40. New York: Academic Press.

Kutzbach, J. E., and Guetter, P. J. (1986). "The influence of changing orbital parameters and surface boundary conditions on climate simulations for the past 18,000 years." *Journal of Atmospheric Sciences* 43, 1726–59.

Lanner, R. M., and Van Devender, T. R. (1974). "Morphology of pinyon pine needles from fossil packrat middens in Arizona." *Forest Science* 20, 207–11.

Lathrop, E. W., and Rowlands, P. G. (1983). "Plant ecology in deserts: An overview." In *Environmental effects of off-road vehicles*. R. H. Webb and H. G. Wilshire, eds., pp. 114–52. New York: Springer-Verlag.

Livingstone, D. A., and Clayton, W. D. (1980). "An altitudinal cline in tropical African grass floras and its paleoecological significance." *Quaternary Research* 13, 392–402.

Lomolino, M. V.; Brown, J. H.; and Davis, R. (1989). "Island biogeography of montane forest mammals in the American Southwest." *Ecology* 70, 180–94.

McLaughlin, S. P. (1986). "Floristic analysis of the southwestern United States." *Great Basin Naturalist* 46, 46–65.

Markgraf, V. (1986). "Plant inertia reassessed." *American Naturalist* 127, 726.

Martin, P. S. (1959). "How many logs make a forest." *Ohio Journal of Science* 59, 221–22.

———. (1963). *The last 10,000 years*. Tucson: University of Arizona Press.

———. (1987). "Late Quaternary extinctions: The promise of TAMS C-14 dating." *Nuclear Instruments and Methods in Physics Research* B29, 179–86.

Martin, P. S., and Mehringer, P. J., Jr. (1965). "Pleistocene pollen analysis and biogeography of the Southwest." In *The late Quaternary of the United States*. H. E. Wright, Jr., ed., pp. 433–51. New Haven: Yale University Press.

Mitchell, J. F. B.; Grahame, N. S.; and Needham, K. J. (1988). "Climate simulations for 9000 years before present: Seasonal variations and effect of the Laurentide ice sheet." *Journal of Geophysical Research* 93, 8283–8303.

Muller, C. H. (1940). "Plant succession in the *Larrea-Flourensia* climax." *Ecology* 21, 206–12.

Patterson, B. D. (1980). "Montane mammalian biogeography in New Mexico." *Southwestern Naturalist* 25, 33–50.

Reveal, J. L. (1979). "Biogeography of the intermountain region." *Mentzelia* 4, 1–92.

Rogers, S. O., and Bendich, A. J. (1985). "Extraction of DNA from milligram amounts of fresh, herbarium and mummified plant tissues." *Plant Molecular Biology* 5, 69–76.

Shreve, F. (1914). *A montane rainforest: A contribution to the physiological plant geography of Jamaica*. Carnegie Institute of Washington Publication 199.

———. (1942). "The desert vegetation of North America." *Botanical Review* 8, 195–246.

Smith, A. G. (1965). "Problems of inertia and threshold related post-glacial habitat changes." *Proceedings of the Royal Society B* 161, 331–42.

Spaulding, W. G.; Leopold, E. B.; and Van Devender, T. R. (1983). "Late Wisconsin paleoecology of the American Southwest." In *Late-Quaternary environments of the United States. Volume 1, The late Pleistocene*, S. C. Porter, ed., pp. 259–93. Minneapolis: University of Minnesota Press.

Stutz, H. C. (1978). "Explosive evolution in perennial *Atriplex* in western North America." *Great Basin Naturalist Memoirs* 2, 161–68.

Teeri, J. A., and Stowe, L. G. (1976). "Climatic patterns and the distribution of C-4 grasses in North America." *Oecologia* 23, 1–12.

Thorne, R. F. (1986). "A historical sketch of the vegetation of the Mohave and Colorado Deserts of the American Southwest." *Annals of the*

Missouri Botanical Garden 73, 642–51.

Tidwell, W. D.; Rushforth, S. R.; and Simper, D. (1972). "Evolution of floras in the Intermountain Region." In *Intermountain Flora*, vol. 1. A. Cronquist, A. H. Holmgren, N. H. Holmgren, and J. L. Reveal, eds., pp. 19–39. New York: Hafner.

Van Devender, T. R.; Thompson, R. S.; and Betancourt, J. L. (1987). "Vegetation history of the deserts of southwestern North America. The nature and timing of the Late Wisconsin-Holocene transition." In *North America and adjacent oceans during the last deglaciation*. W. F. Ruddiman and H. E. Wright, Jr., eds., pp. 323–52. *The Geology of North America*, vol. K-3. Boulder: Geological Society of America.

Van Devender, T. R.; Rea, A. M.; and Smith, M. L. (1985). "The Sangamon interglacial vertebrate fauna from Rancho La Brisca, Sonora, Mexico." *Transactions of the San Diego Society of Natural History* 21, 23–55.

Webb, R. H.; Steiger, J. W.; and Newman, E. V. (1988). *The response of vegetation to disturbance in Death Valley National Monument, California*. U.S. Geological Survey Bulletin 1793.

Webb, T. III (1986). "Is vegetation in equilibrium with climate? How to interpret late Quaternary pollen data." *Vegetatio* 67, 75–91.

Webb, T. III; Cushing, E. J.; and Wright, H. E., Jr. (1983). "Holocene changes in the vegetation of the Midwest." In *Late Quaternary environments of the United States. Volume 2, The Holocene*. H. E. Wright, Jr., ed., pp. 142–65. Minneapolis: University of Minnesota Press.

Wells, P. V. (1977). "Postglacial origin of the present Chihuahuan Desert less than 11,500 years ago." *Transactions of the Symposium on the Biological Resources of the Chihuahuan Desert Region*, pp. 67–83. Washington, D.C.: National Park Service Transactions and Proceedings Series no. 3.

———. (1983a). "Late Quaternary vegetation of the Great Plains." *Transactions of the Nebraska Academy of Sciences* 11, 83–89.

———. (1983b). "Paleobiogeography of montane islands in the Great Basin since the last glaciopluvial." *Ecological Monographs* 53, 373–88.

———. (1987). "Systematics and distribution of pinyons in the late Quaternary." In *Proceedings of the Pinyon-Juniper conference*, pp. 104–8. U.S. Department of Agriculture, Forest Service, Intermountain Research Station, General Technical Report INT-215.

Wells, P. V., and Hunziker, J. H. (1976). "Origin of the creosote bush deserts of southwestern North America." *Annals of the Missouri Botanical Garden* 63, 843–61.

Wells, P. V., and Woodcock, D. (1985). "Full-glacial vegetation of Death Valley, California: Juniper woodland opening to *Yucca* semidesert." *Madroño* 32, 11–23.

Wendorf, F., and Hester, J. J., eds. (1975). *Late Pleistocene environments of the southern High Plains*. Fort Burgwin Research Center Publication 9. Santa Fe: Museum of New Mexico Press.

West, N. E. (1984). "Successional patterns and productivity potentials of pinyon-juniper ecosystems." In *Developing strategies for rangeland management*. National Research Council/National Academy of Sciences, pp. 1301–32. Boulder: Westview Press.

Woodcock, D. (1986). "The late Pleistocene of Death Valley: A climatic reconstruction based on macrofossil data." *Palaeogeography, Palaeoclimatology, and Palaeoecology* 57, 273–83.

Wright, H. E., Jr.; Bent, A. M.; Hansen, B. S.; and Maher, L. J., Jr. (1973). "Present and past vegetation of the Chuska Mountains, northwestern New Mexico." *Geological Society of America Bulletin* 84, 1155–80.

Index

CONTRIBUTORS

Julio L. Betancourt, U.S. Geological Survey, 300 West Congress Street, FB-44, Tucson, AZ 85701.

George L. Bradley, Department of Ecology and Evolutionary Biology, University of Arizona, Tucson, AZ 85721.

Tony L. Burgess, Department of Ecology and Evolutionary Biology, University of Arizona, Tucson, AZ 85721.

Kenneth L. Cole, Indiana Dunes National Lakeshore, 1100 North Mineral Springs Road, Porter, IN 46304.

Lisa K. Croft, Quaternary Research Center and Department of Botany, University of Washington, Seattle, WA 98195.

Nicholas J. Czaplewski, Department of Biological Sciences, Northern Arizona University, Flagstaff, AZ 86011; now at Oklahoma Museum of Natural History, University of Oklahoma, Norman, OK 73019.

Kenneth P. Dial, Department of Biological Sciences, University of Montana, Missoula, MT 59812.

Steven E. Falconer, Department of Anthropology, Arizona State University, Tempe, AZ 85281.

Patricia L. Fall, Department of Geosciences, University of Arizona, Tucson, AZ 85721; now at Department of Geography, Arizona State University, Tempe, AZ 85281.

Robert B. Finley, Jr., 1050 Arapahoe, Apartment 506, Boulder, CO 80302.

W. Eugene Hall, Department of Entomology, University of Arizona, Tucson, AZ 85721.

Cynthia A. Lindquist, Office of Arid Lands Studies, Department of Agriculture, University of Arizona, Tucson, AZ 85721.

Austin Long, Laboratory of Isotope Geochemistry, Department of Geosciences, University of Arizona, Tucson, AZ 85721.

Paul S. Martin, Department of Geosciences, University of Arizona, Tucson, AZ 85721.

Peter J. Mehringer, Jr., Departments of Anthropology and Geology, Washington State University, Pullman, WA 99164-4910.

Desmond J. Nelson, CSIRO Central Australian Laboratory, P.O. Box 2111, Alice Springs, NT 5750, Australia.

Carl A. Olson, Department of Entomology, University of Arizona, Tucson, AZ 85721.

Louis Scott, Department of Botany, University of the Orange Free State, P.O. Box 339, Bloemfontein, 9300 South Africa.

W. Geoffrey Spaulding, Quaternary Research Center and Department of Botany, University of Washington, WA 98195.

Robert S. Thompson, Branch of Paleontology and Stratigraphy, U.S. Geological Survey, Box 25046, Mail Stop 919, Denver Federal Center, Denver, CO 80225.

Laurence J. Toolin, Department of Geosciences, University of Arizona, Tucson, AZ 85721.

Thomas R. Van Devender, Arizona-Sonora Desert Museum, 2021 North Kinney Road, Tucson, AZ 85743.

Terry A. Vaughan, Department of Biological Sciences, Northern Arizona University, Flagstaff, AZ 86011; now at P.O. Box 655, Lake Montezuma, AZ 86432.

Lisa A. Warneke, Laboratory of Isotope Geochemistry, Department of Geosciences, University of Arizona, Tucson, AZ 85721.

Robert H. Webb, U.S. Geological Survey, 300 West Congress Street, FB-44, Tucson, AZ 85701.

Peter E. Wigand, Desert Research Institute, University of Nevada, Reno, NV 89506.

THE EDITORS

Julio L. Betancourt is a physical scientist with the Water Resources Division, U.S. Geological Survey. His interest lies in climatic variability on various time scales and its expression on the hydrology and ecology of arid lands.

Thomas R. Van Devender has published sixty articles with forty-one coauthors on fossils from packrat middens. He serves as research scientist with the Arizona-Sonora Desert Museum in Tucson.

Paul S. Martin has been studying historical biogeography of the Southwest since arriving at the University of Arizona's Desert Laboratory in 1957. Now Emeritus Professor of Geosciences, he comments, "The best thing that happened was the discovery of perishable fossils, especially in fossil packrat middens."